Understanding Society

An Introductory Reader

THIRD EDITION

MARGARET L. ANDERSEN
University of Delaware

KIM A. LOGIO
Saint Joseph's University

HOWARD F. TAYLOR
Princeton University

THOMSON
WADSWORTH

Australia • Brazil • Canada • Mexico • Singapore • Spain
United Kingdom • United States

THOMSON

WADSWORTH

Understanding Society: An Introductory Reader,
Third Edition
Margaret L. Andersen, Kim A. Logio, and Howard F. Taylor

Acquisitions Editor: Chris Caldeira
Assistant Editor: Christina Beliso
Editorial Assistant: Erin Parkins
Marketing Manager: Michelle Williams
Marketing Communications Manager: Linda Yip
Project Manager, Editorial Production: Cheri Palmer
Creative Director: Rob Hugel

Art Director: Caryl Gorska
Print Buyer: Linda Hsu
Permissions Editor: Bob Kauser
Production Service: Dusty Friedman, The Book Company
Cover Designer: Yvo Riezebos
Cover Image: © Digital Vision Ltd./SuperStock
Compositor: Integra

Library of Congress Control Number: 2007938668

ISBN-13: 978-0-495-50430-6
ISBN-10: 0-495-50430-0

Thomson Higher Education
10 Davis Drive
Belmont, CA 94002-3098
USA

For more information about our products, contact us at:
Thomson Learning Academic Resource Center
1-800-423-0563

For permission to use material from this text or product, submit a request online at
http://www.thomsonrights.com.
Any additional questions about permissions can be submitted by e-mail to **thomsonrights@thomson.com.**

Contents

PART III   Socialization and the Life Course

PART IV   Society and Social Interaction

PART V   Groups and Organizations

PART VI Deviance and Crime

PART VII Social Class and Social Stratification

PART VIII Global Stratification

PART XII Social Institutions

Preface

"If you really acquire the sociological perspective, you can never be bored," writes June Jordan, a contemporary African American essayist. We agree, and we present these readings to help students see how fascinating the sociological perspective can be in interpreting human life. This anthology is intended for use in introductory sociology courses. Most of the students in these courses are first- or second-year students, many of whom are not majoring in sociology. We think this anthology that will excite students about the sociological perspective and show what such a perspective can bring to understanding society.

This new edition keeps many of the same themes as the first two editions, but we added thirty-seven new articles that bring new material to the book and reflect some of the developments in society. The table of contents is modeled on the companion text, *Sociology: The Essentials* by Andersen and Taylor, but it can easily be adapted for use with other introductory books or it can be used as a stand-alone text.

For this collection we selected articles that will engage students and show them what sociology can contribute to their understanding of the world. The readings include a variety of perspectives, research methodologies, and current topics. Some have been excerpted to make them more accessible to students. The anthology has a strong focus on diversity, both in the sections on class, race, gender, and age and in other selections throughout the book. We also included many articles that give students a global perspective on various sociological topics. The book features current research in sociology plus some classic readings from sociological theory. Instructors who want to include more such readings in their introductory courses can use the Howard anthology, *Classic Readings in Sociology* (also available from Wadsworth). We developed this book with several themes in mind:

- **Contemporary research:** We wanted students to see examples of strong contemporary research, presented in a fashion that would be accessible to beginning undergraduates. The articles included here feature different styles of sociological research. For example, Christine Williams's article, "The Social Organization of Toy Stores," shows how participant observation can be used to understand situations that students see in everyday life. This can be compared to the use of secondary data in Avis A. Jones-DeWeever and Heidi Hartmann's "Abandoned Before the Storms: Gender, Race, and Class Disparities in the Gulf," which discusses how race, class, and gender influenced the consequences of Hurricane Katrina.

- **Diversity:** In keeping with our understanding that society is increasingly diverse, we selected articles that show the range of experiences people have by virtue of differences in race, gender, class, sexual orientation, and other characteristics (such as age and religion). Numerous articles in the reader focus on African Americans, Native Americans, Latinos, Asian Americans, women, gays and lesbians, Jewish Americans, and people with disabilities, among others. Some of the selections bring a comprehensive analysis of race, class, and gender to the subject at hand, thus adding to students' understanding of how diverse groups experience the social structure of society. As an example, Yen Le Espiritu ("We Don't Sleep Around Like White Girls Do") shows the different patterns of socialization that Filipinas experience regarding gender and sexual identity. Robert Wuthnow ("America and the Challenges of Religious Diversity") explores the significance of religious diversity in American national identity.

- **Global perspective:** We also incorporated a global perspective into the reader, with many sections including articles that broaden students' worldview beyond the borders of the United States. Articles like Arlie Hochschild's essay "The Nanny Chain" will help students see how patterns of domestic help and contemporary immigration link the experiences of U.S. families to families from other nations, whose women are increasingly being employed to provide domestic help for professional workers in the United States. Similarly, Patricia Hill Collins's article, "New Commodities, New Consumers" shows how African American people are used in the capitalist quest for new global markets. Students will be able to see how their own consumption practices engage race and globalization in ways they might not have recognized.

- **Applying sociological knowledge:** Our students commonly ask, "What can you do with a sociological perspective?" We think this is an important question and one with many different answers. Sociologists use their knowledge in a variety of ways: to influence social policy formation, to interpret current events, and to educate people about common misconceptions and stereotypes, to name a few. Many of the new readings have a particularly contemporary appeal because the topics relate directly to changes in students' lives, such as the increasing use of the Internet for social interaction. For example, Sharlene Hess-Biber's article, "Men and Women: Mind and Body," examines the prevalence of eating disorders in a culture where beauty and body ideals are narrowly defined. In our experience, this topic is highly significant to undergraduate readers. Katherine Newman's analysis of school shootings (in "Explaining Rampage School Shootings") will be especially interesting to students, given the massacres at Virginia Tech University in April 2007. The student exercises we have added to this edition also enhance the theme of applying sociological knowledge to everyday life.

- **Classical theory:** We think it is important that introductory students learn about the contributions of classical sociological theorists. Thus, we have kept articles by Weber, Marx and Engels, Du Bois, and Goffman that will showcase some of the most important classics. As with all of the selections, we include discussion questions at the end of each reading to help students think about how such classic pieces are reflected in contemporary issues. For example, Max Weber's argument about the Protestant ethic and the spirit of capitalism is fascinating to think about in the contemporary context of increased consumerism and increased class inequality. Students might ask whether contemporary patterns of wealth and consumption no longer reflect the asceticism and moral calling about which Weber wrote. W. E. B. Du Bois's reflections on double consciousness also continue to be relevant in discussions of race and group perceptions.

NEW TO THE THIRD EDITION

The third edition of *Understanding Society* is organized in fourteen parts, following the outline of most introductory courses. The part titled "Social Institutions" is then subdivided into brief sections on six major social institutions: family, religion, education, work, government and politics, and health care. This organization allows faculty to focus on different institutions in any order they choose.

We significantly revised several of the thematic sections to reflect more current research, particularly the sections on culture, class, gender, sexuality, religion, education, work, and health care.

Several new features have been added. At the conclusion of each article, we include a list of core concepts that are reflected in the article. These concepts are then defined in the glossary at the end of the book.

We also added student exercises at the end of each major section. Through these, students can—on their own or in group projects—apply what they have learned to their observations of social behavior.

Brief introductions before each article place the article in context and help frame students' understanding of the selection. Discussion questions at the end of each reading help students think about the implications of what they have read.

The thirty-seven new readings in the book were selected to engage student interest and reflect the richness of sociological thought. Thus, we included a new piece on immigration, noting how contemporary immigration has changed since waves of immigration in the early twentieth century (Nancy Foner, "Immigrant Women and Work, Then and Now"). Karen Sternheimer's article ("Do Video

Games Kill?") addresses important questions about the link between violent behavior and exposure to violent media content. Michael Messner's research, "Barbie Girls versus Sea Monsters," explores how gender is constructed through children's play. Janny Scott and David Leonhardt's "Shadowy Lines That Still Divide" underscores the dimension of growing inequality in contemporary society. Ruth Rosen ("The Care Crisis"), along with other articles, analyzes the challenge of integrating family and work life. And Sara Mead's article, "The Truth about Boys and Girls," examines the question of whether there is a gender gap in schools that now favors girls.

In sum, with this anthology we hope to capture student interest in sociology, provide interesting research and theory, incorporate the analysis of diversity into the core of the sociological perspective, analyze the increasingly global dimensions of society, and show students how what they learn about sociology can be applied to real issues and problems.

PEDAGOGICAL FEATURES

In addition to the sociological content of this reader, a number of pedagogical features enrich student learning and help instructors teach with the book. Each essay has a **brief introductory paragraph** that identifies the major themes and questions being raised in the article. A short list of **core concepts** at the end of each reading helps students understand how the reading is related to basic ideas in the field of sociology. All of the core concepts are defined in the **glossary** at the end of the book. This will especially help instructors who use this anthology as a stand-alone text. We also follow each article with **discussion questions** that students can use to improve their critical thinking and to reinforce their understanding of the article's major points. Many of these questions could also be used as the basis for class discussion, student papers, or research exercises and projects. And, new to this edition, are the **student exercises** included at the end of each major part. Finally, a **subject/name index** helps students and faculty locate specific topics and authors in the book.

ACKNOWLEDGMENTS

Many people helped in a variety of ways as we revised this anthology. We especially thank Michelle Wilcox for her assistance in preparing the student exercises and the glossary. We also appreciate the support of Linda Keen and Judy Watson, whose work makes a lot of things possible. And this edition is substantially improved based on the recommendations of our reviewers. We give special thanks to Wendy A. Moore, Texas A&M University; Kristin Marsh, University of Mary Washington; Suvarna Cherukuri, Siena College; Catherine

Marrone, SUNY Stony Brook; Edward W. Morris, Ohio University; Gail Murphy-Geiss, Colorado College; and Patricia Gibbs, Foothill College.

We sincerely appreciate the enthusiasm and support provided by Wadsworth's outstanding editorial team. Thank you Chris Caldeira, Tali Beesley, Dusty Friedman, and Michelle Williams for all you do to support this work. And, most especially, we thank Richard, Jim, Nolan, Owen, and Isabel for all the love, smiles, and support they give us every day.

About the Editors

Margaret L. Andersen is the Edward F. and Elizabeth Goodman Rosenberg Professor of Sociology at the University of Delaware, where she also holds joint appointments in Women's Studies and Black American Studies. She is the author of *On Land and On Sea: A Century of Women in the Rosenfeld Collection; Thinking about Women: Sociological Perspectives on Sex and Gender; Race, Class and Gender* (with Patricia Hill Collins); *Sociology: Understanding a Diverse Society* (with Howard F. Taylor); and *Sociology: The Essentials* (with Howard Taylor). She was the 2006 recipient of the American Sociological Association's Jessie Bernard Award and has received the Sociologists for Women in Society's Feminist Lecturer Award. She is the vice president of the American Sociological Association and former president of the Eastern Sociological Society, as well as a recipient of the University of Delaware's Excellence in Teaching Award.

Kim A. Logio is Assistant Professor of Sociology at Saint Joseph's University in Philadelphia, where she teaches courses in research methods, childhood obesity, and women and health. Her research on adolescent body image has appeared in *Violence against Women* and has been presented at the American Sociological Association meetings. Her research on childhood obesity involves urban elementary school students and their understanding of nutrition and food choices. She is also the coauthor of *Adventures in Criminal Justice Research* and is presently working on the fourth edition, with George Dowdall, Earl Babbie, and Fred Halley. She has been heard on National Public Radio discussing childhood obesity. She lives in Delaware with her husband and three children.

Howard F. Taylor is Professor of Sociology at Princeton University. He is the author of *Balance in Small Groups; The IQ Game; Sociology: Understanding a Diverse Society* (with Margaret L. Andersen); and *Sociology: The Essentials* (with Margaret L. Andersen). He is the winner of the Du Bois-Johnson-Frazier Award, given by the American Sociological Association for distinguished research in race and ethnic relations, and has received the Princeton University President's Award for Distinguished Teaching. He is past president of the Eastern Sociological Society and is currently writing a book on *Race, Class, and the Bell Curve in America*.

1

The Sociological Imagination

C. WRIGHT MILLS

First published in 1959, C.Wright Mills' essay, taken from his book The Sociological Imagination, *is a classic statement about the sociological perspective. A man of his times, his sexist language intrudes on his argument, but the questions he posed about the connection between history, social structure, and people's biography (or lived experiences) still resonate today. His central theme is that the task of sociology is to understand how social and historical structures impinge on the lives of different people in society.*

Nowadays men often feel that their private lives are a series of traps. They sense that within their everyday worlds, they cannot overcome their troubles, and in this feeling, they are often quite correct: What ordinary men are directly aware of and what they try to do are bounded by the private orbits in which they live; their visions and their powers are limited to the close-up scenes of job, family, neighborhood; in other milieux, they move vicariously and remain spectators. And the more aware they become, however vaguely, of ambitions and of threats which transcend their immediate locales, the more trapped they seem to feel.

Underlying this sense of being trapped are seemingly impersonal changes in the very structure of continent-wide societies. The facts of contemporary history are also facts about the success and the failure of individual men and women. When a society is industrialized, a peasant becomes a worker; a feudal lord is liquidated or becomes a businessman. When classes rise or fall, a man is employed or unemployed; when the rate of investment goes up or down, a man takes new heart or goes broke. When wars happen, an insurance salesman becomes a rocket launcher; a store clerk, a radar man; a wife lives alone; a child grows up without a father. Neither the life of an individual nor the history of a society can be understood without understanding both.

Yet men do not usually define the troubles they endure in terms of historical change and institutional contradiction. The well-being they enjoy, they do not usually impute to the big ups and downs of the societies in which they live. Seldom aware of the intricate connection between the patterns of their own lives

SOURCE: "The Promise," from *Sociological Imagination* by C. Wright Mills, Copyright 1959, 2000 by Oxford University Press, Inc. Used by permission of Oxford University Press, Inc.

and the course of world history, ordinary men do not usually know what this connection means for the kinds of men they are becoming and for the kinds of history-making in which they might take part. They do not possess the quality of mind essential to grasp the interplay of man and society, of biography and history, of self and world. They cannot cope with their personal troubles in such ways as to control the structural transformations that usually lie behind them. . . .

The sociological imagination enables its possessor to understand the larger historical scene in terms of its meaning for the inner life and the external career of a variety of individuals. It enables him to take into account how individuals, in the welter of their daily experience, often become falsely conscious of their social positions. Within that welter, the framework of modern society is sought, and within that framework the psychologies of a variety of men and women are formulated. By such means the personal uneasiness of individuals is focused upon explicit troubles and the indifference of publics is transformed into involvement with public issues.

The first fruit of this imagination—and the first lesson of the social science that embodies it—is the idea that the individual can understand his own experience and gauge his own fate only by locating himself within his period, that he can know his own chances in life only by becoming aware of those of all individuals in his circumstances. In many ways it is a terrible lesson; in many ways a magnificent one. We do not know the limits of man's capacities for supreme effort or willing degradation, for agony or glee, for pleasurable brutality or the sweetness of reason. But in our time we have come to know that the limits of "human nature" are frighteningly broad. We have come to know that every individual lives, from one generation to the next, in some society; that he lives out a biography, and that he lives it out within some historical sequence. By the fact of his living he contributes, however minutely, to the shaping of this society and to the course of its history, even as he is made by society and by its historical push and shove.

The sociological imagination enables us to grasp history and biography and the relations between the two within society. That is its task and its promise. To recognize this task and this promise is the mark of the classic social analyst. . . .

No social study that does not come back to the problems of biography, of history and of their intersections within a society has completed its intellectual journey. Whatever the specific problems of the classic social analysts, however limited or however broad the features of social reality they have examined, those who have been imaginatively aware of the promise of their work have consistently asked three sorts of questions:

1. What is the structure of this particular society as a whole? What are its essential components, and how are they related to one another? How does it differ from other varieties of social order? Within it, what is the meaning of any particular feature for its continuance and for its change?

2. Where does this society stand in human history? What are the mechanics by which it is changing? What is its place within and its meaning for the development of humanity as a whole? How does any particular feature we

are examining affect, and how is it affected by, the historical period in which it moves? And this period—what are its essential features? How does it differ from other periods? What are its characteristic ways of history-making?

3.   What varieties of men and women now prevail in this society and in this period? And what varieties are coming to prevail? In what ways are they selected and formed, liberated and repressed, made sensitive and blunted? What kinds of "human nature" are revealed in the conduct and character we observe in this society in this period? And what is the meaning for "human nature" of each and every feature of the society we are examining?

Whether the point of interest is a great power state or a minor literary mood, a family, a prison, a creed—these are the kinds of questions the best social analysts have asked. They are the intellectual pivots of classic studies of man in society—and they are the questions inevitably raised by any mind possessing the sociological imagination. For that imagination is the capacity to shift from one perspective to another—from the political to the psychological; from examination of a single family to comparative assessment of the national budgets of the world; from the theological school to the military establishment; from considerations of an oil industry to studies of contemporary poetry. It is the capacity to range from the most impersonal and remote transformations to the most intimate features of the human self—and to see the relations between the two. Back of its use there is always the urge to know the social and historical meaning of the individual in the society and in the period in which he has his quality and his being.

That, in brief, is why it is by means of the sociological imagination that men now hope to grasp what is going on in the world, and to understand what is happening in themselves as minute points of the intersections of biography and history within society. In large part, contemporary man's self-conscious view of himself as at least an outsider, if not a permanent stranger, rests upon an absorbed realization of social relativity and of the transformative power of history. The sociological imagination is the most fruitful form of this self-consciousness. By its use men whose mentalities have swept only a series of limited orbits often come to feel as if suddenly awakened in a house with which they had only supposed themselves to be familiar. Correctly or incorrectly, they often come to feel that they can now provide themselves with adequate summations, cohesive assessments, comprehensive orientations. Older decisions that once appeared sound now seem to them products of a mind unaccountably dense. Their capacity for astonishment is made lively again. They acquire a new way of thinking, they experience a transvaluation of values: in a word, by their reflection and by their sensibility, they realize the cultural meaning of the social sciences.

Perhaps the most fruitful distinction with which the sociological imagination works is between "the personal troubles of milieu" and "the public issues of social structure." This distinction is an essential tool of the sociological imagination and a feature of all classic work in social science.

Troubles occur within the character of the individual and within the range of his immediate relations with others; they have to do with his self and with those limited areas of social life of which he is directly and personally aware.

Accordingly, the statement and the resolution of troubles properly lie within the individual as a biographical entity and within the scope of his immediate milieu—the social setting that is directly open to his personal experience and to some extent his willful activity. A trouble is a private matter: values cherished by an individual are felt by him to be threatened.

Issues have to do with matters that transcend these local environments of the individual and the range of his inner life. They have to do with the organization of many such milieux into the institutions of an historical society as a whole, with the ways in which various milieux overlap and interpenetrate to form the larger structure of social and historical life. An issue is a public matter: some value cherished by publics is felt to be threatened. Often there is a debate about what that value really is and about what it is that really threatens it. This debate is often without focus if only because it is the very nature of an issue, unlike even widespread trouble, that it cannot very well be defined in terms of the immediate and everyday environments of ordinary men. An issue, in fact, often involves a crisis in institutional arrangements, and often too it involves what Marxists call "contradictions" or "antagonisms."

In these terms, consider unemployment. When, in a city of 100,000, only one man is unemployed, that is his personal trouble, and for its relief we properly look to the character of the man, his skills, and his immediate opportunities. But when in a nation of 50 million employees, 15 million men are unemployed, that is an issue, and we may not hope to find its solution within the range of opportunities open to any one individual. The very structure of opportunities has collapsed. Both the correct statement of the problem and the range of possible solutions require us to consider the economic and political institutions of the society, and not merely the personal situation and character of a scatter of individuals.

Consider war. The personal problem of war, when it occurs, may be how to survive it or how to die in it with honor; how to make money out of it; how to climb into the higher safety of the military apparatus; or how to contribute to the war's termination. In short, according to one's values, to find a set of milieux and within it to survive the war or make one's death in it meaningful. But the structural issues of war have to do with its causes; with what types of men it throws up into command; with its effects upon economic and political, family and religious institutions, with the unorganized irresponsibility of a world of nation-states.

Consider marriage. Inside a marriage a man and a woman may experience personal troubles, but when the divorce rate during the first four years of marriage is 250 out of every 1,000 attempts, this is an indication of a structural issue having to do with the institutions of marriage and the family and other institutions that bear upon them.

Or consider the metropolis—the horrible, beautiful, ugly, magnificent sprawl of the great city. For many upper-class people, the personal solution to "the problem of the city" is to have an apartment with private garage under it in the heart of the city, and forty miles out, a house by Henry Hill, garden by Garrett Eckbo, on a hundred acres of private land. In these two controlled environments—with a small staff at each end and a private helicopter connection—most people could solve many of the problems of personal milieux caused by the facts of the city. But all this, however splendid, does not solve the public issues that the structural fact of the city poses. What should be done with this wonderful monstrosity? Break it all up into scattered units, combining residence and work? Refurbish it as it stands? Or, after evacuation, dynamite it and build new cities according to new plans in new places? What should those plans be? And who is to decide and to accomplish whatever choice is made? These are structural issues; to confront them and to solve them requires us to consider political and economic issues that affect innumerable milieux.

In so far as an economy is so arranged that slumps occur, the problem of unemployment becomes incapable of personal solution. In so far as war is inherent in the nation-state system and in the uneven industrialization of the world, the ordinary individual in his restricted milieu will be powerless—with or without psychiatric aid—to solve the troubles this system or lack of system imposes upon him. In so far as the family as an institution turns women into darling little slaves and men into their chief providers and unweaned dependents, the problem of a satisfactory marriage remains incapable of purely private solution. In so far as the overdeveloped megalopolis and the overdeveloped automobile are built-in features of the overdeveloped society, the issues of urban living will not be solved by personal ingenuity and private wealth.

What we experience in various and specific milieux, I have noted, is often caused by structural changes. Accordingly, to understand the changes of many personal milieux we are required to look beyond them. And the number and variety of such structural changes increase as the institutions within which we live become more embracing and more intricately connected with one another. To be aware of the idea of social structure and to use it with sensibility is to be capable of tracing such linkages among a great variety of milieux. To be able to do that is to possess the sociological imagination. . . .

KEY CONCEPTS

issues   sociological imagination   troubles

DISCUSSION QUESTIONS

1. Using either today's newspaper or some other source of news, identify one example of what C. Wright Mills would call an issue. How is this issue reflected in the personal troubles of people it affects? Why would Mills call it a social issue?

2. What are the major historical events that have influenced the biographies of people in your generation? In your parents' generation? What does this tell you about the influence of society and history on biography?

2

The Forest and the Trees

ALLAN G. JOHNSON

Allan Johnson uses the classic example of the forest and the trees as a metaphor to demonstrate that people in society are participating in something larger than themselves. He also argues that the strong cultural belief in individualism blunts the sociological imagination, because it makes you see only individuals, not the social structures that shape diverse group experiences.

As a form of sociological practice, I work with people in corporations, schools, and universities who are trying to deal with issues of diversity. In the simplest sense, diversity is about the variety of people in the world, the varied mix of gender, race, age, social class, ethnicity, religion, and other social characteristics. In the United States and Europe, for example, the workforce is changing as the percentages who are female or from non-European ethnic and racial backgrounds increase and the percentage who are white and male declines.

If the changing mix was all that diversity amounted to, there wouldn't be a problem since in many ways differences make life interesting and enhance creativity. Compared with homogeneous teams, for example, diverse work teams are usually better with problems that require creative solutions. To be sure, diversity brings with it difficulties to be dealt with such as language barriers and different ways of doing things that can confuse or irritate people. But we're the species with the "big brain," the adaptable ones who learn quickly, so learning to get along with people unlike ourselves shouldn't be a problem we can't handle. Like travelers in a strange land, we'd simply learn about one another and make room for differences and figure out how to make good use of them.

As most people know, however, in the world as it is, difference amounts to more than just variety. It's also used as a basis for including some and excluding others, for rewarding some more and others less, for treating some with respect and dignity and some as if they were less than fully human or not even there. Difference is used as a basis for privilege, from reserving for some the simple human dignities that everyone should have, to the extreme of deciding who lives and who dies. Since the workplace is part of the world, patterns of inequality and oppression that permeate the world also show up at work, even though people

SOURCE: Allan G. Johnson. 1997. *The Forest and the Trees: Sociology as Life, Practice, Promise.* Philadelphia: Temple University Press, pp. 7–27. Reprinted with permission.

may like to think of themselves as "colleagues" or part of "the team." And just as these patterns shape people's lives in often damaging ways, they can eat away at the core of a community or an organization, weakening it with internal division and resentment bred and fed by injustice and suffering. . . .

People tend to think of things only in terms of individuals, as if a society or a company or a university were nothing more than a collection of people living in a particular time and place. Many writers have pointed out how individualism affects social life. It isolates us from one another, promotes divisive competition, and makes it harder to sustain a sense of community, of all being "in this together." But individualism does more than affect how we participate in social life. It also affects how we *think* about social life and how we make sense of it. If we think everything begins and ends with individuals—their personalities, biographies, feelings, and behavior—then it's easy to think that social problems must come down to flaws in individual character. If we have a drug problem, it must be because individuals just can't or won't "say no." If there is racism, sexism, heterosexism, classism, and other forms of oppression, it must be because of people who for some reason have the personal "need" to behave in racist, sexist, and other oppressive ways. If evil consequences occur in social life, then it must be because of evil people and their evil ways and motives.

If we think about the world in this way—which is especially common in the United States—then it's easy to see why members of privileged groups get upset when they're asked to look at the benefits that go with belonging to that particular group and the price others pay for it. When women, for example, talk about how sexism affects them, individualistic thinking encourages men to hear this as a personal accusation: "If women are oppressed, then I'm an evil oppressor who wants to oppress them." Since no man wants to see himself as a bad person, and since most men probably don't *feel* oppressive toward women, men may feel unfairly attacked.

In the United States, individualism goes back to the nineteenth century and, beyond that, to the European Enlightenment and the certainties of modernist thinking. It was in this period that the rational mind of the individual person was recognized and elevated to a dominant position in the hierarchy of things, separated from and placed above even religion and God. The roots of individualistic thinking in the United States trace in part to the work of William James who helped pioneer the field of psychology. Later, it was deepened in Europe and the United States by Sigmund Freud's revolutionary insights into the existence of the subconscious and the inner world of individual existence. Over the course of the twentieth century, the individual life has emerged as a dominant framework for understanding the complexities and mysteries of human existence.

You can see this in bookstores and best-seller lists that abound with promises to change the world through "self-help" and individual growth and transformation. Even on the grand scale of societies—from war and politics to international economics—individualism reduces everything to the personalities and behavior of the people we perceive to be "in charge." If ordinary people in capitalist societies feel deprived and insecure, the individualistic answer is that the people who run corporations are "greedy" or the politicians are corrupt and incompetent and otherwise lacking in personal character. The same perspective argues that poverty exists because of the habits, attitudes, and skills of individual poor people, who are blamed for what they supposedly lack as people and told to change if they want anything better for themselves. To make a better world, we think we have to put the "right people" in charge or make better people by liberating human consciousness in a New Age or by changing how children are socialized or by locking up or tossing out or killing people who won't or can't be better than they are. Psychotherapy is increasingly offered as a model for changing not only the inner life of individuals, but also the world they live in. If enough people heal themselves through therapy, then the world will "heal" itself as well. The solution to collective problems such as poverty or deteriorating cities then becomes a matter not of collective solutions but of an accumulation of individual solutions. So, if we want to have less poverty in the world, the answer lies in raising people out of poverty or keeping them from becoming poor, *one person at a time*.

So, individualism is a way of thinking that encourages us to explain the world in terms of what goes on inside individuals and nothing else. We've been able to think this way because we've developed the human ability to be reflexive, which is to say, we've learned to look at ourselves *as selves* with greater awareness and insight than before. We can think about what kind of people we are and how we live in the world, and we can imagine ourselves in new ways. To do this, however, we first have to be able to believe that we exist as distinct individuals apart from the groups and communities and societies that make up our social environment. In other words, the *idea* of the "individual" has to exist before we think about ourselves as individuals, and the idea of the individual has been around for only a few centuries. Today, we've gone far beyond this by thinking of the social environment itself as just a collection of individuals: Society *is* people and people *are* society. To understand social life, all we have to do is understand what makes the individual psyche tick.

If you grow up and live in a society that's dominated by individualism, the idea that society is just people seems obvious. The problem is that this approach ignores the difference between the individual people who participate in social life and the relationships that connect them to one another and to groups and societies. It's true that you can't have a social relationship without people to participate in it and make it happen, but the people and the relationship aren't the same thing. That's why this book's articles plays on the old saying about missing the forest for the trees. In one sense, a forest is simply a collection of individual trees; but it's more than that. It's also a collection of trees that exist in *a particular relation* to one another, and you can't tell what that relation is by just looking at each individual tree. Take a thousand trees and scatter them across the Great Plains of North America, and all you have are a thousand trees. But take those same trees and bring them close together and you have a forest. Same individual trees, but in one case a forest and in another case just a lot of trees.

The "empty space" that separates individual trees from one another isn't a characteristic of any one tree or the characteristics of all the individual trees somehow added together. It's something more than that, and it's crucial to understand the *relationships among* trees that make a forest what it is. Paying attention to that "something more"—whether it's a family or a corporation or an entire society—and how people are related to it is at the heart of sociological practice.

THE ONE THING

If sociology could teach everyone just one thing with the best chance to lead toward everything else we could know about social life, it would, I believe, be this: *We are always participating in something larger than ourselves, and if we want to understand social life and what happens to people in it, we have to understand what it is that we're participating in* and *how we participate in it.* In other words, the key to understanding social life isn't just the forest and it isn't just the trees. It's the forest *and* the trees and how they're related to one another. Sociology is the study of how all this happens.

The "larger" things we participate in are called social systems, and they come in all shapes and sizes. In general, the concept of a system refers to any collection of parts or elements that are connected in ways that cohere into some kind of whole. We can think of the engine in a car as a system, for example, a collection of parts arranged in ways that make the car "go." Or we could think of a language as a system, with words and punctuation and rules for how to combine them into sentences that mean something. We can also think of a family as a system—a collection of elements related to one another in a way that leads us to think of it as a unit. These include things such as the positions of mother, father, wife, husband, parent, child, daughter, son, sister, and brother. Elements also include shared ideas that tie those positions together to make relationships, such as how "good mothers" are supposed to act in relation to children or what a "family" is and what makes family members "related" to one another as kin. If we take the positions and the ideas and other elements, then we can think of what results as a whole and call it a social system.

In similar ways, we can think of corporations or societies as social systems. They differ from one another—and from families—in the kinds of elements they include and how those are arranged in relation to one another. Corporations have positions such as CEOs and stockholders, for example; but the position of "mother" isn't part of the corporate system. People who work in corporations can certainly be mothers in families, but that isn't a position that connects them to a corporation. Such differences are a key to seeing how systems work and produce different kinds of consequences. Corporations are sometimes referred to as "families," for example, but if you look at how families and corporations are actually put together as systems, it's easy to see how unrealistic such notions are. Families don't usually "lay off" their members when times are tough or to boost the bottom line, and they usually don't divide the food on the dinner table according to who's the strongest and best able to grab the lion's share for themselves. But corporations dispense with workers all the time as a way to raise dividends and the value of stock, and top managers routinely take a huge share of each year's profits even while putting other members of the corporate "family" out of work.

What social life comes down to, then, is social systems and how people participate in and relate to them. Note that people *participate* in systems without being *parts* of the systems themselves. In this sense, "father" is a position in my family, and I, Allan, am a person who actually occupies that position. It's a crucial distinction that's easy to lose sight of. It's easy to lose sight of because we're so used to thinking solely in terms of individuals. It's crucial because it means that people aren't systems, and systems aren't people, and if we forget that, we're likely to focus on the wrong thing in trying to solve our problems. . . .

INDIVIDUALISTIC MODELS DON'T WORK

Probably the most important basis for sociological practice is to realize that *the individualistic perspective that dominates current thinking about social life doesn't work.* Nothing we do or experience takes place in a vacuum; everything is always related to a context of some kind. When a wife and husband argue about who'll clean the bathroom, for example, or who'll take care of a sick child when they both work outside the home, the issue is never simply about the two of them even though it may seem that way at the time. We have to ask about the larger context in which this takes place. We might ask how this instance is related to living in a society organized in ways that privilege men over women, in part by not making men feel obliged to share equally in domestic work except when they choose to "help out." On an individual level, he may think she's being a nag; she may think he's being a jerk; but it's never as simple as that. What both may miss is that in a different kind of society, they might not be having this argument in the first place because both might feel obliged to take care of the home and children. In similar ways, when we see ourselves as a unique result of the family we came from, we overlook how each family is connected to larger patterns. The emotional problems we struggle with as individuals aren't due simply to what kind of parents we had, for their participation in social systems—at work, in the community, in society as a whole—shaped them as people, including their roles as mothers and fathers.

An individualistic model is misleading because it encourages us to explain human behavior and experience from a perspective that's so narrow it misses most of what's going on. A related problem is that *we can't understand what goes on in social systems simply by looking at individuals.* In one sense, for example, suicide is a solitary act done by an individual, typically while alone. If we ask why people kill themselves, we're likely to think first of how people feel when they do it—hopeless, depressed, guilty, lonely, or perhaps obliged by honor or duty to sacrifice themselves for someone else or some greater social good. That might explain suicides taken one at a time, but what do we have when we add up all the suicides that happen in a society for a given year? What does that number tell us, and, more importantly, about what? The suicide rate for the entire U.S. population in 1994, for example, was twelve suicides per 100,000 people. If we look inside that number, we find that the rate for males was twenty per 100,000, but the rate for females was only five per 100,000. The rate also differs dramatically by race and country and varies over time. The suicide rate for white males, for example, was 71 percent higher than for black males, and the rate for white females was more than twice that for black females. While the rate in the United States was twelve per 100,000, it was thirty-four per 100,000 in Hungary and only seven per 100,000 in Italy. So, in the United States, males and whites are far more likely than females and blacks to kill themselves; and people in the United States are almost twice as likely as Italians to commit suicide but only one third as likely as Hungarians.

If we use an individualistic model to explain such differences, we'll tend to see them as nothing more than a sum of individual suicides. If males are more likely to kill themselves, then it must be because males are more likely to feel suicidally depressed, lonely, worthless, and hopeless. In other words, the psychological factors that cause individuals to kill themselves must be more common among U.S. males than they are among U.S. females, or more common among people in the United States than among Italians. There's nothing wrong with such reasoning; it may be exactly right *as far as it goes*. But that's just the problem: It doesn't go very far because it doesn't answer the question of *why* these differences exist in the first place. Why, for example, would males be more likely to feel suicidally hopeless and depressed than females, or Hungarians more likely than Italians? Or why would Hungarians who feel suicidally depressed be more likely to go ahead and kill themselves than Italians who feel the same way? To answer such questions, we need more than an understanding of individual psychology. Among other things, we need to pay attention to the fact that words like "female," "white," and "Italian" name positions that people occupy in social systems. This draws attention to how those systems work and what it means to occupy those positions in them.

Sociologically, a suicide rate is a number that describes something about a group or a society, not the individuals who belong to it. A suicide rate of twelve per 100,000 tells us nothing about you or me or anyone else. Each of us either commits suicide during a given year or we don't, and the rate can't tell us who does what. In the same way, how individuals feel before they kill themselves isn't by itself enough to explain why some groups or societies have higher suicide rates than others. Individuals can feel depressed or lonely, but groups and societies can't feel a thing. We could consider that Italians might tend to be less depressed than people in the United States, for example, or that in the United States, people might tend to deal with feelings of depression more effectively than Hungarians. It makes no sense at all, however, to say that the United States is more depressed or lonely than Italy.

While it might work to look at what goes on in individuals as a way to explain why one person commits suicide, this can't explain *patterns* of suicide found in social systems. To do this, we have to look at how people feel and behave *in relation* to systems and how these systems work. We need to ask, for example, how societies are organized in ways that encourage people who participate in them to feel more or less depressed or to respond to such feelings in suicidal or nonsuicidal ways. We need to see how belonging to particular groups shapes people's experience as they participate in social life, and how this limits the alternatives they think they can choose from. What is it about being male or being white that can make suicide a path of least resistance? How, in other words, can we go to the heart of sociological practice to ask how people participate in something larger than themselves and see how this affects the choices they make? How can we see the relationship between people and systems that produces variations in suicide rates or, for that matter, just about everything else that we do and experience, from having sex to going to school to working to dying?

Just as we can't tell what's going on in a system just by looking at individuals, we also can't tell what's going on in individuals just by looking at systems. Something may look like one thing in the system as a whole, but something else entirely when we look at the people who participate in it. If we look at the kind of mass destruction and suffering that war typically causes, for example, an individualistic model suggests a direct link with the "kinds" of people who participate in it. If war produces cruelty, bloodshed, aggression, and conquest, then it must be that the people who participate in it are cruel, bloodthirsty, aggressive people who want to conquer and dominate others. Viewing the carnage and destruction that war typically leaves in its wake, we're likely to ask, "What kind of people could do such a thing?" Sociologically, however, this question misleads us by reducing a social phenomenon to a simple matter of "kinds of people" without looking at the systems those people participate in. Since we're always participating in one system or another, when someone drops a bomb that incinerates thousands of people, we can't explain what happened simply by figuring out "what kind of person would do such a thing." In fact, if we look at what's known about people who fight in wars, they appear fairly normal by most standards and anything but bloodthirsty and cruel. Most accounts portray men in combat, for example, as alternating between boredom and feeling scared out of their wits. They worry much less about glory than they do about not being hurt or killed and getting themselves and their friends home in one piece. For most soldiers, killing and the almost constant danger of being killed are traumatic experiences that leave them forever changed as people. They go to war not in response to some inner need to be aggressive and kill, but because they think it's their duty to go, because they'll go to prison if they dodge the draft, because they've seen war portrayed in books and movies as an adventurous way to prove they're "real men," or because they don't want to risk family and friends rejecting them for not measuring up as true patriots.

People aren't systems, and systems aren't people, which means that social life can produce horrible or wonderful consequences without necessarily meaning that the people who participate in them are horrible or wonderful. Good people participate in systems that produce bad consequences all the time. I'm often aware of this in the simplest situations, such as when I go to buy clothes or food. Many of the clothes sold in the United States are made in sweatshops in cities like Los Angeles and New York and in Third World countries, where people work under conditions that resemble slavery in many respects, and for wages that are so low they can barely live on them. A great deal of the fruit and vegetables in stores are harvested by migrant farm workers who work under conditions that aren't much better. If these workers were provided with decent working conditions and paid a living wage, the price of clothing and food would probably be a lot higher than it is. This means that I benefit directly from the daily mistreatment and exploitation of thousands of people. The fact that I benefit doesn't make me a bad person; but my participation in that system does involve me in what happens to them. . . .

KEY CONCEPTS

diversity   individualism   Enlightenment
social structure

DISCUSSION QUESTIONS

1. Johnson argues that there is a tendency in the United States for people to
    explain everything in individual terms. Using the example of suicide, why
    are individualistic explanations inadequate? What would sociological expla-
    nations emphasize instead?

2. Johnson opens his discussion by noting the diversity that characterizes U.S.
    society. How does he apply the sociological perspective to his understanding
    of diversity and its significance?

3

Not Our Kind of Girl

ELAINE BELL KAPLAN

*Elaine Bell Kaplan's research on African American teen mothers debunks a
number of myths about teenage pregnancy and about African American mothers.
She conducted her research using the method of participant observation. Here she
describes how she did her research, including how her position as both an insider
and an outsider influenced her research project. Her project and its results are
presented in her book,* Not Our Kind of Girl: Unraveling the Myths of
Black Teenage Motherhood *(1995).*

"If we want to solve the problems of the Black community, we have to do
something about illegitimate babies born to teenage mothers." The caller,
who identified himself as Black, was responding to a radio talk show discussion
about the social and economic problems of the Black community. According to

SOURCE: Elaine Bell Kaplan. 1995. *Not Our Kind of Girl: Unraveling the
Myths of Black Teenage Motherhood.* Berkeley: University of California Press,
pp. xviii–xxiii, 10–26, 173–181.

this caller's view, Black teen mothers' children grow up in fatherless households with mothers who have few moral values and little control over their offspring. The boys join gangs; the girls stand a good chance of becoming teen mothers themselves. The caller's perspective captures the popular view of many Americans: that marital status and age-appropriate sexual behavior ensure the well-being of the family and the community. . . .

According to mainstream ideology, men who through hard work have moved up the career ladder and provide their families with decent food on the dinner table, clothes on their backs, and an occasional family vacation have achieved the American Dream. Women's achievements are measured by their marriage and child rearing, done in proper order and at an appropriate age. Teenage girls are expected to replicate these values by refraining from sexual relations before adulthood and marriage.

Certainly, such traditional ideas held sway over the Black community I knew. Two decades ago unmarried teenagers with babies were a rare and unwelcome presence in my Harlem community. These few girls would be subjected to gossip about their lack of morals and stigmatized if they were on welfare. But by the 1980s so many young Black girls were pushing strollers around inner-city neighborhoods that they became an integral part of both the reality and the myth concerning the sexuality of Black underclass culture and Black family values. These Black teenage mothers did not fit in with the American ethic of hard work and strong moral character. . . .

If this conservative ideology is extended to teen mothers, their situation can be explained only as a result of aberrant moral character. If Black adolescent girls fail to achieve, something in their nature prevents them from doing so. As president, Ronald Reagan often urged teenage mothers to "just say no" so that taxpayers would no longer be forced to pay for their sexual behavior. The "Just Say No" slogan invoked by the Reagan and Bush administrations in the 1980s was utilized in the 1990s by both Black and White conservatives in the attempt "to change welfare as we know it." If these politicians have their way, teenage mothers will be shunned, hidden, and ignored.

As I made my way through East Oakland and downtown Richmond to interview teen mothers,* I witnessed a different scenario from the one devised by politicians. Teenage mothers are housed in threatening, drug-infested environments, schooled in jail-like institutions, and obstructed from achieving the American Dream. In our ostensibly open society, teenage mothers are disqualified from full participation and are marked as deviant. Black teenage girls aged fourteen, fifteen, and sixteen—many of them just beginning to show an adolescent interest in wearing makeup, dressing in the latest fashions, and reading teen magazines—are stigmatized. These teen mothers attempt to cope as best they can by redefining their situation in terms that involve the least damage to their self-respect.

---

* Except when necessary for clarification, all teenage mothers and older women who were previously teenage mothers will for the sake of brevity be referred to as teen mothers—a term they use. When appropriate the teen mothers' own mothers will be referred to as adult mothers. All names and places have been changed to protect confidentiality.

Are Black teenage mothers responsible for the socioeconomic problems besetting the Black community, as the radio show caller would have us believe? Do Black teenage mothers have different moral values than most Americans? Do they have babies in order to collect welfare, as politicians suggest? Do the families of Black teenage mothers condone their deviant behavior, as the popular view contends? Or, as William J. Wilson's economic theory suggests, is Black teenage motherhood simply a response to the economic problems of the Black community? Black teenage girls confront a world in which gender norms, poverty, and racism are intertwined. Accordingly, to answer questions about these young mothers, we must sort out a host of complex economic and social problems that pervade their lives. I hope the questions I have asked and the answers given will provide portraits of real teenage mothers involved in real experiences.

The reality of these teenage mothers is that they have had to adopt strategies for survival that seem to them to make sense within their social environment but are as inadequate for them as they were for teenage mothers in the past.

These ethnographic pictures illuminate the way structural contradictions act on psychological well-being and the way people construct and reconstruct their lives in order to cope on a daily basis. One issue that comes through quite clearly in this study, and one that is often overlooked by politicians and various studies on Black teenage mothers, is that these teenagers know what constitutes a successful life. Black teenage mothers...struggle against being considered morally deviant, underclass, and unworthy.

If we are to understand the stories of these teenage mothers and generalize from their experiences in any significant way, we must place them within the current theoretical and political discussions concerning Black teenage mothers. As T. S. Eliot noted long ago, reality is often more troubling than myth.

What is begging for our attention is the fact that adolescence is a time when Black girls, striving for maturity, lose the support of others in three significant ways. First, they are abandoned by the educational system; second, they become mere sexual accompanists for boys and men; third, these problems create a split between the girls and their families and significant others. What is needed to understand the losses, the stresses, and the large and small violences that render such teenage girls incapable of successfully completing their adolescent tasks is a gender, race, and class analysis. When early motherhood is added to these challenges, they become insurmountable. The adolescent mothers I saw were deprived of every resource needed for any human being to function well in our society: education, jobs, food, medical care, a secure place to live, love and respect, the ability to securely connect with others. In addition, these girls were silenced by the insidious and insistent stereotyping of them as promiscuous and aberrant teenage girls....

TALKING TO TEEN MOTHERS

I began my search to understand the rise in Black motherhood by interviewing two teen mothers referred to me by friends. They came to my house early one Saturday morning and stayed for three hours. Although I had prepared a series of general questions, the young women had so much more to say that I was compelled to create a more extensive set. Next, the director of a local family planning center let me attend a teen parent meeting, where I left a letter of introduction inviting those who were interested in my project to contact me. These teen mothers referred me to others. Eventually, I created a list of fifteen teen mothers.

The director of the family planning center also introduced me to Mary Higgins, the director of the Alternative Center in East Oakland. The Center operated with a grant from a large charity organization that allowed it to develop outreach programs geared to the needs of the local teenage population. These programs included an alternative school, day care, self-esteem development, parenting skills training, and personal counseling. Mary in turn introduced me to Ann Getty, a counselor at the center. Through Ann I met Claudia Wilson, a counselor for the Richmond Youth Counseling program. A short time after that meeting, I began to work as a volunteer consultant for the Alternative Center and to attend meetings with counselors and others who visited the center.

Through my contacts at the center, in the autumn of 1985 I met De Vonya Smalls and twenty of the sample of thirty-two teen mothers who participated in this study. The rest of my sample was drawn from other contacts I made in a network of community workers at the Richmond Youth Service Agency and through my work as a volunteer consultant there. The youth agency's counselors introduced me to teenage mothers who lived in the downtown Richmond area. As a consultant, I was able to talk extensively with the adolescents who took part in teen mother programs.

After several months of making contacts, losing some, and making new ones, I was able to pull together the sample of thirty-two teenage mothers. Of this sample, I "hung out" with a core group of seven teen mothers for a period of seven months, including sixteen-year-old De Vonya Smalls. The other six teen mothers who participated were sixteen-year-old Susan Carter, a mother of a two-month-old baby, who was living with her mother and sister in East Oakland; seventeen-year-old Shana Leeds, a mother with a nine-month old baby, who was living with a family friend in downtown Richmond; and eighteen-year-old Terry Parks, a mother of a two-year-old, who was sharing her East Oakland apartment with twenty-year-old Dana Little and her five-year-old son. The group also included twenty-year-old Diane Harris, who had become pregnant at seventeen and within months had exchanged a middle-class lifestyle for that of a welfare mother and was now living in a run-down apartment in East Oakland; Lois Patterson, a twenty-seven-year-old mother of two and long-term welfare recipient, who was living with her extended family in a small, crowded house in East Oakland; and Evie Jenkins, a forty-three-year-old mother of two, who was living on monthly disability insurance in a housing project near downtown Richmond. Like Diane Harris, Evie lost her middle-class status when she became a teenage welfare mother at age seventeen.

I accompanied these women to the Alternative Center, to the welfare office, and to visits with their mothers. Some of the teen mothers could not find private places to talk, so we talked in the back seat of my car, over lunch or dinner in coffee shops, in a shopping mall, at teenage program meetings, or while moving boxes to a new apartment—in other words, anywhere they would let me join them.

Interviewing the teen mothers on a regular basis was difficult: they frequently moved, appointments were missed, telephones were disconnected. One day I tried to call five mothers about planned participant observation sessions only to find all their telephones disconnected. A few mothers were willing to be interviewed because they thought they would benefit in some way. One mother let me interview her because she thought I had access to housing and could get her an apartment. Another thought I would be able to get her into a teen parent program. A few mothers did not bother returning my telephone calls once they discovered I could not pay them.

I did not pay the teen mothers or the others for taking part in these interviews. In exchange for their information, I told the teen mothers about my own family, gave out information about welfare assistance and teen parent programs, and drove them to various stores. I helped De Vonya Smalls move into her first apartment. I went out with the teen mothers to eat Chinese food, shared takeout dinners, and bought potato chips and sodas for, so it seemed, everyone's sisters, brothers, and cousins. I was in some homes so often that the families began to treat me like a friend.

I found myself caught up in the teen mothers' lives more than I had planned. I was able to capture changes in their lives. I watched a teen mother break up with her baby's father. I witnessed DeVonya Smalls and Shana Leeds move in and out of three different homes. I saw Shana Leeds go through the process of applying for AFDC. I sat through long afternoons with Diane Harris discussing her baby's "womanizing" father, only to attend their wedding a few months later.

I also talked to everyone else I could, including the teen mothers' mothers, Black and White teenage girls who were not mothers, teachers, counselors, directors of teen programs, social workers, and Planned Parenthood counselors. Many have definite views about teenage mothers, some representing a more conservative voice than we usually hear in the Black community.

Sadly, most of the teen mothers' fathers and their babies' fathers were not involved in their lives in any significant way. The teen mothers' lack of knowledge about the babies' fathers' whereabouts made it impossible for me to interview the men. The few men who were still involved with the teen mothers refused to be interviewed. The best I could do was to observe some of the dynamics between two teen fathers and mothers.

PERSONAL HISTORIES

The teen mothers' ages ranged from fourteen to forty-three. Seventeen of them were currently teen mothers (aged fourteen to nineteen), and fifteen were older women who had previously been teen mothers (aged twenty to forty-three).

The presence of the two age groups enabled me to appreciate the dynamic quality and long-term effects of teenage pregnancy on the mothers. The current teen mothers brought to the study a "here and now" aspect: I witnessed some of the family drama as it unfolded. The older women brought a sense of history and their reflective skills; the problems of being a teenage mother did not disappear when the teenage mothers became adults. The older women's stories served two goals for this book: to show that the black community has a history of not condoning teenage motherhood, and to locate emerging problems within the structural changes of our society that have affected everyone in recent years. . . .

As a group, the teen mothers' personal histories reveal both common and not so common patterns among teenage mothers. The youngest teen mother was fourteen and the oldest was eighteen at the time of their first pregnancies. Seventeen teen mothers were currently receiving welfare aid. But contrary to the commonly held assumption that welfare mothers beget welfare mothers, only five teen mothers reported that their families had been on welfare for longer than five years. Twenty-four of the teen mothers had grown up in families headed by a single mother—a common pattern among teenage mothers. Thirteen reported that their mothers had been teenage mothers. Unlike other studies that focus on poor teenage mothers, this study also included five middle-class and three working-class teenage mothers whose parents were teachers, civil service managers, or nursing assistants. Nine of the teen mothers were attending high school (of whom six were attending alternative high schools). Several had taken college courses, and two had managed to obtain a college degree.

Along with capturing an ethnographic snapshot of the seven teen mothers, I conducted semistructured interviews in which I asked all the teen mothers specific questions about their experiences before, during, and after their pregnancies. I asked questions about various common perceptions: the idea of passive and promiscuous teenage girls, the role of men in their lives, the notion of strong cultural support for their pregnancies, the concept of extended family support networks, and the idea that teenage mothers have babies in order to receive welfare aid. Each teenage mother was interviewed for two to two and one-half hours. I audiotaped and transcribed all of the interviews.

I transcribed the material verbatim except for names and other identifying markers, which were changed during the transcription. I coded each teen mother on background variables and patterns. I read and reread my fieldnotes, supporting documents, and relevant literature. For this book I chose those quotations that would best represent typical responses, overall categories, and major themes. I used quotations from the core sample of seven as well as from the larger sample of thirty-two to include a wide range of responses.

Whenever possible I have tried to capture the teen mothers' emotional responses to the questions or issues. Often a teen mother would express through a sigh or a laugh feelings about some issue that contradicted her verbal response. For instance, when Terry Parks laughed as she described her feelings about being on welfare, I added a note about her laughter because it indicated to me that she was embarrassed about the subject. Without that notation, I would not have been able to communicate the emotional intensity with which she said the word "welfare" as she talked about her welfare experiences.

THROUGH THE ETHNOGRAPHIC LENS

I use an ethnographic approach to provide an intricate picture of how gender and poverty dictate the lives of these young teenage mothers and how societal gender, race, and class struggles are played out at the personal level. An ethnographic approach can bridge the gap between the sociological discussion of field research and the actual field experience. Studying these women through the lens of ethnography helped me move the teen mothers' personal stories to an objective level of analysis. The ethnographic method allowed the teen mothers to express to me personal information that was close to the heart. The method also allowed me to bring these Black teenage mothers into sociology's purview, to better understand them as persons, to make their voices heard, and to make their lives important to the larger society. The interviews and observations show that Black teenage girls' experiences are structural and troublesome. At all times I have attempted to make these teen mothers' stories real and visible by presenting the teen mothers' own words with as little editing as possible and by revealing their own insights into the interlocking structures of gender, race, and class.

THE INSIDER INTERVIEWER

I could not walk easily into some teen mothers' lives. Being close to the people being interviewed made me both pleased and tense. Being an insider—someone sharing the culture, community, ethnicity, or gender background of the study participants—has its advantages and disadvantages. When the interviewer can identify with the class and ethnic background of the person being interviewed, there is a greater chance of establishing rapport. The person will express a greater range of attitudes and opinions, especially when the opinions to be expressed are somewhat opposed to general public opinion. The situation is more complex when interviewees are asked to reveal information that may serve the researcher's interest but not that of the group involved. "Don't wash dirty linen in public," they remind the researcher.

The most difficult questions I faced, as do most insider interviewers, had to do with the politics of doing interviews in my own community. As an insider I had to decide whether making certain issues public would benefit the group at the same time that it served my research goals. I imagine that these interviews will raise questions. How will the White community perceive Black families if I discuss the conflicts between teen mothers and their mothers, or fathers who refuse to support their children, or the heavy negative sanctioning of these teen mothers by some in the Black community? My work would be taken out of context, several people warned me.

Every Black researcher who works on issues pertaining to her or his community grapples with these questions. We think about the possibility that our findings may contradict what the Black community wants outsiders to know. Some researchers select nonthreatening topics. Others romanticize Black life despite the evidence that life is hard for those on the bottom. And others simply adopt a code of silence, taking a position similar to that of the Black college teacher who in another context made the point to me, "I'm socialized to bear my pain in silence and not go blabbing about my problems to White folks, let alone strangers."

Being an insider did not help me gain the confidence of the teen mothers and others immediately. Most were suspicious of researchers. I lost a chance to interview one group of teen mothers involved in a special school project because the counselors who worked with them did not like the way a White male researcher had treated the teen mothers previously. Indeed, these teen mothers had the right to be suspicious. What these girls and women say about their lives can be used against them by public policy makers, since the Black community is often blamed for its own social and economic situations.

But overall, being a Black woman was helpful, because eventually the teen mothers, realizing we had much in common, stopped being suspicious of me and began to talk candidly of their lives. Occasionally I could not find a babysitter and had to bring my little boy along. I found my son's presence helped reduce the aloofness of my role as researcher and the powerlessness of the teens' position as interview subjects. I was surprised at how helpful my son was in breaking through the first awkward moments. We made him the topic of discussion— mothers can always compare child-care problems. His presence also helped me counter some of the teenagers' tendencies to deny problems. When I talked to De Vonya Smalls about my son's effects on my own schedule, like having to get up at five in the morning instead of at seven, she relaxed and told me about her efforts to study for a test while her baby cried for attention. She also admitted to doing poorly in school.

I decided to study these teenage mothers because Black teenage mothers are not going away, no matter how much we ignore, romanticize, or remain silent about their lives. I strongly disagree with approaches that let the group's code of silence supersede the need to understand the problems and issues of Black teenage mothers. That kind of false ideology only perpetuates the myths about Black teenage motherhood and causes researchers to neglect larger sociological issues or fail to ask pertinent questions about the lives of these mothers. In the name of racial pride, then, we essentially overlook how the larger society shares a great deal of responsibility for these problems. The only way to reduce the number of teenage pregnancies or to improve the lives of teenage mothers is to understand the societal causes by examining the realities of these girls' lives. The time had arrived, as Nate Hare put it, for an end to the unrealistic view of Black lives.

. . . These Black adolescent girls in this study have been abandoned by the educational system and locked out of the job market. They offer a more complex view of Black families than most sociological assessments, in which Black families often appear as either deviant or saintly, but seldom as real as these girls portrayed them. They not only experience this complicated reality but also understand it, and they can vividly describe how they experience these structural constructs and constraints. At the same time, they are deprived of other critical knowledge. Like the adolescent girls in Carol Gilligan's study, these girls are deprived of information about their bodies and sexuality that is critical to their decision making.

The interviews in this study eloquently demonstrate that these young women are not exhibiting lower-class or pathological behavior. . . . Rather, the lives of poor and middle-class Black teenage mothers add complexity to the structural perspective, providing rich evidence that they are not the "other," as the culture-of-poverty literature suggests. . . .

The teen mothers in this study have been abandoned by an educational system that cannot offer them a safe environment conducive to learning, let alone the instruments of learning they need. Their teachers do not have the administrative, community, and financial support necessary to provide uncrowded classrooms in which respect for the other is fostered and excitement about learning is possible. The students do not have up-to-date textbooks, paper supplies, or equipment and training for our computer age. In addition, the schools do not provide girls with an educational environment in which they are perceived to be other than sexual objects. . . .

. . . At an early age, these teen mothers are forsaken by the schools in a variety of ways. They do not have physically safe classrooms in which respect for girls is fostered. They are perceived as and expected to behave as purely sex objects, or they are forgotten and left unprotected because teachers (and men in power) do not believe young Black girls need to concern themselves with acquiring a good education and securing long-term careers; after all, their main concern in life is really having babies and collecting welfare checks. . . .

. . . [W]e have restricted children's and adolescents' knowledge of the maturation process and their sexuality. That children and adolescents, especially girls, are kept ignorant about the developmental process does not foster competence and maturity; rather, it fosters complacency and helplessness. It also puts our adolescent girls at risk, for without knowledge and critical thinking skills, our girls cannot learn to make responsible choices that will ultimately keep them safe.

To make matters more complicated as [these girls were] beginning to develop physically, [they were] facing a perplexing double standard: with . . . development came sanctions against sexual freedom, while young men who were also experiencing adolescent development were granted the right to sexual freedom, from which they inferred the right to access teen girls' bodies. The admonishment to be a "nice girl" collided with the strategies of young men who counted on nice girls to "cave in" to such strategies: real men have sex no matter how they get it. Along with these problems, . . . Black teenage girls began to see little chance of acquiring a decent education, stable employment, marriage, and a comfortable family home—the American Dream. . . .

THE EFFECTS OF RACE AND CLASS

At the time when teenage pregnancies began to increase in the Black community, devastating workplace changes were occurring. Jobs were becoming scarce for many workers, public housing projects were beginning to appear in the community, pushing more people into less space, and mom and pop stores were moving out. Blacks were finding it more and more difficult to find and keep traditional jobs because of the sweeping economic changes occurring everywhere. Many laborers found themselves replaced by cheaper labor. White women began taking service jobs. Jobs ordinarily done by teenagers, such as paper deliveries, baby-sitting, and clerking at local stores, were taken over by adults. Finding themselves with no future, Black girls refocused their attention away from school and onto personal relationships. . . .

Why, then, did these teenage mothers have babies? If we use the village as a metaphor, we must say that in absence of support from the members of a village, in which everyone pools their emotional and economic resources, these teen mothers had babies because they were isolated from society and unwanted by everyone around them. In order for them to grow, they needed to be loved and nourished. Without both, they created love where they could. These interviews confirm De Vonya's revelation that motherhood is a strategy: "I can be a mother if I can't be anything else." . . .

KEY CONCEPTS

American dream   ethnography   ideology
participant observation

DISCUSSION QUESTIONS

1. How did Elaine Bell Kaplan's status as a Black woman give her an insider's view of Black teenage mothers? In what ways did she remain an outsider and how does this affect her research?

2. Why is participant observation a particularly good research method for investigating the questions that Kaplan was asking about Black teen mothers?

4

Promoting Bad Statistics

JOEL BEST

In this article, Joel Best points out how numbers publicly used to describe social problems can be misleading. He shows how advocacy about a given problem can distort accurate, empirical observations. In addition, he emphasizes that statistical information is produced in a social context.

In contemporary society, social problems must compete for attention. To the degree that one problem gains media coverage, moves to the top of politicians' agendas, or becomes the subject of public concern, others will be neglected. Advocates find it necessary to make compelling cases for the importance of particular social problems. They choose persuasive wording and point to disturbing examples, and they usually bolster their case with dramatic statistics.

Statistics have a fetish-like power in contemporary discussions about social problems. We pride ourselves on rational policy making, and expertise and evidence guide our rationality. Statistics become central to the process: numbers evoke science and precision; they seem to be nodules of truth, facts that distill the simple essence of apparently complex social processes. In a culture that treats facts and opinions as dichotomous terms, numbers signify truth—what we call "hard facts." In virtually every debate about social problems, statistics trump "mere opinion."

Yet social problems statistics often involve dubious data. While critics occasionally call some number into question, it generally is not necessary for a statistic to be accurate—or even plausible—in order to achieve widespread acceptance. Advocates seeking to promote social problems often worry more about the processes by which policy makers, the press, and the public come to focus on particular problems, than about the quality of their figures. I seek here to identify some principles that govern this process. They are, if you will, guidelines for creating and disseminating dubious social problems statistics.

Although we talk about facts as though they exist independently of people, patiently awaiting discovery, someone has to produce—or construct—all that we know. Every social statistic reflects the choices that go into producing it. The key choices involve definition and methodology: Whenever we count something, we must first define what it is we hope to count, and then choose the methods by

SOURCE: Joel Best, "Promoting Bad Statistics." *Society*, March/April 2001: 11–15.

which we will go about counting. In general, the press regards statistics as facts, little bits of truth. The human choices behind every number are forgotten; the very presentation of a number gives each claim credibility. In this sense, statistics are like fetishes.

ANY NUMBER IS BETTER THAN NO NUMBER

By this generous standard, a number need not bear close inspection, or even be remotely plausible. To choose an example first brought to light by Christina Hoff Sommers, a number of recent books, both popular and scholarly, have repeated the garbled claim that anorexia kills 150,000 women annually. (The figure seems to have originated from an estimate for the total number of women who are anorexic; only about 70 die each year from the disease.) It should have been obvious that something was wrong with this figure. Anorexia typically affects *young* women. Each year, roughly 8,500 females aged 15–24 die from all causes; another 47,000 women aged 25–44 also die. What are the chances, then, that there could be 150,000 deaths from anorexia each year? But, of course, most of us have no idea how many young women die each year—("It must be a lot. . . . "). When we hear that anorexia kills 150,000 young women per year, we assume that whoever cites the number must know that it is true. It is, after all, a number and therefore presumably factual.

Oftentimes, social problems statistics exist in splendid isolation. When there is only one number, that number has the weight of authority. It is accepted and repeated. People treat the statistic as authoritative because it is a statistic. Often, these lone numbers come from activists seeking to draw attention to neglected social phenomena. One symptom of societal neglect is that no one has bothered to do much research or compile careful records; there often are no official statistics or other sources for more accurate numbers. When reporters cover the story, they want to report facts. When activists have the only available figures, their numbers look like facts, so, in the absence of other numbers, the media simply report the activists' statistics.

Once a number appears in one news report, that story becomes a potential source for everyone seeking information about the social problem; officials, experts, activists, and other reporters routinely repeat figures that appear in press reports.

NUMBERS TAKE ON LIVES OF THEIR OWN

David Luckenbill has referred to this as "number laundering." A statistic's origin—perhaps simply as someone's best guess—is soon forgotten, and through repetition, the figure comes to be treated as a straightforward fact—accurate and authoritative. The trail becomes muddy, and people lose track of the estimate's original source, but they become confident that the number must be correct because it appears everywhere.

It barely matters if critics challenge a number, and expose it as erroneous. Once a number is in circulation, it can live on, regardless of how thoroughly it may have been discredited. Today's improved methods of information retrieval—electronic indexes, full-text databases, and the Internet—make it easier than ever to locate statistics. Anyone who locates a number can, and quite possibly will, repeat it. That annual toll of 150,000 anorexia deaths has been thoroughly debunked, yet the figure continues to appear in occasional newspaper stories. Electronic storage has given us astonishing, unprecedented access to information, but many people have terrible difficulty sorting through what's available and distinguishing good information from bad. Standards for comparing and evaluating claims seem to be wanting. This is particularly true for statistics that are, after all, numbers and therefore factual, requiring no critical evaluation. Why not believe and repeat a number that everyone else uses? Still, some numbers do have advantages.

BIG NUMBERS ARE BETTER THAN LITTLE NUMBERS

Remember: social problems claims must compete for attention; there are many causes and a limited amount of space on the front page of the *New York Times*, Advocates must find ways to make their claims compelling; they favor melodrama—terrible villains, sympathetic, vulnerable victims, and big numbers. Big numbers suggest that there is a big problem, and big problems demand attention, concern, action. They must not be ignored.

Advocates seeking to attract attention to a social problem soon find themselves pressed for numbers. Press and policy makers demand facts ("You say it's a problem? Well, how big a problem is it?"). Activists believe in the problem's seriousness, and they often spend much of their time talking to others who share that belief. They know that the problem is much more serious, much more common than generally recognized ("The cases we know about are only the tip of the iceberg."). When asked for figures, they thus offer their best estimates, educated guesses, guesstimates, ballpark figures, or stabs in the dark. Mitch Snyder, the most visible spokesperson for the homeless in the early 1980s, explained on ABC's "Nightline" how activists arrived at the figure of three million homeless: "Everybody demanded it. Everybody said we want a number. . . . We got on the phone, we made a lot of calls, we talked to a lot of people, and we said, 'Okay, here are some numbers.' They have no meaning, no value." Because activists sincerely believe that the new problem is big and important, and because they suspect that there is a very large dark figure of unreported or unrecorded cases, activists' estimates tend to be high, and to err on the side of exaggeration.

This helps explain the tendency to estimate the scope of social problems in large, suspiciously round figures. There are, we are told, one million victims of elder abuse each year, two million missing children, three million homeless, 60 million functionally illiterate Americans; child pornography may be, depending on your source, a $1 billion or $46 billion industry, and so on. Often, these estimates are the only available numbers.

The mathematician John Allen Paulos argues that innumeracy—the mathematical counterpart to illiteracy—is widespread and consequential. He suggests that innumeracy particularly shapes the way we deal with large numbers. Most of us understand hundreds, even thousands, but soon the orders of magnitude blur into a single category: "It's a lot." Even the most implausible figures can gain widespread acceptance. When missing-children advocates charged that nearly two million children are missing each year, anyone might have done the basic math: there are about 60 million children under 18; if two million are missing, that would be one in 30; that is, every year, the equivalent of one child in every American schoolroom would be missing. A 900-student school would have 30 children missing from its student body each year. To be sure, the press debunked this statistic in 1985, but only four years after missing children became a highly publicized issue and the two-million estimate gained wide circulation. And, of course, having been discredited, the number survives and can still be encountered on occasion.

It is remarkable how often contemporary discussions of social problems make no effort to define what is at issue. Often, we're given a dramatic, compelling example, perhaps a tortured, murdered child, then told that this terrible case is an example of a social problem—in this case, child abuse—and finally given a statistic: "There are more than three million reports of child abuse each year." The example, coupled with the problem's name, seems sufficient to make the definition self-evident. However, definitions cannot always be avoided.

DEFINITIONS: BETTER BROAD THAN NARROW

Because broad definitions encompass more kinds of cases, they justify bigger numbers, and we have already noted the advantages of big numbers. No definition is perfect; there are two principal ways definitions of social problems can be flawed. On the one hand, a definition might be too broad and encompass more than it ought to include. That is, broad definitions tend to identify what methodologists call false positives; they include some cases that arguably ought not to be included as part of the problem. On the other hand, a definition that is too narrow may exclude false negatives, cases that perhaps ought to be included as part of the problem.

In general, activists trying to promote a new social problem view false negatives as more troubling than false positives. Activists often feel frustrated trying to get people concerned about some social condition that has been ignored. The general failure to recognize and acknowledge that something is wrong is part of what the activists want to correct; therefore, they may be especially careful not to make things worse by defining the problem too narrowly. A definition that is too narrow fails to recognize a problem's full extent; in doing so, it helps perpetuate the history of neglecting the problem. Some activists favor definitions broad enough to encompass every case that ought to be included; that is, they promote broad definitions in hopes of eliminating all false negatives.

However, broad definitions may invite criticism. They include cases that not everyone considers instances of social problems; that is, while they minimize false negatives, they do so at the cost of maximizing cases that critics may see as false positives. The rejoinder to this critique returns us to the idea of neglect and the harm it causes. Perhaps, advocates acknowledge, their definitions may seem to be too broad, to encompass cases that seem too trivial to be counted as instances of the social problem. But how can we make that judgment? Here advocates are fond of pointing to terrible examples, to the victim whose one, brief, comparatively mild experience had terrible personal consequences; to the child who, having been exposed to a flasher, suffers a lifetime of devastating psychological consequences. Perhaps, advocates say, other victims with similar experiences suffer less or at least seem to suffer less. But is it fair to define a problem too narrowly to include everyone who suffers? Shouldn't our statistics measure the problem's full extent? While social problems statistics often go unchallenged, critics occasionally suggest that some number is implausibly large, or that a definition is too broad.

DEFENDING NUMBERS BY ATTACKING CRITICS

When activists have generated a statistic as part of a campaign to arouse concern about some social problem, there is a tendency for them to conflate the number with the cause. Therefore, anyone who questions a statistic can be suspected of being unsympathetic to the larger claims, indifferent to the victims' suffering, and so on. *Ad hominem* attack on the motives of individuals challenging numbers is a standard response to statistical confrontations. These attacks allow advocates to refuse to budge; making *ad hominem* arguments lets them imply that their opponents don't want to acknowledge the truth, that their statistics are derived from ideology, rather than methodology. If the advocates' campaign has been reasonably successful, they can argue that there is now widespread appreciation that this is a big, serious problem; after all, the advocates' number has been widely accepted and repeated, surely it must be correct. A fallback stance—useful in those rare cases where public scrutiny leaves one's own numbers completely discredited—is to treat the challenge as meaningless nitpicking. Perhaps our statistics were flawed, the advocates acknowledge, but the precise number hardly makes a difference ("After all, even one victim is too many.").

Similarly, criticizing definitions for being too broad can provoke angry reactions. For advocates, such criticisms seem to deny victim's suffering, minimize the extent of the problem, and by extension endorse the status quo. If broader definitions reflect progress, more sensitive appreciation of the true scope of social problems, then calls for narrowing definitions are retrograde, insensitive refusals to confront society's flaws.

Of course, definitions must be operationalized if they are to lead to statistics. It is necessary to specify how the problem will be measured and the statistic produced. If there is to be a survey, who will be sampled? And how will the questions be worded? In what order will they be asked? How will the responses be coded? Most of what we call social-scientific methodology requires choosing how to measure social phenomena. Every statistic depends upon these choices. Just as advocates' preference for large numbers leads them to favor broad definitions, the desirability of broad definitions shapes measurement choices.

MEASURES: BETTER INCLUSIVE THAN EXCLUSIVE

Most contemporary advocates have enough sociological sophistication to allude to the dark figure—that share of a social problem that goes unreported and unrecorded. Official statistics, they warn, inevitably underestimate the size of social problems. This undercounting helps justify advocates' generous estimates (recall all those references to "the tip of the iceberg"). Awareness of the dark figure also justifies measurement decisions that maximize researchers' prospects for discovering and counting as many cases as possible.

Consider the first federally sponsored National Incidence Studies of Missing, Abducted, Runaway, and Thrownaway Children (NISMART). This was an attempt to produce an accurate estimate for the numbers of missing children. To estimate family abductions (in which a family member kidnaps a child) researchers conducted a telephone survey of households. The researchers made a variety of inclusive measurement decisions: an abduction could involve moving a child as little as 20 feet; it could involve the child's complete cooperation; there was no minimum time that the abduction had to last; those involved may not have considered what happened an abduction; and there was no need that the child's whereabouts be unknown (in most family abductions identified by NISMART, the child was not with someone who had legal custody, but everyone knew where the child was). Using these methods of measurement, a non-custodial parent who took a child for an unauthorized visit, or who extended an authorized visit for an extra night, was counted as having committed a "family abduction." If the same parent tried to conceal the taking or to prevent the custodial parent's contact with the child, the abduction was classified in the most serious ("policy-focal") category. The NISMART researchers concluded that there were 163,200 of these more serious family abductions each year, although evidence from states with the most thorough missing-children reporting systems suggests that only about 9,000 cases per year come to police attention. In other words, the researchers' inclusive measurement choices led to a remarkably high estimate. Media coverage of the family-abduction problem coupled this high figure with horrible examples—cases of abductions lasting years, involving long-term sexual abuse, ending in homicide, and so on. Although most of the episodes identified by NISMART's methods were relatively minor, the press implied that very serious cases were very common ("It's a big number!").

There is nothing atypical about the NISMART example. Advocacy research has become an important source of social problems statistics. Advocates hope research will produce large numbers, and they tend to believe that broad definitions are justified. They deliberately adopt inclusive research measurements that promise to minimize false negatives and generate large numbers. These measurement decisions almost always occur outside public scrutiny and only rarely attract attention. When the media report numbers, percentages, and rates, they almost never explain the definitions and measurements used to produce those statistics.

While many statistics seem to stand alone, occasions do arise when there are competing numbers or contradictory statistical answers to what seems to be the same question. In general, the media tend to treat such competing numbers with a sort of even-handedness.

COMPETING NUMBERS ARE EQUALLY GOOD

Because the media tend to treat numbers as factual, and to ignore definitions and measurement choices, inconsistent numbers pose a problem. Clearly, both numbers cannot be correct. Where a methodologist might try to ask how different advocates arrived at different numbers (in hopes of showing that one figure is more accurate than another, or at least of understanding how the different numbers might be products of different methods), the press is more likely to account for any difference in terms of the competitors' conflicting ideologies or agendas.

Consider the case of the estimates for the crowd size at the 1995 Million Man March. The event's very name set a standard for its success: as the date for the March approached, its organizers insisted that it would attract a million people, while their critics predicted that the crowd would never reach that size. On the day of the march, the organizers announced success: there were, they said, 1.5 to 2 million people present. Alas, the National Park Service Park Police, charged by Congress with estimating the size of demonstrations on the Capitol Mall, calculated that the march drew only 400,000 people (still more than any previous civil rights demonstration). The Park Police knew the Mall's dimensions, took aerial photos, and multiplied the area covered by the crowd by a multiplier based on typical crowd densities. The organizers, like the organizers of many previous demonstrations on the Mall, insisted that the Park Police estimate was far too low. Enter a team of aerial photo analysts from Boston University who eventually calculated that the crowd numbered 837,000 plus or minus 25 percent (i.e., they suggested there might have been a million people in the crowd).

The press covered these competing estimates in standard "he said–she said" style. Few reporters bothered to ask why the two estimates were different. The answer was simple: the BU researchers used a different multiplier. Where the Park Police estimated that there was one demonstrator per 3.6 square feet (actually a fairly densely-packed crowd), the BU researchers calculated that there was a person for every 1.8 square feet (the equivalent of being packed in a crowded elevator). But rather than trying to compare or evaluate the processes by which people arrived at the different estimates, most press reports treated the numbers as equally valid, and implied that the explanation for the difference lay in the motives of those making the estimates.

The March organizers (who wanted to argue that the demonstration had been successful) produced a high number; the Park Police (who, the March organizers insisted, were biased against the March) produced a low one, and the BU scientists (presumably impartial and authoritative) found something in between. The BU estimate quickly found favor in the media: it let the organizers save face (because the BU team conceded the crowd might have reached one million); it seemed to split the difference between the high and low estimates; and it apparently came from experts. There was no effort to judge the competing methods and assumptions behind the different numbers, for example, to ask whether it was likely that hundreds of thousands of men stood packed as close together as the BU researchers imagined for the hours the demonstration lasted.

This example, like those discussed earlier, reveals that public discussions of social statistics are remarkably unsophisticated. Social scientists advance their careers by using arcane inferential statistics to interpret data. The standard introductory undergraduate statistics textbook tends to zip through descriptive statistics on the way to inferential statistics. But it is descriptive statistics— simple counts, averages, percentages, rates, and the like—that play the key role is public discussions over social problems and social policy. And the level of those discussions is not terribly advanced. There is too little critical thinking about social statistics. People manufacture, and other people repeat, dubious figures. While this can involve deliberate attempts to deceive and manipulate, this need not be the case. Often, the people who create the numbers—who, as it were, make all those millions—believe in them. Neither the advocates who create statistics, nor the reporters who repeat them, nor the larger public questions the figures.

What Paulos calls innumeracy is partly to blame—many people aren't comfortable with basic ideas of numbers and calculations. But there is an even more fundamental issue: many of us do not appreciate that every number is a social construction, produced by particular people using particular methods. The naïve, but widespread, tendency is to treat statistics as fetishes, that is, as almost magical nuggets of fact, rather than as someone's efforts to summarize, to simplify complexity. If we accept the statistic as a fetish, then several of the guidelines I have outlined make perfect sense. Any number is better than no number, because the number represents truth. Numbers take on lives of their own because they are true, and their truth justifies their survival. The best way to defend a number is to attack its critics' motives, because anyone who questions a presumably true number must have dubious reasons for doing so. And, when we are confronted with competing numbers, those numbers are equally good, because, after all, they are somehow equivalent bits of truth. At the same time, the guidelines offer those who must produce numbers justifications for favoring big numbers, broad definitions, and inclusive methods. Again, this need not be cynical. Often, advocates are confident that they know the truth, and they approach collecting statistics as a straightforward effort to generate the numbers needed to document what they, after all, know to be true.

Any effort to improve the quality of public discussion of social statistics needs to begin with the understanding that numbers are socially constructed. Statistics are not nuggets of objective fact that we discover; rather, they are people's creations. Every statistic reflects people's decisions to count, their choices of what to count and how to go about counting it, and so on. These choices inevitably shape the resulting numbers.

Public discussions of social statistics need to chart a middle path between naivete (the assumption that numbers are simply true) and cynicism (the suspicion that figures are outright lies told by people with bad motives). This middle path needs to be critical. It needs to recognize that every statistic has to be created, to acknowledge that every statistic is imperfect, yet to appreciate that statistics still offer an essential way of summarizing complex information. Social scientists have a responsibility to promote this critical stance in the public, within the press, and among advocates.

KEY CONCEPTS

anorexia nervosa   debunking   methodology

DISCUSSION QUESTIONS

1. What does Best mean that numbers are social constructions?

2. Find an example in the media of where someone is using numbers to promote concern about a particular social problem. Based on Best's article, what questions would you need to ask to find out if the numbers are accurate?

Applying Sociological Knowledge:
An Exercise for Students

One of the points in this section is how a sociological perspective differs from an individualistic or even psychological perspective. Pick one of the following topics: teen pregnancy, unemployment, child abuse. Compare and contrast what factors might be important to consider when explaining this social issue using a sociological perspective versus a psychological perspective.

5

Size 6: The Western Women's Harem

FATIMA MERNISSI

In this article, Mernissi compares the pressure for American women to fit into small-sized clothing to the enforcement of Muslim women wearing veils to cover their faces. She argues that the male–dominated fashion industry creates cultural expectations that demean women as they age and gain weight. For a woman to wear a size larger than 6 and to be older in appearance is to violate cultural expectations of femininity and will limit her options in society. This article highlights the unique cultural standards in both Western and non–Western societies that work to create and enforce gender inequality.

It was during my unsuccessful attempt to buy a cotton skirt in an American department store that I was told my hips were too large to fit into a size 6. That distressing experience made me realize how the image of beauty in the West can hurt and humiliate a woman as much as the veil does when enforced by the state police in extremist nations such as Iran, Afghanistan, or Saudi Arabia. Yes, that day I stumbled onto one of the keys to the enigma of passive beauty in Western harem fantasies. The elegant saleslady in the American store looked at me without moving from her desk and said that she had no skirt my size. "In this whole big store, there is no skirt for me?" I said. "You are joking." I felt very suspicious and thought that she just might be too tired to help me. I could understand that. But then the saleswoman added a condescending judgment, which sounded to me like an Imam's *fatwa*. It left no room for discussion:

"You are too big!" she said.

"I am too big compared to what?" I asked, looking at her intently, because I realized that I was facing a critical cultural gap here.

"Compared to a size 6," came the saleslady's reply.

Her voice had a clear-cut edge to it that is typical of those who enforce religious laws. "Size 4 and 6 are the norm," she went on, encouraged by my bewildered look. "Deviant sizes such as the one you need can be bought in special stores."

That was the first time that I had ever heard such nonsense about my size. In the Moroccan streets, men's flattering comments regarding my particularly

SOURCE: Reprinted with permission of Pocket Books, a Division of Simon & Schuster Adult Publishing Group, from *Scheherazade Goes West* by Fatima Mernissi. Copyright © 2001 by Fatima Mernissi. All Rights Reserved. New York: Pocket Books, pp. 208–220.

generous hips have for decades led me to believe that the entire planet shared their convictions. It is true that with advancing age, I have been hearing fewer and fewer flattering comments when walking in the medina, and sometimes the silence around me in the bazaars is deafening. But since my face has never met with the local beauty standards, and I have often had to defend myself against remarks such as *zirafa* (giraffe), because of my long neck, I learned long ago not to rely too much on the outside world for my sense of self-worth. In fact, paradoxically, as I discovered when I went to Rabat as a student, it was the self-reliance that I had developed to protect myself against "beauty blackmail" that made me attractive to others. My male fellow students could not believe that I did not give a damn about what they thought about my body. "You know, my dear," I would say in response to one of them, "all I need to survive is bread, olives, and sardines. That you think my neck is too long is your problem, not mine."

In any case, when it comes to beauty and compliments, nothing is too serious or definite in the medina, where everything can be negotiated. But things seemed to be different in that American department store. In fact, I have to confess that I lost my usual self-confidence in that New York environment. Not that I am always sure of myself, but I don't walk around the Moroccan streets or down the university corridors wondering what people are thinking about me. Of course, when I hear a compliment, my ego expands like a cheese soufflé, but on the whole, I don't expect to hear much from others. Some mornings, I feel ugly because I am sick or tired; others, I feel wonderful because it is sunny out or I have written a good paragraph. But suddenly, in that peaceful American store that I had entered so triumphantly, as a sovereign consumer ready to spend money, I felt savagely attacked. My hips, until then the sign of a relaxed and uninhibited maturity, were suddenly being condemned as a deformity.

"And who decides the norm?" I asked the saleslady, in an attempt to regain some self-confidence by challenging the established rules. I never let others evaluate me, if only because I remember my childhood too well. In ancient Fez, which valued round-faced plump adolescents, I was repeatedly told that I was too tall, too skinny, my cheekbones were too high, my eyes were too slanted. My mother often complained that I would never find a husband and urged me to study and learn all that I could, from storytelling to embroidery, in order to survive. But I often retorted that since "Allah had created me the way I am, how could he be so wrong, Mother?" That would silence the poor woman for a while, because if she contradicted me, she would be attacking God himself. And this tactic of glorifying my strange looks as a divine gift not only helped me to survive in my stuffy city, but also caused me to start believing the story myself. I became almost self-confident. I say almost, because I realized early on that self-confidence is not a tangible and stable thing like a silver bracelet that never changes over the years. Self-confidence is like a tiny fragile light, which goes off and on. You have to replenish it constantly.

"And who says that everyone must be a size 6?" I joked to the saleslady that day, deliberately neglecting to mention size 4, which is the size of my skinny twelve-year-old niece.

At that point, the saleslady suddenly gave me an anxious look. "The norm is everywhere, my dear," she said. "It's all over, in the magazines, on television, in the ads. You can't escape it. There is Calvin Klein, Ralph Lauren, Gianni Versace, Giorgio Armani, Mario Valentino, Salvatore Ferragamo, Christian Dior, Yves Saint-Laurent, Christian Lacroix, and Jean-Paul Gaultier. Big department stores go by the norm." She paused and then concluded, "If they sold size 14 or 16, which is probably what you need, they would go bankrupt."

She stopped for a minute and then stared at me, intrigued. "Where on earth do you come from? I am sorry I can't help you. Really, I am." And she looked it too. She seemed, all of a sudden, interested, and brushed off another woman who was seeking her attention with a cutting, "Get someone else to help you, I'm busy." Only then did I notice that she was probably my age, in her late fifties. But unlike me, she had the thin body of an adolescent girl. Her knee-length, navy blue, Chanel dress had a white silk collar reminiscent of the subdued elegance of aristocratic French Catholic schoolgirls at the turn of the century. A pearl-studded belt emphasized the slimness of her waist. With her meticulously styled short hair and sophisticated makeup, she looked half my age at first glance.

"I come from a country where there is no size for women's clothes," I told her. "I buy my own material and the neighborhood seamstress or craftsman makes me the silk or leather skirt I want. They just take my measurements each time I see them. Neither the seamstress nor I know exactly what size my skirt is. We discover it together in the making. No one cares about my size in Morocco as long as I pay taxes on time. Actually, I don't know what my size is, to tell you the truth."

The saleswoman laughed merrily and said that I should advertise my country as a paradise for stressed working women. "You mean you don't watch your weight?" She inquired, with a tinge of disbelief in her voice. And then, after a brief moment of silence, she added in a lower register, as if talking to herself: "Many women working in highly paid fashion-related jobs could lose their positions if they didn't keep to a strict diet."

Her words sounded so simple, but the threat they implied was so cruel that I realized for the first time that maybe "size 6" is a more violent restriction imposed on women than is the Muslim veil. Quickly I said good-bye so as not to make any more demands on the saleslady's time or involve her in any more unwelcome, confidential exchanges about age-discriminating salary cuts. A surveillance camera was probably watching us both.

Yes, I thought as I wandered off, I have finally found the answer to my harem enigma. Unlike the Muslim man, who uses space to establish male domination by excluding women from the public arena, the Western man manipulates time and light. He declares that in order to be beautiful, a woman must look fourteen years old. If she dares to look fifty, or worse, sixty, she is beyond the pale. By putting the spotlight on the female child and framing her as the ideal of beauty, he condemns the mature woman to invisibility. In fact, the modern Western man enforces Immanuel Kant's nineteenth-century theories: To be beautiful, women have to appear childish and brainless. When a woman looks mature and self-assertive, or allows her hips to expand, she is condemned as ugly. Thus, the walls of the European harem separate youthful beauty from ugly maturity.

These Western attitudes, I thought, are even more dangerous and cunning than the Muslim ones because the weapon used against women is time. Time is less visible, more fluid than space. The Western man uses images and spotlights to freeze female beauty within an idealized childhood, and forces women to perceive aging—that normal unfolding of the years—as a shameful devaluation. "Here I am, transformed into a dinosaur," I caught myself saying aloud as I went up and down the rows of skirts in the store, hoping to prove the saleslady wrong—to no avail. This Western time-defined veil is even crazier than the space-defined one enforced by the Ayatollahs.

The violence embodied in the Western harem is less visible than in the Eastern harem because aging is not attacked directly, but rather masked as an aesthetic choice. Yes, I suddenly felt not only very ugly but also quite useless in that store, where, if you had big hips, you were simply out of the picture. You drifted into the fringes of nothingness. By putting the spotlight on the prepubescent female, the Western man veils the older, more mature woman, wrapping her in shrouds of ugliness. This idea gives me the chills because it tattoos the invisible harem directly onto a woman's skin. Chinese foot-binding worked the same way: Men declared beautiful only those women who had small, childlike feet. Chinese men did not force women to bandage their feet to keep them from developing normally—all they did was to define the beauty ideal. In feudal China, a beautiful woman was the one who voluntarily sacrificed her right to unhindered physical movement by mutilating her own feet, and thereby proving that her main goal in life was to please men. Similarly, in the Western world, I was expected to shrink my hips into a size 6 if I wanted to find a decent skirt tailored for a beautiful woman. We Muslim women have only one month of fasting, Ramadan, but the poor Western woman who diets has to fast twelve months out of the year. "*Quelle horreur*," I kept repeating to myself while looking around at the American women shopping. All those my age looked like youthful teenagers.

According to the writer Naomi Wolf, the ideal size for American models decreased sharply in the 1990s. "A generation ago, the average model weighed 8 percent less than the average American woman, whereas today she weighs 23 percent less. . . . The weight of Miss America plummeted, and the average weight of Playboy Playmates dropped from 11 percent below the national average in 1970 to 17 percent below it in eight years."[1] The shrinking of the ideal size, according to Wolf, is one of the primary reasons for anorexia and other health-related problems: "Eating disorders rose exponentially, and a mass of neurosis was promoted that used food and weight to strip women of . . . a sense of control."[2]

Now, at last, the mystery of my Western harem made sense. Framing youth as beauty and condemning maturity is the weapon used against women in the West just as limiting access to public space is the weapon used in the East. The objective remains identical in both cultures: to make women feel unwelcome, inadequate, and ugly.

The power of the Western man resides in dictating what women should wear and how they should look. He controls the whole fashion industry, from cosmetics to underwear. The West, I realized, was the only part of the world where women's fashion is a man's business. In places like Morocco, where you design your own clothes and discuss them with craftsmen and -women, fashion is your own business. Not so in the west. As Naomi Wolf explains in *The Beauty Myth*, men have engineered a prodigious amount of fetish-like, fashion-related paraphernalia: "Powerful industries—the $33-billion-a-year diet industry, the $20-billion cosmetic industry, the $300-million cosmetic surgery industry, and the $7-billion pornography industry—have arisen from the capital made out of unconscious anxieties, and are in turn able, through their influence on mass culture, to use, stimulate, and reinforce the hallucination in a rising economic spiral."[3]

But how does the system function? I wondered. Why do women accept it?

Of all the possible explanations, I like that of the French sociologist, Pierre Bourdieu, the best. In his latest book, *La Domination Masculine*, he proposes something he calls "*la violence symbolique*": "Symbolic violence is a form of power which is hammered directly on the body, and as if by magic, without any apparent physical constraint. But this magic operates only because it activates the codes pounded in the deepest layers of the body."[4] Reading Bourdieu, I had the impression that I finally understood Western man's psyche better. The cosmetic and fashion industries are only the tip of the iceberg, he states, which is why women are so ready to adhere to their dictates. Something else is going on on a far deeper level. Otherwise, why would women belittle themselves spontaneously? Why, argues Bourdieu, would women make their lives more difficult, for example, by preferring men who are taller or older than they are? "The majority of French women wish to have a husband who is older and also, which seems consistent, bigger as far as size is concerned," writes Bourdieu.[5] Caught in the enchanted submission characteristic of the symbolic violence inscribed in the mysterious layers of the flesh, women relinquish what he calls "les signes ordinaries de la hiérarchie sexuelle," the ordinary signs of sexual hierarchy, such as old age and a larger body. By so doing, explains Bourdieu, women spontaneously accept the subservient position. It is this spontaneity Bourdieu describes as magic enchantment.[6]

Once I understood how this magic submission worked, I became very happy that the conservative Ayatollahs do not know about it yet. If they did, they would readily switch to its sophisticated methods, because they are so much more effective. To deprive me of food is definitely the best way to paralyze my thinking capabilities.

Both Naomi Wolf and Pierre Bourdieu come to the conclusion that insidious "body codes" paralyze Western women's abilities to compete for power, even though access to education and professional opportunities seem wide open, because the rules of the game are so different according to gender. Women enter the power game with so much of their energy deflected to their physical appearance that one hesitates to say the playing field is level. "A cultural fixation on female thinness is not an obsession about female beauty," explains Wolf. It is "an obsession about female obedience. Dieting is the most potent political sedative in women's history; a quietly mad population is a tractable one."[7] Research, she contends, "confirmed what most women know too well—that concern with weight leads to a 'virtual collapse of self-esteem and sense of effectiveness' and that ... prolonged and periodic caloric restriction' resulted in a distinctive personality whose traits are passivity, anxiety, and emotionality."[8] Similarly, Bourdieu, who focuses more on how this myth hammers its inscriptions onto the flesh itself, recognizes that constantly reminding women of their physical appearance destabilizes them emotionally because it reduces them to exhibited objects. "By confining women to the status of symbolical objects to be seen and perceived by the other, masculine domination . . . puts women in a state of constant physical insecurity. . . . They have to strive ceaselessly to be engaging, attractive, and available."[9] Being frozen into the passive position of an object whose very existence depends on the eye of its beholder turns the educated modern Western woman into a harem slave.

"I thank you, Allah, for sparing me the tyranny of the 'size 6 harem,'" I repeatedly said to myself while seated on the Paris-Casablanca flight, on my way back home at last. "I am so happy that the conservative male elite does not know about it. Imagine the fundamentalists switching from the veil to forcing women to fit size 6."

How can you stage a credible political demonstration and shout in the streets that your human rights have been violated when you cannot find the right skirt?

NOTES

1. Naomi Wolf, *The Beauty Myth: How Images of Beauty Are Used Against Women* (New York: Anchor Books, Doubleday, 1992), p. 185.

2. Ibid., p. 11.

3. Ibid., p. 17.

4. Pierre Bourdieu: "La force symbolique est une forme de pouvoir qui s'exerce sur les corps, directement, et comme par magie, en dehors de toute contraine physique, mais cette magie n'opère qu'en s'appuyant sur des dispositions déposées, tel des ressorts, au plus profond des corps." In *La Domination Masculine* (Paris: Editions du Seuil, 1998), op. cit. p. 44.

   Here I would like to thank my French editor, Claire Delannoy, who kept me informed of the latest debates on women's issues in Paris by sending me Bourdieu's book and many others. Delannoy has been reading this manuscript since its inception in 1996 (a first version was published in Casablanca by Edition Le Fennec in 1998 as "Êtes-Vous Vacciné Contre le Harem").

5. *La Domination Masculine*, op. cit., p. 41.

6. Bourdieu, op.cit., p. 42.

7. Wolf, op. cit., p. 187.

8. Wolf, quoting research carried out by S. C. Woolly and O. W. Woolly, op. cit., pp. 187–188.

9. Bourdieu, *La Domination Masculine*, p. 73.

KEY CONCEPTS

cultural relativism   culture shock   ethnocentrism

DISCUSSION QUESTIONS

1. How is the shopping experience described in this article similar to an experience you may have had? What cultural understanding do you need to have before shopping for clothing, toys, or some other consumer product?

2. How do the cultural expectations for men's fashion compare and contrast to those for women? Do men experience a similarly limiting definition of fashionable?

6

Do Video Games Kill?

KAREN STERNHEIMER

In this article, Sternheimer summarizes the argument that video games contribute to the violence among youth and could be partly responsible for school shootings, like the ones in Paducah, Kentucky, and Littleton, Colorado. She argues that much of the media coverage on this connection overlooks research that shows little direct relationship between video violence and violent behavior. Also, the public fear of juvenile violence is misguided and grounded in racist assumptions.

As soon as it was released in 1993, a video game called *Doom* became a target for critics. Not the first, but certainly one of the most popular first-person shooter games, *Doom* galvanized fears that such games would teach kids to kill. In the years after its release, *Doom* helped video gaming grow into a multibillion dollar industry, surpassing Hollywood box-office revenues and further fanning public anxieties.

Then came the school shootings in Paducah, Kentucky; Springfield, Oregon; and Littleton, Colorado. In all three cases, press accounts emphasized that the shooters loved *Doom*, making it appear that the critics' predictions about video games were coming true.

But in the ten years following *Doom*'s release, homicide arrest rates fell by 77 percent among juveniles. School shootings remain extremely rare; even during the 1990s, when fears of school violence were high, students had less than a 7 in

SOURCE: From Karen Sternheimer. "Do Video Games Kill?" *Contexts* 6, no. 1 (Winter 2007): 13–17. The American Sociological Association. Used by Permission. All rights reserved.

10 million chance of being killed at school. During that time, video games became a major part of many young people's lives, few of whom will ever become violent, let alone kill. So why is the video game explanation so popular?

CONTEMPORARY FOLK DEVILS

In 2000 the FBI issued a report on school rampage shootings, finding that their rarity prohibits the construction of a useful profile of a "typical" shooter. In the absence of a simple explanation, the public symbolically linked these rare and complex events to the shooters' alleged interest in video games, finding in them a catchall explanation for what seemed unexplainable—the white, middle-class school shooter. However, the concern about video games is out of proportion to their actual threat.

Politicians and other moral crusaders frequently create "folk devils," individuals or groups defined as evil and immoral. Folk devils allow us to channel our blame and fear, offering a clear course of action to remedy what many believe to be a growing problem. Video games, those who play them, and those who create them have become contemporary folk devils because they seem to pose a threat to children.

Such games have come to represent a variety of social anxieties: about youth violence, new computer technology, and the apparent decline in the ability of adults to control what young people do and know. Panics about youth and popular culture have emerged with the appearance of many new technologies. Over the past century, politicians have complained that cars, radio, movies, rock music, and even comic books caused youth immorality and crime, calling for control and sometimes censorship.

Acting on concerns like these, politicians often engage in battles character-ized as between good and evil. The unlikely team of Senators Joseph Lieberman, Sam Brownback, Hillary Rodham Clinton, and Rick Santorum introduced a bill in March 2005 that called for $90 million to fund studies on media effects. Lieberman commented, "America is a media-rich society, but despite the flood of information, we still lack perhaps the most important piece of information—what effect are media having on our children?" Regardless of whether any legislation passes, the senators position themselves as protecting children and benefit from the moral panic they help to create.

CONSTRUCTING CULPABILITY

Politicians are not the only ones who blame video games. Since 1997, 199 news-paper articles have focused on video games as a central explanation for the Paducah, Springfield, and Littleton shootings. This helped to create a groundswell of fear that schools were no longer safe and that rampage shootings could happen wherever there were video games. The shootings legitimated existing concerns about the new medium and about young people in general. Headlines such as "Virtual Realities Spur School Massacres" (*Denver Post*, July 27, 1999), "Days of Doom" (*Pittsburgh*

Post-Gazette, May 14, 1999), "Bloodlust Video Games Put Kids in the Crosshairs" (*Denver Post*, May 30, 1999), and "All Those Who Deny Any Linkage between Violence in Entertainment and Violence in Real Life, Think Again" (*New York Times*, April 26, 1999) insist that video games are the culprit.

These headlines all appeared immediately after the Littleton shooting, which had the highest death toll and inspired most (176) of the news stories alleging a video game connection. Across the country, the press attributed much of the blame to video games specifically, and to Hollywood more generally. The *Pittsburgh Post-Gazette* article "Days of Doom" noted that "eighteen people have now died at the hands of avid *Doom* players." The *New York Times* article noted above began, "By producing increasingly violent media, the entertainment industry has for decades engaged in a lucrative dance with the devil," evoking imagery of a fight against evil. It went on to construct video games as a central link: "The two boys apparently responsible for the massacre in Littleton, Colo., last week were, among many other things, accomplished players of the ultra-violent video game *Doom*. And Michael Carneal, the 14-year-old boy who opened fire on a prayer group in a Paducah, Ky., school foyer in 1997, was also known to be a video-game expert."

Just as many stories insisted that video games deserved at least partial blame, editorial pages around the country made the connection as well:

> President Bill Clinton is right. He said this shooting was no isolated incident, that Kinkel and other teens accused of killing teachers and fellow students reflect a changing culture of violence on television and in movies and video games. (*Cleveland Plain Dealer*, May 30, 1998)

> The campaign to make Hollywood more responsible . . . should proceed full speed ahead. (*Boston Herald*, April 9, 2000)

> Make no mistake, Hollywood is contributing to a culture that feeds on and breeds violence. . . . When entertainment companies craft the most shocking video games and movies they can, peddle their virulent wares to an impressionable audience with abandon, then shrug off responsibility for our culture of violence, they deserve censure. (*St. Louis Post-Dispatch*, April 12, 2000)

The video game connection took precedence in all these news reports. Some stories mentioned other explanations, such as the shooters' social rejection, feelings of alienation at school, and depression, but these were treated mostly as minor factors compared with video games. Reporters gave these other reasons far less attention than violent video games, and frequently discussed them at the end of the articles.

The news reports typically introduce experts early in the stories who support the video game explanation. David Grossman, a former army lieutenant described as a professor of "killology," has claimed that video games are "murder simulators" and serve as an equivalent to military training. Among the 199 newspaper articles published, 17 of them mentioned or quoted Grossman. Additionally, an attorney who has filed several lawsuits against video game producers wrote an article for the *Denver Post* insisting that the games are to blame. By contrast, only seven articles identified sociologists as experts. Writers routinely presented alternative explanations as rebuttals but rarely explored them in depth.

REPORTING ON RESEARCH

By focusing so heavily on video games, news reports downplay the broader social contexts. While a handful of articles note the roles that guns, poverty, families, and the organization of schools may play in youth violence in general, when reporters mention research to explain the shooters' behavior, the vast majority of studies cited concern media effects, suggesting that video games are a central cause.

Since the early days of radio and movies, investigators have searched for possible effects—typically negative—that different media may have on audiences, especially children. Such research became more intense following the rise in violent crime in the United States between the 1960s and early 1990s, focusing primarily on television. Several hundred studies asked whether exposure to media violence predicts involvement in actual violence.

Although often accepted as true—one scholar has gone so far as to call the findings about the effects of media violence on behavior a "law"—this body of research has been highly controversial. One such study fostered claims that television had led to more than 10,000 murders in the United States and Canada during the 20th century. This and many other media-effects studies rely on correlation analysis, often finding small but sometimes statistically significant links between exposure to media violence and aggressive behavior.

But such studies do not demonstrate that media violence causes aggressive behavior, only that the two phenomena exist together. Excluding a host of other factors (such as the growing unrest during the civil rights and antiwar movements, and the disappearance of jobs in central cities) may make it seem that a direct link exists between the introduction of television and homicides. In all likelihood any connection is incidental.

It is equally likely that more aggressive people seek out violent entertainment. Aggression includes a broad range of emotions and behaviors, and is not always synonymous with violence. Measures of aggression in media-effects research have varied widely, from observing play between children and inanimate objects to counting the number of speeding tickets a person received. Psychologist Jonathan Freedman reviewed every media-violence study published in English and concluded that "the majority of studies produced evidence that is inconsistent or even contradicts" the claim that exposure to media violence causes real violence.

Recently, video games have become a focus of research. Reviews of this growing literature have also been mixed. A 2001 meta-analysis in *Psychological Science* concluded that video games "will increase aggressive behavior," while a similar review published that same year in a different journal found that "it is not possible to determine whether video game violence affects aggressive behavior." A 2005 review found evidence that playing video games improves spatial skills and reaction times, but not that the games increase aggression.

The authors of the *Psychological Science* article advocate the strong-effects hypothesis. Two of their studies were widely reported on in 2000, the year after the Columbine High School shootings, with scant critical analysis. But their research was based on college undergraduates, not troubled teens, and it measured aggression in part by subjects' speed in reading "aggressive" words on a computer screen or blasting opponents with sound after playing a violent video game. These measures do not approximate the conditions the school shooters experienced, nor do they offer much insight as to why they, and not the millions of other players, decided to acquire actual weapons and shoot real people.

Occasionally reporters include challenges like this in stories containing media-effects claims, but news coverage usually refers to this body of research as clear, consistent, and conclusive. "The evidence, say those who study violence in culture, is unassailable: Hundreds of studies in recent decades have revealed a direct correlation between exposure to media violence—now including video games—and increased aggression," said the *New York Times* (April 26, 1999). The *Boston Herald* quoted a clinical psychologist who said, "Studies have already shown that watching television shows with aggressive or violent content makes children more aggressive" (July 30, 2000). The psychologist noted that video game research is newer, but predicted that "in a few years, studies will show that video games increase a child's aggression even more than violent TV shows." News reports do not always use academic sources to assess the conclusiveness of media effects research. A *Pittsburgh Post-Gazette* story included a quote by an attorney, who claimed, "Research on this has been well-established" (May 14, 1999).

It is no accident that media-effects research and individual explanations dominate press attempts to explain the behavior of the school shooters. Although many politicians are happy to take up the cause against video games, popular culture itself suggests an apolitical explanation of violence and discourages a broader examination of structural factors. Focusing on extremely rare and perhaps unpredictable outbursts of violence by young people discourages the public from looking closely at more typical forms of violence against young people, which is usually perpetrated by adults.

The biggest problem with media-effects research is that it attempts to decontextualize violence. Poverty, neighborhood instability, unemployment, and even family violence fall by the wayside in most of these studies. Ironically, even mental illness tends to be overlooked in this psychologically oriented research. Young people are seen as passive media consumers, uniquely and uniformly vulnerable to media messages.

MISSING MEDIA STUDIES

News reports of the shootings that focus on video games ignore other research on the meanings that audiences make from media culture. This may be because its qualitative findings are difficult to turn into simple quotations or sound bites. Yet in seeking better understanding of the role of video games in the lives of the shooters and young people more generally, media scholars could have added much to the public debate.

For instance, one study found that British working-class boys boast about how many horror films they have seen as they construct their sense of masculinity by appearing too tough to be scared. Another study examined how younger boys talk about movies and television as a way to manage their anxieties and insecurities regarding their emerging sense of masculinity. Such studies illustrate why violent video games may appeal to many young males.

Media scholars have also examined how and why adults construct concerns about young people and popular culture. One such study concluded that some adults use their condemnation of media as a way to produce cultural distinctions that position them above those who enjoy popular culture. Other researchers have found that people who believe their political knowledge is superior to that of others are more likely to presume that media violence would strongly influence others. They have also found that respondents who enjoy television violence are less likely to believe it has a negative effect.

Just as it is too simplistic to assert that video game violence makes players more prone to violence, news coverage alone, however dramatic or repetitive, cannot create consensus among the public that video games cause youth violence. Finger-wagging politicians and other moralizers often alienate as many members of the public as they convert. In an ironic twist, they might even feed the antiauthoritarian appeal that may draw players of all ages to the games.

The lack of consensus does not indicate the absence of a moral panic, but reveals contradictory feelings toward the target group. The intense focus on video games as potential creators of violent killers reflects the hostility that some feel toward popular culture and young people themselves. After adult rampage shootings in the workplace (which happen more often than school shootings), reporters seldom mention whether the shooters played video games. Nor is an entire generation portrayed as potential killers.

AMBIVALENCE ABOUT JUVENILE JUSTICE

The concern in the late 1990s about video games coincided with a growing ambivalence about the juvenile justice system and young offenders. Fears about juvenile "super-predators," fanned by former Florida Representative Bill McCollom's 1996 warning that we should "brace ourselves" against the coming storm of young killers, made the school shootings appear inevitable. McCollom and other politicians characterized young people as a "new breed," uniquely dangerous and amoral.

These fears were produced partially by the rise in crime during the late 1980s and early 1990s, but also by the so-called echo boom that produced a large generation of teens during the late 1990s. Demographic theories of crime led policymakers to fear that the rise in the number of teen males would bring a parallel rise in crime. In response, virtually every state changed its juvenile justice laws during the decade. They increased penalties, imposed mandatory minimum sentences, blended jurisdiction with criminal courts, and made it easier to transfer juvenile cases to adult criminal courts.

So before the first shot was fired in Paducah, politicians warned the public to be on the lookout for killer kids. Rather than being seen as tragic anomalies, these high-profile incidents appeared to support scholarly warnings that all kids posed an increasing threat. Even though juvenile (and adult) crime was in sharp decline by the late nineties, the intense media coverage contributed to the appearance of a new trend.

Blaming video games meant that the shooters were set aside from other violent youth, frequently poor males of color, at whom our get-tough legislation has been targeted. According to the National Center for Juvenile Justice, African-American youth are involved in the juvenile justice system more than twice as often as whites. The video game explanation constructs the white, middle-class shooters as victims of the power of video games, rather than fully culpable criminals. When boys from "good" neighborhoods are violent, they seem to be harbingers of a "new breed" of youth, created by video games rather than by their social circumstances. White, middle-class killers retain their status as children easily influenced by a game, victims of an allegedly dangerous product. African-American boys, apparently, are simply dangerous.

While the news media certainly asked what role the shooters' parents may have played, the press tended to tread lightly on them, particularly the Kinkels of Springfield, Oregon, who were their son's first murder victims. Their middle-class, suburban, or rural environments were given little scrutiny. The white school shooters did more than take the lives of their classmates; their whiteness and middle-class status threatened the idea of the innocence and safety of suburban America.

In an attempt to hold more than just the shooters responsible, the victims' families filed lawsuits against film producers, Internet sites, and video game makers. Around the same time, Congress made it more difficult to sue gun manufacturers for damages. To date, no court has found entertainment producers liable for causing young people to commit acts of violence. In response to a lawsuit following the Paducah shootings, a Kentucky circuit judge ruled that "we are loath to hold that ideas and images can constitute the tools for a criminal act," and that product liability law did not apply because the product did not injure its consumer. The lawsuit was dismissed, as were subsequent suits filed after the other high-profile shootings.

GAME OVER?

Questions about the power of media and the future of the juvenile justice system persist. In March 2005, the U.S. Supreme Court ruled that juvenile executions were unconstitutional. This ruling represents an about-face in the 25-year trend toward toughening penalties for young offenders. While many human rights and children's advocates praised this decision, it was sharply criticized by those who believe that the juvenile justice system is already too lenient. Likewise, critics continue to target video games, as their graphics and plot capabilities grow more complex and at times more disturbing. Meanwhile, youth crime rates continue to decline. If we want to understand why young people, particularly in middle-class or otherwise stable environments, become homicidal, we need to look beyond the games they play. While all forms of media merit critical analysis, so do the supposedly "good" neighborhoods and families that occasionally produce young killers.

KEY CONCEPTS

mass media   moral panic   popular culture

DISCUSSION QUESTIONS

1. Consider the programs you watched as a child. Are violent images present in these programs? Does watching television "desensitize" young people to violence?

2. How is American society a violent one? Think of all the ways in which the culture of America reinforces the presence of violence.

7

September 11, 2001

Mass Murder and Its Roots in the Symbolism of American Consumer Culture

GEORGE RITZER

Ritzer discusses the terrorist acts of September 11, 2001, as attacks against American culture. He explains how McDonald's fast-food chain, credit cards, large shopping malls, and discount stores are symbols of America's consumption culture that affront the rest of the world regularly. This discussion offers insight into why other nations feel such animosity toward the United States.

. . . On September 11, 2001, the terrorists not only killed thousands of innocent people and destroyed buildings of various sorts, they also sought to destroy (and in one case succeeded) major symbols of America's preeminent position in the globalization process: The World Trade Center was a symbol of America's global hegemony in the economic realm, and the

SOURCE: George Ritzer. "September 11, 2001: Mass Murder and Its Roots in the Symbolism of American Consumer Culture." In *McDonaldization: The Reader*, edited by George Ritzer, 199–212. Thousand Oaks, CA: Pine Forge Press, 2002.

Pentagon is obviously the icon of its military preeminence around the world. In addition, there is a widespread belief that the fourth plane, the one that crashed in Pennsylvania, was headed for the symbol of American political power—the White House. Obviously, the common element in all these targets is that they are, among other things, cultural icons, with the result that the terrorist attacks can be seen as assaults on American culture. (This is not, of course, to deny the very material effects on people, buildings, the economy, and so on.) Furthermore, although symbols, jobs, businesses, and lives were crippled or destroyed, the main objective was symbolic—the demonstration that the most important symbols of American culture were not only vulnerable but could be, and were, badly hurt or destroyed. The goal was to show the world that the United States was not an invulnerable superpower but that it could be assaulted successfully by a small number of terrorists. One implication was that if such important symbols could be attacked successfully, nothing in the United States (as well as in U.S. interests around the world) was safe from the wrath of terrorists. Thus, we are talking about an assault on, among other things, culture—an assault designed to have a wide-ranging impact throughout the United States and the world.

In emphasizing culture, I am not implying that economic, political, and military issues (to say nothing of the loss of life) are less important. Indeed, these domains are encompassed, at least in part, under the broad heading of culture and attacks on cultural icons. Clearly, many throughout the world are angered by a variety of things about the United States, especially its enormous economic, political, and military influence and power. In fact, this essay will focus on one aspect of the economy—consumption—and its role in producing hostility to the United States. Again, this is just one factor in the creation of this hostility, but it is certainly worthy of further discussion. Others, with greater expertise in those areas, will certainly be analyzing the military, political, and other economic dimensions and implications of the events of September 11, 2001.

By focusing on America's role in consumption and its impact around the world, I am not condoning the terrorist attacks (they are among the most heinous of acts in human history) or blaming the United States for those attacks. Rather, my objective is to discuss one set of reasons that people in many different countries loathe (while a far larger number of people love) the United States. Indeed, it is a truism that, often, love and hate coexist in the same people. However, needless to say, those involved in these terrorist acts had nothing but hatred for the United States.

CONSUMPTION

American hegemony throughout the world is most visible and, arguably, of greatest significance, in economic and cultural terms, in the realm of consumption. On a day-to-day basis in much of the world, people are far more likely to be confronted by American imperialism in the realm of consumption than they are in other economic domains. . . .

Rather than focus on consumption in general, I will discuss three of its aspects of greatest concern to me: fast-food restaurants, credit cards, and "cathedrals of consumption" (for example, discounters such as Wal-Mart). Before getting to these, it is important to point out that they, and many other components of our consumer culture, are not only physical presences throughout the world, they are media presences by way of television, movies, the Internet, and so on. Furthermore, even in those countries where these phenomena are not yet material realities, they are already media presences. As a result, their impact is felt even though they have not yet entered a particular country, and in those countries where they already exist, their impact is increased because they are also media presences.

I want to focus on the ways in which, from the perspective of those in other nations and cultures, fast-food restaurants, credit cards, and cathedrals of consumption bring with them (a) an American way of doing business, (b) an American way of consuming, and (c) American cultural icons....

FAST-FOOD RESTAURANTS

... McDonald's is today a global corporation with restaurants in nearly 150 nations throughout the world, and we can expect expansion into other nations in the coming years.

As it moves into each new nation, it brings with it a variety of American ways of doing business. In fact, in more recent years, the impact of its ways of doing business were surely felt long before the restaurant chain itself became a physical presence. McDonald's has been such a resounding success, and has offered so many important business innovations, that business leaders in other nations were undoubtedly incorporating many of its ideas almost from the inception of the chain. Of course, the major business innovation here is the franchise system. Although the franchise system predated McDonald's by many years . . . , the franchise system came of age with the development of the McDonald's chain. Kroc made a number of innovations in franchising . . . that served to make it a far more successful system. The central point, given the interest of this essay, is that this system has been adopted, adapted, and modified by all sorts of businesses not only in the United States but throughout the world. In the case of franchise systems in other nations, they are doing business in a way that is similar to, if not identical with, comparable American franchises.

The fact that indigenous businesses (e.g., Russkoye Bistro in Russia, Nirulas in India) are conducting their business based, at least to a larger degree, on an American business model is not visible to most people. However, they are affected in innumerable ways by the ways in which these franchises operate. Thus, day-to-day behaviors are influenced by all this, even if consumers are unaware of these effects.

What is far more obvious, even to consumers, is that people are increasingly consuming like Americans. This is clear not only in American chains in other countries but in indigenous clones of those chains. In terms of the former, in Japan, to take one example, McDonald's has altered long-standing traditions about how people are expected to eat. Thus, although eating while standing has long been taboo, in McDonald's restaurants, many Japanese eat just that way. Similarly, long expected not to touch food with their hands or drink directly from containers, many Japanese are doing just that in McDonald's and elsewhere. Much the same kind of thing is happening in indigenous clones of American fast-food restaurants in Japan such as Mos Burger.

These and many other changes in the way people consume are obvious, and they affect the way people live their lives on a daily basis. Just as many Japanese may resent these incursions into, and changes in, the ways in which they have traditionally conducted their everyday lives, those in many other cultures are likely to have their own wide-ranging set of resentments. However, these changes involve much more than transformations in the way people eat. The "McDonaldization thesis" involves far more than restaurants; universities . . . , churches . . . , and museums, among many other settings, can be seen as becoming McDonaldized. Almost no sector of society is immune from McDonaldization, and this means that innumerable aspects of people's everyday lives are transformed by it.

. . . McDonald's itself has become such an icon, as has its "golden arches," Ronald McDonald, and many of its products—Big Mac, Egg McMuffin, and so on. Other fast-food chains have brought with them their own icons—Burger King's Whopper, Colonel Sanders of Kentucky Fried Chicken, and so on. These icons are accepted, even embraced, by most, but others are likely to be angered by them. For example, traditional Japanese foods such as sushi and rice are being replaced, at least for some, by Big Macs and "supersized" French fried potatoes. Because food is such a central part of any culture, such a transformation is likely to enrage some. More important, perhaps, is the ubiquity of the McDonald's restaurant, especially its golden arches, throughout so many nations of the world. To many in other societies, these are not only important symbols in themselves but have become symbols of the United States and, in some cases, even more important than more traditional symbols (such as the American embassy and the flag). In fact, there have been a number of incidents in recent years in which protests against the United States and its actions have taken the form of actions against the local McDonald's restaurants. To some, a McDonald's restaurant, especially when it is placed in some traditionally important locale, represents an affront, a "thumb in the eye," to the society and its culture. It is also perceived as a kind of "Trojan Horse," and the view is that hidden within its bright and attractive wrappings and trappings are all manner of potential threats to local culture. Insulted by, and fearful of, such "foreign" entities, a few react by striking out at them and the American culture and business world that stands behind them.

CREDIT CARDS

. . . The credit card represents an American way of doing business, especially a reliance on the extension of credit to maintain and to increase sales. Many nations have been dominated, and some still are, by "cash-and-carry" business. Businesses have typically been loathe to grant credit, and when they do, it has usually not been for large amounts of money. When credit was granted, strong collateral was required. This is in great contrast to the credit card industry, which has granted billions of dollars in credit with little or no collateral. In these and many other ways, traditional methods of doing business are being threatened and eroded by the incursion of the credit card. . . .

. . . [C]redit cards are perceived as playing a key role in the development and expansion of consumer culture, a role characterized by hyperconsumption. Although many are overjoyed to be deeply immersed in consumer culture and others would dearly love to be so involved, still others are deeply worried by it on various grounds. One of the concerns, felt not only in the United States but perhaps even more elsewhere in the world, is the degree to which immersion in the seeming superficialities of consumption and fashion represents a threat, if not an affront, to deep-seated cultural and religious values. For example, many have viewed modern consumption as a kind of religion, and I have described malls and other consumption settings as cathedrals of consumption. As such, they can be seen as alternatives, and threats, to conventional religions in many parts of the world. At the minimum, the myriad attractions of consumption and a day at the mall serve as powerful alternatives to visiting one's church, mosque, or synagogue.

Finally, credit cards in general, to say nothing of the major brands—Visa and MasterCard (as well as the "charge card" and its dominant brand, *American Express*)—are seen as major icons of American culture. While these icons are similar to, say, McDonald's and its golden arches, there is something quite unique and powerful about credit cards. Although one who lives outside the United States may encounter a McDonald's and its arches every day, or maybe every few days, a Visa credit card, for example, is *always* with those who have one. It is always there in one's wallet, and it is probably a constant subconscious reality. Furthermore, one is continually reminded of it every time one passes a consumption site, especially one that has the logo of the credit card on its door or display window. Even without the latter, the mere presence of a shop and its goods is a reminder that one possesses a credit card and that the shop can be entered and goods can be purchased. The credit card is a uniquely powerful cultural icon because it is with cardholders all the time and they are likely to be reminded of it continually.

CATHEDRALS OF CONSUMPTION

Cathedrals of consumption, many of which are also American innovations, are increasing presences elsewhere in the world. There is a long list of these cathedrals of consumption . . ., but let us focus on two—shopping malls and discounters, especially Wal-Mart. American-style fully enclosed shopping malls are springing up all over the world . . . . Most of these are indigenous developments, but the model is the American mall. . . .

These, of course, represent American ways of doing business. In the case of the mall, this involves the concentration of businesses in a single setting devoted to them. In the case of discounters, it represents the much greater propensity of American businesses (in comparison with their peers around the world) to compete on a price basis and to offer consumers deep discounts. Although appealing to many people, resentment may develop not only because these represent American rather than indigenous business practices but because they pose threats to local businesses. As in the United States, still more resentment is likely to be generated because small local shops are likely to be driven out of business by the development of a mall or the opening of a Wal-Mart on the outskirts of town.

Again, more obvious is the way consumers are led to alter their behaviors as a result of these developments. For example, instead of walking or bicycling to local shops, increasing numbers are more likely to drive to the new and very attractive malls and discounters. This can also lead to movement toward the increasing American reality that such trips are not just about shopping; such settings have become *destinations* where people spend many hours wandering from shop to shop, having lunch, and even seeing a movie or having a drink. Consumption sites have become places to while away days, and as such, they pose threats to alternative public sites, such as parks, zoos, and museums. In the end, malls and discounters are additional and very important contributors to the development of hyperconsumption and all the advantages and problems associated with it. Settings such as a massive shopping mall with a huge adjacent parking lot and a large Wal-Mart with its parking lot are abundantly obvious to people, as are the changes they help to create in the way natives consume.

Settings such as these are perceived as American cultural icons. Wal-Mart may be second only to McDonald's in terms of the association of consumption sites with things American, and the suburban mall is certainly broadly perceived in a similar way. Furthermore, malls are likely to house a number of other cultural icons, such as McDonald's, the Gap, and so on. And still further, the latter are selling yet other icons in the form of products such as Big Macs, blue jeans, Nike shoes, and so on. Many of those icons will be taken from the malls and eaten, worn, and otherwise displayed in public. Their impact is amplified because their well-known logos and names are likely to be plastered all over these products. Again, there is an "in-your-face" quality to all of this, and although many will be led to want these things, others will react negatively to the ubiquity of these emblems of America and its consumer culture and that these emblems tend to supplant indigenous symbols.

The argument here is that the recent terrorist attacks can be seen as assaults on American cultural icons—specifically the World Trade Center (business and consumption), the Pentagon (military), and potentially, the White House (political)....

Although I have focused on three American cultural icons in this essay and their worldwide proliferation, in vast portions of the world they are of minimal importance or completely nonexistent. Even where they are not physical presences, however, they are known through movies, television, magazines, and newspapers, and even by word of mouth. Thus, their influence throughout the world far exceeds their material presence in the world....

CONCLUSION

Wars are always about culture, at least to some degree, but the one we have embarked on seems to reek with cultural symbolism. We live in an era—the era of globalization—in which not only cultural products and the businesses that sell them but also the symbols that go to their essence are known throughout the world. In fact, some—Nike and Tommy Hilfiger come to mind—are *nothing but symbols*; they manufacture nothing (except symbols). Although we deeply mourn the loss of life, the terrorists were after more. Surely, they wanted to kill people, but mainly because their deaths represented symbolically the fact that America could be made to bleed. But they also wanted to destroy some of the American symbols best known throughout the world; in destroying them, they were, they thought, symbolically destroying the United States. Interestingly, the initial response from the United States was largely symbolic—American flags were displayed everywhere; the sounds of the national anthem wafted through the air on a regular basis; red, white, and blue ribbons were wrapped around trees and telephone poles; and so on. Of course, the response soon went beyond symbols: Missiles have been launched and bombs dropped, special forces are in action, and people are dying. However, should the accused mastermind of the terrorist acts—Osama Bin Laden—be caught or killed, he will quickly become an even greater cultural icon than he is at the moment (his likeness already adorns T-shirts in Pakistan and elsewhere). Destroying his body may be satisfying to some, but it may well create a greater cultural problem for the United States, one that might translate into still more American citizens killed and structures destroyed. . . .

KEY CONCEPTS

bureaucracy	consumerism	McDonaldization

DISCUSSION QUESTIONS

1. Think back to the days immediately following September 11, 2001. What cultural symbols did Americans display as a sign of unity against terrorists? What are some other examples of American culture that some nations may take issue with and find offensive?

2. What are the issues to consider when the United States attempts to bring democracy to other nations? How do we balance bringing mostly welcomed American products and practices to a foreign culture and possibly destroying native cultural practices in the process?

8

Men and Women: Mind and Body

SHARLENE HESSE-BIBER

In this chapter from her book The Cult of Thinness, *Hesse-Biber explains how culture dictates thinness as part of femininity and a way to control women's bodies. In ancient Chinese culture, the practice of foot binding served to control women and girls. In Victorian England, corsets were worn to control women's bodies. She draws the link between these practices and current trends in dieting and plastic surgery. The "cult of thinness" is an example of the cultural control over women.*

You know guys just sort of think you're stupid. Sure, I get a lot of attention being blonde and female, but what society really thinks of women is that intelligence doesn't even figure in. Just body and face. The time my mom asked me what I was going to be when I grew up, I said I'd put myself on the cover of a magazine. For men, it's the money they make, not how they look. Nobody cares if a guy is a good father or even a good person. He's the president of this company or that, or the head doctor of such and such a hospital. It's his job that counts.

—TRACY, COLLEGE SOPHOMORE

Our culture judges a man primarily in terms of how powerful, ambitious, aggressive, and dominant he is in the worlds of thought and action. These are qualities more of the mind than the body. A woman, on the other hand, is judged almost entirely in terms of her appearance, her attractiveness to men, and her ability to keep the species going. Her sexual body can even be quite dangerous to a man on his way to success. She becomes a temptress distracting him from his true work and the pursuit of rationality, knowledge, and power.

The split between mind and body is a central idea in Western culture. It often frames our perceptions of what it means to be feminine and masculine.[1] Well-known studies conducted by Paul Rosenkrantz[2] and his colleagues in the 1960s gave college students a list of extreme personality traits and their opposites: very passive versus very active; very illogical versus very logical; very vain about appearance versus uninterested in appearance, and so on. The students were asked to ascribe each trait to either males or females, and to rank each trait's social

SOURCE: Sharlene Hesse-Biber. 2007. *The Cult of Thinness,* New York: Oxford University Press, pp. 32–56.

desirability. The findings from this study (and from others) showed that traits most often associated with competence and social desirability were assigned to men, and those associated with having an "emotional" life were assigned to women. Men were viewed as more independent than women, more logical, more direct, more self-confident, and more ambitious. Women were seen as being more gentle than men, more soft-spoken, more talkative, and more tender.

The problem with a dichotomy is that it provides no middle ground. If one trait is positive, then its opposite is negative—if a man is strong, then a woman is weak. But such a narrow view of human behavior ignores the reality that there are a range of traits common to both sexes.[3]

The college-age women I interviewed fully understood this mind and body gender stereotyping. Angela said:

> My body is the most important thing. It's like that's all I ever had
> because that's all everyone ever said about me. My mother would say
> that I am smart and stuff, but really they focused on my looks. And even
> my doctor enjoys my looks. He used to make me walk across the room
> to check my spine and he'd comment on how cute I walked, that I
> wiggled. Why comment on it at all?

She noted how difficult it is for women to be acknowledged for "male" attributes: "I think men are shocked when they see a gorgeous, really intelligent woman. You'll always hear about how 'gorgeous.' You don't hear or very rarely hear, 'She's a very intelligent woman.'" . . .

HISTORICAL ROOTS OF THE MIND/BODY DICHOTOMY

. . . Cultural rules have controlled women's bodies throughout history.[4] Our Anglo-American legal institutions, for example, created laws based on the biological differences between men and women. U.S. laws enacted at the turn of the century regulated the number of hours per day or week that women were allowed to work. Other laws prevented them from working at night or in certain occupations like mining or smelting. This "protective legislation" drove women out of certain jobs, while it gained reduced hours for men in those occupations where women remained. In addition, domestic relations laws (those concerning property rights, pension benefits, maternity leave policies, etc.) have reinforced the idea that women are reproducers (the reproductive body) and men are breadwinners (the rational mind).[5]

For those of us embedded in a particular society, it is sometimes difficult to see how culture controls women's bodies. In the United States today, women who diet or have their breasts enlarged and their tummies tucked regard this as an exercise of free will. But if we compare these practices with two historical examples, one from ancient China and the other from the Victorian era, we may gain a new perspective.

ANCIENT CHINA AND THE PRACTICE
OF FOOT BINDING

For 1,000 years, the Confucian philosophy of hierarchical patriarchal authority formed the basis of Chinese society. Males were considered higher than females, and the old had authority over the young. As Confucius, born in 551 B.C., wrote, "Women are, indeed, human beings, but they are of a lower state than men."[6]

Foot binding, one of the most dramatic examples of control over such "lower beings," originated around the tenth century, with court dancers who wrapped their feet to imitate pointed, sickle moons (not unlike toe shoes in Western ballet). The Chinese court and the upper class had always prized small feet in women: Now they copied this practice and took it to extremes. It became an important symbol of high status within Chinese society. In time, it filtered down to the masses as well.[7]

This custom lasted more than 1,000 years and served to virtually cripple women in the name of beauty and femininity. Little girls had their toes bent under into the sole, with the heel and toes bowed forcibly together, and wrapped tightly. The bones eventually broke and the foot could no longer grow. This severely deformed clubfoot, only a few inches long, became known as the "lotus" or "lily" foot. Walking on these stumps was painful, if not impossible.

One young girl described suffering intensely:

> I was inflicted with the pain of foot binding when I was seven. I was an active child but from then on my free and optimistic nature vanished. Mother consulted references in order to select an auspicious day for it. She shut the bedroom door, boiled water, and from a box withdrew binding, shoes, knife, needle and thread. "Today is a lucky day," she said. "If bound today, your feet will never hurt; if bound tomorrow, they will." She then bent my toes toward the plantar [the mid-region of the sole] with a binding cloth ten feet long and two inches wide, doing the right foot first, and then the left. She ordered me to walk, but the pain proved unbearable.
>
> That night, mother wouldn't let me remove the shoes. My feet felt on fire and I couldn't sleep, mother struck me for crying. On the following days, I tried to hide but was forced to walk on my feet. Mother hit me on my hands and feet for resisting. Mother would remove the bindings and wipe the blood and pus which dripped from my feet. She told me that only with removal of the flesh could my feet become slender.
>
> . . . Every two weeks, I changed to new shoes. After changing more than ten pairs of shoes, my feet were reduced to a little over four inches. I had been binding for a month when my younger sister started; when no one was around we would weep together.[8]

Why did women, especially mothers, continue a custom that inflicted such suffering?[9]

The bound foot, a symbol of feminine beauty, represented a woman's only prospects in life. Inheritance laws of Chinese society were male dominated. Unable to inherit property or pass on the ancestral name, a girl was an economic liability until she left to join her husband's family. A good match with prosperous in-laws offered a girl's parents a chance to recoup their investment.[10] While the bride's parents could gain social and economic status, the groom's parents acquired another source of labor, both productive and reproductive. Upper-class women did not work outside their homes, but even though foot-bound, they still performed household chores, made handicrafts, and attended to the needs of their extended families.[11] Even peasant families who dreamed of marrying into a higher class crippled their daughters accordingly. Only the poorest female field workers escaped.[12]

This custom lasted as long as it did because it reinforced the patriarchal authority of Chinese society. Foot binding supported society's belief in the dichotomy between mind and body, and supported dualistic thinking (men were superior and women inferior; men were valued and women devalued). Women followed a strict line of obedience, first to fathers, then to husbands, and finally to sons upon the death of the husband.[13]

> The minds of footbound women were as contracted as their feet. Daughters were taught to cook, supervise the household, and embroider shoes for the golden Lotus. Intellectual and physical restriction had the usual male justification. Women were perverse and sinful, lewd and lascivious, if left to develop naturally. The Chinese believed that being born a woman was payment for evils committed in a previous life. Footbinding was designed to spare a woman the disaster of another such incarnation.[14]

Foot binding also restricted female mobility and sexuality, which are potential sources of power and resistance for women in preindustrial societies. As Susan Greenhalgh notes in her research on women in Old China:

> The greatest threat to the family system came from the women because they married *in* from an *outside* family. Women marrying into the patriarchal family could disrupt its stability by offering dissenting opinions about the allocation of labor and goods within the family, or by simply refusing to accept patterns of authority and interaction already established, and returning to their natal homes.[15]

Along with their bound feet, girls had to embrace a set of personality traits: "chaste and yielding, calm and upright"; "not talkative yet agreeable" in speech; "restrained and exquisite" in appearance and demeanor; engaged in work which demonstrated her skills in "handiwork and embroidery."[16] Upper-class Chinese women, who spent a great deal of time and energy (and pain) on this feminine ideal, were unlikely to challenge the established order.

Sexuality, another threat, also had to be channeled. Binding limited the possibility of extramarital affairs, since the bound woman could not leave the home unaided. Finally, as a prominent aspect of feminine identity, the bound foot was an important part of sexual rituals—even a focal point of sexual excitement. "The tiny and fragile appearance of the foot aroused in the male a combination of lust and pity. He longed to touch it, and being allowed to do so meant that the woman was his."[17]

Foot binding in ancient China seems grotesque and cruel, but it can help us understand the current Cult of Thinness. Foot binding reflected the economic and social power structure of a patriarchal society, which defined women in terms of their bodies—"commodities" for domestic economic exchange and social control.

When societies make the transition from medieval or traditional political-economic systems, like ancient China, to modern systems, power within that society shifts from centralized authority to various institutional powers, including those controlled by patriarchal interests.[18] A dominant force in modern society, capitalism in its early form relied on the *external* control of women's bodies, as in the practice of corseting.[19] But over time this control has become more *internal*, through self-imposed body practices and rituals. Because modern women are also consumers, this focus on their bodies has created many multibillion-dollar industries.

THE RISE OF CONSUMER CULTURE
AND THE PRACTICE OF CORSETING

Women wore the corset in England, the United States, and Western Europe for most of the nineteenth century, when the rise of capitalism led to women's emerging roles as both consumer and commodity.

The image of the Victorian woman was, in part, a response to the dramatic changes that accompanied industrialization. Vast economic expansion created a large, prosperous middle class. Work became segregated from the home, where the middle-class woman was expected to stay, supported by a well-to-do husband.

She was also expected to be a fragile, pure creature, submissive to her spouse and subservient to domestic needs. Her decorative value, a quality that embraced her beauty, her character, and her temperament, defined her. This ideal, later referred to as the "cult of true womanhood," demanded uncompromising virtue.[20] Like the rich woman of ancient China, the Victorian woman became a prized showpiece, evidence of her husband's wealth. As managers of hearth and home, middle- and upper-class wives became the chief consumers of early capitalists' products.[21]

Instead of the clubbed "lotus" foot, the important symbols of beauty and status for women were paleness, fragility, and a tightly cinched waist. The waist had a special erotic significance symbolizing passivity, dependence, and, more perversely, bondage.[22] The proud husband encircling his wife's waist in his broad hands, notes one researcher, demonstrated power and control.[23] According to one French beauty writer, Pauline Mariette:

> The waist is the most essential and principal part of the woman's body, with respect to the figure . . . . The bee, the wasp . . . those are the beings whose graceful and slender waist is always given as the point of comparison . . . . The waist gives woman her jauntiness, the pride of her appearance, the delicacy and grandeur of her gait, the unconstraint and delight of her pose.[24]

To attain such an ideal, a tightly laced undergarment reinforced with whalebone, and later steel, constricted women's waists for many hours a day.[25] This pressure often caused pain and distorted the internal organs. One zealous mother laced her daughter's stays too tightly and killed her at the age of 20:[26] ". . . her ribs had grown into her liver, and that her other entrails were much hurt by being crushed together with her stays, which her mother had ordered to be twitched so straight that it often brought tears into her eyes whilst the maid was dressing her."[27]

The following description is more common:

> I ordered a pair of stays, made very strong and filled with stiff bone, measuring only fourteen inches round the waist. With the assistance of my maid, I managed to lace my waist to eighteen inches. At night I slept in my corset. The next day my maid got my waist to seventeen inches, and so on, an inch every day, until she got them to meet. For the first few days, the pain was very great, but as soon as the stays were laced close, and I had worn them so for a few days, I began to care nothing about it, and in a month or so I would not have taken them off on any account. For I quite enjoyed the sensation, and when I let my husband see me in a dress to fit I was amply repaid for my trouble.[28]

A tight corset prevented women from moving around very much, which tended to make them dependent and submissive.[29] Like the family furniture, a wife was considered a possession. Indeed, many were nearly as immobile as furniture. "Accounts of nineteenth-century house fires reveal that women occasionally went up in flames with the household goods because of immobilizing corsets and skirts too full to run in."[30]

Why did women corset themselves so willingly? Like the women of ancient China, their identities and rewards depended on their body image. One historian commented: "In an age when alternatives to marriage for women were grim and good husbands scarce, the pressures to conform to the submissive ideal that men demanded were enormous."[31]

Patriarchal interests, which characterized women primarily as wives, mothers, and decorative objects, complemented an economy relying more and more on domestic consumption.[32] Capitalism motivated producers to create new needs and exploit new markets, most of which centered around the body and its functioning. Advertising was also crucial in helping to define women as the primary consumers. It promoted insecurity by encouraging women "to adopt a critical attitude toward body, self and life style." They rushed to purchase the latest household items, which were important symbols of being a good wife and mother. They flocked to buy beauty products, which were signs of a woman's femininity and ability to hold on to her man. As long as a woman viewed her body as an object, she was controllable and profitable.[33]

THE ORIGINS OF THE CULT OF THINNESS: FROM EXTERNAL TO INTERNAL BODY CONTROL

Nineteenth-century industrialization and mass production influenced body image in general for both sexes. In her book *Never Too Thin*, social historian Roberta Seid points out that for women, beauty was becoming democratized as ready-made clothing introduced the idea of standard sizes.[34] The machine age promoted a streamlined aesthetic. "While ... slenderness had been associated with sickness and fragility, now many health authorities cautioned against overeating and excess weight."[35] New studies related obesity to premature mortality. By the turn of the century, technological innovation, efficiency, and economic growth reinforced the ideal of a slender body, providing metaphors for the desired human body. "To be as efficient, as effective, as economical, as beautiful as the sleek new machines, as the rationalized workplace.... It was these ... developments that forged the society we know today and that established the framework for our prejudice against fat."[36] In general, Seid notes that men were not bound by the pressure to look slender. A new male image, the "self-made" man, arose with the coming of the Industrial Revolution and the Protestant ethic. The "self-made" man strives for upward mobility through hard work, ability, and thrifty ways, not his physical appearance.[37] In fact men's clothing remained relatively unchanged over many decades.[38]

Twentieth-century capitalism included the diet, beauty, cosmetic, fitness, and health industries. Along with modern patriarchy, it continues to control women into the twenty-first century through pressures to be thin.[39] Increasingly, modern woman achieves this new ideal not through the purchase of a corset or girdle, but through self-directed dieting and exercise. Taken to the extreme, this self-direction becomes an eating disorder.[40]

The shift from external to internal control was part of the ideology of "women's independence." As the nineteenth century drew to a close, middle-class women were increasingly involved in social reform, volunteer activities, and work outside the home such as teaching and nursing.[41] By 1870 more were entering college, and by the 1890s they were beginning to compete with men in such professions as law, medicine,[42] and journalism. A new interest in physical fitness led doctors to prescribe women's tennis, golf, swimming, horseback riding, and bicycling. Dancing was the rage. Suffragettes were marching for the right to vote. Women were becoming more physically mobile, and abandoning their tight corsets.[43]

Of course, the 1920s women's movement threatened the traditional view of how men and women focused their lives, and a predictable backlash followed. Just as women were demanding more "space" and equality, the culture's standards of attractiveness demanded that they shrink.[44] A slender female body, achieved though dieting, became the dominant image for most of the twentieth century.[45]

In effect, patriarchal and consumer interests co-opted and harnessed women's newfound independence. In order to attain the ultra-slender ideal, women began to purchase diet products and to spend enormous amounts of time and energy on their bodies.[46] These activities continue to divert economic and emotional capital away from other investments women might make, like political activism, education, and careers, which might empower women and change their thinking about mind and body.

Even during the first wave of feminism, the slim, youthful, albeit rather sexless "flapper" of the 1920s became the most important symbol of American beauty. She had a straight, boyish figure, and exposed her slender legs. As one historian notes, it was a trivialized image:

> On the one hand, she indicated a new freedom in sensual expression by shortening her skirts and discarding her corsets. On the other hand, she bound her breasts, ideally had a small face and lips . . . and expressed her sensuality not through eroticism, but through constant, vibrant movement. . . . The name "flapper" itself bore overtones of the ridiculous. Drawing from a style of flapping galoshes popular among young women before the war, it connoted irrelevant movement.[47]

Psychiatrist John A. Ryle observed that cases of anorexia nervosa increased during the Flapper Era, which he attributed to "the spreading of the slimming fashion" and "the more emotional lives of the younger generation since the War."[48] Researchers at a 1926 New York Academy of Science conference on adult weight reported an outbreak of eating problems, which they linked to a "psychic contagion." One physician described a significant increase in the number of women whose dramatic weight reduction led to mental breakdown and hospitalization.[49]

However, "the flapper along with the entire exuberant culture of the 1920s vanished into the abyss of the Depression and then the consuming preoccupations of the Second World War."[50] Hemlines fell in the 1930s, and the defined waist returned. The ideal woman of the 1930s still had plenty of curves, but overall she remained slim.

The late 1940s and 1950s saw a temporary interruption of a long-term trend toward slenderness. Political and social reaction after World War II drove many white middle-class women out of their war effort jobs and back to their kitchens[51] in a period of "resurgent Victorianism."[52] As the economy switched back to domestic production, it urged women again to focus on a consumer role. Young men used the G.I. Bill to pay for educations and buy first homes. The "family wage" was enough to support a family, and it also justified why women should be paid less. While women still went to college, their numbers in the professions declined and many opted to marry upon leaving school. Economic expansion and the rise of suburbia created the white middle-class housewife.[53]

To complete the picture of domestic bliss, American fashion in the 1950s revived the hourglass figure. The girdle, long full skirts, and even crinolines came back to create a silhouette not unlike the Victorian lady. Hollywood provided a busty new feminine image, first personified by Marilyn Monroe, and later carried to extremes by the "Mammary Goddesses" like Jayne Mansfield.[54]

THE ULTRA-SLENDER IDEAL

Within a decade, however, thin was back in. This time, the super-slim body ideal met and merged with other social influences. These forces included a new feminist movement and changes in women's roles, the increasing power of the media, and rampant consumerism. As Seid says:

> The imperative to be thin became monolithic as fashion's decrees were reinforced and pushed by all cultural authorities—the health industry, the federal government, employers, teachers, religious leaders, and parents until the concept became so self sustaining, so internalized that no reinforcement was necessary.[55]

The women's movement of the 1960s offered alternative visions to the "happy housewife" of the 1950s. Women began to close the gap in higher education, and their numbers in the labor force, with and without children, increased dramatically. The contraceptive revolution gave women some increased control over their own fertility.[56]

Yet as women gained economic, social, and political resources with which to chart their own destinies, the pressure to shrink in body size returned.[57] The media began to play a dominant role in this pressure. In the 1960s, films were no longer the most important influence in defining beauty. Instead, television, the American fashion industry, and women's magazines became the arbiters of image. Fashion photography, showcasing clothing, demanded stick-thin bodies.[58] In the mid-1960s, a 17-year-old, 5′ 6″, 97-pound British model entered the American fashion scene. Her name was Twiggy. She became an instant celebrity and many young women began to emulate her. Understandably, researchers point to this decade as the era of marked increase in eating disorders.[59]

Of course, another major fashion influence, the Barbie doll, had already arrived. Writing in the magazine *Smithsonian*, Doug Stewart notes, "If all the Barbies sold since 1959 were laid head to heeled foot ... they would circle the earth three and a half times."[60] CNN reports that "more than a billion Barbies have been sold in 150 countries." CNN news reporter Aaron Brown offers this account:

> Barbie was invented in 1959 and then it was revolutionary. Ruth Handler insisted that her doll have breasts. Baby dolls had dominated the market until Barbie came along. Someone once figured out that had the original Barbie been human, she would have been about 5′ 6″ and her figure would have been 39-18-33. Barbie was an immediate sensation.

The company that sold the doll, Mattel, became an instant success story. Ruth Handler and a Barbie model even rang the closing bell at the New York Stock Exchange.[61]

Barbie demonstrates that while roles can change over time, one may never find relief from the Cult of Thinness. "She was a model in 1959, a career girl in 1963, a surgeon in 1973 and an aerobics instructor in 1984."[62] Her body image includes exaggerated breasts, impossibly long legs, nonexistent hips, and a waist tinier than a Victorian lady's. This is the "perfect figure" presented to little girls. Barbie's 2006 line includes dolls with such names as "Hard Rock Café Barbie Doll" and "Peppermint Obsession Barbie Doll," as well as "Maiko Barbie Doll" from Mattel's "World Culture/Dolls of the World Barbie Collection." The Mattel company describes her as having "white makeup, a traditional hairstyle, white socks and Japanese sandals, she's the picture of femininity."[63] The most current Barbies sold online, for example through Target stores, features them in professional dress; however, their femininity is always assured, as in the case of the new "Barbie Pet Doctor," which Mattel describes wearing a "cute printed lab coat."[64] If I were a bit more cynical, I might wonder who really *is* the pet.

Women's magazines also contribute to the obsession with image, fashion, and thinness. One researcher points out

> that women's magazines collectively comprise a social institution which serves to foster and maintain a cult of femininity. This cult is manifested both as a social group to which all these born females can belong, and as a set of practices and beliefs: rites and rituals, sacrifices and ceremonies, whose periodic performance reaffirms a common femininity and shared group membership. In promoting a cult of femininity, these journals are not merely reflecting the female role in society; they are also supplying one source of definitions of, and socialization into, that role.[65]

Why are women who have gained some economic independence still expressing their self-reliance and inner control through these body rituals? An opposing view suggests that dieting and physical fitness are not methods for the subordination of women, but ways that women can feel powerful. After all, for many women, feeling fat means feeling powerless. However, by investing time, money, and energy on attaining a thin body, women may be substituting a momentary sense of power for "real authority." Some feminists take the argument even further, pointing out that being overweight is, itself, a way of expressing power. In *Fat is a Feminist Issue*, Susie Orbach notes that being overweight is one way to say "no" to feeling powerless. A fat person defies Western notions of beauty and challenges, in Orbach's words, "the ability of culture to turn women into mere products."[66] Kim Chernin says that in a feminist age, men feel drawn to and perhaps less threatened by women with childish bodies because "there is something less disturbing about the vulnerability and helplessness of a small child, and something truly disturbing about the body and mind of a mature woman."[67]

The fact remains that regardless of their economic worth, women are socialized to rely on their "natural" resources—beauty, charm, and nurturance—to attract the opposite sex.[68] The stakes of physical attractiveness for women are high, since appearance, including body weight, affects social success.[69] Research studies suggest that college-age women experience even a few extra pounds as a major problem in their lives. Women report more dissatisfaction with their weight and body image than men, and in fact many women willingly embrace the mind/body dichotomy, partly because the woman who invests herself in her body often reaps enormous rewards and benefits. Ignoring investments in one's body can mean the loss of both self-esteem and social status.[70]

I do not want to imply that all women are enslaved by bodily concerns (or that all men are the "enemy"). Throughout history, many women have found ways to resist or alter controlling social practices. For example, the exaggeratedly corseted Victorian figure drew attention to the waist and enhanced the bosom, and some women began to use it to express their sexuality. In time, political conservatives reacted to the way women subverted the intent of corseting. They decried the practice as a sign of loose morals.[71]

It would be hard to portray all women as "victims." Women frequently collude in promoting body rituals. Like the mothers who bound their daughters' feet or tightened their corsets, today's mom may recommend the latest diet and fitness club to her daughter. Many women try to cut deals with the system.... For other young women, working out in the gym may build the confidence they need to compete with men in the work world.

In a way, women's bodies are cultural artifacts, continually molded by history and culture.[72] Subjected to such pressures, the "natural body" gets lost. What replaces it may be the bewigged eighteenth-century countess, the wasp-waisted Victorian housewife, the leggy flapper, or the waif modeling Calvin Klein jeans. All are bodily reflections of the play of power within a society.[73]

NOTES

1. See Michelle Zimbalist Rosaldo, "Women, Culture and Society: A Theoretical Overview," in *Women, Culture and Society*, ed. Michelle Zimbalist Rosaldo and Louise Lamphere (Palo Alto, CA.: Stanford University Press, 1974). See also: Shirley B. Ortner, "Is Female to Male as Nature Is to Culture?" in *Women, Culture and Society*, ed. Michelle Zimbalist Rosaldo and Louise Lamphere (Palo Alto, CA: Stanford University Press, 1974) p. 67–88. Both these authors note that women are symbolized as closer to nature and men are more closely identified with culture. Male activities are given preference over female activities.

2. Paul Rosenkrantz, Susan Vogel, Helen Bee, and Donald Broverman, "Sex-Role Stereotypes and Self-Concepts in College Students," *Journal of Consulting and Clinical Psychology*, 32 (1968): 287–291. See also: P.A. Smith and E. Midlarksy, "Empirically

Derived Conceptions of Femaleness and Maleness: A Current View," *Sex Roles*, 12, no. 3/4 (1985): 313–328; R.J. Canter and B.E. Meyerowitz, "Sex-Role Stereotypes: Self-Reports of Behavior," *Sex Roles*, 10, no. 3/4 (1984): 293–306.

3. See: Nancy Jay, "Gender and Dichotomy," *Feminist Studies*, 7, no. 1 (1981): 37–56. See also: G. Lloyd, *The Man of Reason: "Male" and "Female" in Western Philosophy* (London: Methuen, 1984); Eleanor Maccoby and Carol Nagy Jacklin, *The Psychology of Sex Differences* (Palo Alto, CA.: Stanford University Press, 1974); Marion Lowe, "Social Bodies: The Interaction of Culture and Women's Biology," in *Biological Woman: The Convenient Myth*, ed. R. Hubbard, M.S. Henifin, and B. Fried, pp. 91–116 (Cambridge, MA.: Schenkman Publishing, 1982). See also: C.F. Epstein, *Deceptive Distinctions: Sex, Gender, and the Social Order* (New Haven, CT: Yale University Press, and New York: Russell Sage Foundation, 1988), chapter 4.

4. Ruth Berman notes that the dualist rationalism of Aristotle and Plato demonstrates how society's rulers limit and distort the understanding of even profound thinkers in their desire to maintain the status quo in their self-interest. Leaders have historically and currently used the practice of invoking an apparently natural hierarchy of human worth to justify widely disparate economic and social conditions. Ruth Berman, "From Aristotle's Dualism to Materialist Dialectics: Feminist Transformation of Science and Society," in *Gender/Body/Knowledge: Feminist Reconstructions of Being and Knowing*, ed. Alison M. Jagger and Susan R. Bordo (New Brunswick, NJ: Rutgers University Press, 1989), 224–255; *Body/Politics: Women and the Discourses of Science*, ed. Mary Jacobus, Evelyn Fox Keller and Sally Shuttleworth (New York: Routledge, 1990). . . .

5. See: Susan Lehrer, *Origins of Protective Labor Legislation for Women: 1905–1925*. (Albany: State University of New York Press, 1987). See also: Zillah R. Eisenstein, *The Female Body and the Law*. (Berkeley: University of California Press, 1988).

6. James W. Bashford, *China: An Interpretation* (New York: Abingdon Press, 1961), 128. Cited in Susan Greenhalgh, "Bound Feet, Hobbled Lives: Women in Old China," *Frontiers*, 2, no. 1 (1977): 17–21.

7. See: Wolfram Eberhard, "Introduction" in Howard S. Levy, *Chinese Footbinding: The History of a Curious Erotic Custom* (New York: Walton Rawls Publisher, 1966), 15–19; Susan Greenhalgh, "Bound Feet, Hobbled Lives: Women in Old China," *Frontiers*, 2, no. 1 (1977): 17–21.

8. Howard S. Levy, *Chinese Footbinding: The History of a Curious Erotic Custom* (New York: Walton Rawls publisher, 1966), 26–27.

9. C. Fred Blake notes that a mother who bound her daughter's feet considered it a sign of caring. Blake notes: "The 'tradition' could not have passed from mothers to daughters if not for mothers' credibility as 'caring.' The conundrum of a mothers' care consciously causing her daughter excruciating pain is contained in a single word, *teng*, which . . . refers to 'hurting,' 'caring,' or a conflation of both in the same breath" (682). See: C. Fred Blake, "Foot-binding in Neo-Confucian China and the Appropriation of Female Labor," *Signs: Journal of Women in Culture and Society*, 19, no. 3 (1994): 676–712.

10. Fei Hsiao-tung, *China's Gentry: Essays in Rural-Urban Relations* (Chicago: University of Chicago Press, 1953), 32, 84; Chow Yung-ten, *Social Mobility in China: Status Careers among the Gentry in a Chinese Community* (New York: Atherton Press, 1966). Both cited in Susan Greenhalgh, "Bound Feet, Hobbled Lives: Women in Old China," *Frontiers*, 2, no. 1 (1977): 7–21.

11. Anthropologist C. Fred Blake notes, however, that by binding women's feet Chinese society "masked" the real contribution of women's labor to the overall economy: "The material contributions that women made to the family were indeed substantial. They included women's traditional handiwork—making items like clothes and shoes—as well as their biological contributions in making sons for the labor-intensive economy" (700). He notes that binding women's feet, a symbol of their labor power, made it easier for the family system to take over their labor power (707–708). See: C. Fred Blake, "Foot-binding in Neo-Confucian China and the Appropriation of Female Labor," *Signs: Journal of Women in Culture and Society*, 19, no. 3 (1994): 676–712.

12. Anthropologist C. Fred Blake notes that "'big feet' of ordinary women were demeaned as clumsy and crude and as a disaster to the natural foundation—the productivity—of the civilized world" (693). C. Fred Blake, "Foot-binding in Neo-Confucian China and the Appropriation of Female Labor," *Signs: Journal of Women in Culture and Society*, 19, no. 3 (1994): 676–712.

13. Susan Greenhalgh, "Bound Feet, Hobbled Lives: Women in Old China," *Frontiers*, 2, no. 1 (1977): 7–21. (See especially p. 12.)

14. See: Andrea Dworkin, *Women Hating* (New York: Dutton, 1974), 103–104. . . .

15. Susan Greenhalgh, "Bound Feet, Hobbled Lives: Women in Old China," *Frontiers*, 2, no. 1 (1977): 13.

16. See: Susan Greenhalgh, "Bound Feet, Hobbled Lives: Women in Old China," *Frontiers*, 2, no. 1 (1977): 12. See also: Florence Ayscough, *Chinese Women Yesterday and Today* (Boston: Houghton Mifflin, 1937); C. Fred Blake, "Footbinding in Neo-Confucian China and the Appropriation of Female Labor," *Signs: Journal of Women in Culture and Society*, 19, no. 3 (1994): 685.

17. Howard S. Levy, *Chinese Footbinding: The History of a Curious Erotic Custom* (New York: Walton Rawls Publisher, 1966), 32. For an extensive discussion of the erotic nature of the bound foot, see: Bernard Rudofsky, *The Kimono Mind* (New York: Doubleday, 1965).

18. Michel Foucault, *Discipline and Punish: The Birth of the Prison*, trans. Alan Sheridan (New York: Pantheon Books, 1977).

19. It is important to point out that males were also subjected to body rituals and practices. Early capitalism with its mechanization of production needed disciplined bodies to do the mundane work routines in the early factory system. Using the body as a central arena of disciplinary power and control allowed nineteenth-century capitalism to operate efficiently and profitably. As Dreyfus and Rabinow note: "Without the insertion of disciplined, orderly individuals into the machinery of production, the new demands of capitalism would have been stymied." Hubert L. Dreyfus and P. Rabinow, *Michel Foucault: Beyond Structuralism and Hermeneutics* (Chicago: University of Chicago Press, 1983), 135. A worker's body was equated with that of a machine (whereas women workers were considered a lower-paid category compared with males). In his approach known as "Taylorism," Frederick Taylor, founder of "scientific management," envisioned a worker as part of the machinery of production. Through the application of scientific principles, specifically the technique of "time and motion studies," he proposed to ascertain how to get the most efficiency out of a given worker, ignoring some of the humanistic aspects of work. See: Frederick Taylor, *Principles of Scientific Management* (New York: W.W. Norton, 1967).

20. Ann Gordon, Mari Jo Buhle, and Nancy Schrom, "Women in American Society: An Historical Contribution," *Radical America*, 5, no. 4 (July–August 1971): 3–66.

21. Kathryn Weibel, *Mirror, Mirror: Images of Women Reflected in Popular Culture* (New York: Anchor Books, 1977), 176–177. Kathryn Weibel argues that the separation in roles fostered by the Industrial Revolution was reflected in the disparity in comfort and ornamentation between men's and women's clothes. The Industrial Revolution generated a larger, relatively wealthy, middle class of men. As has been the case historically, wives were expected to display the wealth of their husbands, becoming more "ornamented" and more "stuffed-looking" as middle class wealth increased during the nineteenth century.

22. Fashion historian Valerie Steele notes that "the vast majority of women of all classes wore corsets and the degree of tightness varied according to design of dress, social occasion and age, personality and figure of the individual woman." See: Valerie Steele, *Fashion and Eroticism: Ideals of Feminine Beauty from the Victorian Era to the Jazz Age* (New York: Oxford University Press, 1985), 162. See also: Helen E. Roberts, "The Exquisite Slave: The Role of Clothes in the Making of the Victorian Woman," *Signs: Journal of Women in Culture and Society*, 2, no. 3 (1977): 554–569.

23. Kathryn Weibel, *Mirror, Mirror: Images of Women Reflected in Popular Culture* (New York: Anchor Books, 1977), 180. While establishing links between social or cultural influences and illness is difficult, it has been suggested that a form of anemia, known as chlorosis, reflected the cultural repression women experienced during the Victorian Era. Joan Brumberg notes that chlorosis, an illness characterized by weakness, fainting, and passivity was widespread among young women in the United States dating from 1870 to 1920. See: Joan J. Brumberg, "Chlorotic Girls, 1870–1920: A Historical Perspective on Female Adolescence," *Child Development*, 53, no. 6 (1982): 1468–1477. Other researchers suggest that rates of classical conversion hysteria may be another example of the importance of cultural pressures, in this case an environment where sexual repression and dependency were primary characterizations of women's role. Donald M. Schwartz, Michael G. Thompson, and Craig L. Johnson, "Anorexia Nervosa and Bulimia: The Socio-Cultural Context," *International Journal of Eating Disorders*, 1, no. 30 (1982): 20–36.

24. Pauline Mariette, *L'Art de la Toilette* (Paris: Librairie Centrale, 1866), 40–41. Cited in Valerie Steele, *Fashion and Eroticism: Ideals of Feminine Beauty from the Victorian Era to the Jazz Age* (New York: Oxford University Press, 1985), 108.

25. Lawrence Stone compares corset makers with those who practice orthodontia. He notes that corset makers were "the affluent equivalents of the orthodontists of the late twentieth-century America, who also cater for a real need as well as a desire for perfection in a certain area thought to be important for success in life." See Lawrence Stone, *The Family, Sex and Marriage in England 1500–1800* (New York, Harper & Row, 1977). As cited in William Bennett and Joel Gurin, *The Dieter's Dilemma: Eating Less and Weighing More* (New York: Basic Books, 1982), 183. . . .

26. William Bennett and Joel Gurin, *The Dieter's Dilemma: Eating Less and Weighing More* (New York: Basic Books, 1982), 183.

27. Lawrence Stone, *The Family, Sex and Marriage in England 1500–1800* (New York: Harper & Row, 1977). As cited in William Bennett and Joel Gurin, *The Dieter's Dilemma: Eating Less and Weighing More* (New York: Basic Books, 1982), 183.

28. *Englishwoman's Domestic Magazine*, 3d ser. 4 (1868): 54. Cited in Helen E. Roberts,

"The Exquisite Slave: The Role of Clothes in the Making of the Victorian Woman," *Signs: Journal of Women in Culture and Society*, 2, no. 3 (1977): 564.

29. See: Lorna Duffin, "The Conspicuous Consumptive: Woman as an Invalid," in *The Nineteenth Century Woman, Her Cultural and Physical World*, ed. Sara Delamont and Lorna Duffin (London: Croom Helm, 1978), 26–56; Barbara Ehrenreich and Deirdre English, *For Her Own Good: 150 Years of the Experts' Advice to Women* (Garden City, NY: Anchor Books, 1979); Helen E. Roberts, "The Exquisite Slave: The Role of Clothes in the Making of the Victorian Woman," *Signs: Journal of Women in Culture and Society*, 2, no. 3 (1977): 554–569; Thorstein Veblen, *The Theory of the Leisure Class* (New York: Random House, Modern Library, original work published 1899). . . .

30. Kathryn Weibel, *Mirror, Mirror: Images of Women Reflected in Popular Culture* (New York: Anchor Books, 1977), 179.

31. Helen E. Roberts, "The Exquisite Slave: The Role of Clothes in the Making of the Victorian Woman," *Signs: Journal of Women in Culture and Society*, 2, no. 3 (1977): 564.

32. It is important to note that patriarchal interests were also threatened during early industrialism. Capitalism challenged patriarchal power by separating the home from the workplace. Ehrenreich and English (1979) note: "The household was left with only the most biological activities—eating, sex, sleeping, the care of small children . . . birth, dying and the care of the sick and aged." Furthermore, "It was now possible for a woman to enter the market herself and exchange her labor for the means of survival." See: Barbara Ehrenreich and Deidre English, *For Her Own Good: 150 Years of the Experts' Advice to Women* (Garden City, NY: Anchor Books, 1979), 10, 13, 27.

33. Mike Featherstone, "The Body in Consumer Culture," *Theory, Culture and Society*, 1, no. 2 (1982): 20. See also: Stuart Ewen, *Captains of Consciousness: Advertising and the Roots of the Consumer Culture* (New York: McGraw Hill, 1976); Joseph Hansen and Evelyn Reed, *Cosmetics, Fashions and the Exploitation of Women* (New York: Pathfinder Press, 1986); Heidi Hartmann, "Capitalism, Patriarchy and Job Segregation by Sex," *Signs: Journal of Women in Culture and Society*, 1, no. 3 (1976): 137–169.

34. Roberta Seid provides an excellent detailed historical analysis of American society's movement toward slenderness. See especially: chapter 5, Roberta Pollack Seid, *Never Too Thin: Why Women Are at War with Their Bodies* (New York: Prentice-Hall Press, 1989); see also: Stuart Ewen and Elizabeth Ewen, *Channels of Desire: Mass Images and the Shaping of the American Consciousness* (New York: McGraw-Hill, 1982).

35. Ibid, 85.

36. Ibid, 83.

37. Ibid, 115.

38. Ibid, 115.

39. Barbara Ehrenreich and Deidre English, *For Her Own Good: 150 Years of the Experts Advice to Women* (Garden City, NY: Anchor Books, 1979); Stuart Ewen, *Captains of Consciousness: Advertising and the Roots of Consumer Culture* (New York: McGraw-Hill, 1976); Joseph Hansen and Evelyn Reed, *Cosmetics, Fashions and the Exploitation of Women* (New York: Pathfinder Press, 1986); Heidi Hartmann, "Capitalism, Patriarchy, and Job Segregation by Sex," *Signs: Journal of Women in Culture and*

Society, 1, no. 3 (1976): 137–169; Brett Silverstein, *Fed Up! The Food Forces That Make You Fat, Sick and Poor* (Boston: South End Press, 1984).

40. Sharlene Hesse-Biber, "Women, Weight and Eating Disorders: A Socio-Cultural and Political-Economic Analysis," *Women's Studies International Forum*, 14, no. 3 (1991): 173–191; Susan Bordo, *Unbearable Weight: Feminism, Western Culture and the Body* (Berkeley University of California Press, 1993).

41. Mary Frank Fox and Sharlene Hesse-Biber, *Women at Work* (Palo Alto CA: Mayfield Publishing, 1984), 19.

42. In the late 1800s, a large number of women entered medicine. The percentage of women in the 1893–1894 medical graduating classes in the Boston area was 23.7%. Women accounted for 17% of Boston's medical community. But, by the end of World War I, the numbers of women dropped off. See: Augusta Greenblatt, "Women in Medicine," *National Forum: The Phi Beta Kappa Journal*, 61, no. 4, (1981): 10–11.

43. For a detailed discussion of this transition, see: Lois Banner, *American Beauty*, (Chicago: The University of Chicago Press, 1983); and William Bennett and Joel Gurin, *The Dieter's Dilemma: Eating Less and Weighing More* (New York: Basic Books, 1982), especially chapter 7, titled "The Century of Svelte."

44. See: Susie Orbach, *Hunger Strike: The Anorectic's Struggle as a Metaphor of Our Age* (New York: W.W. Norton, 1986), 75.

45. Roberta Pollack Seid, *Never Too Thin: Why Women Are at War with Their Bodies* (New York: Prentice Hall, 1989).

46. Ibid.

47. See: Lois W. Banner, *American Beauty* (New York: Knopf, 1983), 279.

48. John A. Ryle, "Discussion of Anorexia Nervosa," *Proceedings of the Royal Society of Medicine*, 32 (1939): 735–737. It is important to note that documented cases of anorexia nervosa were cited in the medical literature well before this time. See: William Gull, "Anorexia Nervosa (Apepsia Hysterica, Anorexia Hysterica)," *Transactions of the Clinical Society of London*, 7 (1974): 22–28; and (English language translation) Ernest-Charles Lasegue, "On Hysterical Anorexia," *Medical Times and Gazette*, no. 2, September 6, 1873, 22–28. pp. 265–266; September 27, 1873, pp. 367–369....

49. "Weight Reduction Linked to the Mind," *New York Times*, p. 6, February 24, 1926.

50. Richard A. Gordon, *Anorexia and Bulimia: Anatomy of a Social Epidemic* (Cambridge, MA: Basil Blackwell, 1990), 78.

51. See: William Bennett and Joel Gurin, *The Dieter's Dilemma: Eating Less and Weighing More* (New York: Basic Books, 1982), 207.

52. Lois W. Banner, *American Beauty* (New York: Knopf, 1983), 283.

53. This is amply documented in Betty Friedan, *The Feminine Mystique* (New York: Norton, 1963).

54. Historian Lois Banner notes that during the 1950s women's sports suffered a setback. Banner states: "With few exceptions, the kind of acclaim accorded to individual women sports stars in the 1920s and 1930s no longer existed, and the commercial women's swimming and basketball teams popular in these earlier decades faded from view.... In high schools and colleges, women's athletics similarly came to occupy a modest position vis-à-vis men's sports." Lois W. Banner, *American Beauty* (New

York: Knopf, 1983), 285. During the 1950s, there was evidence of the flapper in the image of Debbie Reynolds and Sandra Dee. Sandra Dee played the popular Gidget, who was portrayed as looking for a husband and not serious about a career. See: Lois W. Banner, *American Beauty* (New York: Knopf, 1983), 283.

55. Roberta Pollack Seid, *Never Too Thin: Why Women Are at War with Their Bodies* (New York: Prentice Hall, 1989), 257.

56. Mary Frank Fox and Sharlene Hesse-Biber, *Women at Work* (Palo Alto, CA: Mayfield Publishing, 1984).

57. An empirical test of this theory on changing body image comes from a study by Silverstein, Perdue, Peterson, Vogel, and Fantini (1986). They studied the standards of bodily attractiveness across time and note that over the course of the twentieth century, as the proportion of American women who worked in the professions or who graduated from college increased, the standard of bodily attractiveness became less curvaceous. They note that this occurred especially in the 1920s and during the 1960s. Thinness may be considered a sign of conforming to a constricting feminine image (like corseting), whereas greater weight may convey a strong, powerful image. See: Brett Silverstein, Lauren Perdue, Barbara Peterson, Linda Vogel, and Deborah A. Fantini, "Possible Causes of the Thin Standard of Bodily Attractiveness for Women," *International Journal of Eating Disorders*, 5, no. 5 (1986): 135–144.

58. Lois W. Banner, *American Beauty* (New York: Knopf, 1983), 266–287.

59. Allan Mazur, "U.S. Trends in Feminine Beauty and Overadaptation," *Journal of Sex Research*, 22, no. 3 (1986): 281–303.

60. Doug Stewart, "In the Cutthroat World of Toy Sales, Child's Play is Serious Business," *Smithsonian*, 20, no. 9 (December, 1989,): 80.

61. Aaron Brown, "Look Back at Those Who Passed in 2002." *CNN NewsNight*, December 31, 2002, http://transcripts.cnn.com/TRANSCRIPTS/0212/31/asb.00.html

62. Doug Stewart, "In the Cutthroat World of Toy Sales, Child's Play is Serious Business," *Smithsonian*, 20, no. 9 (December, 1989): 72–84.

63. Mattel, Inc. "Doll Showcase: 2006 Line" from Barbie B Collector, 2005, http://collectdolls.about.com/ About, Inc. a part of the New York Times, Company. 2006.

64. Target. "Dolls + Accessories: Barbie, Target.com (2005).

65. Marjorie Ferguson, *Forever Feminine: Women's Magazines and the Cult of Femininity* (London: Heinemann, 1983), 184.

66. Susie Orbach, *Fat Is a Feminist Issue* (New York: Berkeley Press, 1978), 21.

67. See: Kim Chernin, *The Obsession: Reflections on the Tyranny of Slenderness* (New York: Harper & Row, 1981), 110.

68. Pauline B. Bart, "Emotional and Social Status of the Older Woman," in *No Longer Young: The Older Woman in America: Occasional Papers in Gerontology.* Ann Arbor: University of Michigan, Institute of Gerontology, 1975, 321; Daniel BarTal and Leonard Saxe, "Physical Attractiveness and Its Relationship to Sex-Role Stereotyping," *Sex Roles*, 2, no. 2 (1976): 123–133; Peter Blumstein and Pepper W. Schwartz, *American Couples: Money, Work and Sex* (New York: William Morrow, 1983); Glen H. Elder "Appearance and Education in Marriage Mobility," *American Sociological*

Review, 34 (1969): 519–533; Susan Sontag, "The Double Standard of Aging," *Saturday Review*, 55, no. 39 (1972): 29–38.

69. Sharlene Hesse-Biber, Alan Clayton-Matthews, and John Downey, "The Differential Importance of Weight among College Men and Women," *Genetic, Social and General Psychology Monographs*, 113, no. 4 (1987): 511–538.

70. R. Pingitore, B. Spring, and D. Garfield, "Gender Differences in Body Satisfaction. "*Obesity Research*, 5, no. 5, (1997): 402–409. See: M. Tiggemann, "Gender Differences in the Interrelationships between Weight Dissatisfaction, Restraint, and Self-esteem," *Sex Roles*, 30, no. 5–6 (1994): 319–330. The authors point out that feeling fat is related to women's self-esteem.

71. David Kunzle, "Dress Reform as Antifeminism: A Response to Helene E. Roberts' 'The Exquisite Slave: The Role of Clothes in the Making of the Victorian Woman," *Signs: Journal of Women in Culture and Society*, 2, no. 3 (1977): 570–579. . . .

72. Barbara Ehrenreich and Deirde English, *For Her Own Good: 150 Years of the Experts' Advice to Women* (Garden City, NY: Anchor Books); Zillah R. Eisenstein, *The Female Body and the Law* (Berkeley: The University of California Press, 1988); Emily Martin, *The Woman in the Body: A Cultural Analysis of Reproduction* (Boston: Beacon Press, 1987); Helena Michie, *The Flesh Made Word: Female Figures and Women's Bodies* (New York: Oxford University Press, 1987); Gayle Rubin, "The Traffic in Women," in *Toward an Anthropology of Women*, ed. Rayna R. Reiter (New York: Monthly Review Press, 1975), 157–210; Bryan S. Turner, *The Body and Society* (New York: Basil Blackwell, 1984); Susan Bordo, "Anorexia Nervosa: Psychopathology as the Crystallization of Culture," in *Feminism and Foucault: Reflections on Resistance*, ed. Irene Diamond and Lee Quinby (Boston: Northeastern University Press, 1988), 87–117.

73. Michel Foucault, *Discipline and Punish: The Birth of Prison*. trans. Alan Sheridan. (New York: Pantheon Books, 1977).

KEY CONCEPTS

cult   cultural hegemony   custom

dominant culture

DISCUSSION QUESTIONS

1. What types of fashion trends can you think of today that are common among women and girls that are, or have the potential to be, harmful?

2. How can we balance between the cultural expectation to be thin, the importance of being happy with your own body, and the negative health consequences of being overweight? How does society contribute to each of these concerns?

Applying Sociological Knowledge:
An Exercise for Students

Take a look around your campus and make note of any differences in clothing that people are wearing that you would associate with cross-cultural differences. What issues would you face in your family, community, or friendship network were you to adopt that style of dress? What do these issues teach you about how cultural expectations shape social norms and group conformity?

9

Leaving Home for College: Expectations for Selective Reconstruction of Self

DAVID KARP, LYNDA LYTLE HOLMSTROM, AND PAUL S. GRAY

Many young adults leave home for the first time when they go away to college. This article addresses the changes young students go through when they leave high school and their family home. The authors discuss how personal changes in identity and perceptions of self are more significant than the geographical move to college.

In their important and much discussed critique of American culture, *Habits of the Heart* (1984), Robert Bellah and his colleagues remark that American parents are of two minds about the prospect of their children leaving home. The thought that their children will leave is difficult, but perhaps more troublesome is the thought that they might not. In contrast to many cultures, American parents place great emphasis on their children establishing independence at a relatively early age. Still, as Bellah's wry comment suggests, they are deeply ambivalent about their children leaving home. The data presented in this paper, part of a larger project on family dynamics during the year that a child applies for admission to college, show that such ambivalence is shared by the children. Our goal here is to document some of the social psychological complexities of achieving independence in America by analyzing the perspectives of 23 primarily upper-middle-class high school seniors as they moved through the college application process and contemplated leaving home.[1]

Of course, a great deal has been written about the internal conflict that surrounds any significant personal change (most obviously, Erik Erikson 1963, 1968, 1974, 1980; see also Manaster 1977; O'Mally 1995). Although researchers have attended to the phenomenon of "incompletely launched young adults" (Heer, Hodge, and Felson 1985; Grigsby and McGowan 1986; Schnaiberg and Goldenberg 1989), little has been written about how relatively sheltered, middle- to upper-middle-class children think about "leaving the nest." Leaving home for college is perhaps among the greatest changes that the economically comfortable students we interviewed have thus far encountered in their lives. For them, going to college carries great significance as a coming-of-age moment, in part because it has been long anticipated and not to do

SOURCE: © 1998 by the Society for the Study of Symbolic Interaction.
Reprinted from *Symbolic Interaction*. Vol. 21, No. 3, pp. 253–276 by permission.

so would be unacceptable from a normative stand point. Literature on students who "beat the odds" by going to college suggests that this is also an important transition for them, but one carrying fundamentally different meanings. Unlike the middle- or upper-middle-class students we interviewed, who are trying, at the least, to retain their class position, students arriving at college from less privileged backgrounds must confront wholly new cultural worlds (Rodriguez 1982; Smith 1993; Hooks 1993).

The 23 students with whom we were able to complete interviews simply assumed, as did their parents, that they would go to college.[2] Among the 30 sets of parents, all but four individuals had attended college (and two received some different training beyond high school). All of the adults, however, felt strongly about the necessity of college attendance for their children. One of the four who did not go to college, a self-made and extraordinarily successful entrepreneur, did offer some reservation about the utility of an education in the rough and tumble "real world." Even so, both he and his wife were highly invested in getting their son into a prestigious college. While all of the children knew their parents' expectations and fully expected to meet them, we did speak with two students who had some misgivings about whether they really wanted to go to college. Like their counterparts in our sample, these students knew they would go, but still entertained private doubts about their interest in and motivation for college work.[3]

What does it mean to become independent of one's parents, family, and high school friendship groups? As Anna Freud noted, "few situations in life are more difficult to cope with than the attempts of adolescent children to liberate themselves" (Bassoff 1988, p.xi). Young people are ambivalent regarding independence; it is hard to break away. Their ambivalence embodies both symbolic and pragmatic dimensions. Symbolically, independence is the desired outcome of a necessary process of differentiation (Blos 1962). The task for adolescents is "to find their own way in the world and develop confidence that they are strong enough to survive outside the protective family circle" (Bassoff 1988, p. 3). To establish their own identity and sense of purpose, ". . . they need to wrench themselves away from those who threaten their developing selfhood" (Bassoff 1988, p. 3; see also Campbell, Adams, and Dobson 1984; Katchadourian and Boli 1994). However, independence also has a pragmatic side. In college, young people can "start over"; they can make new friends, establish intimate relationships, and develop the skills and knowledge to help them become self-supporting adults. "But the truth is that they are not sure they can take care of themselves or that they want to be left alone" (Bassoff 1988, p. 3). . . .

IDENTITY AFFIRMATION, IDENTITY RECONSTRUCTION, AND IDENTITY DISCOVERY

While the students in this study anticipated college as a time during which they would maintain, refine, build upon, and elaborate certain of their identities, they also anticipated negotiating some fundamental identity changes. The students saw college as the time for discovering who they *really* were. They anticipated finding wholly new and permanent life identities during the college years. In addition, they believed that going to college provides a unique opportunity to consciously establish some new identities. Repeatedly, students described the importance of going away to college in terms of an opportunity to discard disliked identities while making a variety of "fresh starts." Their words suggest that college-bound students look forward to re-creating themselves in a context far removed (often geographically, but always symbolically) from their family, high school, and community. The immediately following sections attend, in turn, to how upper-middle-class high school seniors (1) anticipate change, (2) strategize about solidifying certain identities, [and] (3) evaluate identities they wish to escape....

Anticipating Change

Along with such turning points as marriage, having children, and making an occupational commitment, it is plain that leaving for college is self-consciously understood as a dramatic moment of personal transformation. The students with whom we spoke all saw leaving home as a critical juncture in their lives. One measure of consensus in the way our 23 respondents interpreted the meaning of leaving home is the similarity of their words. Students used nearly identical phrases in describing the transition to college as the time to "move on," "discover who I really am," to "start over," to "become an adult," to "become independent," to "begin a new life." The students, moreover, explicitly saw going to college as the "next stage" of their lives....

While all the students interviewed recognized the need for change and were looking forward to it, their certainty about the appropriateness of moving on did not prevent them from feeling anxiety and ambivalence about the transition to college. Theirs is an anticipation composed of optimism, excitement, anxiety, and sometimes fear.

> [I'm] starting the rest of my life. I mean, deciding what I'm going to do and figuring out my future. I mean, that's one thing I'm looking forward to, but it's also one thing I'm not looking forward to. I have mixed feelings about that. It's exciting to figure out your future. In another sense it's scary to have all of the responsibility. (White male attending a public school)

These comments suggest that the prospect of leaving home generates an anticipatory socialization process characterized by multiple and sometimes contradictory feelings and emotions. Students long for independence, anticipate the excitement that accompanies all fresh starts, but worry about their ability to fully meet the challenge....

Affirming Who I Really Am

The one concrete and critical choice that college-bound students must make is which school, in fact, to attend. This decision is often an agonizing one for both students and their parents and involves very high levels of "emotion work" (Hochschild 1983). The significance of making the college choice and the anxiety that it occasions go well beyond questions of money, course curricula, or the physical amenities of the institutions themselves. What makes the decision so difficult is that the students know they are choosing the context in which their new identities will be established, . . . The fateful issue in the minds of the students is whether people with their identity characteristics and aspirations will be able to flourish. Consequently, it is not surprising that the most consistent and universal pattern in our data is the effort expended by students to find a school where "a person like me" will feel comfortable. . . .

In the most global way, prospective students were searching for a place where the students seemed friendly. On several occasions, students remarked that they were turned on or off to a school because their "tour guide" was either really nice or not friendly enough. . . .

In contrast to the students-like-me theme, an interesting sub-set of seniors expressed a strong interest in diversity. These students not only wanted to meet new people, but different kinds of new people. Students who wanted diversity were excited at the prospect of meeting people different from themselves as a critical learning experience. It is important to note that it was primarily the minority students we interviewed who looked for diversity as they contemplated colleges. An Asian student put it this way:

> The more mixed the better. I think interaction with other ethnic and racial groups is very healthy. If possible, I would not mind having, you know, like an Afro-American roommate. I'd love to. (Asian male attending a public school) . . .

While the statements immediately above illustrate that students make careful assessments about the goodness of fit between certain aspects of themselves and the character of different colleges, a dominant theme in the interviews concerned change. Students repeatedly commented that, during their college years, they expected their identities to shift in two fundamental ways. First, they anticipated discovering "who I am" in the broadest sense. Second, they saw college as providing a fresh start because they could discard some of their disliked, sticky identities, often acquired as early as grade school.

Creating the Person I Want to Be

. . . Seen in terms of Erving Goffman's (1959) dramaturgical model of interaction, going to college provides a new stage and audience, together allowing for new identity performances. Goffman notes (1959, p. 6) that "When an individual appears before others his actions will influence the definition of the situation which they come to have. Sometimes the individual will act in a thoroughly calculating manner, expressing himself in a given way solely in order to give the kind of impression to others that is likely to evoke from them a specific response he is concerned to obtain." To the extent that such impression-management is most centrally dependent upon information control, leaving home provides an unparalleled opportunity to abandon labels that have most contributed to disliked and unshakable identities. When students speak of college as providing a fresh start, they have in mind the possibility of fashioning new roles and identities. Going to college promises the chance to edit, to revise, to re-write certain parts of their biographies.

> It's sort of like starting a new life. I'll have connections to the past, but I'm obviously starting with a clean slate. . . . Because no one cares how you did in your high school after you're in college. So everyone's equal now. (White male attending a public high school) . . .

As students described their hopes about college, the theme of "fresh starts" was almost universally voiced. . . . Leaving home, friends, and community offers students the possibility to jettison identities which are the product of others' consistent definitions of them over many years. Going to college provides a unique opportunity to display new identities consistent with the person they wish to become.

The data presented thus far are meant to convey the symbolic weightiness of the transition from high school to college. Every student with whom we spoke saw leaving home as a critical biographical moment. They see it as a definitive life stage when their capacity for independence will be fully tested for the first time. Some have had a taste of independence at summer camps and the like, but the transition to college is viewed as the "real thing." Their words, we have been suggesting, indicate that they see strong connections among leaving home, gaining independence, achieving adult status, and transforming their identities. Students carefully attempt to pick a college where they will fit in, thus indicating the importance of retaining and consolidating certain parts of their identities (see Shreier 1991). In addition, they believe that they will discover, in a holistic sense, who they "really" are during the college years.

WILL THEY MISS ME?

. . . The family is a social system in which roles are interconnected and interdependent. When a child goes off to college, the system is disturbed and the family will try to adapt to the new circumstances. College-bound seniors worry about this process of adaptation. They speculate that their remaining siblings will miss them, or will be left to face the unremitting attentiveness and concern of parents. They also wonder about prospective changes in their parents' marital relationship. In particular, they are concerned for their mothers, whom they identify as being more invested than their fathers in keeping the family system *status quo ante*. Finally, and most significantly, these late adolescents manifest insecurity about their place in the family, especially now that they are leaving. Several of them remarked ruefully, "I should hope they feel some grief [laughter]." "I think they'll be lonelier. I hope they will." "They'll miss me, I hope. . . . I hope they feel my presence being gone. . . . They don't have to be, like, mourning my departure, but just a little bit would be nice." It's not that they actually want their parents and siblings to suffer, but missing them would be proof positive that their membership in the family was valued, and that their future place in the family system is assured, in spite of their changing addresses. . . .

In many of our conversations, it appeared that the worst thing about going away to college was that the young people would no longer be able to participate in many aspects of family life. However, perhaps no issue symbolizes the worry associated with leaving home as powerfully as pending decisions over space in the household. How quickly one's bedroom is claimed by other members of the family is, for many of these students, a commentary on the fragility of their position. Although Silver (1996) points out that both the home room and college dorm room are used to symbolically affirm family relations, our conversations with students were more focused on the meanings they attached to their bedrooms at home. One senior said, "They always joke around and they say, 'Oh, we're going to make your room into a den.'" . . .

Some of the seniors are beginning to understand that the nature of relations with their parents will be altered forever. They will have much more discretion concerning what to reveal about themselves, and therefore much more control over the impression they choose to give their parents. As one young woman put it, "I will experience a lot of things without them there, so that they won't know that they've happened . . . [unless] I tell them or if they can see a difference in me." Others expressed shared anxieties about personal transformations and the consequent stability of their place in the family constellation. . . .

What are we to make of these worries, speculations, and musings? College-bound young adults genuinely want to remain attached to their families, even as they are yearning for true independence. Getting into college is understood as a point of departure which has the potential to alter fundamentally their relationship with their family. However, in spite of their worries, most students see the transition to college as a good thing—a positive transformation with life-long consequences. They cannot predict precisely how their relations with parents and siblings will change, but they know for sure that they have initiated a process that will alter the character of these primary relationships. Such knowledge is plainly implicated in the calculus of ambivalence they feel about leaving home:

> It's like, if you want to be treated like an adult, you have to act like an adult. If you want to be treated like a child, act like a child. If you want to be treated like an adult the rest of your life, you've got to start sometime. (White male attending a public high school)

"You've got to start sometime." That, of course, is exactly what they are doing as they embark on their great adventure of self-discovery, into college first and hopefully, thereby, toward full adulthood.

NOTES

1. We used father's occupation as a proxy for social class. We characterized our sample as predominantly upper-middle class. A sampling of the types of father's occupations that warrant this description includes: physician, lawyer, professor, administrator, and architect. A few occupations were either higher or lower in status.

2. Either because we could not reach them or because they declined to be interviewed, we did not speak to eight of the 31 students originally included in our sample. The number is 31 because one of the 30 families had twins.

3. One student, who declined to be interviewed, did not complete the college application process during his senior year in high school. He was the only student in our sample who did not anticipate attending college in the year following high school graduation.

REFERENCES

Basoff, Evelyn. 1988. *Mothers and Daughters: Loving and Letting Go.* New York: Penguin Books.

Bellah, Robert, Richard Madsen, William Sullivan, Ann Swidler, and Steven Tipton. 1985. *Habits of the Heart: Individualism and Commitment in American Life.* Berkeley: University of California Press.

Blos, Peter. 1962. *On Adolescence: A Psychoanalytic Interpretation.* New York: Free Press.

Campbell, Eugene, Gerald Adams, and William Dobson. 1984. "Familial Correlates of Identity Formation in Late Adolescence: A Study of the Predictive Utility of Connectedness and Individuality in Family Relations." *Journal of Youth and Adolescence* 13: 509–525.

Erikson, Erik. 1963. *Childhood and Society*, 2nd ed. New York: W. W. Norton.

———. 1968. *Identity: Youth and Crisis.* New York: W. W. Norton.

———. 1974. *Dimensions of a New Identity.* New York: W. W. Norton.

———. 1980. *Identity and the Life Cycle.* New York: W. W. Norton.

Goffman, Erving. 1959. *The Presentation of Self in Everyday Life.* Garden City, NY: Doubleday Anchor.

Grigsby, Jill, and Jill McGowan. 1986. "Still in the Nest: Adult Children Living with Their Parents." *Sociology and Social Research* 70: 146–148.

Heer, David, Robert Hodge, and Marcus Felson. 1985. "The Cluttered Nest: Evidence that Young Adults Are More Likely to Live at Home Now Than in the Recent Past." *Sociology and Social Research* (69): 436–441.

Hochschild, Arlie. 1983. *The Managed Heart: Commercialization of Human Feeling.* Berkeley: University of California Press.

Hooks, Bell. 1993. "Keeping Close to Home: Class and Education." Pp. 99–111 in *Working-Class Women in the Academy*, edited by Michelle Tokarczyk and Elizabeth Fay. Amherst, MA: The University of Massachusetts Press.

Katchadourian, Herant, and John Boli. 1994. *Cream of the Crop: The Impact of Elite Education in the Decade After College.* New York: Basic Books.

Manaster, Guy. 1977. *Adolescent Development and the Life Tasks.* Boston: Allyn and Bacon.

O'Mally, Dawn. 1995. *Adolescent Development: Striking a Balance Between Attachment and Autonomy.* Ph.D. dissertation, Department of Psychology, Harvard University, Cambridge, MA.

Rodriguez, Richard. 1982. *Hunger of Memory: The Education of Richard Rodriguez.* Boston: David R. Godine.

Schnaiberg, Allan, and Sheldon Goldenberg. 1998. "From Empty Nest to Crowded Nest: The Dynamics of Incompletely-Launched Young Adults." *Social Problems* 36: 251–269.

Schreier, Barbara. 1991. *Fitting In: Four Generations of College Life*. Chicago: Chicago Historical Society.

Silver, Ira. 1996. "Role Transitions, Objects, and Identity." *Symbolic Interaction* 19: 1–20.

Smith, Patricia. 1993. "Grandma Went to Smith, All Right, But She Went from Nine to Five: A Memoir." Pp. 126–139 in *Working-Class Women in the Academy,* edited by Michelle Tokarczyk and Elizabeth Fay, Amherst, MA: The University of Massachusetts Press.

KEY CONCEPTS

anticipatory
socialization

identity

rite of passage

DISCUSSION QUESTIONS

1. What changes did you go through (or are you going through) during your first year of college? How do these changes influence your self-identity and how others perceive you?

2. When you go home for vacations and visits, how does home feel different now that you have lived away? What feels the same?

10

Barbie Girls Versus Sea Monsters

Children Constructing Gender

MICHAEL A. MESSNER

In this article, Messner analyzes the gender differences among preschool soccer teams. His analysis uncovers how young boys and girls "do gender" when they interact and play with and among one another. He also discusses

SOURCE: Michael A. Messner. "Barbie Girls versus Sea Monsters: Children Constructing Gender." *Gender & Society* 14, no. 6 (December 2000): 765–784.

how the youth soccer league is structured in gendered ways. Finally, the research shows that popular culture icons, like Barbie dolls, provide symbols of gendered expectations for children.

In the past decade, studies of children and gender have moved toward greater levels of depth and sophistication (e.g., Jordan and Cowan 1995; McGuffy and Rich 1999; Thorne 1993). In her groundbreaking work on children and gender, Thorne (1993) argued that previous theoretical frameworks, although helpful, were limited: The top-down (adult-to-child) approach of socialization theories tended to ignore the extent to which children are active agents in the creation of their worlds—often in direct or partial opposition to values or "roles" to which adult teachers or parents are attempting to socialize them. Developmental theories also had their limits due to their tendency to ignore group and contextual factors while overemphasizing "the constitution and unfolding of *individuals* as boys or girls" (Thorne 1993, 4). In her study of grade school children, Thorne demonstrated a dynamic approach that examined the ways in which children actively construct gender in specific social contexts of the classroom and the playground. Working from emergent theories of performativity, Thorne developed the concept of "gender play" to analyze the social processes through which children construct gender. Her level of analysis was not the individual but "*group life*—with social relations, the organization and meanings of social situations, the collective practices through which children and adults create and recreate gender in their daily interactions" (Thorne 1993, 4).

A key insight from Thorne's research is the extent to which gender varies in salience from situation to situation. Sometimes, children engage in "relaxed, cross sex play"; other times—for instance, on the playground during boys' ritual invasions of girls' spaces and games—gender boundaries between boys and girls are activated in ways that variously threaten or (more often) reinforce and clarify these boundaries. However, these varying moments of gender salience are not free-floating; they occur in social contexts such as schools and in which gender is formally and informally built into the division of labor, power structure, rules, and values (Connell 1987).

The purpose of this article is to use an observation of a highly salient gendered moment of group life among four- and five-year-old children as a point of departure for exploring the conditions under which gender boundaries become activated and enforced. I was privy to this moment as I observed my five-year-old son's first season (including weekly games and practices) in organized soccer. Unlike the long-term, systematic ethnographic studies of children conducted by Thorne (1993) or Adler and Adler (1998), this article takes one moment as its point of departure. I do not present this moment as somehow "representative" of what happened throughout the season; instead, I examine this as an example of what Hochschild (1994, 4) calls "magnified moments," which are "episodes of heightened importance, either epiphanies, moments of intense glee or unusual insight, or moments in which things go intensely but meaningfully wrong. In either case, the moment stands out; it is metaphorically rich, unusually elaborate and often echoes [later]." A magnified moment in daily life offers a window into the social construction of reality. It presents researchers with an opportunity to excavate gendered meanings and processes through an analysis of institutional and cultural contexts. The single empirical observation that serves as the point of departure for this article was made during a morning. Immediately after the event, I recorded my observations with detailed notes. I later slightly revised the notes after developing the photographs that I took at the event.

I will first describe the observation—an incident that occurred as a boys' four- and five-year-old soccer team waited next to a girls' four- and five-year-old soccer team for the beginning of the community's American Youth Soccer League (AYSO) season's opening ceremony. I will then examine this moment using three levels of analysis.

> *The interactional level*: How do children "do gender," and what are the contributions and limits of theories of performativity in understanding these interactions?
>
> *The level of structural context*: How does the gender regime, particularly the larger organizational level of formal sex segregation of AYSO, and the concrete, momentary situation of the opening ceremony provide a context that variously constrains and enables the children's interactions?
>
> *The level of cultural symbol*: How does the children's shared immersion in popular culture (and their differently gendered locations in this immersion) provide symbolic resources for the creation, in this situation, of apparently categorical differences between the boys and the girls?

Although I will discuss these three levels of analysis separately, I hope to demonstrate that interaction, structural context, and culture are simultaneous and mutually intertwined processes, none of which supersedes the others.

BARBIE GIRLS VERSUS SEA MONSTERS

It is a warm, sunny Saturday morning. Summer is coming to a close, and schools will soon reopen. As in many communities, this time of year in this small, middle- and professional-class suburb of Los Angeles is marked by the beginning of another soccer season. This morning, 156 teams, with approximately 1,850 players ranging from 4 to 17 years old, along with another 2,000 to 3,000 parents, siblings, friends, and community dignitaries have gathered at the local high school football and track facility for the annual AYSO opening ceremonies. Parents and children wander around the perimeter of the track to find the assigned station for their respective teams. The coaches muster their teams and chat with parents. Eventually, each team will march around the track, behind their new team banner, as they are announced over the loudspeaker system and are applauded by the crowd. For now though, and for the next 45 minutes to an hour, the kids, coaches, and parents must stand, mill around, talk, and kill time as they await the beginning of the ceremony.

The Sea Monsters is a team of four- and five-year-old boys. Later this day, they will play their first-ever soccer game. A few of the boys already know each other from preschool, but most are still getting acquainted. They are wearing their new uniforms for the first time. Like other teams, they were assigned team colors—in this case, green and blue—and asked to choose their team name at their first team meeting, which occurred a week ago. Although they preferred "Blue Sharks," they found that the name was already taken by another team and settled on "Sea Monsters." A grandmother of one of the boys created the spiffy team banner, which was awarded a prize this morning. As they wait for the ceremony to begin, the boys inspect and then proudly pose for pictures in front of their new award-winning team banner. The parents stand a few feet away— some taking pictures, some just watching. The parents are also getting to know each other, and the common currency of topics is just how darned cute our kids look, and will they start these ceremonies soon before another boy has to be escorted to the bathroom?

Queued up one group away from the Sea Monsters is a team of four- and five-year-old girls in green and white uniforms. They too will play their first game later today, but for now, they are awaiting the beginning of the opening ceremony. They have chosen the name "Barbie Girls," and they also have a spiffy new team banner. But the girls are pretty much ignoring their banner, for they have created another, more powerful symbol around which to rally. In fact, they are the only team among the 156 marching today with a team float—a red Radio Flyer wagon base, on which sits a Sony boom box playing music, and a 3-foot-plus-tall Barbie doll on a rotating pedestal. Barbie is dressed in the team colors— indeed, she sports a custom-made green-and-white cheerleader-style outfit, with the Barbie Girls' names written on the skirt. Her normally all-blonde hair has been streaked with Barbie Girl green and features a green bow, with white polka dots. Several of the girls on the team also have supplemented their uniforms with green bows in their hair.

The volume on the boom box nudges up and four or five girls begin to sing a Barbie song. Barbie is now slowly rotating on her pedestal, and as the girls sing more gleefully and more loudly, some of them begin to hold hands and walk around the float, in sync with Barbie's rotation. Other same-aged girls from other teams are drawn to the celebration and, eventually, perhaps a dozen girls are singing the Barbie song. The girls are intensely focused on Barbie, on the music, and on their mutual pleasure.

As the Sea Monsters mill around their banner, some of them begin to notice, and then begin to watch and listen as the Barbie Girls rally around their float. At first, the boys are watching as individuals, seemingly unaware of each other's shared interest. Some of them stand with arms at their sides, slack-jawed, as though passively watching a television show. I notice slight smiles on a couple of their faces, as though they are drawn to the Barbie Girls' celebratory fun. Then, with side-glances, some of the boys begin to notice each other's attention on the Barbie Girls. Their faces begin to show signs of distaste. One of them yells out, "NO BARBIE!" Suddenly, they all begin to move—jumping up and down, nudging and bumping one other—and join into a group chant: "NO BARBIE! NO BARBIE! NO BARBIE!" They now appear to be every bit as gleeful as the girls, as they laugh, yell, and chant against the Barbie Girls.

The parents watch the whole scene with rapt attention. Smiles light up the faces of the adults, as our glances sweep back and forth, from the sweetly celebrating Barbie Girls to the aggressively protesting Sea Monsters. "They are SO different!" exclaims one smiling mother approvingly. A male coach offers a more in-depth analysis: "When I was in college," he says, "I took these classes from professors who showed us research that showed that boys and girls are the same. I believed it, until I had my own kids and saw how different they are." "Yeah," another dad responds, "Just look at them! They are so different!"

The girls, meanwhile, show no evidence that they hear, see, or are even aware of the presence of the boys who are now so loudly proclaiming their opposition to the Barbie Girls' songs and totem. They continue to sing, dance, laugh, and rally around the Barbie for a few more minutes, before they are called to reassemble in their groups for the beginning of the parade.

After the parade, the teams reassemble on the infield of the track but now in a less organized manner. The Sea Monsters once again find themselves in the general vicinity of the Barbie Girls and take up the "NO BARBIE!" chant again. Perhaps put out by the lack of response to their chant, they begin to dash, in twos and threes, invading the girls' space, and yelling menacingly. With this, the Barbie Girls have little choice but to recognize the presence of the boys—some look puzzled and shrink back, some engage the boys and chase them off. The chasing seems only to incite more excitement among the boys. Finally, parents intervene and defuse the situation, leading their children off to their cars, homes, and eventually to their soccer games.

THE PERFORMANCE OF GENDER

In the past decade, especially since the publication of Judith Butler's highly influential *Gender Trouble* (1990), it has become increasingly fashionable among academic feminists to think of gender not as some "thing" that one "has" (or not) but rather as situationally constructed through the performances of active agents. The idea of gender as performance analytically foregrounds the agency of individuals in the construction of gender, thus highlighting the situational fluidity of gender: here, conservative and reproductive, there, transgressive and disruptive. Surely, the Barbie Girls versus Sea Monsters scene described above can be fruitfully analyzed as a moment of crosscutting and mutually constitutive gender performances: The girls—at least at first glance—appear to be performing (for each other?) a conventional four- to five-year-old version of emphasized femininity. At least on the surface, there appears to be nothing terribly transgressive here. They are just "being girls," together. The boys initially are unwittingly constituted as an audience for the girls' performance but quickly begin to perform (for each other?—for the girls, too?) a masculinity that constructs itself in opposition to Barbie, and to the girls, as not feminine. They aggressively confront—first through loud verbal chanting, eventually through bodily invasions— the girls' ritual space of emphasized femininity, apparently with the intention of disrupting its upsetting influence. The adults are simultaneously constituted as an adoring audience for their children's performances and as parents who perform for each other by sharing and mutually affirming their experience-based narratives concerning the natural differences between boys and girls.

In this scene, we see children performing gender in ways that constitute themselves as two separate, opposed groups (boys vs. girls) and parents performing gender in ways that give the stamp of adult approval to the children's performances of difference, while constructing their own ideological narrative that naturalizes this categorical difference. In other words, the parents do not seem to read the children's performances of gender as social constructions of gender. Instead, they interpret them as the inevitable unfolding of natural, internal differences between the sexes....

The parents' response to the Barbie Girls versus Sea Monsters performance suggests one of the main limits and dangers of theories of performativity. Lacking an analysis of structural and cultural context, performances of gender can all too easily be interpreted as free agents' acting out the inevitable surface manifestations of a natural inner essence of sex difference. An examination of structural and cultural contexts, though, reveals that there was nothing inevitable about the girls' choice of Barbie as their totem, nor in the boys' response to it.

THE STRUCTURE OF GENDER

In the entire subsequent season of weekly games and practices, I never once saw adults point to a moment in which boy and girl soccer players were doing the *same* thing and exclaim to each other, "Look at them! They are *so similar!*" The actual similarity of the boys and the girls, evidenced by nearly all of the kids' routine actions throughout a soccer season—playing the game, crying over a skinned knee, scrambling enthusiastically for their snacks after the games, spacing out on a bird or a flower instead of listening to the coach at practice—is a key to understanding the salience of the Barbie Girls versus Sea Monsters moment for gender relations. In the face of a multitude of moments that speak to similarity, it was this anomalous Barbie Girls versus Sea Monsters moment—where the boundaries of gender were so clearly enacted—that the adults seized to affirm their commitment to difference. It is the kind of moment—to use Lorber's (1994, 37) phrase—where "believing is seeing," where we selectively "see" aspects of social reality that tell us a truth that we prefer to believe, such as the belief in categorical sex difference. No matter that our eyes do not see evidence of this truth most of the rest of the time.

In fact, it was not so easy for adults to actually "see" the empirical reality of sex similarity in everyday observations of soccer throughout the season. That is due to one overdetermining factor: an institutional context that is characterized by informally structured sex segregation among the parent coaches and team managers, and by formally structured sex segregation among the children. The structural analysis developed here is indebted to Acker's (1990) observation that organizations, even while appearing "gender neutral," tend to reflect, re-create, and naturalize a hierarchical ordering of gender....

Adult Divisions of Labor and Power

There was a clear—although not absolute—sexual division of labor and power among the adult volunteers in the AYSO organization. The Board of Directors consisted of 21 men and 9 women, with the top two positions—commissioner and assistant commissioner—held by men. Among the league's head coaches, 133 were men and 23 women. The division among the league's assistant coaches was similarly skewed. Each team also had a team manager who was responsible for organizing snacks, making reminder calls about games and practices, organizing team parties and the end-of-the-year present for the coach. The vast majority of team managers were women. A common slippage in the language of coaches and parents revealed the ideological assumptions underlying this position: I often noticed people describe a team manager as the "team mom." In short, as Table 1 shows, the vast majority of the time, the formal authority of the head coach and assistant coach was in the hands of a man, while the backup, support role of team manager was in the hands of a woman.

These data illustrate Connell's (1987, 97) assertion that sexual divisions of labor are interwoven with, and mutually supportive of, divisions of power and authority among women and men. They also suggest how people's choices to volunteer for certain positions are shaped and constrained by previous institutional practices. There is no formal AYSO rule that men must be the leaders, women the supportive followers. And there are, after all, *some* women coaches and *some* men team managers. So, it may appear that the division of labor among adult volunteers simply manifests an accumulation of individual choices and preferences. When analyzed structurally, though, individual men's apparently free choices to volunteer disproportionately for coaching jobs, alongside individual women's apparently free choices to volunteer disproportionately for team manager jobs, can be seen as a logical collective result of the ways that the institutional structure of sport has differentially constrained and enabled women's and men's previous options and experiences (Messner 1992). Since boys and men have had far more opportunities to play organized sports and thus to gain skills and knowledge, it subsequently appears rational for adult men to serve in positions of knowledgeable authority, with women serving in a support capacity (Boyle and McKay 1995). Structure—in this case, the historically constituted division of labor and power in sport—constrains current practice. In turn, structure becomes an object of practice, as the choices and actions of today's parents re-create divisions of labor and power similar to those that they experienced in their youth.

T A B L E 1 **Adult Volunteers as Coaches and Team Managers, by Gender (in percentages) (*N* = 156 teams)**

	Head Coaches	Assistant Coaches	Team Managers
Women	15	21	86
Men	85	79	14

The Children: Formal Sex Segregation

As adult authority patterns are informally structured along gendered lines, the children's leagues are formally segregated by AYSO along lines of age and sex. In each age-group, there are separate boys' and girls' leagues. The AYSO in this community included 87 boys' teams and 69 girls' teams. Although the four- to five-year-old boys often played their games on a field that was contiguous with games being played by four- to five-year-old girls, there was never a formal opportunity for cross-sex play. Thus, both the girls' and the boys' teams could conceivably proceed through an entire season of games and practices in entirely homosocial contexts. In the all-male contexts that I observed throughout the season, gender never appeared to be overtly salient among the children, coaches, or parents. It is against this backdrop that I might suggest a working hypothesis about structure and the variable salience of gender: The formal sex segregation of children does not, in and of itself, make gender overtly salient. In fact, when children are absolutely segregated, with no opportunity for cross-sex interactions, gender may appear to disappear as an overtly salient organizing principle. How-ever, when formally sex-segregated children are placed into immediately con-tiguous locations, such as during the opening ceremony, highly charged gendered interactions between the groups (including invasions and other kinds of border work) become more possible.

Although it might appear to some that formal sex segregation in children's sports is a natural fact, it has not always been so for the youngest age-groups in AYSO. As recently as 1995, when my older son signed up to play as a five-year-old, I had been told that he would play in a coed league. But when he arrived to his first practice and I saw that he was on an all-boys team, I was told by the coach that AYSO had decided this year to begin sex segregating all age-groups, because "during half-times and practices, the boys and girls tend to separate into separate groups. So the league thought it would be better for team unity if we split the boys and girls into separate leagues." I suggested to some coaches that a similar dynamic among racial ethnic groups (say, Latino kids and white kids clustering as separate groups during halftimes) would not similarly result in a decision to create racially segregated leagues. That this comment appeared to fall on deaf ears illustrates the extent to which many adults' belief in the need for sex segregation—at least in the context of sport—is grounded in a mutually agreed-upon notion of boys' and girls' "separate worlds," perhaps based in ideologies of natural sex difference.

The gender regime of AYSO, then, is structured by formal and informal sexual divisions of labor and power. This social structure sets ranges, limits, and possibilities for the children's and parents' interactions and performances of gender, but it does not determine them. Put another way, the formal and informal gender regime of AYSO made the Barbie Girls versus Sea Monsters moment possible, but it did not make it inevitable. It was the agency of the children and the parents within that structure that made the moment happen. But why did this moment take on the symbolic forms that it did? How and why do the girls, boys, and parents construct and derive meanings from this moment, and how can we interpret these meanings? These questions are best grappled within in the realm of cultural analysis.

THE CULTURE OF GENDER

The difference between what is "structural" and what is "cultural" is not clear-cut. For instance, the AYSO assignment of team colors and choice of team names (cultural symbols) seem to follow logically from, and in turn reinforce, the sex segregation of the leagues (social structure). These cultural symbols such as team colors, uniforms, songs, team names, and banners often carried encoded gendered meanings that were then available to be taken up by the children in ways that constructed (or potentially contested) gender divisions and boundaries.

Team Names

Each team was issued two team colors. It is notable that across the various age-groups, several girls' teams were issued pink uniforms—a color commonly recognized as encoding feminine meanings—while no boys' teams were issued pink uniforms. Children, in consultation with their coaches, were asked to choose their own team names and were encouraged to use their assigned team colors as cues to theme of the team name (e.g., among the boys, the "Red Flashes," the "Green Pythons," and the blue-and-green "Sea Monsters"). When I analyzed the team names of the 156 teams by age-group and by sex, three categories emerged:

1. Sweet names: These are cutesy team names that communicate small stature, cuteness, and/or vulnerability. These kinds of names would most likely be widely read as encoded with feminine meanings (e.g., "Blue Butterflies," "Beanie Babes," "Sunflowers," "Pink Flamingos," and "Barbie Girls").

2. Neutral or paradoxical names: Neutral names are team names that carry no obvious gendered meaning (e.g., "Blue and Green Lizards," "Team Flubber," "Galaxy," "Blue Ice"). Paradoxical names are girls' team names that carry mixed (simultaneously vulnerable *and* powerful) messages (e.g., "Pink Panthers," "Flower Power," "Little Tigers").

3. Power names: These are team names that invoke images of unambiguous strength, aggression, and raw power (e.g., "Shooting Stars," "Killer Whales," "Shark Attack," "Raptor Attack," and "Sea Monsters").

. . . [A]cross all age-groups of boys, there was only one team name coded as a sweet name—"The Smurfs," in the 10- to 11-year-old league. Across all age categories, the boys were far more likely to choose a power name than anything else, and this was nowhere more true than in the youngest age-groups, where 35 of 40 (87 percent) of boys' teams in the four-to-five and six-to-seven age-groups took on power names. A different pattern appears in the girls' team name choices, especially among the youngest girls. Only 2 of the 12 four- to five-year-old girls' teams chose power names, while 5 chose sweet names and 5 chose neutral/paradoxical names. At age six to seven, the numbers begin to tip toward the boys' numbers but still remain different, with half of the girls' teams now choosing power names. In the middle and older girls' groups, the sweet names all but disappear, with power names dominating, but still a higher proportion of neutral/paradoxical names than among boys in those age-groups.

Barbie Narrative versus Warrior Narrative

How do we make sense of the obviously powerful spark that Barbie provided in the opening ceremony scene described above? Barbie is likely one of the most immediately identifiable symbols of femininity in the world. More conservatively oriented parents tend to happily buy Barbie dolls for their daughters, while perhaps deflecting their sons' interest in Barbie toward more sex-appropriate "action toys." Feminist parents, on the other hand, have often expressed open contempt—or at least uncomfortable ambivalence—toward Barbie. This is because both conservative and feminist parents see dominant cultural meanings of emphasized femininity as condensed in Barbie and assume that these meanings will be imitated by their daughters. Recent developments in cultural studies, though, should warn us against simplistic readings of Barbie as simply conveying hegemonic messages about gender to unwitting children (Attfield 1996; Seiter 1995). In addition to critically analyzing the cultural values (or "preferred meanings") that may be encoded in Barbie or other children's toys, feminist scholars of cultural studies point to the necessity of examining "reception, pleasure, and agency," and especially "the fullness of reception contexts" (Walters 1999, 246). The Barbie Girls versus Sea Monsters moment can be analyzed as a "reception context," in which differently situated boys, girls, and parents variously used Barbie to construct pleasurable intergroup bonds, as well as boundaries between groups. . . .

. . . Indeed, as the Barbie Girls rallied around Barbie, their obvious pleasure did not appear to be based on a celebration of quiet passivity (as feminist parents might fear). Rather, it was a statement that they—the Barbie Girls—were here in this public space. They were not silenced by the boys' oppositional chanting. To the contrary, they ignored the boys, who seemed irrelevant to their celebration. And, when the boys later physically invaded their space, some of the girls responded by chasing the boys off. In short, when I pay attention to what the girls *did* (rather than imposing on the situation what I *think* Barbie "should" mean to the girls), I see a public moment of celebratory "girl power."

And this may give us better basis from which to analyze the boys' oppositional response. First, the boys may have been responding to the threat of displacement they may have felt while viewing the girls' moment of celebratory girl power. Second, the boys may simultaneously have been responding to the fears of feminine pollution that Barbie had come to symbolize to them. But why might Barbie symbolize feminine pollution to little boys? A brief example from my older son is instructive. When he was about three, following a fun day of play with the five-year-old girl next door, he enthusiastically asked me to buy him a

Barbie like hers. He was gleeful when I took him to the store and bought him one. When we arrived home, his feet had barely hit the pavement getting out of the car before an eight-year-old neighbor boy laughed at and ridiculed him: "A *Barbie*? Don't you know that Barbie is a *girl's toy*?" No amount of parental intervention could counter this devastating peer-induced injunction against boys' playing with Barbie. My son's pleasurable desire for Barbie appeared almost overnight to transform itself into shame and rejection. The doll ended up at the bottom of a heap of toys in the closet, and my son soon became infatuated, along with other boys in his preschool, with Ninja Turtles and Power Rangers. . . .

By kindergarten, most boys appear to have learned—either through experiences similar to my son's, where other male persons police the boundaries of gender-appropriate play and fantasy and/or by watching the clearly gendered messages of television advertising—that Barbie dolls are not appropriate toys for boys (Rogers 1999, 30). To avoid ridicule, they learn to hide their desire for Barbie, either through denial and oppositional/pollution discourse and/or through sublimation of their desire for Barbie into play with male-appropriate "action figures" (Pope et al. 1999). In their study of a kindergarten classroom, Jordan and Cowan (1995, 728) identified "warrior narratives . . . that assume that violence is legitimate and justified when it occurs within a struggle between good and evil" to be the most commonly agreed-upon currency for boys' fantasy play. They observe that the boys seem commonly to adapt story lines that they have seen on television. Popular culture—film, video, computer games, television, and comic books—provides boys with a seemingly endless stream of Good Guys versus Bad Guys characters and stories—from cowboy movies, Superman and Spiderman to Ninja Turtles, Star Wars, and Pokémon—that are available for the boys to appropriate as the raw materials for the construction of their own warrior play. . . .

A cultural analysis suggests that the boys' and the girls' previous immersion in differently gendered cultural experiences shaped the likelihood that they would derive and construct different meanings from Barbie—the girls through pleasurable and symbolically empowering identification with "girl power" narratives; the boys through oppositional fears of feminine pollution (and fears of displacement by girl power?) and with aggressively verbal, and eventually physical, invasions of the girls' ritual space. The boys' collective response thus constituted them differently, *as boys*, in opposition to the girls' constitution of themselves *as girls*. An individual girl or boy, in this moment, who may have felt an inclination to dissent from the dominant feelings of the group (say, the Latina Barbie Girl who, her mother later told me, did not want the group to be identified with Barbie, or a boy whose immediate inner response to the Barbie Girls' joyful celebration might be to join in) is most likely silenced into complicity in this powerful moment of border work.

What meanings did this highly gendered moment carry for the boys' and girls' teams in the ensuing soccer season? Although I did not observe the Barbie Girls after the opening ceremony, I did continue to observe the Sea Monsters' weekly practices and games. During the boys' ensuing season, gender never reached this "magnified" level of salience again—indeed, gender was rarely raised verbally or performed overtly by the boys. On two occasions, though, I observed the coach jokingly chiding the boys during practice that "if you don't watch out, I'm going to get the Barbie Girls here to play against you!" This warning was followed by gleeful screams of agony and fear, and nervous hopping around and hugging by some of the boys. Normally, though, in this sex-segregated, all-male context, if boundaries were invoked, they were not boundaries between boys and girls but boundaries between the Sea Monsters and other boys' teams, or sometimes age boundaries between the Sea Monsters and a small group of dads and older brothers who would engage them in a mock scrimmage during practice. But it was also evident that when the coach was having trouble getting the boys to act together, as a group, his strategic and humorous invocation of the dreaded Barbie Girls once again served symbolically to affirm their group status. They were a team. They were the boys.

CONCLUSION

The overarching goal of this article has been to take one empirical observation from everyday life and demonstrate how a multilevel (interactionist, structural, cultural) analysis might reveal various layers of meaning that give insight into the everyday social construction of gender. This article builds on observations made by Thorne (1993) concerning ways to approach sociological analyses of children's worlds. The most fruitful approach is not to ask why boys and girls are so different but rather to ask how and under what conditions boys and girls constitute themselves as separate, oppositional groups. Sociologists need not debate whether gender is "there"—clearly, gender is always already there, built as it is into the structures, situations, culture, and consciousness of children and adults. The key issue is under what conditions gender is activated as a salient organizing principle in social life and under what conditions it may be less salient. These are important questions, especially since the social organization of categorical gender difference has always been so clearly tied to gender hierarchy (Acker 1990; Lorber 1994). In the Barbie Girls versus Sea Monsters moment, the performance of gendered boundaries and the construction of boys' and girls' groups as categorically different occurred in the context of a situation systematically structured by sex segregation, sparked by the imposing presence of a shared cultural symbol that is saturated with gendered meanings, and actively supported and applauded by adults who basked in the pleasure of difference, reaffirmed.

I have suggested that a useful approach to the study of such "how" and "under what conditions" questions is to employ multiple levels of analysis. At the most general level, this project supports the following working propositions.

Interactionist theoretical frameworks that emphasize the ways that social agents "perform" or "do" gender are most useful in describing how groups of people actively create (or at times disrupt) the boundaries that delineate seemingly categorical differences between male persons and female persons. In this case, we saw how the children and the parents interactively performed gender in a way that constructed an apparently natural boundary between the two separate worlds of the girls and the boys.

Structural theoretical frameworks that emphasize the ways that gender is built into institutions through hierarchical sexual divisions of labor are most useful in explaining under what conditions social agents mobilize variously to disrupt or to affirm gender differences and inequalities. In this case, we saw how the sexual division of labor among parent volunteers (grounded in their own histories in the gender regime of sport), the formal sex segregation of the children's leagues, and the structured context of the opening ceremony created conditions for possible interactions between girls' teams and boys' teams.

Cultural theoretical perspectives that examine how popular symbols that are injected into circulation by the culture industry are variously taken up by differently situated people are most useful in analyzing how the meanings of cultural symbols, in a given institutional context, might trigger or be taken up by social agents and used as resources to reproduce, disrupt, or contest binary conceptions of sex difference and gendered relations of power. In this case, we saw how a girls' team appropriated a large Barbie around which to construct a pleasurable and empowering sense of group identity and how the boys' team responded with aggressive denunciations of Barbie and invasions. . . .

. . . The eventual interactions between the boys and the girls were made possible—although by no means fully determined—by the structure of the gender regime and by the cultural resources that the children variously drew on.

On the other hand, the gendered division of labor in youth soccer is not seamless, static, or immune to resistance. One of the few woman head coaches, a very active athlete in her own right, told me that she is "challenging the sexism" in AYSO by becoming the head of her son's league. As post–Title IX women increasingly become mothers and as media images of competent, heroic female athletes become more a part of the cultural landscape for children, the gender regimes of children's sports may be increasingly challenged (Dworkin and Messner 1999). Put another way, the dramatically shifting opportunity structure and cultural imagery of post–Title IX sports have created opportunities for new kinds of interactions, which will inevitably challenge and further shift institutional structures. Social structures simultaneously constrain and enable, while agency is simultaneously reproductive and resistant.

REFERENCES

Acker, Joan. 1990. Hierarchies, jobs, bodies: A theory of gendered organizations. *Gender & Society* 4:139–58.

Adler, Patricia A., and Peter Adler. 1998. *Peer power: Preadolescent culture and identity*. New Brunswick, NJ: Rutgers University Press.

Attfield, Judy. 1996. Barbie and Action Man: Adult toys for girls and boys, 1959–93. In *The gendered object*, edited by Pat Kirkham, 80–89. Manchester, UK, and New York: Manchester University Press.

Boyle, Maree, and Jim McKay. 1995. "You leave your troubles at the gate": A case study of the exploitation of older women's labor and "leisure" in sport. *Gender & Society* 9:556–76.

Butler, Judith. 1990. *Gender trouble: Feminism and the subversion of identity*. New York and London: Routledge.

Connell, R.W. 1987. *Gender and power*. Stanford, CA: Stanford University Press.

Dworkin, Shari L., and Michael A. Messner. 1999. Just do . . . what?: Sport, bodies, gender. In *Revisioning gender*, edited by Myra Marx Ferree, Judith Lorber, and Beth B. Hess, 341–61. Thousand Oaks, CA: Sage.

Hochschild, Arlie Russell. 1994. The commercial spirit of intimate life and the abduction of feminism: Signs from women's advice books. *Theory, Culture & Society* 11:1–24.

Jordan, Ellen, and Angela Cowan. 1995. Warrior narratives in the kindergarten classroom: Renogotiating the social contract? *Gender & Society* 9:727–43.

Lorber, Judith. 1994. *Paradoxes of gender*. New Haven, CT, and London: Yale University Press.

McGuffy, C. Shawn, and B. Lindsay Rich. 1999. Playing in the gender transgression zone: Race, class and hegemonic masculinity in middle childhood. *Gender & Society* 13:608–27.

Messner, Michael A. 1992. *Power at play: Sports and the problem of masculinity*. Boston: Beacon.

Pope, Harrison G., Jr., Roberto Olivarda, Amanda Gruber, and John Borowiecki. 1999. Evolving ideals of male body image as seen through action toys. *International Journal of Eating Disorders* 26:65–72.

Rogers, Mary F. 1999. *Barbie culture*. Thousand Oaks, CA: Sage.

Seiter, Ellen. 1995. *Sold separately: Parents and children in consumer culture*. New Brunswick, NJ: Rutgers University Press.

Thorne, Barrie. 1993. *Gender play: Girls and boys in school*. New Brunswick, NJ: Rutgers University Press.

Walters, Suzanna Danuta. 1999. Sex, text, and context: (In) between feminism and cultural studies. In *Revisioning gender*, edited by Myra Marx Ferree, Judith Lorber, and Beth B. Hess, 222–57. Thousand Oaks, CA: Sage.

KEY CONCEPTS

doing gender   gender segregation   gender socialization

DISCUSSION QUESTIONS

1. Think back to a time when you may have played in organized groups (sports, camps, or some other activity). How is the example of the soccer league described in this article similar to what you may have experienced? How is it different?

2. What are some popular children's toys today? How do they socialize children into gendered roles? Are there toys today that construct gender differently than when you were a child?

11

"We Don't Sleep Around Like White Girls Do"

Family, Culture, and Gender in Filipina American Lives

YEN LE ESPIRITU

In this article Le Espiritu summarizes research on Filipino families in San Diego. The socialization of girls growing up in immigrant Filipino families emphasizes the roles of chastity, nurturance, and family-oriented responsibilities. Unlike the "moral flaws" of White daughters, Filipinas are expected to act in ways that are gender-appropriate and consistent with their Asian culture.

I want my daughters to be Filipino especially on sex. I always emphasize to them that they should not participate in sex if they are not married. We are also Catholic. We are raised so that we don't engage in going out with men while we are not married. And I don't like it to happen to my daughters as if they have no values. I don't like them to grow up that way, like the American girls.
FILIPINA IMMIGRANT MOTHER

I found that a lot of the Asian American friends of mine, we don't date like white girls date. We don't sleep around like white girls do. Everyone is really mellow at dating because your parents were constraining and restrictive.
SECOND-GENERATION FILIPINA DAUGHTER

. . . **[G]**ender is a key to immigrant identity and a vehicle for racialized immigrants to assert cultural superiority over the dominant group. In immigrant communities, culture takes on a special significance: not only does it form a lifeline to the home country and a basis for

SOURCE: Yen Le Espiritu. "We Don't Sleep Around Like White Girls Do: Family, Culture, and Gender in Filipina American Lives." *Signs: Journal of Women in Culture and Society* 26, no. 2 (2001).

group identity in a new country, it is also a base from which immigrants stake their political and sociocultural claims on their new country (Eastmond 1993, 40). For Filipino immigrants, who come from a homeland that was once a U.S. colony, cultural reconstruction has been especially critical in the assertion of their presence in the United States—a way to counter the cultural Americanization of the Philippines, to resist the assimilative and alienating demands of U.S. society, and to reaffirm to themselves their self-worth in the face of colonial, racial, class, and gendered subordination. Before World War II, Filipinos were barred from becoming U.S. citizens, owning property, and marrying whites. They also encountered discriminatory housing policies, unfair labor practices, violent physical encounters, and racist as well as anti-immigrant discourse.[1] While blatant legal discrimination against Filipino Americans is largely a matter of the past, Filipinos continue to encounter many barriers that prevent full participation in the economic, social, and political institutions of the United States (Azores-Gunner 1986–87; Cabezas, Shinagawa, and Kawaguchi 1986–87; Okamura and Agbayani 1997). Moreover, the economic mobility and cultural assimilation that enables white ethnics to become "unhyphenated whites" is seldom extended to Filipino Americans (Espiritu 1994). Like other Asians, the Filipino is "always seen as an immigrant, as the 'foreigner-within,' even when born in the United States" (Lowe 1996, 5). Finally, although Filipinos have been in the United States since the middle of the 1700s and Americans have been in the Philippines since at least the late 1800s, U.S. Filipinos—as racialized nationals, immigrants, and citizens—are "still practically an invisible and silent minority" (San Juan 1991, 117). Drawing from my research on Filipino American families in San Diego, California, I explore . . . the ways racialized immigrants claim through gender the power denied them by racism.

My epigraphs, quotations of a Filipina immigrant mother and a second-generation Filipina daughter, suggest that the virtuous Filipina daughter is partially constructed on the conceptualization of white women as sexually immoral. This juxtaposition underscores the fact that femininity is a relational category, one that is co-constructed with other racial and cultural categories. These narratives also reveal that women's sexuality and their enforced "morality" are fundamental to the structuring of social inequalities. Historically, the sexuality of racialized women has been systematically demonized and disparaged by dominant or oppressor groups to justify and bolster nationalist movements, colonialism, and/or racism. But as these narratives indicate, racialized groups also criticize the morality of white women as a strategy of resistance—a means of asserting a morally superior public face to the dominant society. . . .

But this strategy is not without costs. The elevation of Filipina chastity (particularly that of young women) has the effect of reinforcing masculinist and patriarchal power in the name of a greater ideal of national/ethnic self-respect. Because the control of women is one of the principal means of asserting moral superiority, young women in immigrant families face numerous restrictions on their autonomy, mobility, and personal decision making. . . .

STUDYING FILIPINOS IN SAN DIEGO

San Diego, California has long been a favored area of settlement for Filipinos and is today the third-largest U.S. destination for Filipino immigrants (Rumbaut 1991, 220).... San Diego has been a primary area of settlement for Filipino navy personnel and their families since the early 1900s. As in other Filipino communities along the Pacific Coast, the San Diego community grew dramatically in the twenty-five years following passage of the 1965 Immigration Act. New immigration contributed greatly to the tripling of San Diego county's Filipino American population from 1970 to 1980 and its doubling from 1980 to 1990. In 1990, nearly 96,000 Filipinos resided in the county. Although they made up only 4 percent of the county's general population, they constituted close to 50 percent of the Asian American population (Espiritu 1995). Many post-1965 Filipino immigrants have come to San Diego as professionals—most conspicuously as health care workers. A 1992 analysis of the socioeconomic characteristics of recent Filipino immigrants in San Diego indicated that they were predominantly middle-class, college-educated, and English-speaking professionals who were more likely to own than rent their homes (Rumbaut 1994). At the same time, about two-thirds of the Filipinos surveyed indicated that they had experienced racial and ethnic discrimination (Espiritu and Wolf, forthcoming).

The information on which this article is based comes mostly from in-depth interviews that I conducted with almost one hundred Filipinos in San Diego.[2] Using the "snowball" sampling technique, ... I chose participants ... through a network of Filipino American contacts whom the first group of respondents trusted. To capture the diversity within the Filipino American community, I sought and selected respondents of different backgrounds and with diverse viewpoints. The sample is about equally divided between first-generation immigrants (those who came to the United States as adults) and Filipinas/os who were born and/or raised in the United States.... They included poor working-class immigrants who barely eked out a living, as well as educated professionals who thrived in middle- and upper-calss suburban neighborhoods. However, the class status of most was much more ambiguous.... Reflecting the prominence of the U.S. Navy in San Diego, more than half of my respondents were affiliated with or had relatives affiliated with the U.S. Navy.

My tape-recorded interviews, conducted in English, ranged from three to ten hours each and took place in offices, coffee shops, and homes. My questions were open-ended and covered three general areas: family and immigration history, ethnic identity and practices, and community development among San Diego's Filipinos.... Even without prompting, young Filipinas almost always recounted stories of restrictive gender roles and gender expectations, particularly of parental control over their whereabouts and sexuality.

I believe that my own personal and social characteristics influenced the actual process of data collection, the quality of the materials that I gathered, and my analysis of them. As a Vietnam-born woman who immigrated to the United States at the age of twelve, I came to the research project not as an "objective" outsider but as a fellow Asian immigrant who shared some of the life experiences of my respondents. During the fieldwork process, I . . . actively shared with my informants my own experiences of being an Asian immigrant woman: of being perceived as an outsider in U.S. society, of speaking English as a second language, of being a woman of color in a racialized patriarchal society, and of negotiating intergenerational tensions within my own family. . . . These shared experiences enable me to bring to the work a comparative perspective that is implicit, intuitive, and informed by my own identities and positionalities—and with it a commitment to approach these subjects with both sensitivity and rigor. . . .

CONSTRUCTING THE DOMINANT GROUP: THE MORAL FLAWS OF WHITE AMERICANS

. . . While much has been written on how whites have represented the (im)morality of people of color (Collins 1991; Marchetti 1993; Hamamoto 1994), there has been less critical attention to how people of color have represented whites. . . . I argue that female morality—defined as women's dedication to their families and sexual restraint—is one of the few sites where economically and politically dominated groups can construct the dominant group as other and themselves as superior. Because womanhood is idealized as the repository of tradition, the norms that regulate women's behaviors become a means of determining and defining group status and boundaries. As a consequence, the burdens and complexities of cultural representation fall most heavily on immigrant women and their daughters. . . .

Family-Oriented Model Minorities: "White Women Will Leave You"

. . . Many of my respondents constructed their "ethnic" culture as principled and "American" culture as deviant. Most often, this morality narrative revolves around family life and family relations. When asked what set Filipinos apart from other Americans, my respondents—of all ages and class backgrounds—repeatedly contrasted close-knit Filipino families to what they perceived to be the more impersonal quality of U.S. family relations. . . . "Americans" are characterized as lacking in strong family ties and collective identity, less willing to do the work of family and cultural maintenance, and less willing to abide by patriarchal norm in husband/wife relations. . . .

Implicit in negative depictions of U.S. families as uncaring, selfish, and distant is the allegation that white women are not as dedicated to their families as Filipina women are to theirs. Several Filipino men who married white women recalled being warned by their parents and relatives that "white women will leave you." . . . For some Filipino men, perceived differences in attitudes about women's roles between Filipina and non-Filipina women influenced their marital choice. A Filipino American navy man explained why he went back to the Philippines to look for a wife:

My goal was to marry a Filipina. I requested to be stationed in the Philippines to get married to a Filipina. I'd seen the women here and basically they are spoiled. They have a tendency of not going along together with their husband. They behave differently. They chase the male, instead of the male, the normal way of the traditional way is for the male to go after the female. They have sex without marrying. They want to do their own things. So my idea was to go back home and marry somebody who has never been here. I tell my son the same thing: if he does what I did and finds himself a good lady there, he will be in good hands.

Another man who had dated mostly white women in high school recounted that when it came time for him to marry, he "looked for the kind of women" he met while stationed in the Philippines: "I hate to sound chauvinistic about marriages, but Filipinas have a way of making you feel like you are a king. They also have that tenderness, that elegance. And we share the same values about family, education, religion, and raising children.". . .

Racialized Sexuality and (Im)Morality:
"In America, . . . Sex Is Nothing"

Sexuality, as a core aspect of social identity, is fundamental to the structuring of gender inequality (Millett 1970). Sexuality is also a salient marker of otherness and has figured prominently in racist and imperialist ideologies (Gilman 1985; Stoler 1991). Historically, the sexuality of subordinate groups—particularly that of racialized women—has been systematically stereotyped by the dominant groups.[3] At stake in these stereotypes is the construction of women of color as morally lacking in the areas of sexual restraint and traditional morality. Asian women—both in Asia and in the United States—have been racialized as sexually immoral, and the "Orient"—and its women—has long served as a site of European male-power fantasies, replete with lurid images of sexual license, gynecological aberrations, and general perversion (Gilman 1985, 89). In colonial Asia in the nineteenth and early twentieth centuries, for example, female sexuality was a site for colonial rulers to assert their moral superiority and thus their supposed natural and legitimate right to rule. The colonial rhetoric of moral superiority was based on the construction of colonized Asian women as subjects of sexual desire and fulfillment and European colonial women as the paragons of virtue and the bearers of a redefined colonial morality (Stoler 1991). The discourse of morality has also been used to mark the "unassimilability" of Asians in the United States. At the turn of the twentieth century, the public perception of Chinese women as disease-ridden, drug-addicted prostitutes served to underline the depravity of "Orientals" and played a decisive role in the eventual passage of exclusion laws against all Asians (Mazumdar 1989, 3–4). The stereotypical view that all Asian women were prostitutes, first formed in the 1850s, persisted. Contemporary American popular culture continues to endow

Asian women with an excess of "womanhood," sexualizing them but also impugning their sexuality (Espiritu 1997, 93).

Filipinas—both in the Philippines and in the United States—have been marked as desirable but dangerous "prostitutes" and/or submissive "mail-order brides" (Halualani 1995; Egan 1996). These stereotypes emerged out of the colonial process, especially the extensive U.S. military presence in the Philippines. Until the early 1990s, the Philippines, at times unwillingly, housed some of the United States's largest overseas airforce and naval bases (Espiritu 1995, 14). Many Filipino nationalists have charged that "the prostitution problem" in the Philippines stemmed from U.S. and Philippine government policies that promoted a sex industry—brothels, bars, and massage parlors—for servicemen stationed or on leave in the Philippines. During the Vietnam War, the Philippines was known as the "rest and recreation" center of Asia, hosting approximately ten thousand U.S. servicemen daily (Coronel and Rosca 1993; Warren 1993). In this context, *all* Filipinas were racialized as sexual commodities, usable and expendable. A U.S.-born Filipina recounted the sexual harassment she faced while visiting Subic Bay Naval Station in Olongapo City:

> One day, I went to the base dispensary. . . . I was dressed nicely, and as I walked by the fire station, I heard catcalls and snide remarks being made by some of the firemen. . . . I was fuming inside. The next thing I heard was, "How much do you charge?" I kept on walking. "Hey, are you deaf or something? How much do you charge? You have a good body." That was an incident that I will never forget. (Quoted in Espiritu 1995, 77)

The sexualized racialization of Filipina women is also captured in Marianne Valanueva's short story "Opportunity" (1991). As the protagonist, a "mail-order bride" from the Philippines, enters a hotel lobby to meet her American fiancé, the bellboys snicker and whisper *puta* (whore): a reminder that US. economic and cultural colonization in the Philippines always forms a backdrop to any relations between Filipinos and Americans (Wong 1993, 53).

Cognizant of the pervasive hypersexualization of Filipina women, my respondents, especially women who grew up near military bases, were quick to denounce prostitution, to condemn sex laborers, and to declare (unasked) that they themselves did not frequent "that part of town." As one Filipina immigrant said,

> Growing up [in the Philippines], I could never date an American because my dad's concept of a friendship with an American is with a G.I. The only reason why my dad wouldn't let us date an American is that people will think that the only way you met was because of the base. I have never seen the inside of any of the bases because we were just forbidden to go there.

Many of my respondents also distanced themselves culturally from the Filipinas who serviced US. soldiers by branding them "more Americanized" and "more Westernized." In other words, these women were sexually promiscuous because they had assumed the sexual mores of white women. This characterization allows my respondents to symbolically disown the Filipina "bad girl" and, in so doing, to uphold the narrative of Filipina sexual virtuosity and white female sexual promiscuity. In the following narrative, a mother who came to the United States in her thirties contrasted the controlled sexuality of women in the Philippines with the perceived promiscuity of white women in the United States:

> In the Philippines, we always have chaperon when we go out. When we go to dances, we have our uncle, our grandfather, and auntie all behind us to make sure that we behave in the dance hall. Nobody goes necking outside. You don't even let a man put his hand on your shoulders. When you were brought up in a conservative country, it is hard to come here and see that it is all freedom of speech and freedom of action. Sex was never mentioned in our generation. I was thirty already when I learned about sex. But to the young generation in America, sex is nothing.

Similarly, another immigrant woman criticized the way young American women are raised: "Americans are so liberated. They allow their children, their girls, to go out even when they are still so young." In contrast, she stated that, in "the Filipino way, it is very important, the value of the woman, that she is a virgin when she gets married."

The ideal "Filipina," then, is partially constructed on the community's conceptualization of white women. She is everything that they are not: she is sexually modest and dedicated to her family; they are sexually promiscuous and uncaring. Within the context of the dominant culture's pervasive hypersexualization of Filipinas, the construction of the "ideal" Filipina—as family-oriented and chaste—can be read as an effort to reclaim the morality of the community. This effort erases the Filipina "bad girl," ignores competing sexual practices in the Filipino communities, and uncritically embraces the myth of "Oriental femininity." Cast as the embodiment of perfect womanhood and exotic femininity, Filipinas (and other Asian women) in recent years have been idealized in U.S. popular culture as more truly "feminine" (i.e., devoted, dependent, domestic) and therefore more desirable than their more modern, emancipated sisters (Espiritu 1997, 113). Capitalizing on this image of the "superfemme," mail-order bride agencies market Filipina women as "'exotic, subservient wife imports' for sale and as alternatives for men sick of independent 'liberal' Western women" (Halualani 1995, 49; see also Ordonez 1997, 122).

Embodying the moral integrity of the idealized ethnic community, immigrant women, particularly young daughters, are expected to comply with male-defined criteria of what constitute "ideal" feminine virtues. While the sexual behavior of adult women is confined to a monogamous, heterosexual context, that of young women is denied completely (see Dasgupta and DasGupta 1996, 229–31). . . .

THE CONSTRUCTION(S) OF THE "IDEAL" FILIPINA: "BOYS ARE BOYS AND GIRLS ARE DIFFERENT"

... Although details vary, young women of various groups and across space and time—for example, second-generation Chinese women in San Francisco in the 1920s (Yung 1995), U.S.-born Italian women in East Harlem in the 1930s (Orsi 1985), young Mexican women in the Southwest during the interwar years (Ruiz 1992), and daughters of Caribbean and Asian Indian immigrants on the East Coast in the 1990s (Dasgupta and DasGupta 1996; Waters 1996)—have identified strict parental control on their activities and movements as the primary source of intergenerational conflict. Recent studies of immigrant families also identify gender as a significant determinant of parent-child conflict, with daughters more likely than sons to be involved in such conflicts and instances of parental derogation (Rumbaut and Ima 1988; Woldemikael 1989; Matute-Bianchi 1991; Gibson 1995).

Although immigrant families have always been preoccupied with passing on their native culture, language, and traditions to both male and female children, it is daughters who have the primary burden of protecting and preserving the family. Because sons do not have to conform to the image of an "ideal" ethnic subject as daughters do, they often receive special day-to-day privileges denied to daughters (Haddad and Smith 1996, 22–24; Waters 1996, 75–76). This is not to say that immigrant parents do not place undue expectations on their sons; rather, these expectations do not pivot around the sons' sexuality or dating choices.[4] In contrast, parental control over the movement and action of daughters begins the moment they are perceived as young adults and sexually vulnerable. It regularly consists of monitoring their whereabouts and forbidding dating (Wolf 1997). For example, the immigrant parents I interviewed seldom allowed their daughters to date, to stay out late, to spend the night at a friend's house, or to take an out-of-town trip.

Many of the second-generation women I spoke to complained bitterly about these parental restrictions. They particularly resented what they saw as gender inequity in their families: the fact that their parents placed far more restrictions on their activities and movements than on their brothers'. Some decried the fact that even their younger brothers had more freedom than they did. "It was really hard growing up because my parents would let my younger brothers do what they wanted but I didn't get to do what I wanted even though I was the oldest. I had a curfew and my brothers didn't. I had to ask if I could go places and they didn't. My parents never even asked my brothers when they were coming home." ...

When questioned about this double standard, parents generally responded by explaining that "girls are different":

I have that Filipino mentality that boys are boys and girls are different. Girls are supposed to be protected, to be clean. In the early years, my daughters have to have chaperons and curfews. And they know that they have to be virgins until they get married. The girls always say that is not fair. What is the difference between their brothers and them? And my answer always is, "In the Philippines, you know, we don't do that. The girls stay home. The boys go out." It was the way that I was raised. I still want to have part of that culture instilled in my children. And I want them to have that to pass on to their children.

Even among self-described Western-educated and "tolerant" parents, many continue to ascribe to "the Filipino way" when it comes to raising daughters. As one college-educated father explains,

> Because of my Western education, I don't raise my children the way my parents raised me. I tended to be a little more tolerant. But at times, especially in certain issues like dating, I find myself more towards the Filipino way in the sense that I have only one daughter so I tended to be a little bit stricter. So the double standard kind of operates: it's alright for the boys to explore the field but I tended to be overly protective of my daughter. My wife feels the same way because the boys will not lose anything, but the daughter will lose something, her virginity, and it can be also a question of losing face, that kind of thing.

Although many parents discourage or forbid dating for daughters, they still fully expect these young women to fulfill their traditional roles as women: to marry and have children. A young Filipina recounted the mixed messages she received from her parents:

> This is the way it is supposed to work: Okay, you go to school. You go to college. You graduate. You find a job. Then you find your husband, and you have children. That's the whole time line. But my question is, if you are not allowed to date, how are you supposed to find your husband? They say "no" to the whole dating scene because that is secondary to your education, secondary to your family. They do push marriage, but at a later date. So basically my parents are telling me that I should get married and I should have children but that I should not date.

. . . The restrictions on girls' movement sometimes spill over to the realm of academics. Dasgupta and DasGupta (1996, 230) recount that in the Indian American community, while young men were expected to attend faraway competitive colleges, many of their female peers were encouraged by their parents to go to the local colleges so that they could live at or close to home. Similarly, Wolf (1997, 467) reports that some Filipino parents pursued contradictory tactics with their children, particularly their daughters, by pushing them to achieve academic excellence in high school but then "pulling the emergency brake" when they contemplated college by expecting them to stay at home, even if it meant going to a less competitive college, or not going at all. . . .

I argue that these parental restrictions are attempts to construct a model of Filipina womanhood that is chaste, modest, nurturing, and family-oriented. Women are seen as responsible for holding the cultural line, maintaining racial boundaries, and marking cultural difference. This is not to say that parent-daughter conflicts exist in all Filipino immigrant families. Certainly, Filipino parents do not respond in a uniform way to the challenges of being racial-ethnic minorities, and I met parents who have had to change some of their ideas and practices in response to their inability to control their children's movements and choices:

> I have three girls and one boy. I used to think that I wouldn't allow my daughters to go dating and things like that, but there is no way I could do that. I can't stop it. It's the way of life here in America.... (Professional Filipino immigrant father)

> My children are born and raised here, so they do pretty much what they want. They think they know everything. I can only do so much as a parent.... When I try to teach my kids things, they tell me that I sound like an old record: They even talk back to me sometimes.... The first time my daughter brought her boyfriend to the house, she was eighteen years old. I almost passed away, knocked out.... (Working-class Filipino immigrant mother)

These narratives call attention to the shifts in the generational power caused by the migration process and to the possible gap between what parent say they want for their children and their ability to control the young....

SANCTIONS AND REACTIONS: "THAT IS NOT WHAT A DECENT FILIPINO GIRL SHOULD DO"

I do not wish to suggest that immigrant communities are the only ones in which parents regulate their daughters' mobility and sexuality. Feminist scholars have long documented the construction, containment, and exploitation of women's sexuality in various societies (Maglin and Perry 1996). We also know that the cultural anxiety over unbounded female sexuality is most apparent with regard to adolescent girls (Tolman and Higgins 1996, 206). The difference is in the ways immigrant and nonimmigrant families sanction girls' sexuality. To control sexually assertive girls nonimmigrant parents rely on the gender-based good girl/bad girl dichotomy in which "good girls" are passive, threatened sexual objects while "bad girls" are active, desiring sexual agents (Tolman and Higgins 1996).... This good girl/bad girl cultural story conflates femininity with sexuality, increases women's vulnerability to sexual coercion, and justifies women's containment in the domestic sphere.

Immigrant families, though, have an additional strategy: they can discipline their daughters as racial/national subjects as well as gendered ones. That is, as self-appointed guardians of "authentic" cultural memory, immigrant parents can attempt to regulate their daughters' independent choices by linking them to cultural ignorance or betrayal. As both parents and children recounted, young women who disobeyed parental strictures were often branded "non-ethnic," "untraditional," "radical," "selfish," and "not caring about the family." Female sexual choices were also linked to moral degeneracy, defined in relation to a narrative of a hegemonic white norm. Parents were quick to warn their daughters about "bad" Filipinas who had become pregnant outside marriage.[5] As in the case of "bar girls" in the

Philippines, Filipina Americans who veered from acceptable behaviors were deemed "Americanized"—as women who have adopted the sexual mores and practices of white women. As one Filipino immigrant father described "Americanized" Filipinas: "They are spoiled because they have seen the American way. They go out at night. Late at night. They go out on dates. Smoking. They have sex without marrying."

From the perspective of the second-generation daughters, these charges are stinging. The young women I interviewed were visibly pained—with many breaking down and crying—when they recounted their parents' charges. This deep pain, stemming in part from their desire to be validated as Filipina, existed even among the more "rebellious" daughters. One twenty-four-year-old daughter explained:

> My mom is very traditional. She wants to follow the Filipino customs, just really adhere to them, like what is proper for a girl, what she can and can't do, and what other people are going to think of her if she doesn't follow that way. When I pushed these restrictions, when I rebelled and stayed out later than allowed, my mom would always say, "That is not what a decent Filipino girl should do. You should come home at a decent hour. What are people going to think of you?" And that would get me really upset, you know, because I think that my character is very much the way it should be for a Filipina. I wear my hair long, I wear decent makeup. I dress properly, conservative. I am family oriented. It hurts me that she doesn't see that I am decent, that I am proper and that I am not going to bring shame to the family or anything like that.

This narrative suggests that even when parents are unable to control the behaviors of their children, their (dis)approval remains powerful in shaping the emotional lives of their daughters (see Wolf 1997). Although better-off parents can and do exert greater controls over their children's behaviors than do poorer parents (Wolf 1992; Kibria 1993), I would argue that all immigrant parents—regardless of class background— possess this emotional hold on their children. Therein lies the source of their power: As immigrant parents, they have the authority to determine if their daughters are "authentic" members of their racial-ethnic community. . . .

Faced with parental restrictions on their mobility, young Filipinas struggle to gain some control over their own social lives, particularly over dating. In many cases, daughters simply misinform their parents of their whereabouts or date without their parents' knowledge. They also rebel by vowing to create more egalitarian relationships with their own husbands and children. A thirty-year-old Filipina who is married to a white American explained why she chose to marry outside her culture:

> In high school, I dated mostly Mexican and Filipino. It never occurred to me to date a white or black guy. I was not attracted to them. But as I kept growing up and my father and I were having all these conflicts, I knew that if I married a Mexican or a Filipino, [he] would be exactly like my father. And so I tried to date anyone that would not remind me of my dad. A lot of my Filipina friends that I grew up with had similar experiences. So I knew that it wasn't only me. I was determined to marry a white person because he would treat me as an individual.[6]

Another Filipina who was labeled "radical" by her parents indicated that she would be more open-minded in raising her own children: "I see myself as very traditional in upbringing but I don't see myself as constricting on my children one day and I wouldn't put the gender roles on them. I wouldn't lock them into any particular way of behaving." It is important to note that even as these Filipinas desired new gender norms and practices for their own families, the majority hoped that their children would remain connected to Filipino culture.

My respondents also reported more serious reactions to parental restrictions, recalling incidents of someone they knew who had run away, joined a gang, or attempted suicide. A Filipina high-school counselor relates that most of the Filipinas she worked with "are really scared because a lot of them know friends that are pregnant and they all pretty much know girls who have attempted suicide." A 1995 random survey of San Diego public high schools conducted by the Federal Centers for Disease Control and Prevention (CDC) found that, in comparison with other ethnic groups, female Filipino students had the highest rates of seriously considering suicide (45.6 percent) as well as the highest rates of actually attempting suicide (23 percent) in the year preceding the survey. In comparison, 33.4 percent of Latinas, 26.2 percent of white women, and 25.3 percent of black women surveyed said they had suicidal thoughts (Lau 1995).

CONCLUSION

Mainstream American society defines white middle-class culture as the norm and whiteness as the unmarked marker of others' difference (Frankenberg 1993). In this article, I have shown that many Filipino immigrants use the largely gendered discourse of morality as one strategy to decenter whiteness and to locate themselves above the dominant group, demonizing it in the process. Like other immigrant groups, Filipinos praise the United States as a land of significant economic opportunity but simultaneously denounce it as a country inhabited by corrupted and individualistic people of questionable morals. In particular, they criticize American family life, American individualism, and American women (see Gabbacia 1994, 113). Enforced by distorting powers of memory and nostalgia, this rhetoric of moral superiority often leads to patriarchal calls for a cultural "authenticity" that locates family honor and national integrity in the group's female members. Because the policing of women's bodies is one of the main means of asserting moral superiority, young women face numerous restrictions on their autonomy, mobility, and personal decision making. This practice of cultural (re)construction reveals how deeply the conduct of private life can be tied to larger social structures.

The construction of white Americans as the "other" and American culture as deviant serves a dual purpose: It allows immigrant communities both to reinforce patriarchy through the sanctioning of women's (mis)behavior and to present an unblemished, if not morally superior, public face to the dominant society. Strong in family values, heterosexual morality, and a hierarchical family structure, this public face erases the Filipina "bad girl" and ignores competing (im)moral practices in the Filipino communities. Through the oppression of Filipina women and the denunciation of white women's morality, the immigrant community attempts to exert its moral superiority over the dominant Western culture and to reaffirm to itself its self-worth in the face of economic, social, political, and legal subordination. In other words, the immigrant community uses restrictions on women's lives as one form of resistance to racism. This form of cultural resistance, however, severely restricts the lives of women, particularly those of the second generation, and it casts the family as a potential site of intense conflict and oppressive demands in immigrant lives.

NOTES

1. Cordova 1983; Sharma 1984; Scharlin and Villanueva 1992; Jung 1999.

2. My understanding of Filipino American lives is also based on the many conversations I have had with my Filipino American students at the University of California, San Diego, and with Filipino American friends in the San Diego area and elsewhere.

3. Writing on the objectification of black women, Patricia Hill Collins (1991) argues that popular representations of black females—mammy, welfare queen, and Jezebel—all pivot around their sexuality, either desexualizing or hypersexualizing them. Along the same line, Native American women have been portrayed as sexually excessive (Green 1975), Chicana women as "exotic and erotic" (Mirande 1980), and Puerto Rican and Cuban women as "tropical bombshells, . . . sexy, sexed and interested" (Tafolla 1985, 39).

4. The relationship between immigrant parents and their sons deserves an article of its own.

5. According to a 1992 health assessment report of Filipinos in San Francisco, Filipino teens have the highest pregnancy rates among all Asian groups and, in 1991, the highest rate of increase in the number of births as compared with all other racial or ethnic groups (Tiongson 1997, 257).

6. The few available studies on Filipino American intermarriage indicate a high rate relative to other Asian groups. In 1980, Filipino men in California recorded the highest intermarriage rate among all Asian groups, and Filipina women had the second-highest rate, after Japanese American women (Agbayani-Siewert and Revilla 1995, 156).

REFERENCES

Azores-Gunter, Tania Fortunata M. 1986–87. "Educational Attainment and Upward Mobility: Prospects for Filipino Americans." *Amerasia Journal* 13(1): 39–52.

Cabezas, Amado, Larry H. Shinagawa, and Gary Kawaguchi. 1986–87. "New Inquiries into the Socioeconomic Status of Pilipino Americans in California." *Amerasia Journal* 13(1): 1–21.

Collins, Patricia Hill. 1991. *Black Feminist Thought: Knowledge, Consciousness, and the Politics of Empowerment.* New York: Routledge.

Cordova, Fred. 1983. *Filipinos. Forgotten Asian Americans, a Pictorial Essay, 1763–1963.* Dubuque, Iowa: Kendall/Hunt.

Coronel, Sheila, and Ninotchka Rosca. 1993. "For the Boys: Filipinas Expose Years of Sexual Slavery by the U.S. and Japan." *Ms.*, November/December, 10–15.

Dasgupta, Shamita Das, and Sayantani DasGupta. 1996. "Public Face, Private Space: Asian Indian Women and Sexuality." In *"Bad Girls/Good Girls": Women, Sex, and Power in the Nineties*, ed. Nan Bauer Maglin and Donna Perry, 226–43. New Brunswick, N.J.: Rutgers University Press.

Eastmond, Marita. 1993. "Reconstructing Life: Chilean Refugee Women and the Dilemmas of Exile." In *Migrant Women: Crossing Boundaries and Changing Identities*, ed. Gina Buijs, 35–53. Oxford: Berg.

Egan, Timothy. 1996. "Mail-Order Marriage, Immigrant Dreams and Death." *New York Times*, May 26, 12.

Espiritu, Yen Le. 1994. "The Intersection of Race, Ethnicity, and Class: The Multiple Identities of Second Generation Filipinos." *Identities* 1(2–3):249–73.

———. 1995. *Filipino American Lives.* Philadelphia: Temple University Press.

———. 1997. *Asian American Women and Men: Labor, Laws, and Love.* Thousand Oaks, Calif: Sage.

Espiritu, Yen Le, and Diane L. Wolf. 2001. "The Paradox of Assimilation: Children of Filipino Immigrants in San Diego." In *Ethnicities: Children of Immigrants in America*, ed. Rubén G. Rumbaut and Alejandro Portes. Berkeley: University of California Press; New York: Russell Sage Foundation.

Gibson, Margaret A. 1995. "Additive Acculturation as a Strategy for School Improvement." In *California's Immigrant Children: Theory, Research, and Implications for Educational Policy*, ed. Ruben Rumbaut and Wayne A. Cornelius, 77–105. La Jolla: Center for U.S.–Mexican Studies, University of California, San Diego.

Gilman, Sander L. 1985. *Difference and Pathology: Stereotypes of Sexuality, Race, and Madness.* Ithaca, N.Y.: Cornell University Press.

Green, Rayna. 1975. "The Pocahontas Perplex: The Image of Indian Women in American Culture." *Massachusetts Review* 16(4):698–714.

Haddad, Yvonne Y., and Jane I. Smith. 1996. "Islamic Values among American Muslims." In *Family and Gender among American Muslims: Issues Facing Middle Eastern Immigrants and Their Descendants*, ed. Barbara C. Aswad and Barbara Bilge, 19–40. Philadelphia: Temple University Press.

Halualani, Rona Tamiko. 1995. "The Intersecting Hegemonic Discourses of an Asian Mail-Order Bride Catalog: Pilipina 'Oriental Butterfly' Dolls for Sale." *Women's Studies in Communication* 18(1):45–64.

Hamanoto, Darrell Y. 1994. *Monitored Peril: Asian Americans and the Politics of Representation.* Minneapolis: University of Minnesota Press.

Jung, Moon-Kie. 1999. "No Whites: No Asians: Race, Marxism and Hawaii's Pre-emergent Working Class." *Social Science History* 23(3):357–93.

Kibria, Nazli. 1993: *Family Tightrope: The Changing Lives of Vietnamese Immigrant Community.* Princeton, N.J.: Princeton University Press.

Lau, Angela. 1995. "Filipino Girls Think Suicide at Number One Rate." *San Diego Union-Tribune*, February 11, A-1.

Lowe, Lisa. 1996. *Immigrant Acts: On Asian American Cultural Politics*. Durham, N.C.: Duke University Press.

Maglin, Nan Bauer, and Donna Perry. 1996. "Introduction." In *"Bad Girls/Good Girls": Women, Sex, and Power in the Nineties*, ed. Nan Bauer Maglin and Donna Perry, xiii–xxvi. New Brunswick, N.J.: Rutgers University Press: .

Marchetti, Gina. 1993. *Romance and the "Yellow Peril": Race, Sex, and Discursive Strategies in Hollywood Fiction*. Berkeley: University of California Press.

Matute-Bianchi, Maria Eugenia. 1991. "Situational Ethnicity and Patterns of School Performance among Immigrant and Nonimmigrant Mexican-Descent Students." In *Minority Status and Schooling: A Comparative Study of Immigrant and Involuntary Minorities*, ed. Margaret A. Gibson and John U. Ogbu, 205–47. New York: Garland.

Mazumdar, Suchetta. 1989. "General Introduction: A Woman-Centered Perspective on Asian American History." In *Making Waves: An Anthology by and about Asian American Women*, ed. Asian Women United of California, 1–22. Boston: Beacon.

Millett, Kate. 1970. *Sexual Politics*. Garden City, N.Y.: Doubleday.

Mirande, Alfredo. 1980. "The Chinano Family: A Reanalysis of Conflicting Views?" In *Rethinking Marriage, Child Rearing, and Family Organization*, ed. Arlene S. Skolnick and Jerome H. Skolnick, 479–93. Berkeley: University of California Press.

Okamura, Jonathan, and Amefil Agbayani. 1997. "*Pamantasan*: Filipino American Higher Education." In *Filipino Americans: Transformation and Identity*, ed. Maria P. Root, 183–97. Thousand Oaks, Calif.: Sage.

Ordonez, Raquel Z. 1997. "Mail-Order Brides: An Emerging Community." In *Filipino Americans: Transformation and Identity*, ed. Maria P. Root, 121–42. Thousand Oaks, Calif: Sage.

Orsi, Robert Anthony. 1985. *The Madonna of 115th Street: Faith and Community in Italian Harlem, 1880–1950*. New Haven, Conn.: Yale University Press.

Ruiz, Vicki L. 1992. "The Flapper and the Chaperone: Historical Memory among Mexican-American Women." In *Seeking Common Ground: Multidisciplinary Studies*, ed. Donna Gabbacia. Westport, Conn.: Greenwood.

Rumbaut, Ruben. 1991. "Passages to America: Perspectives on the New Immigration." In *America at Century's End*, ed. Alan Wolfe, 208–44. Berkeley: University of California Press.

———. 1994. "The Crucible Within: Ethnic Identity, Self-Esteem, and Segmented Assimilation among Children of Immigrants." *International Migration Review* 28(4):748–94.

Rumbaut, Ruben, and Kenji Ima. 1988. *The Adaptation of Southeast Asian Refugee Youth: A Comparative Study*. Washington, D.C.: U.S. Office of Refugee Resettlement.

San Juan, E., Jr. 1991. "Mapping the Boundaries: The Filipino Writer in the U.S." *Journal of Ethnic Studies* 19(1):117–31.

Scharlin, Craig, and Lilia V. Villanueva. 1992. *Philip Vera Cruz: A Personal History of Filipino Immigrants and the Farmworkers Movement*. Los Angeles: University of California, Los Angeles Labor Center, Institute of Labor Relations, and Asian American Studies Center.

Sharma, Miriam. 1984. "Labor Migration and Class Formation among the Filipinos in Hawaii, 1906–46." In *Labor Immigration under Capitalism: Asian Workers in the United States before World War II*, ed. Lucie Cheng and Edna Bonacich, 579–611. Berkeley: University of California Press.

Stoler, Ann Laura. 1991. "Carnal Knowledge and Imperial Power: Gender, Race, and Morality in Colonial Asia." In *Gender at the Crossroads of Knowledge: Feminist Anthropology in the Postmodern Era*, ed. Micaela di Leonardo, 51–104. Berkeley: University of California Press.

Tafolla, Carmen. 1985. *To Split a Human: Mitos, Machos y la Mujer Chicana*. San Antonio, Tex.: Mexican American Cultural Center.

Tiongson, Antonio T., Jr. 1997. "Throwing the Baby out with the Bath Water." In *Filipino Americans: Transformation and Identity*, ed. Maria P. Root, 257–71. Thousand Oaks, Calif.: Sage.

Tolman, Deborah L., and Tracy E. Higgins. 1996. "How Being a Good Girl Can Be Bad for Girls." In *"Bad Girls/Good Girls": Women, Sex, and Power in the Nineties*, ed. Nan Bauer Maglin and Donna Perry, 205–25. New Brunswick, N.J.: Rutgers University Press.

Villanueva, M. 1991. *Ginseng and Other Tales from Manila*. Corvallis, Oreg.: Calyx.

Warren, Jenifer. 1993. "Suit Asks Navy to Aid Children Left in Philippines." *Los Angeles Times*, March 5, A3.

Waters, Mary C. 1996. "The Intersection of Gender, Race, and Ethnicity in Identity Development of Caribbean American Teens." In *Urban Girls: Resisting Stereotypes, Creating Identities*, ed. Bonnie J. Ross Leadbeater and Niobe Way, 65–81. New York: New York University Press.

Woldemikael, T. M. 1939. *Becoming Black American: Haitians and American Institutions in Evanston, Illinois*. New York: AMS Press.

Wolf, Diane L. 1992. *Factory Daughters: Gender, Household Dynamics, and Rural Industrialization in Java*. Berkeley: University of California Press.

———. 1997. "Family Secrets: Transnational Struggles among Children of Filipino Immigrants." *Sociological Perspectives* 40(3):457–82.

Wong, Sau-ling. 1993. *Reading Asian American Literature: From Necessity to Extravagance*. Princeton, N.J.: Princeton University Press.

Yung, Judy. 1995. *Unbound Feet: A Social History of Chinese Women in San Francisco*. Berkeley: University of California Press.

KEY CONCEPTS

family	gender identity	patriarchy

DISCUSSION QUESTIONS

1. Think about your own household rules while growing up. Did your parents restrict your activities in ways that reflect social expectations? Were you taught that you "represent" your family? How about representing an entire race-ethnic group?

2. How are boys and girls socialized differently with regard to sexuality? What is acceptable and expected for men? For women?

12

Age Matters

TONI M. CALASANTI AND KATHLEEN F. SLEVIN

This piece explains how the study of older people and aging has been absent from traditional feminist scholarship. The authors describe how age relations involve a system of inequality that keeps old people disadvantaged. Feminist research can only advance if being old is recognized as another social location that experiences marginalization and devaluation. Calasanti and Slevin call for the further study and understanding of ageism as a system of inequality.

An inadvertent but pernicious ageism burdens much of feminist scholarship and activism. It stems from failing to study old people on their own terms and from failing to theorize age relations—the system of inequality, based on age, that privileges the not-old at the expense of the old (Calasanti 2003). Some feminists mention age-based oppression but treat it as a given—an "et cetera" on a list of oppressions, as if to indicate that we already know what it is. As a result, feminist work suffers, and we engage in our own oppression.

This [article] urges a shift in how feminist scholars approach the study of inequalities by demonstrating how and why age matters.... We briefly discuss the omission of old age and age relations from feminism and point to some of the ways in which ageism permeates feminist work and the wider society. We then outline more clearly what we mean by age relations and how these form an oppressive system....

THE BIAS OF MIDDLE AGE

Feminists consider age but rarely old people or age relations. Most focus on young adult or middle-aged women and on girls. Some attend to Sontag's notion of the "double standard of aging," by which women suffer scorn and exclusion as they grow old—"a humiliating process of gradual sexual disqualification" (Sontag 1972: 102). But even studies of women "of a certain age" (Sontag 1972: 99) focus on middle age—a time when physical markers such as menopause, wrinkles, and the like emerge and care work for old people begins to occupy women's time.

SOURCE: Toni M. Calasanti and Kathleen F. Slevin. 2006. *Age Matters: Realigning Feminist Thinking*. New York: Routledge, pp. 1–17.

Even though feminists have contributed to the literature on bodies, discussion of old bodies is sorely lacking (Laz 2003) beyond discussions of menopause. And while a handful of scholars in their fifties or sixties have done important work on age oppression, such literature "primarily refers not to deep old age but to the late middle years, roughly equating to fifties to seventies, and to the processes and experiences of aging rather than old age itself" (Twigg 2004: 62). That is, feminist scholarship on aging bodies has generally not been concerned with the "Fourth Age"—a time qualitatively different from the Third Age in that it is marked by serious infirmity. Care-work research tends not to examine *old* women who give or *receive* care. Old age, as a political location, has been ignored. . . .

Feminists exclude old people both in their choice of research questions and in their theoretical approaches. They often write or say "older" rather than "old" to avoid the negativity of the latter. They may see old age as a social construction and take it as a sign of women's inequality that they are denigrated as "old" before men are, but we do not often question the stigma affixed to old age. We don't ask why it seems denigrating to label someone "old." Feminists have analyzed how terms related to girls and women, such as "sissy" and "girly," are used to put men and boys down and reinforce women's inferiority. Yet we have not considered the *age* relations that use these terms to keep old and young groups in their respective places.

The absence of a feminist critique of ageism and age relations furthers the oppression that old people face, especially those marginalized at the intersections of multiple hierarchies. For example, by accepting the cultural dictate to "age successfully" (e.g., Friedan 1993) that underlies the "new gerontology" (Holstein and Minkler 2003), feminists reinforce ageism. Developed by Rowe and Kahn, the notion of successful aging was meant to displace the view of old age as a time of disease and decline with a "vigorous emphasis on the potential for and indeed the likelihood of a healthy and engaged old age" (Holstein and Minkler 2003: 787). Successful aging requires maintenance of the activities popular among the middle-agers privileged with money and leisure time. Thus, staying fit, or at least appearing fit, is highly valued social capital. In this sense, successful aging means not aging, not being "old," or, at the very least, not looking old. The body has become central to identity and to aging, and the maintenance of its youthful appearance has become a lifelong project that requires increasing levels of work.

Many of the age-resisting cultural practices are the purview of women. Successful aging assumes a "feminine" aspect in the ideal that the good elderly woman be healthy, slim, discreetly sexy, and independent (Ruddick 1999). Suffice it to say, our constructions of old age contain little that is positive. Fear of and disgust with growing old are widespread; people stigmatize it and associate it with personal failure, with "letting yourself go." Furthermore, class, gender, and racial biases embedded in these standards of middle age emphasize control over and choice about aging. We see advertising images of old people playing golf or tennis, traveling, sipping wine in front of sunsets, and strolling (or jogging) on the beaches of upscale resorts. Such pursuits, and the consumption depicted in ads for posh retirement communities, assume a sort of active lifestyle available only to a select group (McHugh 2000): men, whose race and class make them most likely to be able to afford it, and their spouses.

Cruikshank noted the "almost inescapable" judgment that old women's bodies are unattractive, but we know little about how old women endure this rejection (Cruikshank 2003: 147). Thus, though reporting on women who have aged "successfully" (e.g., Friedan 1993) might help negate ageist stereotypes of old women as useless or unhappy, it remains ageist in that it reinforces these standards of middle age. In light of the physical changes that occur as they age, then, many old people must develop strategies to preserve their "youthfulness" so that they will not be seen as old. As a result, old people and their bodies have become subject to a kind of discipline to activity. Those who are chronically impaired, or who prefer to be contemplative, are considered to be "problem" old people (Katz 2000; Holstein 1999; Holstein and Minkler 2003). Those who remain "active" are "not old"; those who are less active are "old" and thus less valuable. . . .

Age categories have real consequences, and bodies—old bodies—matter. They have a material reality along with their social interpretation (Laz 2003). Old people are *not*, in fact, just like middle-aged persons but only older. They are different. And as is the case with other forms of oppression, we must acknowledge and accept these differences, even see them as valuable. We must distinguish between age resistance and age denial (Twigg 2004: 63), and to do so, we must theorize the age relations that underlie the devaluation of old age.

AGE RELATIONS

. . . The first assertion, that societies are organized on the basis of age, is widely documented by scholars in aging studies. Age is a master status characteristic that defines individuals as well as groups (Hendricks 2003). Societies proscribe appropriate behaviors and obligations based on age. The second and third aspects of age relations speak more directly to issues of power and how and why such age-based organization matters for life chances. Old age not only exacerbates other inequalities but also is a social location in its own right, conferring a loss of power for all those designated as "old" regardless of their advantages in other hierarchies.

When feminists explore power relations such as those based on gender, we point to systematic differences between women and men (recognizing that other power relations come into play). In theorizing age relations, then, we also posit systematic differences between being, for instance, an *old* woman and a *young* woman. This position does not deny the importance of life-course and aging processes but instead posits discrimination and exclusion based on age—across lines of such inequalities as race, ethnicity, sexuality, class, or gender. The point at which one becomes "old" varies with these other inequalities. Once reached, old age brings losses of authority and status. Old age is a unique time of life and not simply an additive result of events occurring over the life course. Those who are perceived to be old are marginalized and lose power; they are subjected to violence (such as elder abuse) and to exploitation and cultural imperialism (Laws 1995). They suffer inequalities in distributions of authority, status, and money, and these inequalities are seen to be natural, and thus beyond dispute. Next, we briefly discuss how old people experience these inequalities.

Loss of Power

Old people lose authority and autonomy. For instance, doctors treat old patients differently than younger clients, more often withholding information, services, and treatment of medical problems (Robb, Chen, and Haley 2002). On one hand, doctors often take the complaints of old people less seriously than younger clients, attributing them to "old age" (Quadagno 1999). On the other hand, old age has been biomedicalized—a process whereby the outcomes of social factors are defined as medical or personal problems to be alleviated by medical intervention. Old people lose their ability to make decisions about their bodies and undergo drug therapies rather than other curative treatments (Wilson 2000; Estes and Binney 1991).

Workplace Issues and Marginalization

Ageism costs old people in the labor market both status and money. Although the attitudes and beliefs of employers are certainly implicated (e.g., Encel 1999), often ageism is more subtly incorporated into staffing and recruitment policies, career structures, and retirement policies (Bytheway 1995). The inability to earn money in later life means that most old people must rely on others—family or the state. And when we consider the economic dependence and security of old people, the oppressive nature of age relations become apparent. The fiscal policies and welfare retrenchment in many Western countries provide one lens on the discrimination faced by old people as they increasingly face cutbacks. As Wilson (2000: 9) noted, "Economic policies are often presented as rational and inevitable but, given the power structure of society, these so-called inevitable choices usually end up protecting younger age groups and resulting in unpleasant outcomes for those in later life (cuts in pensions or charges for health care)." Demographic projections about aging populations are often used to justify such changes, even though relevant evidence is often lacking. Furthermore, neither the public nor decision makers seem willing to consider counterevidence, such as cross-cultural comparisons that reveal little relationship between the percentage of social spending on old persons and their percentage within the overall population (Wilson 2000). Predictions of dire consequences attendant on an aging population are similarly unrelated. Indeed, with only 12.4 percent of its population age sixty-five years and older, the United States ranks thirty-seventh among countries with at least 10 percent of their population age sixty-five years and older, well below the almost 19 percent of the top three countries (Italy, Japan, and Greece) (Federal Interagency Forum on Aging Related Statistics 2004).

Decreases in income, erosion of pensions, and proposals to "reform" Social Security are not the only ways old people are marginalized when they leave the labor market. Laws (1995: 115) suggested that labor-market participation shapes identity—such that participation in waged labor "is a crucial element of citizenship, in the definition of social worthiness, and in the development of a subject's self-esteem." In conjunction with the sort of cultural denigration we describe next, the lack of labor-market participation encourages young people to see old people as "other" and not fully deserving of citizenship rights (Wilson 2000: 161). Such disenfranchisement may be informal (rather than based in laws), but it is real nonetheless (Laws 1995), as seen in the previous policy discussion.

Wealth and Income

In the contemporary United States, many people believe that many old people hold vast economic resources—an assertion that is certainly counter to claims that old people lose status or money in later life. However, the greatest inequalities in terms of income and wealth exist among old people, such that many are quite poor (Pampel 1998). The small number of old people with tremendous wealth is offset by the vast majority who rely on Social Security to stay above the poverty line. In concrete terms, Social Security—whose monthly payments averaged $1,013 for men and $764 for women in 2003—provides more than half of all income received for two-thirds of old people in the United States; indeed it amounts to almost half of income for four-fifths. Even more, it composes 90 percent or more of all income for a full one-third of older people and 100 percent of income for more than one-fifth (22 percent). Reliance on these payments is high for all but the richest quintile of old people, whose earnings and pensions add more income than does Social Security. Overall economic dependence of old people on this state-administered program is thus quite high, and more still when we realize that, even with Social Security, about one-fifth of old minority men and more than a fourth of old minority women fall below the age-adjusted poverty line (Social Security Administration 2004; Federal Interagency Forum on Aging Related Statistics 2004).

The poverty line provides an example of the differential treatment of old people. The poverty threshold is higher for old people than the rest of the population. In 2003 an old person's income had to be below $8,825—compared to $9,573 for those younger than sixty-five—to be officially designated "poor" (DeNavas-Walt, Proctor, and Mills 2004). Notably, most of the public is unaware of this fact. Poverty thresholds are calculated based on estimates of costs for nutritionally adequate diets, and because of their slower metabolism, old people need fewer calories than younger people. Thus, old people are assumed to need less money than those younger than sixty-five, despite their high medical expenses. As a result, official statistics greatly underestimate the number of old persons who are poor.

Cultural Devaluation

Finally, old people are subject to a "cultural imperialism" exemplified by "the emphasis on youth and vitality that undermines the positive contributions of older people" (Laws 1995: 113). The reality that being old, in and of itself, is a position of low status is apparent in the burgeoning antiaging industry (including the new field of "longevity medicine"), which is estimated to gross between $27 billion and $43 billion a year (with the expectation of a rise to $64 billion by

2007), depending on how expansive a definition one uses (Mehlman, Binstock, Juengst, Ponsaran, and Whitehouse 2004; U.S. Senate Special Committee on Aging 2001; *Dateline NBC*, March 6, 2001). Besides ingesting nutritional supplements and testosterone or human growth hormones, increasing numbers of people spend hours at the gym, undergo cosmetic surgery, and use lotions, creams, and hair dyes to erase the physical markers of age. Such is the equation of old age with disease and physical and mental decline that visible signs of aging serve to justify limitation of the rights and authority of old people. Many view old age as a "natural" part of life with unavoidable decrements—an equation apparent in the medical doctors' treatment of symptoms as "just old age" rather than as signs of illness or injury that merit care. The equation of aging with a natural order justifies ageism. . . .

Age relations differ from other power relations in that one's group membership shifts over time. As a result, one can experience both aspects of age relations—advantage and disadvantage—during a lifetime. Although other social locations can be malleable, such dramatic shifts in status remain uncommon. Few change racial or gender identities, but we all grow old or die first. Intersecting inequalities affect when this (becoming old) occurs, but the fact remains that where individuals stand in relation to old age *must* change (Calasanti and Slevin 2001).

REFERENCES

Bytheway, B. (1995) *Ageism.* Buckingham, UK: Open University Press.

Calasanti, T.M. (2003) "Theorizing Age Relations," in *The Need for Theory: Critical Approaches to Social Gerontology,* ed. S. Biggs, A. Lowenstein, and J. Hendricks, 199–218. Amityville, NY: Baywood.

Calasanti, T.M., and Slevin, K.F. (2001) *Gender, Social Inequalities, and Aging.* Walnut Creek, CA: AltaMira Press.

Cruikshank, M. (2003) *Learning to Be Old.* New York: Rowman and Littlefield.

Dateline NBC. March 6, 2001.

DeNavas-Walt, C., Proctor, B.D., and Mills, R.J. (2004) "Income, Poverty, and Health Insurance Coverage in the United States: 2003," in *Current Population Reports, P60–226.* Washington, DC: U.S. Census Bureau, U.S. Government Printing Office.

Encel, S. (1999) "Age Discrimination in Employment in Australia," *Ageing International* 25:69–84.

Estes, C.L., and Binney, E.A. (1991) "The Biomedicalization of Aging: Dangers and Dilemmas," in *Critical Perspectives on Aging: The Political and Moral Economy of Growing Old,* ed. M. Minkler and C.L. Estes, 117–34. Amityville, NY: Baywood.

Federal Interagency Forum on Aging Related Statistics. (2004) *Older Americans 2004: Key Indicators of Well-Being.* Washington, DC: U.S. Government Printing Office.

Friedan, B. (1993) *The Fountain of Age.* New York: Simon and Schuster.

Hendricks, J. (2003) "Structure and Identity—Mind the Gap: Toward a Personal Resource Model of Successful Aging," in *The Need for Theory: Critical Approaches to Social Gerontology*, ed. S. Biggs, A. Lowenstein, and J. Hendricks, 63–87. Amityville, NY: Baywood.

Holstein, M.B. (1999) "Women and Productive Aging: Troubling Implications," in *Critical Gerontology: Perspectives from Political and Moral Economy*, ed. M. Minkler and C.L. Estes, 359–73. Amityville, NY: Baywood.

Holstein, M.B., and Minkler, M. (2003) "Self, Society, and the 'New Gerontology,'" *The Gerontologist* 43 (6) 787–96.

Katz, S. (2000) "Busy Bodies: Activity, Aging, and the Management of Everyday Life," *Journal of Aging Studies* 14:135–52.

Laws, G. (1995) "Understanding Ageism: Lessons from Feminism and Postmodernism," *The Gerontologist* 35 (1) 112–18.

Laz, C. (2003) "Age Embodied," *Journal of Aging Studies* 17:503–19.

McHugh, K. (2000) "The 'Ageless Self'? Emplacement of Identities in Sun-Belt Retirement Communities," *Journal of Aging Studies* 14:103–15.

Mehlman, M.J., Binstock, R.H., Juengst, E.T., Ponsaran, R.S., and Whitehouse, P.J. (2004) "Anti-aging Medicine: Can Consumers Be Better Protected?" *The Gerontologist* 44 (3) 304–10.

Pampel, F.C. (1998) *Aging, Social Inequality, and Public Policy.* Thousand Oaks, CA: Pine Forge Press.

Quadagno, J.S. (1999) *Aging and the Life Course.* Boston: McGraw-Hill.

Robb, C., Chen, H., and Haley, W.E. (2002) "Ageism in Mental Health Care: A Critical Review," *Journal of Clinical Geropsychology* 8 (1): 1–12.

Ruddick, S. (1999) "Virtues and Age," in *Mother Time: Women, Aging and Ethics*, ed. M.U. Walker, 45–60. Lanham, MD: Rowman and Littlefield.

Social Security Administration. (2004) *Fast Facts and Figures about Social Security, 2004.* Washington, DC: U.S. Government Printing Office.

Sontag, S. (1972) "The Double Standard of Aging," in Saturday Review of the Society, ed. M. Rainbolt and J. Fleetwood, 55:29–38. Reprinted in *On the Contrary: Essays by Men and Women*, Albany, NY: SUNY Press, 1983, 99–112.

Twigg, J. (2004) "The Body, Gender, and Age: Feminist Insights in Social Gerontology," *Journal of Aging Studies* 18 (1): 59–73.

U.S. Senate Special Committee on Aging. (2001) *Swindlers, Hucksters and Snake Oil Salesman: Hype and Hope Marketing Anti-aging Products to Seniors* (Serial No. 107–14). Washington, DC: U.S. Government Printing Office.

Wilson, G. (2000) *Understanding Old Age.* Thousand Oaks, CA: Sage.

KEY CONCEPTS

ageism   age prejudice   age stratification

DISCUSSION QUESTIONS

1. Consider popular television programs today. Are there any older characters consistently present on television? If so, how are older characters portrayed? What roles do they have in the program?

2. What issues are important to aging Americans? How are these issues addressed by the government? How are they addressed by communities?

Applying Sociological Knowledge:
An Exercise for Students

Socialization occurs in many different contexts. Pick one of the following situations that involves entry to a new role: becoming a college student, becoming a parent, getting your first job. What changes do you experience in this new role? How are the expectations associated with your new status communicated to you? Are these formally or informally communicated? What does this teach you about the socialization process?

13

The Presentation of Self in Everyday Life

ERVING GOFFMAN

Erving Goffman likens social interaction to a "con game," in which we are consistently trying to put forward a certain impression or "self" in order to get something from others. Although many will not see human behavior so cynically, Goffman's analysis sheds light on how people try to manage the impression that others have of them.

When an individual plays a part he implicitly requests his observers to take seriously the impression that is fostered before them. They are asked to believe that the character they see actually possesses the attributes he appears to possess, that the task he performs will have the consequences that are implicitly claimed for it, and that, in general, matters are what they appear to be. In line with this, there is the popular view that the individual offers his performance and puts on his show "for the benefit of other people." It will be convenient to begin a consideration of performances by turning the question around and looking at the individual's own belief in the impression of reality that he attempts to engender in those among whom he finds himself.

At one extreme, one finds that the performer can be fully taken in by his own act; he can be sincerely convinced that the impression of reality which he stages is the real reality. When his audience is also convinced in this way about the show he puts on—and this seems to be the typical case—then for the moment at least, only the sociologist or the socially disgruntled will have any doubts about the "realness" of what is presented.

At the other extreme, we find that the performer may not be taken in at all by his own routine. This possibility is understandable, since no one is in quite as good an observational position to see through the act as the person who puts it on. Coupled with this, the performer may be moved to guide the conviction of his audience only as a means to other ends, having no ultimate concern in the conception that they have of him or of the situation. When the individual has no belief in his own act and no ultimate concern with the beliefs of his audience, we may call him cynical, reserving the term "sincere" for individuals who believe in the impression fostered by their own performance. It should be understood that the cynic, with all his professional disinvolvement, may obtain unprofessional pleasures from his masquerade, experiencing a kind of gleeful spiritual aggression

SOURCE: Erving Goffman. 1959. *The Presentation of Self in Everyday Life.* Garden City, NY: Anchor Doubleday, pp. 17–27.

from the fact that he can toy at will with something his audience must take seriously.

It is not assumed, of course, that all cynical performers are interested in deluding their audiences for purposes of what is called "self-interest" or private gain. A cynical individual may delude his audience for what he considers to be their own good, or for the good of the community, etc. For illustrations of this we need not appeal to sadly enlightened showmen such as Marcus Aurelius or Hsun Tzǔ. We know that in service occupations practitioners who may otherwise be sincere are sometimes forced to delude their customers because their customers show such a heartfelt demand for it. Doctors who are led into giving placebos, filling station attendants who resignedly check and recheck tire pressures for anxious women motorists, shoe clerks who sell a shoe that fits but tell the customer it is the size she wants to hear—these are cynical performers whose audiences will not allow them to be sincere. Similarly, it seems that sympathetic patients in mental wards will sometimes feign bizarre symptoms so that student nurses will not be subjected to a disappointingly sane performance. So also, when inferiors extend their most lavish reception for visiting superiors, the selfish desire to win favor may not be the chief motive; the inferior may be tactfully attempting to put the superior at ease by simulating the kind of world the superior is thought to take for granted.

I have suggested two extremes: an individual may be taken in by his own act or be cynical about it. These extremes are something a little more than just the ends of a continuum. Each provides the individual with a position which has its own particular securities and defenses, so there will be a tendency for those who have traveled close to one of these poles to complete the voyage. Starting with lack of inward belief in one's role, the individual may follow the natural movement described by Park:

> It is probably no mere historical accident that the word person, in its first meaning, is a mask. It is rather a recognition of the fact that everyone is always and everywhere, more or less consciously, playing a role . . . It is in these roles that we know each other; it is in these roles that we know ourselves.[1]

In a sense, and in so far as this mask represents the conception we have formed of ourselves—the role we are striving to live up to—this mask is our truer self, the self we would like to be. In the end, our conception of our role becomes second nature and an integral part of our personality. We come into the world as individuals, achieve character, and become persons.[2] . . .

Front, then, is the expressive equipment of a standard kind intentionally or unwittingly employed by the individual during his performance. For preliminary purposes, it will be convenient to distinguish and label what seem to be the standard parts of front.

First, there is the "setting," involving furniture, décor, physical layout, and other background items which supply the scenery and stage props for the spate of human action played out before, within, or upon it. . . .

If we take the term "setting" to refer to the scenic parts of expressive equipment, one may take the term "personal front" to refer to the other items of expressive equipment, the items that we most intimately identify with the performer himself and that we naturally expect will follow the performer wherever he goes. As part of personal front we may include: insignia of office or rank; clothing; sex, age, and racial characteristics; size and looks; posture; speech patterns; facial expressions; bodily gestures; and the like. Some of these vehicles for conveying signs, such as racial characteristics, are relatively fixed and over a span of time do not vary for the individual from one situation to another. On the other hand, some of these sign vehicles are relatively mobile or transitory, such as facial expression, and can vary during a performance from one moment to the next. . . .

In addition to the fact that different routines may employ the same front, it is to be noted that a given social front tends to become institutionalized in terms of the abstract stereotyped expectations to which it gives rise, and tends to take on a meaning and stability apart from the specific tasks which happen at the time to be performed in its name. The front becomes a "collective representation" and a fact in its own right.

When an actor takes on an established social role, usually he finds that a particular front has already been established for it. Whether his acquisition of the role was primarily motivated by a desire to perform the given task or by a desire to maintain the corresponding front, the actor will find that he must do both.

Further, if the individual takes on a task that is not only new to him but also unestablished in the society, or if he attempts to change the light in which his task is viewed, he is likely to find that there are already several well-established fronts among which he must choose. Thus, when a task is given a new front we seldom find that the front it is given is itself new.

NOTES

1.   Robert Ezra Park, *Race and Culture* (Glencoe, IL: The Free Press, 1950), p. 249.
2.   Ibid., 250.

KEY CONCEPTS

dramaturgical model   presentation of self

DISCUSSION QUESTIONS

1.   How many "selves" do you think you could "play" or "do" in order to accomplish something with another person? Discuss two such selves and try to get them to be quite different from each other.

2.   "All the world's a stage," wrote William Shakespeare. So might Goffman have said this. How does his analysis of the presentation of self in everyday life suggest that life is a drama where we all play our parts?

14

Code of the Street

ELIJAH ANDERSON

Elijah Anderson's study of interaction on the street shows the vast array of implicit "codes" of behavior or rules that guide street interaction. His analysis helps explain the complexity of street interaction and provides a sociological explanation of street violence.

In some of the most economically depressed and drug- and crime-ridden pockets of the city, the rules of civil law have been severely weakened, and in their stead a "code of the street" often holds sway. At the heart of this code is a set of prescriptions and proscriptions, or informal rules, of behavior organized around a desperate search for respect that governs public social relations, especially violence, among so many residents, particularly young men and women. Possession of respect—and the credible threat of vengeance—is highly valued for shielding the ordinary person from the interpersonal violence of the street. In this social context of persistent poverty and deprivation, alienation from broader society's institutions, notably that of criminal justice, is widespread. The code of the street emerges where the influence of the police ends and personal responsibility for one's safety is felt to begin, resulting in a kind of "people's law," based on "street justice." This code involves a quite primitive form of social exchange that holds would-be perpetrators accountable by promising an "eye for an eye," or a certain "payback" for transgressions. In service to this ethic, repeated displays of "nerve" and "heart" build or reinforce a credible reputation for vengeance that works to deter aggression and disrespect, which are sources of great anxiety on the inner-city street. . . .

In approaching the goal of painting an ethnographic picture of these phenomena, I engaged in participant-observation, including direct observation, and conducted in-depth interviews. Impressionistic materials were drawn from various social settings around the city, from some of the wealthiest to some of the most economically depressed, including carryouts, "stop and go" establishments, laundromats, taverns, playgrounds, public schools, the Center City indoor mall known as the Gallery, jails, and public street corners. In these settings I encountered a wide variety of people—

SOURCE: Elijah Anderson. 1999. *Code of the Street.* New York: W. W. Norton, pp. 9–11, 32–34, 312–317. Reprinted with permission.

adolescent boys and young women (some incarcerated, some not), older men, teen-age mothers, grandmothers, and male and female schoolteachers, black and white, drug dealers, and common criminals. To protect the privacy and confidentiality of my subjects, names and certain details have been disguised. . . .

Of all the problems besetting the poor inner-city black community, none is more pressing than that of interpersonal violence and aggression. This phenomenon wreaks havoc daily on the lives of community residents and increasingly spills over into downtown and residential middle-class areas. Muggings, burglaries, carjackings, and drug-related shootings, all of which may leave their victims or innocent bystanders dead, are now common enough to concern all urban and many suburban residents.

The inclination to violence springs from the circumstances of life among the ghetto poor—the lack of jobs that pay a living wage, limited basic public services (police response in emergencies, building maintenance, trash pickup, lighting, and other services that middle-class neighborhoods take for granted), the stigma of race, the fallout from rampant drug use and drug trafficking, and the resulting alienation and absence of hope for the future. Simply living in such an environment places young people at special risk of falling victim to aggressive behavior. Although there are often forces in the community that can counteract the negative influences—by far the most powerful is a strong, loving, "decent" (as inner-city residents put it) family that is committed to middle-class values—the despair is pervasive enough to have spawned an oppositional culture, that of "the street," whose norms are often con-sciously opposed to those of mainstream society. These two orientations—decent and street—organize the community socially, and the way they coexist and interact has important consequences for its residents, particularly for children growing up in the inner city. Above all, this environment means that even youngsters whose home lives reflect mainstream values—and most of the homes in the community do—must be able to handle themselves in a street-oriented environment.

This is because the street culture has evolved a "code of the street," which amounts to a set of informal rules governing interpersonal public behavior, particu-larly violence. The rules prescribe both proper comportment and the proper way to respond if challenged. They regulate the use of violence and so supply a rationale allowing those who are inclined to aggression to precipitate violent encounters in an approved way. The rules have been established and are enforced mainly by the street-oriented; but on the streets the distinction between street and decent is often irrelevant. Everybody knows that if the rules are violated, there are penalties. Knowl-edge of the code is thus largely defensive, and it is literally necessary for operating in public. Therefore, though families with a decency orientation are usually opposed to the values of the code, they often reluctantly encourage their children's familiarity with it in order to enable them to negotiate the inner-city environment.

At the heart of the code is the issue of respect—loosely defined as being treated "right" or being granted one's "props" (or proper due) or the deference one deserves. However, in the troublesome public environment of the inner city, as people increasingly feel buffeted by forces beyond their control, what one deserves in the way of respect becomes ever more problematic and uncertain. This situation in turn further opens up the issue of respect to sometimes intense interpersonal negotiation, at times resulting in altercations. In the street culture, especially among young people, respect is viewed as almost an external entity, one that is hard-won but easily lost—and so must constantly be guarded. The rules of the code in fact provide a framework for negotiating respect. With the right amount of respect, individuals can avoid being bothered in public. This security is important, for if they *are* bothered, not only may they face physical danger, but they will have been disgraced or "dissed" (disrespected). Many of the forms dissing can take may seem petty to middle-class people (maintaining eye contact for too long, for example), but to those invested in the street code, these actions, a virtual slap in the face, become serious indications of the other person's intentions. Consequently, such people become very sensitive to advances and slights, which could well serve as a warning of imminent physical attack or confrontation.

The hard reality of the world of the street can be traced to the profound sense of alienation from mainstream society and its institutions felt by many poor inner-city black people, particularly the young. The code of the street is actually a cultural adaptation to a profound lack of faith in the police and the judicial system—and in others who would champion one's personal security. The police, for instance, are most often viewed as representing the dominant white society and as not caring to protect inner-city residents. When called, they may not respond, which is one reason many residents feel they must be prepared to take extraordinary measures to defend themselves and their loved ones against those who are inclined to aggression. Lack of police accountability has in fact been incorporated into the local status system: the person who is believed capable of "taking care of himself" is accorded a certain deference and regard, which translates into a sense of physical and psychological control. The code of the street thus emerges where the influence of the police ends and where personal responsibility for one's safety is felt to begin. Exacerbated by the proliferation of drugs and easy access to guns, this volatile situation results in the ability of the street-oriented minority (or those who effectively "go for bad") to dominate the public spaces. . . .

The attitudes and actions of the wider society are deeply implicated in the code of the street. Most people residing in inner-city communities are not totally invested in the code; it is the significant minority of hard-core street youth who maintain the code in order to establish reputations that are integral to the extant social order. Because of the grinding poverty of the communities these people inhabit, many have—or feel they have—few other options for expressing themselves. For them the standards and rules of the street code are the only game in town.

And as was indicated above, the decent people may find themselves caught up in problematic situations simply by being at the wrong place at the wrong time, which is why a primary survival strategy of residents here is to "see but don't see." The extent to which some children—particularly those who through upbringing have become most alienated and those who lack strong and conventional social support—experience, feel, and internalize racist rejection and contempt from mainstream society may strongly encourage them to express contempt for the society in turn. In dealing with this contempt and rejection, some youngsters consciously invest themselves and their considerable mental resources in what amounts to an oppositional culture, a part of which is the code of the street. They do so to preserve themselves and their own self-respect. Once they do, any respect they might be able to garner in the wider system pales in comparison with the respect available in the local system; thus they often lose interest in even attempting to negotiate the mainstream system.

At the same time, many less alienated young people have assumed a street-oriented demeanor as way of expressing their blackness while really embracing a much more moderate way of life; they, too, want a nonviolent setting in which to live and one day possibly raise a family. These decent people are trying hard to be part of the mainstream culture, but the racism, real and perceived, that they encounter helps legitimize the oppositional culture and, by extension, the code of the street. On occasion they adopt street behavior; in fact, depending on the demands of the situation, many people attempt to codeswitch, moving back and forth between decent and street behavior. . . .

In addition, the community is composed of working-class and very poor people since those with the means to move away have done so, and there has also been a proliferation of single-parent households in which increasing numbers of kids are being raised on welfare. The result of all this is that the inner-city community has become a kind of urban village, apart from the wider society and limited in terms of resources and human capital. Young people growing up here often receive only the truncated version of mainstream society that comes from television and the perceptions of their peers. . . .

According to the code, the white man is a mysterious entity, a part of an enormous monolithic mass of arbitrary power, in whose view black people are insignificant. In this system and in the local social context, the black man has very little clout; to salvage something of value, he must outwit, deceive, oppose, and ultimately "end-run" the system.

Moreover, he cannot rely on this system to protect him; the responsibility is his, and he is on his own. If someone rolls on him, he has to put his body, and often his life, on the line. The physicality of manhood thus becomes extremely important. And urban brinksmanship is observed and learned as a matter of course. . . .

Urban areas have experienced profound structural economic changes, as deindustrialization—the movement from manufacturing to service and high-tech—and the growth of the global economy have created new economic conditions. Job opportunities increasingly go abroad to Singapore, Taiwan, India, and Mexico, and to nonmetropolitan America, to satellite cities like King of Prussia, Pennsylvania. Over the last fifteen years, for example, Philadelphia has lost 102,500 jobs, and its manufacturing employment has declined by 53 percent. Large numbers of inner-city people, in particular, are not adjusting effectively to the new economic reality. Whereas low-wage jobs—especially unskilled and low-skill factory jobs—used to exist simultaneously with poverty and there was hope for the future, now jobs simply do not exist, the present economic boom notwithstanding. These dislocations have left many inner-city people unable to earn a decent living. More must be done by both government and business to connect inner-city people with jobs.

The condition of these communities was produced not by moral turpitude but by economic forces that have undermined black, urban, working-class life and by a neglect of their consequences on the part of the public. Although it is true that persistent welfare dependency, teenage pregnancy, drug abuse, drug dealing, violence, and crime reinforce economic marginality, many of these behavioral problems originated in frustrations and the inability to thrive under conditions of economic dislocation. This in turn leads to a weakening of social and family structure, so children are increasingly not being socialized into mainstream values and behavior. In this context, people develop profound alienation and may not know what to do about an opportunity even when it presents itself. In other words, the social ills that the companies moving out of these neighborhoods today sometimes use to justify their exodus are the same ones that their corporate predecessors, by leaving, helped to create.

Any effort to place the blame solely on individuals in urban ghettos is seriously misguided. The focus should be on the socioeconomic structure, because it was structural change that caused jobs to decline and joblessness to increase in many of these communities. But the focus also belongs on the public policy that has radically threatened the well-being of many citizens. Moreover, residents of these communities lack good education, job training, and job networks, or connections with those who could help them get jobs. They need enlightened employers able to understand their predicament and willing to give them a chance. Government, which should be assisting people to adjust to the changed economy, is instead cutting what little help it does provide. . . .

The emergence of an underclass isolated in urban ghettos with high rates of joblessness can be traced to the interaction of race prejudice, discrimination, and the effects of the global economy. These factors have contributed to the profound social isolation and impoverishment of broad segments of the inner-city black population. Even though the wider society and economy have been experiencing accelerated prosperity for almost a decade, the fruits of it often miss the truly disadvantaged isolated in urban poverty pockets.

In their social isolation an oppositional culture, a subset of which is the code of the street, has been allowed to emerge, grow, and develop. This culture is essentially one of accommodation with the wider society, but different from past efforts to accommodate the system. A larger segment of people are now not simply isolated but ever more profoundly alienated from the wider society and its institutions. For instance, in conducting the fieldwork for this book, I visited numerous inner-city schools, including elementary, middle, and high schools, located in areas of concentrated poverty. In every one, the so-called oppositional culture was well entrenched. In one elementary school, I learned from interviewing kindergarten, first-grade, second-grade, and fourth-grade teachers that through the first grade, about a fifth of the students were invested in the code of the street; the rest are interested in the subject matter and eager to take instruction from the teachers—in effect, well disciplined. By the fourth grade, though, about three-quarters of the students have bought into the code of the street or the oppositional culture.

As I have indicated throughout this work, the code emerges from the school's impoverished neighborhood, including overwhelming numbers of single-parent homes, where the fathers, uncles, and older brothers are frequently incarcerated—so frequently, in fact, that the word "incarcerated" is a prominent part of the young child's spoken vocabulary. In such communities there is not only a high rate of crime but also a generalized diminution of respect for law. As the residents go about meeting the exigencies of public life, a kind of people's law results, . . . Typically, the local streets are, as we saw, tough and dangerous places where people often feel very much on their own, where they themselves must be personally responsible for their own security, and where in order to be safe and to travel the public spaces unmolested, they must be able to show others that they are familiar with the code—that physical transgressions will be met in kind.

In these circumstances the dominant legal codes are not the first thing on one's mind; rather, personal security for self, family, and loved ones is. Adults, dividing themselves into categories of street and decent, often encourage their children in this adaptation to their situation, but at what price to the children and at what price to wider values of civility and decency? As the fortunes of the inner city continue to decline, the situation becomes ever more dismal and intractable. . . .

KEY CONCEPTS

deindustrialization norms urban underclass

DISCUSSION QUESTIONS

1. List several ways that subtle or nonverbal behavior becomes important "on the street."
2. What specific ways does Anderson see street behavior as stemming from social structural conditions for African Americans?

15

The Impact of Internet Communications on Social Interaction

THOMAS WELLS BRIGNALL III AND THOMAS VAN VALEY

The increased use of the Internet has altered the way young people interact. This article applies Goffman's theory about the presentation of self to social interaction that takes place on the Internet. The "rules" for social interaction

SOURCE: Thomas Wells Brignall III and Thomas Van Valey. 2005.
"The Impact of Internet Communications on Social Interaction." *Sociological Spectrum* 25: 335–348.

and the skill that develops when you engage in face-to-face interaction are changed when communicating online. The authors suggest that young people today are cyberkids and experience social interaction and education differently because of the Internet.

INTRODUCTION

The Internet is clearly on the way to becoming an integral tool of business, communication, and popular culture in the United States and in other parts of the world. The Internet is also being presented as a pedagogical tool for much of public education. However, its extraordinary growth is not without concern. Of particular relevance is the issue of the potential impact of the Internet and especially computer-mediated communications on the nature and quality of social interaction among cyberkids.

According to NetValue, people who chat online are among the heaviest users of the Internet (2002, p.1). Furthermore, 20% of female chatters are teenage girls. NetValue, Nielsen, and eMarketer (Ramsey, 2000) all conclude that while teenagers are not yet the dominant demographic of Internet users, their usage is growing very rapidly. Regardless of country of origin, recent data on Internet usage indicates that young people are becoming some of the heaviest users of the Internet (Nielsen 2002, p. 1). According to Nielsen (2002), 74% of United States residents between the ages of 12 and 18 are now using the Internet.

Among Internet users between 12 and 18, 35% spend 31–60 minutes per day online, and 44% spend more than an hour per day online. Indeed, almost 4% of these cyberkids spend four or more hours online each day. America Online reported in their national survey of more than 6,700 teens and parents of teens that "fifty-six percent of teens (aged 18 to 19) prefer Internet to the telephone" (Pastore 2002, p. 1). The survey also reported that in order to keep up their communications with friends, more than 81% of teens use e-mail, while 70% use instant messaging.

Certainly, one can argue that online communication is not yet the dominant form of communication among young people. However, there is little question that the phenomenon of online communication among teens and children is growing rapidly. Moreover, it is not hard to imagine a time in the near future when elementary and secondary school students may spend several hours a day doing schoolwork, communicating with friends, teachers, and family, and seeking entertainment, all via the computer and the Internet. It also follows, therefore, that substantial portions of students' experiences with interpersonal communications (especially with persons not already known to them) are likely to be computer-mediated. If children and teenagers are already using computers as a significant form of education, communication, and entertainment, it may well be that less time is being spent having face-to-face interactions with peers.

If the amount of time spent in face-to-face interactions among youth is shrinking, there may be significant consequences for their development of social skills and their presentation of self. Several decades ago, Goffman (1959, 1967) suggested that individuals who lack the normative communication, cultural, and civility skills in a society would find it difficult to interact with others successfully. At the time, Goffman was referring to the variety of visual and auditory cues that occur in face-to-face communications. Today, with computer-mediated communications, it is possible that none of the cues that Goffman wrote about may be present during online communication.

Some authors have already suggested that online behavior is different from offline behavior (Rheingold 1993; Postman 1992; Jones 1995; Miller 1996). Examples of such differences in behavior include individuals who are willing to misrepresent themselves by feigning a different gender, skin color, sexual orientation, physical condition, or age. Other differences in observed behaviors include the open display of group norm violations such as aggressive behavior, racism, sexism, homophobia, personal attacks, harassment, and a tendency for individuals to quickly abandon groups and conversations, refusing to deal with issues they find difficult to immediately resolve. If these authors are correct, it is critical that the specific elements of interpersonal communication involved in computer-mediated interactions be identified and compared with face-to-face interactions. This article examines the nature of Internet communications that take place among young people (particularly the elements of the presentation of self that do not occur, or occur with limited frequency), and suggests some potential consequences for the education and socialization of our youth.

THEORIES OF SOCIAL INTERACTION

In *Asylums* (1961), Goffman described how small rituals have replaced the big rituals that occurred in traditional interpersonal relations. "What remains are brief rituals one individual performs for and to another, attesting to civility and good will on the performer's part and to the recipient's possession of a small patrimony of sacredness" (Goffman 1961, p. 63). However, the choice of the particular set of rules an individual chooses to follow derives from requirements established in social encounters. Therefore, an individual's concept of self is shaped by the sum of the social interactions in which that individual engages.

Goffman (1967) further argues that the self is the actor in an ongoing play that responds to the judgments of others. While each individual has more than one role, he or she is saved from role strain by "audience segregation," that is, by employing multiple roles in their interactions with others. Because an individual can play a unique role in each social situation, that individual can effectively be a different person in each situation without contradiction. However, audience segregation regularly breaks down, and an individual can present a role incompatible with ones presented on other occasions. In these situations, role strain occurs.

Children learn how to cope with such role strains through their own social experience and by watching others navigate social interactions that involve contradictory or competing roles. Indeed, Goffman argues that coping with role strain and developing impression management are necessary skills for the success of individuals in everyday social interactions. Children also must learn how to manage "front stage" and "back stage" behavior. The front stage is open to judgment by an audience, where the back stage is a place where actors can discuss, polish, or refine their performances without revealing themselves to the same audience. However, because there are multiple layers of front stages and back stages, individuals learn how much they can reveal to other characters.

All front stage roles contain a number of elements that are visual or auditory in nature including physical appearance (e.g., demographic characteristics, and physical features, such as size, make up, hairstyle, posture), manner of speaking (e.g., the use of standard vs. slang dialect, accents, regional vocabulary choices, and voice inflections), and the use of various props (e.g., clothing, car, and food preferences). Together, these elements help to create the role that is presented to others. However, individuals in face-to-face interactions not only present these more obvious indicators of their roles, they also give out more subtle cues such as posture, hand gestures, tone of voice, movement in a conversation, eye contact, and levels of social formality. Moreover, many of these indicators of role (both the obvious and the subtle) vary widely across groups.

However, in order to maintain a positive on-going relationship in any difficult face-to-face circumstance, an individual must learn the appropriate socialization rituals. Knowing these rituals and being able to play a proper front stage role is crucial in order for an individual to get along with others. Indeed, the appearance of getting along with others is sometimes far more important than whether individuals actually like one another. Once again, it is difficult for individuals to succeed if they lack the proper social skills of the various groups with which they interact. Only with practice, will individuals develop and learn to improve their interaction skills, and develop a better presentation when communicating to individuals via their front and back stages. . . .

ONLINE INTERACTION

From the beginning of the Internet, it was clear that the interaction taking place online was a new form of social interaction. What was not clear, and still is to be determined, are the consequences of this new form of social interaction. The Internet itself is neither negative nor positive. It is inanimate, an object or a tool that can be used in various ways. To reify the Internet and suggest that it is somehow inherently liberating or enslaving is misleading. Nevertheless, it is important to look at how interaction on the Internet differs from other forms of interaction. These differences may play havoc with traditional social interaction rituals. . . .

Because online interaction is different, it is entirely possible that children with high levels of online interactions would adopt different techniques of social interaction. Several authors such as LaRose, Eastin, and Greeg (2001) and Schmitz (1997) cultivate the notion that for many individuals, online communication helps facilitate face-to-face communications with current relationships and sometimes with new relationships. Boyd and Walther (2002) even argue that Internet social support is superior to face-to-face social support. According to them, online social support offers benefits that face-to-face social networks cannot: anonymity, constant access to better quality expertise, and enhanced modes of expression, with less chance of embarrassment and without incurring an obligation to the support provider. However, Spears and Lea (1992) argue that the absence of social and contextual cues undermine the perception of leadership, status, and power, and leads to reduced impact of social norms and therefore to deregulated, anti-normative behavior.

THE IMPACT OF INTERNET COMMUNICATIONS

Our fundamental position is that online social interaction is one form of role-play, and thus an element in the development of the self. Moreover, if a substantial amount of communication is accomplished online, either at home, at school, or elsewhere, children and cyberkids are likely to develop the skills necessary for online interaction, but they are also likely to lack some of the skills that are involved in face-to-face interaction.

Tapscott (1998) claims that the children of the Internet generation are already way ahead of their parents and many other adults (who do not yet understand what the Internet is or how it works). However, Tapscott and others may have made a crucial mistake in their logic. They have forgotten that political and economic power are in the hands of adults. Therefore, the members of the cyberkid generation must understand, adapt, and modify their interactions in order to get along in society.

Moreover, if students develop unique interaction rituals based on online communication without enough experience or understanding of traditional face-to-face interaction rituals, the likelihood for friction and/or conflict to occur undoubtedly increases. Such youths may be perceived as rude, insolent, disconnected, spoiled, or apathetic. It can be argued that new cultural and social phenomena have typically produced tensions between the generations. However, not having the skills or the knowledge of how to communicate with people who have different values and attitudes complicates the traditional struggles between youth and their elders. Furthermore, such rifts will not only be between the young and the old, but between any groups that are different from one another. . . .

The skills and lessons of socialization that students need to learn in order to cope in everyday life, however, cannot be manufactured by computer simulations or video games (at least not yet). Classrooms with computers hooked up to the Internet predispose students to work as individuals rather than as members of any social group. Although students can interact with others when they are on the Internet, they are often able to choose with whom they wish to interact and how they want to manage it. Even if the students are interacting with fellow classmates while online, is it not reasonable that the use of a computer in mediating those interactions will alter the interaction rituals in which they engage? . . .

DISCUSSION

Given that we know so little about the nature of the social interactions that take place over the Internet, not only but especially by youth, it is clear that much research is needed. One obvious arena would include studies of young children followed over time to determine if there are recognizable online interaction rituals and what consequences they may have on the social development of the children. There is a growing body of research on this general issue—whether computer-mediated communications "displace" more traditional face-to-face forms of communication. A special issue of *IT & Society* recently focused on it, reporting the results of a number of time-diary studies in the United States and elsewhere. However, the conclusions are mixed. Some studies have found reduced levels of time spent with friends and family (Nie and Hillygus 2002; Pronovost 2002; Nie and Erbring 2002). Others have either reported no difference (Robinson et al. 2002; Gershny 2002) or increased levels of sociability among Internet users (Neustadl and Robinson 2002; Cummings 2002).

We agree with Wellman and Gulia (1999) that "the internet . . . is not a separate reality. People bring to their online interactions such baggage as their gender, stage in the life cycle, cultural milieu, socioeconomic status, and offline connections with others" (p. 3). Online and face-to-face social interactions share some elements, and many individuals interact using both forms of communication.

This discussion is not about the members of the current generations who have grown up with the Internet. It is about future generations where the Internet is used as a primary source of communication, the focal center of school-work and research, the main source of entertainment, and the primary medium for the development of contemporary issues. If the strength of the Internet of the future is the fact that individuals can choose with whom they want to interact, then it may also be one of the Internet's weaknesses when it comes to the development of social interaction skills. The demands of learning to get along with others are likely to become drowned out by self-interested pursuits. The possibility of a narrow world perspective seems certain for those individuals who choose to isolate themselves from people and ideas with whom they feel uncomfortable. If the easiest solution to avoid dissonance is to avoid situations that produce it, then the potential for an unrealistic social process is high. We are not opposing or supporting computer-mediated communications or the Internet in schools. We are simply suggesting that it is important to assess the social impacts of the Internet while the frequency of use is still relatively low. Once they become ubiquitous, the possibility of such research has been forever lost.

REFERENCES

Boyd, S. and Joseph, B. Walther. 2002. "Attraction to Computer-Mediated Social Support." Pp. 153–188 in *Communication Technology and Society: Audience Adoption and Uses*, edited by C. A. Lin and D. Atkin. Cresskill, NJ: Hampton Press.

Cummings, J., L. Sproull, and S. Kiesler. 2002. "Beyond Hearing: Where Real World and Online Support Meet." *Group Dynamics: Theory, Research, and Practices*, 6(1):78–88.

Gershney, J. 2002. "Mass Media, Leisure, and Home IT: A Panel Time-Diary Approach." Pp. 53–66 retrieved June 8, 2002 from http://www.IT and Society.org.

Goffman, E. 1959. *The Presentation of Self in Everyday Life*. New York: Anchor.

Goffman, E. 1961. *Asylums: Essays on the Social Situation of Mental Patients and Other Inmates*. New York: Anchor.

Goffman, E. 1967. *Interaction Ritual: Essays on Face to Face Behavior*. New York: Pantheon Books.

Jones, G. S. 1995. *CyberSociety: Computer-Mediated Communication and Community*. Thousand Oaks, CA: Sage Publications.

LaRose, R., M. S. Eastin, and J. Gregg. 2001. "Reformulating the Internet Paradox: Social Cognitive Explanations of Internet Use and Depression." Retrieved January 30, 2003 from www.behavior.net/JOB/v1n1/paradox.html.

Miller, E. S. 1996. *Civilizing Cyberspace: Policy, Power, and The Information Superhighway*. New York: ACM Press.

NetValue. (2002). "Internet use patterns." Edited by R. A. Cole Retrieved December, 5, 2002 from www.netvalue.com/corp/actionnaires/index.htm.

Neustadl, A. and J. P. Robinson. 2002. "Media Use Differences Between Internet Users and Nonusers in the General Social Survey." Pp. 100–120. Retrieved June 8, 2002 from http://www.IT and Society.org

Nie, N. H. and L. Erbring. (2002). "Internet and Society: A Preliminary Report." *IT and Society* 1(1):275–283.

Nie, N. H. and D. S. Hillygus. (2002). "The Impact of Internet Use on Sociability: Time-diary Findings." *IT and Society* 1(1):1–20.

Nielsen Net Ratings. (2002). "Internet User Growth." Retrieved September 19, 2002 from http://www.nielsen-netratings.com/news.jsp.

Pastore, M. (2002) "Internet Key to Communication Among Youth." Retrieved January 30, 2003 from http://cyberatlas.internet.com/big_picture/demographics/article/0,5901_961881,00.html.

Postman, N. (1992). *Technopoly*. New York: Vintage Books.

Pronovost, G. (2002). "The Internet and Time Displacement: A Canadian Perspective." *IT and Society* 1(1):44–53.

Ramsey, G. (2000). *The E-demographics and Usage Patterns Report: September 2000.* Retrieved October 23, 2000 from http://www.emarketer.com/ereports/ecommerce _b2b/.

Rheingold, H. (1993). *The Virtual Community: Homesteading on the Electronic Frontier.* Cambridge, Massachusetts: Addison-Wesley Publishing Company Reading .

Robinson, J. P., M. Kestnbaum, A. Neustadl, and A. Alvavez. 2002. "The Internet and other Uses of Time." Pp. 244–262 in The Internet in Everyday Life, edited by B. Wellman and C. Haythornthwaite. Malden, MA: Blackwell Publishing.

Schmitz, J. (1997). "Structural Relations, Electronic Media, and Social Change: The Public Electronic Network and the Homeless." Pp. 80–101 in *Virtual Culture: Identity and Communication in Cybersociety*, edited by S. G. Jones London: Sage.

Spears, R. and M. Lea. (1992). "Social Influence and the Influence of the "Social" in Computer-mediated Communication." Pp. 30–65 in *Contexts of Computer-mediated Communication*, edited by M. Lea. London: Harvester-Wheatsheaf.

Tapscott, D. (1998). *Growing Up Digital: The Rise of the Net Generation.* New York: McGraw-Hill Trade.

Wellman, B. and M. Gulia. 1999. "Net Surfers Don't Ride Alone: Virtual Community as Cummunity." Pp. 331–367 in *Networks in the Global Village*, edited by Barry Wellman. Boulder; Co: Westview Press, 1999.

KEY CONCEPTS

ritual   role strain   social interaction

DISCUSSION QUESTIONS

1. How have Facebook, MySpace, and other online communication sites changed the way you interact with friends? In your opinion, is the nature of the friendship changed when communicating online as opposed to in-person?

2. What "rules" do you think exist when e-mailing? For example, is there a difference in the way you address a professor in an e-mail compared to the way you address a friend or classmate? Consider e-mail etiquette. Is there such a thing? Should there be?

Applying Sociological Knowledge:
An Exercise for Students

For a 48-hour period, keep a detailed log of every form of technology that you use for social interaction (including various forms of communication, networking, scheduling, etc.). How is this different from or similar to face-to-face interaction? Try to imagine life without some of these technologies. How do you think they have changed the character of social interaction? Can you also imagine what technologies might shape social interaction in the future?

16

Clique Dynamics

PATRICIA ADLER AND PETER ADLER

Patricia Adler and Peter Adler take a look at clique formation and friendship groupings in schools. In their study of children's friendship groups, they analyze how cliques can generate tremendous power and influence over clique members.

A dominant feature of children's lives is the clique structure that organizes their social world. The fabric of their relationships with others, their levels and types of activity, their participation in friendships, and their feelings about themselves are tied to their involvement in, around, or outside the cliques organizing their social landscape. Cliques are, at their base, friendship circles, whose members tend to identify each other as mutually connected. Yet they are more than that; cliques have a hierarchical structure, being dominated by leaders, and are exclusive in nature, so that not all individuals who desire membership are accepted. They function as bodies of power within grades, incorporating the most popular individuals, offering the most exciting social lives, and commanding the most interest and attention from classmates.... As such they represent a vibrant component of the preadolescent experience, mobilizing powerful forces that produce important effects on individuals.

The research on cliques is cast within the broader literature on elementary school children's friendship groups. A first group of such works examines independent variables that can have an influence on the character of children's friendship groups. A second group looks at the features of children's inter- and intra-group relations. A third group concentrates on the behavioral dynamics specifically associated with cliques. Although these studies are diverse in their focus, they identify several features as central to clique functioning without thoroughly investigating their role and interrelation: boundary maintenance and definitions of membership (exclusivity); a hierarchy of popularity (status stratification and differential power), and relations between in-groups and out-groups (cohesion and integration).

In this [essay] we look at these dynamics and their association, at the way clique leaders generate and maintain their power and authority (leadership, power/dominance), and at what it is that influences followers to comply so readily with clique leaders' demands (submission). These interactional dynamics

SOURCE: Patricia Adler and Peter Adler. 1998. *Peer Power: Preadolescent Culture and Identity.* New Brunswick, NJ: Rutgers University Press, pp. 56–69. Reprinted with permission.

are not intended to apply to all children's friendship groups, only those (populated by one-quarter to one-half of the children) that embody the exclusive and stratified character of cliques.

TECHNIQUES OF INCLUSION

The critical way that cliques maintained exclusivity was through careful membership screening. Not static entities, cliques irregularly shifted and evolved their membership, as individuals moved away or were ejected from the group and others took their place. In addition, cliques were characterized by frequent group activities designed to foster some individuals' inclusion (while excluding others). Cliques had embedded, although often unarticulated, modes for considering and accepting (or rejecting) potential new members. These modes were linked to the critical power of leaders in making vital group decisions. Leaders derived power through their popularity and then used it to influence membership and social stratification within the group. This stratification manifested itself in tiers and subgroups within cliques composed of people who were hierarchically ranked into levels of leaders, followers, and wannabes. Cliques embodied systems of dominance, whereby individuals with more status and power exerted control over others' lives.

Recruitment

Initial entry into cliques often occurred at the invitation or solicitation of clique members.... Those at the center of clique leadership were the most influential over this process, casting their votes for which individuals would be acceptable or unacceptable as members and then having other members of the group go along with them. If clique leaders decided they liked someone, the mere act of their friendship with that person would accord them group status and membership....

Potential members could also be brought to the group by established members who had met and liked them. The leaders then decided whether these individuals would be granted a probationary period of acceptance during which they could be informally evaluated. If the members liked them, the newcomers would be allowed to remain in the friendship circle, but if they rejected them, they would be forced to leave.

Tiffany, a popular, dominant girl, reflected on the boundary maintenance she and her best friend Diane, two clique leaders, had exercised in fifth grade:

Q: *Who defines the boundaries of who's in or who's out?*

TIFFANY: Probably the leader. If one person might like them, they might introduce them, but if one or two people didn't like them, then they'd start to get everyone up. Like in fifth grade, there was Dawn Bolton and she was new. And the girls in her class that were in our clique liked her, but Diane and I didn't like her, so we kicked her out. So then she went to the other clique, the Emily clique....

Application

A second way for individuals to gain initial membership into a clique occurred through their actively seeking entry.... Several factors influenced the likelihood that a person would be accepted as a candidate for inclusion, as Darla, a popular fourth-grade girl described: "Coming in, it's really hard coming in, it's like really hard, even if you are the coolest person, they're still like, 'What is *she* doing [exasperated]?' You can't be too pushy, and like I don't know, it's really hard to get in, even if you can. You just got to be there at the right time, when they're nice, in a nice mood."

According to Rick, a fifth-grade boy who was in the popular clique but not a central member, application for clique entry was more easily accomplished by individuals than groups. He described the way individuals found routes into cliques: "It can happen any way. Just you get respected by someone, you do something nice, they start to like you, you start doing stuff with them. It's like you just kind of follow another person who is in the clique back to the clique, and he says, 'Could this person play?' So you kind of go out with the clique for a while and you start doing stuff with them, and then they almost like invite you in. And then soon after, like a week or so, you're actually in. It all depends.... But you can't bring your whole group with you, if you have one. You have to leave them behind and just go in on your own."

Successful membership applicants often experienced a flurry of immediate popularity. Because their entry required clique leaders' approval, they gained associational status.

Friendship Realignment

Status and power in a clique were related to stratification, and people who remained more closely tied to the leaders were more popular. Individuals who wanted to be included in the clique's inner echelons often had to work regularly to maintain or improve their position.

Like initial entry, this was sometimes accomplished by people striving on their own for upward mobility. In fourth grade, Danny was brought into the clique by Mark, a longtime member, who went out of his way to befriend him. After joining the clique, however, Danny soon abandoned Mark when Brad, the clique leader, took an interest in him. Mark discussed the feelings of hurt and abandonment this experience left him with: "I felt really bad, because I made friends with him when nobody knew him and nobody liked him, and I put all my friends to the side for him, and I brought him into the group, and then he dumped me. He was my friend first, but then Brad wanted him.... He moved up and left me behind, like I wasn't good enough anymore."

The hierarchical structure of cliques, and the shifts in position and relationships within them, caused friendship loyalties within these groups to be less reliable than they might have been in other groups. People looked toward those above them and were more susceptible to being wooed into friendship with individuals more popular than they. When courted by a higher-up, they could easily drop their less popular friends....

Ingratiation

Currying favor with people in the group, like previous inclusionary endeavors, can be directed either upward (supplication) or downward (manipulation)....
Note that children often begin their attempts at entry into groups with low-risk tactics; they first try to become accepted by more peripheral members, and only later do they direct their gaze and inclusion attempts toward those with higher status. The children we observed did this as well, making friendly overtures toward clique followers and hoping to be drawn by them into the center.

The more predominant behavior among group members, however, involved currying favor with the leader to enhance their popularity and attain greater respect from other group members. One way they did this was by imitating the style and interests of the group leader. Marcus and Adam, two fifth-grade boys, described the way borderline people would fawn on their clique and its leader to try to gain inclusion:

MARCUS: Some people would just follow us around and say, "Oh yeah, whatever he says, yeah, whatever his favorite kind of music is, is my favorite kind of music."

ADAM: They're probably in a position then they want to be more in because if they like what we like, then they think more people will probably respect them. Because if some people in the clique think this person likes their favorite groups, say it's REM, or whatever, so it's say Bud's [the clique leader's], this person must know what we like in music and what's good and what's not, so let's tell him that he can come up and join us after school and do something.

Fawning on more popular people not only was done by outsiders and peripherals but was common practice among regular clique members, even those with high standing. Darla, a second-tier fourth-grade girl, ... described how, in fear, she used to follow the clique leader and parrot her opinions: "I was never mean to the people in my grade because I thought Denise might like them and then I'd be screwed. Because there were some people that I hated that she liked and I acted like I loved them, and so I would just be mean to the younger kids, and if she would even say, 'Oh she's nice,' I'd say, 'Oh yeah, she's really nice!' " Clique members, then, had to stay abreast of the leader's shifting tastes and whims if they were to maintain status and position in the group. Part of their membership work involved a regular awareness of the leader's fads and fashions, so that they would accurately align their actions and opinions with the current trends in timely manner....

TECHNIQUES OF EXCLUSION

Although inclusionary techniques reinforced individuals' popularity and prestige while maintaining the group's exclusivity and stratification, they failed to contribute to other, essential, clique features such as cohesion and integration, the management of in-group and out-group relationships, and submission to clique leadership. These features are rooted, along with further sources of domination and power, in cliques' exclusionary dynamics.

Out-Group Subjugation

When they were not being nice to try to keep outsiders from straying too far from their realm of influence, clique members predominantly subjected outsiders to exclusion and rejection. They found sport in picking on these lower-status individuals. As one clique follower remarked, "One of the main things is to keep picking on unpopular kids because it's just fun to do." [Sociologist] Eder . . . notes that this kind of ridicule, where the targets are excluded and not enjoined to participate in the laughter, contrasts with teasing, where friends make fun of each other in a more lighthearted manner but permit the targets to remain included in the group by also jokingly making fun of themselves. Diane, a clique leader in fourth grade, described the way she acted toward outsiders: "Me and my friends would be mean to the people outside of our clique. Like, Eleanor Dawson, she would always try to be friends with us, and we would be like, 'Get away, ugly.'"

Interactionally sophisticated clique members not only treated outsiders badly but managed to turn others in the clique against them. Parker and Gottman . . . observe that one of the ways people do this is through gossip. Diane recalled the way she turned all the members of her class, boys as well as girls, against an outsider: "I was always mean to people outside my group like Crystal, and Sally Jones; they both moved schools. . . . I had this gummy bear necklace, with pearls around it and gummy bears. She [Crystal] came up to me one day and pulled my necklace off. I'm like, 'It was my favorite necklace,' and I got all of my friends, and all the guys even in the class, to revolt against her. No one liked her. That's why she moved schools, because she tore my gummy bear necklace off and everyone hated her. They were like, 'That was mean. She didn't deserve that. We hate you.'" . . .

In-Group Subjugation

Picking on people within the clique's confines was another way to exert dominance. More central clique members commonly harassed and were mean to those with weaker standing. Many of the same factors prompting the ill treatment of outsiders motivated high-level insiders to pick on less powerful insiders. Rick, a fifth-grade clique follower, articulated the systematic organization of downward harassment: "Basically the people who are the most popular, their life outside in the playground is picking on other people who aren't as popular, but are in the group. But the people just want to be more popular so they stay in the group, they just kind of stick with it, get made fun of, take it. . . . They come back everyday, you do more ridicule, more ridicule, more ridicule, and they just keep taking it because they want to be more popular, and they actually like you but you don't like them. That goes on a lot, that's the main thing in the group. You make fun of someone, you get more popular, because insults is what they like, they like insults."

The finger of ridicule could be pointed at any individual but the leader. It might be a person who did something worthy of insult, it might be someone who the clique leader felt had become an interpersonal threat, or it might be someone singled out for no apparent reason. . . . Darla, the second tier fourth grader discussed earlier, described the ridicule she encountered and her feelings of mortification when the clique leader derided her hair: "Like I remember, she embarrassed me so bad one day. Oh my God, I wanted to kill her! We were in music class and we were standing there and she goes, 'Ew! what's all that shit in your hair?' in front of the whole class. I was so embarrassed, 'cause, I guess I had dandruff or something."

Often, derision against insiders followed a pattern, where leaders started a trend and everyone followed it. This intensified the sting of the mockery by compounding it with multiple force. Rick analogized the way people in cliques behaved to the links on a chain: "Like it's a chain reaction, you get in a fight with the main person, then the person right under him will not like you, and the person under him won't like you, and et cetera, and the whole group will take turns against you. A few people will still like you because they will do their own thing, but most people will do what the person in front of them says to do, so it would be like a chain reaction. It's like a chain; one chain turns, and the other chain has to turn with them or else it will tangle."

Compliance

Going along with the derisive behavior of leaders or other high-status clique members could entail either active or passive participation. Active participation occurred when instigators enticed other clique members to pick on their friends. For example, leaders would often come up with the idea of placing phony phone calls to others and would persuade their followers to do the dirty work. They might start the phone call and then place followers on the line to finish it, or they might pressure others to make the entire call, thus keeping one step distant from becoming implicated, should the victim's parents complain.

Passive participation involved going along when leaders were mean and manipulative, as when Trevor submissively acquiesced in Brad's scheme to convince Larry that Rick had stolen his money. Trevor knew that Brad was hiding the money the whole time, but he watched while Brad whipped Larry into a frenzy, pressing him to deride Rick, destroy Rick's room and possessions, and threaten to expose Rick's alleged theft to others. It was only when Rick's mother came home, interrupting the bedlam, that she uncovered the money and stopped Larry's onslaught. The following day at school, Brad and Trevor could scarcely contain their glee. As noted earlier, Rick was demolished by the incident and cast out by the clique; Trevor was elevated to the status of Brad's best friend by his coconspiracy in the scheme. . . .

Stigmatization

Beyond individual incidents of derision, clique insiders were often made the focus of stigmatization for longer periods of time. Unlike outsiders who commanded less enduring interest, clique members were much more involved in picking on their friends, whose discomfort more readily held their attention. Rick noted that the duration of this negative attention was highly variable: "Usually at certain times, it's just a certain person you will pick on all the time, if they do something wrong. I've been picked on for a month at a time, or a week, or a day, or just a couple of minutes, and then they will just come to respect you again." When people became the focus of stigmatization, as happened to Rick, they were rejected by all their friends. The entire clique rejoiced in celebrating their disempowerment. They would be made to feel alone whenever possible. Their former friends might join hands and walk past them through the play yard at recess, physically demonstrating their union and the discarded individual's aloneness.

Worse than being ignored was being taunted. Taunts ranged from verbal insults to put-downs to singsong chants. Anyone who could create a taunt was favored with attention and imitated by everyone. Even outsiders, who would not normally be privileged to pick on a clique member, were able to elevate themselves by joining in on such taunting. . . .

The ultimate degradation was physical. Although girls generally held themselves to verbal humiliation of their members, the culture of masculinity gave credence to boys' injuring each other. . . . Fights would occasionally break out in which boys were punched in the ribs or stomach, kicked, or given black eyes. When this happened at school, adults were quick to intervene. But after hours or on the school bus boys could be hurt. Physical abuse was also heaped on people's homes or possessions. People spit on each other or others' books or toys, threw eggs at their family's cars, and smashed pumpkins in front of their house.

Expulsion

While most people returned to a state of acceptance following a period of severe derision . . . this was not always the case. Some people became permanently excommunicated from the clique. Others could be cast out directly, without undergoing a transitional phase of relative exclusion. Clique members from any stratum of the group could suffer such a fate, although it was more common among people with lower status.

When Davey, mentioned earlier, was in sixth grade, he described how expulsion could occur as a natural result of the hierarchical ranking, where a person at the bottom rung of the system of popularity was pushed off. He described the ordinary dynamics of clique behavior:

Q: *How do clique members decide who they are going to insult that day?*

DAVEY: It's just basically everyone making fun of everyone. The small people making fun of smaller people, the big people making fun of the small people. Nobody is really making fun of people bigger than them because they can get rejected, because then they can say, "Oh yes, he did this and that, this and that, and we shouldn't like him anymore." And everybody else says, "Yeah, yeah, yeah," 'cause all the lower people like him, but all the higher people don't. So the lowercase people just follow the highercase people. If one person is doing something wrong, then they will say, "Oh yeah, get out, good-bye." . . .

KEY CONCEPTS

clique   ingroup   outgroup

DISCUSSION QUESTIONS

1. Take a look at your own friendship group in school. Which of the processes of both inclusion and exclusion do you observe?
2. What forms of negative sanction, or punishment, do the more powerful high status clique members deliver to others? List some, noting how they differ in severity.

17

Sexual Assault on Campus

A Multilevel, Integrative Approach to Party Rape

ELIZABETH A. ARMSTRONG, LAURA HAMILTON,
AND BRIAN SWEENEY

In this article the authors discuss their research of a "party dorm" at a large university. The research uncovers patterns of gendered behavior that contribute to the risk of sexual assault during parties, specifically at fraternities. College women are struggling to find a balance between having fun and being in danger. The authors argue that parties are structured in such a way that puts men in control. Strategies that teach women to simply be careful fall short in their efforts to prevent sexual assault.

A 1997 National Institute of Justice study estimated that between one-fifth and one-quarter of women are the victims of completed or attempted rape while in college (Fisher, Cullen, and Turner 2000). College women "are at

SOURCE: Elizabeth A. Armstrong, Laura Hamilton, and Brian Sweeney.
"Sexual Assault on Campus: A Multilevel, Integrative Approach to Party
Rape." *Social Problems* 53 (2006): 483–499.

greater risk for rape and other forms of sexual assault than women in the general population or in a comparable age group" (Fisher et al. 2000:iii). At least half and perhaps as many as three-quarters of the sexual assaults that occur on college campuses involve alcohol consumption on the part of the victim, the perpetrator, or both (Abbey et al. 1996; Sampson 2002). The tight link between alcohol and sexual assault suggests that many sexual assaults that occur on college campuses are "party rapes." A recent report by the U.S. Department of Justice defines party rape as a distinct form of rape, one that "occurs at an off-campus house or on- or off-campus fraternity and involves ... plying a woman with alcohol or targeting an intoxicated woman" (Sampson 2002:6). While party rape is classified as a form of acquaintance rape, it is not uncommon for the woman to have had no prior interaction with the assailant, that is, for the assailant to be an in-network stranger (Abbey et al. 1996).

Colleges and universities have been aware of the problem of sexual assault for at least 20 years, directing resources toward prevention and providing services to students who have been sexually assaulted. Programming has included education of various kinds, support for *Take Back the Night* events, distribution of rape whistles, development and staffing of hotlines, training of police and administrators, and other efforts. Rates of sexual assault, however, have not declined over the last five decades (Adams-Curtis and Forbes 2004:95; Bachar and Koss 2001; Marine 2004; Sampson 2002:1).

Why do colleges and universities remain dangerous places for women in spite of active efforts to prevent sexual assault? While some argue that "we know what the problems are and we know how to change them" (Adams-Curtis and Forbes 2004:115), it is our contention that we do not have a complete explanation of the problem. To address this issue we use data from a study of college life at a large midwestern university and draw on theoretical developments in the sociology of gender (Connell 1987, 1995; Lorber 1994; Martin 2004; Risman 1998, 2004). Continued high rates of sexual assault can be viewed as a case of the reproduction of gender inequality—a phenomenon of central concern in gender theory.

We demonstrate that sexual assault is a predictable outcome of a synergistic intersection of both gendered and seemingly gender neutral processes operating at individual, organizational, and interactional levels. The concentration of homogenous students with expectations of partying fosters the development of sexualized peer cultures organized around status. Residential arrangements intensify students' desires to party in male-controlled fraternities. Cultural expectations that partygoers drink heavily and trust party-mates become problematic when combined with expectations that women be nice and defer to men. Fulfilling the role of the partier produces vulnerability on the part of women, which some men exploit to extract non-consensual sex. The party scene also produces fun, generating student investment in it. Rather than criticizing the party scene or men's behavior, students blame victims. By revealing mechanisms that lead to the persistence of sexual assault and outlining implications for policy, we hope to encourage colleges and universities to develop fresh approaches to sexual assault prevention.

APPROACHES TO COLLEGE SEXUAL ASSAULT

Explanations of high rates of sexual assault on college campuses fall into three broad categories. The first tradition, a psychological approach that we label the "individual determinants" approach, views college sexual assault as primarily a consequence of perpetrator or victim characteristics such as gender role attitudes, personality, family background, or sexual history (Flezzani and Benshoff 2003; Forbes and Adams-Curtis 2001; Rapaport and Burkhart 1984). While "situational variables" are considered, the focus is on individual characteristics (Adams-Curtis and Forbes 2004; Malamuth, Heavey, and Linz 1993). For example, Antonia Abbey and associates (2001) find that hostility toward women, acceptance of verbal pressure as a way to obtain sex, and having many consensual sexual partners distinguish men who sexually assault from men who do not. Research suggests that victims appear quite similar to other college women (Kalof 2000), except that white women, prior victims, first-year college students, and more sexually active women are more vulnerable to sexual assault (Adams-Curtis and Forbes 2004; Humphrey and White 2000).

The second perspective, the "rape culture" approach, grew out of second wave feminism (Brownmiller 1975; Buchward, Fletcher, and Roth 1993; Lottes 1997; Russell 1975; Schwartz and DeKeseredy 1997). In this perspective, sexual assault is seen as a consequence of widespread belief in "rape myths," or ideas about the nature of men, women, sexuality, and consent that create an environment conducive to rape. For example, men's disrespectful treatment of women is normalized by the idea that men are naturally sexually aggressive. Similarly, the belief that women "ask for it" shifts responsibility from predators to victims (Herman 1989; O'Sullivan 1993). This perspective initiated an important shift away from individual beliefs toward the broader context. However, rape supportive beliefs alone cannot explain the prevalence of sexual assault, which requires not only an inclination on the part of assailants but also physical proximity to victims (Adams-Curtis and Forbes 2004:103).

A third approach moves beyond rape culture by identifying particular contexts—fraternities and bars—as sexually dangerous (Humphrey and Kahn 2000; Martin and Hummer 1989; Sanday 1990, 1996; Stombler 1994). Ayres Boswell and Joan Spade (1996) suggest that sexual assault is supported not only by "a generic culture surrounding and promoting rape," but also by characteristics of the "specific settings" in which men and women interact (p. 133). Mindy Stombler and Patricia Yancey Martin (1994) illustrate that gender inequality is institutionalized on campus by "formal structure" that supports and intensifies an already "high-pressure heterosexual peer group" (p. 180). This perspective grounds sexual assault in organizations that provide opportunities and resources.

We extend this third approach by linking it to recent theoretical scholarship in the sociology of gender. Martin (2004), Barbara Risman (1998; 2004), Judith Lorber (1994) and others argue that gender is not only embedded in individual selves, but also in cultural rules, social interaction, and organizational arrangements. This integrative perspective identifies mechanisms at each level that contribute to the reproduction of gender inequality (Risman 2004).

Socialization processes influence gendered selves, while cultural expectations reproduce gender inequality in interaction. At the institutional level, organizational practices, rules, resource distributions, and ideologies reproduce gender inequality. Applying this integrative perspective enabled us to identify gendered processes at individual, interactional, and organizational levels that contribute to college sexual assault.

Risman (1998) also argues that gender inequality is reproduced when the various levels are "all consistent and interdependent" (p. 35). Processes at each level depend upon processes at other levels. Below we demonstrate how interactional processes generating sexual danger depend upon organizational resources and particular kinds of selves. We show that sexual assault results from the intersection of processes at all levels.

We also find that not all of the processes contributing to sexual assault are explicitly gendered. For example, characteristics of individuals such as age, class, and concern with status play a role. Organizational practices such as residence hall assignments and alcohol regulation, both intended to be gender neutral, also contribute to sexual danger. Our findings suggest that apparently gender neutral social processes may contribute to gender inequality in other situations.

METHOD

Data are from group and individual interviews, ethnographic observation, and publicly available information collected at a large midwestern research university. Located in a small city, the school has strong academic and sports programs, a large Greek system, and is sought after by students seeking a quintessential college experience. Like other schools, this school has had legal problems as a result of deaths associated with drinking. In the last few years, students have attended a sexual assault workshop during first-year orientation. Health and sexuality educators conduct frequent workshops, student volunteers conduct rape awareness programs, and *Take Back the Night* marches occur annually.

The bulk of the data presented in this paper were collected as part of ethnographic observation during the 2004–05 academic year in a residence hall identified by students and residence hall staff as a "party dorm." While little partying actually occurs in the hall, many students view this residence hall as one of several places to live in order to participate in the party scene on campus. This made it a good place to study the social worlds of students at high risk of sexual assault—women attending fraternity parties in their first year of college. The authors and a research team were assigned to a room on a floor occupied by 55 women students (51 first-year, 2 second-year, 1 senior, and 1 resident assistant [RA]). We observed on evenings and weekends throughout the entire academic school year. We collected in-depth background information via a detailed nine-page survey that 23 women completed and conducted interviews with 42 of the women (ranging from 1¼ to 2½ hours). All but seven of the women on the floor completed either a survey or an interview.

With at least one-third of first-year students on campus residing in "party dorms" and one-quarter of all undergraduates belonging to fraternities or sororities, this social world is the most visible on campus. As the most visible scene on campus, it also attracts students living in other residence halls and those not in the Greek system. Dense pre-college ties among the many in-state students, class and race homogeneity, and a small city location also contribute to the dominance of this scene. Of course, not all students on this floor or at this university participate in the party scene. To participate, one must typically be heterosexual, at least middle class, white, American-born, unmarried, childless, traditional college age, politically and socially mainstream, and interested in drinking. Over three-quarters of the women on the floor we observed fit this description.

There were no non-white students among the first and second year students on the floor we studied. This is a result of the homogeneity of this campus and racial segregation in social and residential life. African Americans (who make up 3 to 5% of undergraduates) generally live in living-learning communities in other residence halls and typically do not participate in the white Greek party scene. We argue that the party scene's homogeneity contributes to sexual risk for white women. We lack the space and the data to compare white and African American party scenes on this campus, but in the discussion we offer ideas about what such a comparison might reveal.

We also conducted 16 group interviews (involving 24 men and 63 women) in spring 2004. These individuals had varying relationships to the white Greek party scene on campus. Groups included residents of an alternative residence hall, lesbian, gay, and bisexual students, feminists, re-entry students, academically-focused students, fundamentalist Christians, and sorority women. Eight group interviews were exclusively women, five were mixed in gender composition, and three were exclusively men. The group interviews covered a variety of topics, including discussions of social life, the transition to college, sexual assault, relationships, and the relationship between academic and social life. Participants completed a shorter version of the survey administered to the women on the residence hall floor. From these students we developed an understanding of the dominance of this party scene.

We also incorporated publicly available information about the university from informal interviews with student affairs professionals and from teaching (by all authors) courses on gender, sexuality, and introductory sociology. Classroom data were collected through discussion, student writings, e-mail correspondence, and a survey that included questions about experiences of sexual assault. . . .

EXPLAINING PARTY RAPE

We show how gendered selves, organizational arrangements, and interactional expectations contribute to sexual assault. We also detail the contributions of processes at each level that are not explicitly gendered. We focus on each level in turn, while attending to the ways in which processes at all levels depend upon and reenforce others. We show that fun is produced along with sexual assault, leading students to resist criticism of the party scene.

Selves and Peer Culture in the Transition from High School to College

Student characteristics shape not only individual participation in dangerous party scenes and sexual risk within them but the development of these party scenes. We identify individual characteristics (other than gender) that generate interest in college partying and discuss the ways in which gendered sexual agendas generate a peer culture characterized by high-shakes competition over erotic status.

Non-Gendered Characteristics Motivate Participation in Party Scenes
Without individuals available for partying, the party scene would not exist. All the women on our floor were single and childless, as are the vast majority of undergraduates at this university; many, being upper-middle class, had few responsibilities other than their schoolwork. Abundant leisure time, however, is not enough to fuel the party scene. Media, siblings, peers, and parents all serve as sources of anticipatory socialization (Merton 1957). Both partiers and non-partiers agreed that one was "supposed" to party in college. This orientation was reflected in the popularity of a poster titled "What I Really Learned in School" that pictured mixed drinks with names associated with academic disciplines. As one focus group participant explained,

> You see these images of college that you're supposed to go out and have fun and drink, drink lots, party and meet guys. [You are] supposed to hook up with guys, and both men and women try to live up to that. I think a lot of it is girls want to be accepted into their groups and guys want to be accepted into their groups.

Partying is seen as a way to feel a part of college life. Many of the women we observed participated in middle and high school peer cultures organized around status, belonging, and popularity (Eder 1985; Eder, Evans, and Parker 1995; Milner 2004). Assuming that college would be similar, they told us that they wanted to fit in, be popular, and have friends. . . .

Peer Culture as Gendered and Sexualized   Partying was also the primary way to meet men on campus. The floor was locked to non-residents, and even men living in the same residence hall had to be escorted on the floor. The women found it difficult to get to know men in their classes, which were mostly mass lectures. They explained to us that people "don't talk" in class. Some complained they lacked casual friendly contact with men, particularly compared to the mixed-gender friendship groups they reported experiencing in high school.

Meeting men at parties was important to most of the women on our floor. The women found men's sexual interest at parties to be a source of self-esteem and status. They enjoyed dancing and kissing at parties, explaining to us that it proved men "liked" them. This attention was not automatic, but required the skillful deployment of physical and cultural assets (Stombler and Padavic 1997; Swidler 2001). Most of the party-oriented women on the floor arrived with appropriate gender presentations and the money and know-how to preserve and refine them. While some more closely resembled the "ideal" college party girl (white, even features, thin but busty, tan, long straight hair, skillfully made-up, and well-dressed in the latest youth styles), most worked hard to attain this presentation. They regularly straightened their hair, tanned, exercised, dieted, and purchased new clothes. . . .

The psychological benefits of admiration from men in the party scene were such that women in relationships sometimes felt deprived. One woman with a serious boyfriend noted that she dressed more conservatively at parties because of him, but this meant she was not "going to get any of the attention." She lamented that no one was "going to waste their time with me" and that, "this is taking away from my confidence." Like most women who came to college with boyfriends, she soon broke up with him.

Men also sought proof of their erotic appeal. As a woman complained, "Every man I have met here has wanted to have sex with me!" . . . The women found that men were more interested than they were in having sex. These clashes in sexual expectations are not surprising: men derived status from securing sex (from high-status women), while women derived status from getting attention (from high-status men). These agendas are both complementary and adversarial: men give attention to women en route to getting sex, and women are unlikely to become interested in sex without getting attention first.

University and Greek Rules, Resources, and Procedures

Simply by congregating similar individuals, universities make possible hetero-sexual peer cultures. The university, the Greek system, and other related organizations structure student life through rules, distribution of resources, and procedures (Risman 2004).

Sexual danger is an unintended consequence of many university practices intended to be gender neutral. The clustering of homogeneous students intensi-fies the dynamics of student peer cultures and heightens motivations to party. Characteristics of residence halls and how they are regulated push student party-ing into bars, off-campus residences, and fraternities. While factors that increase the risk of party rape are present in varying degrees in all party venues (Boswell and Spade 1996), we focus on fraternity parties because they were the typical party venue for the women we observed and have been identified as particularly unsafe (see also Martin and Hummer 1989; Sanday 1990). Fraternities offer the most reliable and private source of alcohol for first-year students excluded from bars and house parties because of age and social networks.

University Practices as Push Factors   The university has latitude in how it enforces state drinking laws. Enforcement is particularly rigorous in residence halls. We observed RAs and police officers (including gun-carrying peer police) patrolling the halls for alcohol violations. Women on our floor were "documented" within the first week of school for infractions they felt were minor. Sanctions are severe—a $300 fine, an 8-hour alcohol class, and probation for a year. As a consequence, students engaged in only minimal, clandestine alcohol consumption in their rooms. In comparison, alcohol flows freely at fraternities.

The lack of comfortable public space for informal socializing in the residence hall also serves as a push factor. A large central bathroom divided our floor. A sterile lounge was rarely used for socializing. There was no cafeteria, only a convenience store and a snack bar in a cavernous room furnished with big-screen televisions. Residence life sponsored alternatives to the party scene such as "movie night" and special dinners, but these typically occurred early in the evening. Students defined the few activities sponsored during party hours (e.g., a midnight trip to Wal-Mart) as uncool.

Intensifying Peer Dynamics The residence halls near athletic facilities and Greek houses are known by students to house affluent, party-oriented students. White, upper-middle class, first-year students who plan to rush request these residence halls, while others avoid them. One of our residents explained that "everyone knows what [the residence hall] is like and people are dying to get in here. People just think it's a total party or something." Students of color tend to live elsewhere on campus. As a consequence, our floor was homogenous in terms of age, race, sexual orientation, class, and appearance. . . .

The homogeneity of the floor intensified social anxiety, heightening the importance of partying for making friends. Early in the year, the anxiety was palpable on weekend nights as women assessed their social options by asking where people were going, when, and with whom. One exhausted floor resident told us she felt that she "needed to" go out to protect her position in a friendship group. At the beginning of the semester, "going out" on weekends was virtually compulsory. By 11 p.m. the floor was nearly deserted.

Male Control of Fraternity Parties The campus Greek system cannot operate without university consent. The university lists Greek organizations as student clubs, devotes professional staff to Greek-oriented programming, and disbands fraternities that violate university policy. Nonetheless, the university lacks full authority over fraternities; Greek houses are privately owned and chapters answer to national organizations and the Interfraternity Council (IFC) (i.e., a body governing the more than 20 predominantly white fraternities).

Fraternities control every aspect of parties at their houses: themes, music, transportation, admission, access to alcohol, and movement of guests. Party themes usually require women to wear scant, sexy clothing and place women in subordinate positions to men. During our observation period, women attended parties such as "Pimps and Hos," "Victoria's Secret," and "Playboy Mansion"—the last of which required fraternity members to escort two scantily-clad dates. Other recent themes included: "CEO/Secretary Ho," "School Teacher/Sexy Student," and "Golf Pro/Tennis Ho."

Some fraternities require pledges to transport first-year students, primarily women, from the residence halls to the fraternity houses. From about 9 to 11 p.m. on weekend nights early in the year, the drive in front of the residence hall resembled a rowdy taxi-stand, as dressed-to-impress women waited to be carpooled to parties in expensive late-model vehicles. By allowing party-oriented first-year women to cluster in particular residence halls, the university made them easy to find. One fraternity member told us this practice was referred to as "dorm-storming."

Transportation home was an uncertainty. Women sometimes called cabs, caught the "drunk bus," or trudged home in stilettos. Two women indignantly described a situation where fraternity men "wouldn't give us a ride home." The women said, "Well, let us call a cab." The men discouraged them from calling the cab and eventually found a designated driver. The women described the men as "just dicks" and as "rude."

Fraternities police the door of their parties, allowing in desirable guests (first-year women) and turning away others (unaffiliated men). Women told us of abandoning parties when male friends were not admitted. They explained that fraternity men also controlled the quality and quantity of alcohol. Brothers served themselves first, then personal guests, and then other women. Non-affiliated and unfamiliar men were served last, and generally had access to only the least desirable beverages. The promise of more or better alcohol was often used to lure women into private spaces of the fraternities.

Fraternities are constrained, though, by the necessity of attracting women to their parties. Fraternities with reputations for sexual disrespect have more success recruiting women to parties early in the year. One visit was enough for some of the women. A roommate duo told of a house they "liked at first" until they discovered that the men there were "really not nice."

The Production of Fun and Sexual Assault in Interaction

Peer culture and organizational arrangements set up risky partying conditions, but do not explain *how* student interactions at parties generate sexual assault. At the interactional level we see the mechanisms through which sexual assault is produced. As interactions necessarily involve individuals with particular characteristics and occur in specific organizational settings, all three levels meet when interactions take place. Here, gendered and gender neutral expectations and routines are intricately woven together to create party rape. Party rape is the result of fun situations that shift—either gradually or quite suddenly—into coercive situations. Demonstrating how the production of fun is connected with sexual assault requires describing the interactional routines and expectations that enable men to employ coercive sexual strategies with little risk of consequence. . . .

Cultural expectations of partying are gendered. Women are supposed to wear revealing outfits, while men typically are not. As guests, women cede control of turf, transportation, and liquor. Women are also expected to be grateful for men's hospitality, and as others have noted, to generally be "nice" in ways that men are not (Gilligan 1982; Martin 2003; Phillips 2000; Stombler and Martin 1994; Tolman 2002). The pressure to be deferential and gracious may be intensified by men's older age and fraternity membership. The quandary for women, however, is that fulfilling the gendered role of partier makes them vulnerable to sexual assault.

Women's vulnerability produces sexual assault only if men exploit it. Too many men are willing to do so. Many college men attend parties looking for casual sex. A student in one of our classes explained that "guys are willing to do damn near anything to get a piece of ass." A male student wrote the following description of parties at his (non-fraternity) house:

> Girls are continually fed drinks of alcohol. It's mainly to party but my roomies are also aware of the inhibition-lowering effects. I've seen an old roomie block doors when girls want to leave his room; and other times I've driven women home who can't remember much of an evening yet sex did occur. Rarely if ever has a night of drinking for my roommate ended without sex. I know it isn't necessarily and assuredly sexual assault, but with the amount of liquor in the house I question the amount of consent a lot.

Another student—after deactivating—wrote about a fraternity brother "telling us all at the chapter meeting about how he took this girl home and she was obviously too drunk to function and he took her inside and had sex with her." Getting women drunk, blocking doors, and controlling transportation are common ways men try to prevent women from leaving sexual situations. Rape culture beliefs, such as the belief that men are "naturally" sexually aggressive, normalize these coercive strategies. Assigning women the role of sexual "gatekeeper" relieves men from responsibility for obtaining authentic consent, and enables them to view sex obtained by undermining women's ability to resist it as "consensual" (e.g., by getting women so drunk that they pass out)....

We heard many stories of negative experiences in the party scene including at least one account of a sexual assault in every focus group that included heterosexual women. Most women who partied complained about men's efforts to control their movements or pressure them to drink. Two of the women on our floor were sexually assaulted at a fraternity party in the first week of school—one was raped. Later in the semester, another woman on the floor was raped by a friend. A fourth woman on the floor suspects she was drugged; she became disoriented at a fraternity party and was very ill for the next week.

Party rape is accomplished without the use of guns, knives, or fists. It is carried out through the combination of low level forms of coercion—a lot of liquor and persuasion, manipulation of situations so that women cannot leave, and sometimes force (e.g., by blocking a door, or using body weight to make it difficult for a woman to get up). These forms of coercion are made more effective by organizational arrangements that provide men with control over how partying happens and by expectations that women let loose and trust their party-mates. This systematic and effective method of extracting non-consensual sex is largely invisible, which makes it difficult for victims to convince anyone—even themselves—that a crime occurred. Men engage in this behavior with little risk of consequences.

Student Responses and the Resiliency of the Party Scene

The frequency of women's negative experiences in the party scene poses a problem for those students most invested in it. Finding fault with the party scene potentially threatens meaningful identities and lifestyles. The vast majority of heterosexual encounters at parties are fun and consensual. Partying provides a chance to meet new people, experience and display belonging, and to enhance social position. Women on our floor told us that they loved to flirt and be admired, and they displayed pictures on walls, doors, and websites commemorating their fun nights out.

The most common way that students—both women and men—account for the harm that befalls women in the party scene is by blaming victims. By attributing bad experiences to women's "mistakes," students avoid criticizing the party scene or men's behavior within it. Such victim-blaming also allows women to feel that they can control what happens to them. The logic of victim-blaming suggests that sophisticated, smart, careful women are safe from sexual assault. Only "immature," "naïve," or "stupid" women get in trouble. When discussing the sexual assault of a friend, a floor resident explained that:

> She somehow got like sexually assaulted . . . by one of our friends' old roommates. All I know is that kid was like bad news to start off with. So, I feel sorry for her but it wasn't much of a surprise for us. He's a shady character.

Another floor resident relayed a sympathetic account of a woman raped at knife point by a stranger in the bushes, but later dismissed party rape as nothing to worry about "'cause I'm not stupid when I'm drunk." Even a feminist focus group participant explained that her friend who was raped "made every single mistake and almost all of them had to with alcohol. . . . She got ridiculed when she came out and said she was raped." These women contrast "true victims" who are deserving of support with "stupid" women who forfeit sympathy (Phillips 2000). Not only is this response devoid of empathy for other women, but it also leads women to blame themselves when they are victimized (Phillips 2000).

Sexual assault prevention strategies can perpetuate victim-blaming. Instructing women to watch their drinks, stay with friends, and limit alcohol consumption implies that it is women's responsibility to avoid "mistakes" and their fault if they fail. Emphasis on the precautions women should take—particularly if not accompanied by education about how men should change their behavior—may also suggest that it is natural for men to drug women and take advantage of them. Additionally, suggesting that women should watch what they drink, trust partymates, or spend time alone with men asks them to forgo full engagement in the pleasures of the college party scene. . . .

Opting Out   While many students find the party scene fun, others are more ambivalent. Some attend a few fraternity parties to feel like they have participated in this college tradition. Others opt out of it altogether. On our floor, 44 out of the 51 first-year students (almost 90%) participated in the party scene. Those on the floor who opted out worried about sexual safety and the consequences of engaging in illegal behavior. For example, an interviewee who did not drink was appalled by the fraternity party transport system. She explained that:

> All those girls would stand out there and just like, no joke, get into these big black Suburbans driven by frat guys, wearing like seriously no clothes, piled on top of each other. This could be some kidnapper taking you all away to the woods and chopping you up and leaving you there. How dumb can you be?

In her view, drinking around fraternity men was "scary" rather than "fun."

Her position was unpopular. She, like others who did not party, was an outsider on the floor. Partiers came home loudly in the middle of the night, threw up in the bathrooms, and rollerbladed around the floor. Socially, the others simply did not exist. A few of our "misfits" successfully created social lives outside the floor. The most assertive of the "misfits" figured out the dynamics of the floor in the first weeks and transferred to other residence halls.

However, most students on our floor lacked the identities or network connections necessary for entry into alternative worlds. Life on a large university campus can be overwhelming for first-year students. Those who most needed an alternative to the social world of the party dorm were often ill-equipped to actively seek it out. They either integrated themselves into partying or found themselves alone in their rooms, microwaving frozen dinners and watching television. A Christian focus group participant described life in this residence hall: "When everyone is going out on a Thursday and you are in the room by yourself and there are only two or three other people on the floor, that's not fun, it's not the college life that you want."

DISCUSSION AND IMPLICATIONS

We have demonstrated that processes at individual, organizational, and interactional levels contribute to high rates of sexual assault. Some individual level characteristics that shape the likelihood of a sexually dangerous party scene developing are not explicitly gendered. Party rape occurs at high rates in places that cluster young, single, party-oriented people concerned about social status. Traditional beliefs about sexuality also make it more likely that one will participate in the party scene and increase danger within the scene. This university contributes to sexual danger by allowing these individuals to cluster.

However, congregating people is not enough, as parties cannot be produced without resources (e.g., alcohol and a viable venue) that are difficult for underage students to obtain. University policies that are explicitly gender-neutral—such as the policing of alcohol use in residence halls—have gendered consequences. This policy encourages first-year students to turn to fraternities to party. Only fraternities, not sororities, are allowed to have parties, and men structure parties in ways that control the appearance, movement, and behavior of female guests. Men also control the distribution of alcohol and use its scarcity to engineer social interactions. The enforcement of alcohol policy by both university and Greek organizations transforms alcohol from a mere beverage into an unequally distributed social resource.

Individual characteristics and institutional practices provide the actors and contexts in which interactional processes occur. We have to turn to the interactional level, however, to understand *how* sexual assault is generated. Gender neutral expectations to "have fun," lose control, and trust one's party-mates become problematic when combined with gendered interactional expectations. Women are expected to be "nice" and to defer to men in interaction. This expectation is intensified by men's position as hosts and women's as grateful guests. The heterosexual script, which directs men to pursue sex and women to play the role of gatekeeper, further disadvantages women, particularly when virtually *all* men's methods of extracting sex are defined as legitimate. . . .

Our analysis also provides a framework for analyzing the sources of sexual risk in non-university partying situations. Situations where men have a home turf advantage, know each other better than the women present know each other, see the women as anonymous, and control desired resources (such as alcohol or drugs) are likely to be particularly dangerous. Social pressures to "have fun," prove one's social competency, or adhere to traditional gender expectations are also predicted to increase rates of sexual assault within a social scene.

This research has implications for policy. The interdependence of levels means that it is difficult to enact change at one level when the other levels remain unchanged. . . . Without change in institutional arrangements, efforts to change cultural beliefs are undermined by the cultural commonsense generated by encounters with institutions. Efforts to educate about sexual assault will not succeed if the university continues to support organizational arrangements that facilitate and even legitimate men's coercive sexual strategies. Thus, our research implies that efforts to combat sexual assault on campus should target all levels, constituencies, and processes simultaneously. Efforts to educate both men and women should indeed be intensified, but they should be reinforced by changes in the social organization of student life.

Researchers focused on problem drinking on campus have found that reduction efforts focused on the social environment are successful (Berkowitz 2003:21). Student body diversity has been found to decrease binge drinking on campus (Weschsler and Kuo 2003); it might also reduce rates of sexual assault. Existing student heterogeneity can be exploited by eliminating self-selection into age-segregated, white, upper-middle class, heterosexual enclaves and by working to make residence halls more appealing to upper-division students. Building more aesthetically appealing housing might allow students to interact outside of alcohol-fueled party scenes. Less expensive plans might involve creating more living-learning communities, coffee shops, and other student-run community spaces.

While heavy alcohol use is associated with sexual assault, not all efforts to regulate student alcohol use contribute to sexual safety. Punitive approaches sometimes heighten the symbolic significance of drinking, lead students to drink more hard liquor, and push alcohol consumption to more private and thus more dangerous spaces. Regulation inconsistently applied—e.g., heavy policing of residence halls and light policing of fraternities—increases the power of those who can secure alcohol and host parties. More consistent regulation could decrease the value of alcohol as a commodity by equalizing access to it.

Sexual assault education should shift in emphasis from educating women on preventative measures to educating both men and women about the coercive behavior of men and the sources of victim-blaming. Mohler-Kuo and associates (2004) suggest, and we endorse, a focus on the role of alcohol in sexual assault. Education should begin before students arrive on campus and continue throughout college. It may also be most effective if high-status peers are involved in disseminating knowledge and experience to younger college students.

Change requires resources and cooperation among many people. Efforts to combat sexual assault are constrained by other organizational imperatives. Student investment in the party scene makes it difficult to enlist the support of even those most harmed by the state of affairs. Student and alumni loyalty to partying (and the Greek system) mean that challenges to the party scene could potentially cost universities tuition dollars and alumni donations. Universities must contend with Greek organizations and bars, as well as the challenges of internal coordination. Fighting sexual assault on all levels is critical, though, because it is unacceptable for higher education institutions to be sites where women are predictably sexually victimized.

REFERENCES

Abbey, Antonia, Pam McAuslan, Tina Zawacki, A. Monique Clinton, and Philip Buck. 2001. "Attitudinal, Experiential, and Situational Predictors of Sexual Assault Perpetration." *Journal of Interpersonal Violence* 16:784–807.

Abbey, Antonia, Lisa Thomson Ross, Donna McDuffie, and Pam McAuslan. 1996. "Alcohol and Dating Risk Factors for Sexual Assault among College Women." *Psychology of Women Quarterly* 20:147–69.

Adams-Curtis, Leah and Gordon Forbes. 2004. "College Women's Experiences of Sexual Coercion: A Review of Cultural, Perpetrator, Victim, and Situational Variables." *Trauma, Violence, and Abuse: A Review Journal* 5:91–122.

Bachar, Karen and Mary Koss. 2001. "From Prevalence to Prevention: Closing the Gap between What We Know about Rape and What We Do." Pp. 117–42 in *Sourcebook on Violence against Women*, edited by C. Renzetti, J. Edleson, and R. K. Bergen. Thousand Oaks, CA: Sage.

Berkowitz, Alan. 2003. "How Should We Talk about Student Drinking—And What Should We Do about It?" *About Campus* May/June:16–22.

Boswell, A. Ayres and Joan Z. Spade. 1996. "Fraternities and Collegiate Rape Culture: Why Are Some Fraternities More Dangerous Places for Women?" *Gender & Society* 10:133–47.

Brownmiller, Susan. 1975. *Against Our Will: Men, Women, and Rape*. New York: Bantam Books.

Buchward, Emilie, Pamela Fletcher, and Martha Roth, eds. 1993. *Transforming a Rape Culture*. Minneapolis, MN: Milkweed Editions.

Connell, R. W. 1987. *Gender and Power*. Palo Alto, CA: Stanford University Press.

———. 1995. *Masculinities*. Berkeley, CA: University of California Press.

Eder, Donna. 1985. "The Cycle of Popularity: Interpersonal Relations among Female Adolescents." *Sociology of Education* 58:154–65.

Eder, Donna, Catherine Evans, and Stephen Parker. 1995. *School Talk: Gender and Adolescent Culture*. New Brunswick, NJ: Rutgers University Press.

Fisher, Bonnie, Francis Cullen, and Michael Turner. 2000. "The Sexual Victimization of College Women." Washington, DC: National Institute of Justice and the Bureau of Justice Statistics.

Flezzani, James and James Benshoff. 2003. "Understanding Sexual Aggression in Male College Students: The Role of Self-Monitoring and Pluralistic Ignorance." *Journal of College Counseling* 6:69–79.

Forbes, Gordon and Leah Adams-Curtis. 2001. "Experiences with Sexual Coercion in College Males and Females: Role of Family Conflict, Sexist Attitudes, Acceptance of Rape Myths, Self-Esteem, and the Big-Five Personality Factors." *Journal of Interpersonal Violence* 16:865–89.

Gilligan, Carol. 1982. *In a Different Voice: Psychological Theory and Women's Development*. Cambridge, MA: Harvard University Press.

Herman, Diane. 1989. "The Rape Culture." Pp. 20–44 in *Women: A Feminist Perspective*, edited by J. Freeman. Mountain View, CA: Mayfield.

Humphrey, John and Jacquelyn White. 2000. "Women's Vulnerability to Sexual Assault from Adolescence to Young Adulthood." *Journal of Adolescent Health* 27:419–24.

Humphrey, Stephen and Arnold Kahn. 2000. "Fraternities, Athletic Teams, and Rape: Importance of Identification with A Risky Group." *Journal of Interpersonal Violence* 15:1313–22.

Kalof, Linda. 2000. "Vulnerability to Sexual Coercion among College Women: A Longitudinal Study." *Gender Issues* 18:47–58.

Lorber, Judith. 1994. *Paradoxes of Gender*. New Haven, CT: Yale University Press.

Lottes, Ilsa L. 1997. "Sexual Coercion among University Students: A Comparison of the United States and Sweden." *Journal of Sex Research* 34:67–76.

Malamuth, Neil, Christopher Heavey, and Daniel Linz. 1993. "Predicting Men's Antisocial Behavior against Women: The Interaction Model of Sexual Aggression." Pp. 63–98 in *Sexual Aggression: Issues in Etiology, Assessment, and Treatment*, edited by G. N. Hall, R. Hirschman, J. Graham, and M. Zaragoza. Washington, D.C.: Taylor and Francis.

Marine, Susan. 2004. "Waking Up from the Nightmare of Rape." *The Chronicle of Higher Education*. November 26, p. B5.

Martin, Karin. 2003. "Giving Birth Like a Girl." *Gender & Society*. 17:54–72.

Martin, Patricia Yancey. 2004. "Gender as a Social Institution." *Social Forces* 82:1249–73.

Martin, Patricia Yancey and Robert A. Hummer. 1989. "Fraternities and Rape on Campus." *Gender & Society* 3:457–73.

Merton, Robert. 1957. *Social Theory and Social Structure*. New York: Free Press.

Milner, Murray. 2004. *Freaks, Geeks, and Cool Kids: American Teenagers, Schools, and the Culture of Consumption.* New York: Routledge.

Mohler-Kuo, Meichun, George W. Dowdall, Mary P. Koss, and Henry Weschler. 2004. "Correlates of Rape While Intoxicated in a National Sample of College Women." *Journal of Studies on Alcohol* 65:37–45.

O'Sullivan, Chris. 1993. "Fraternities and the Rape Culture." Pp. 23–30 in *Transforming a Rape Culture,* edited by E. Buchward, P. Fletcher, and M. Roth. Minneapolis, MN: Milkweed Editions.

Phillips, Lynn. 2000. *Flirting with Danger: Young Women's Reflections on Sexuality and Domination.* New York: New York University.

Rapaport, Karen and Barry Burkhart. 1984. "Personality and Attitudinal Characteristics of Sexually Coercive College Males." *Journal of Abnormal Psychology* 93:216–21.

Risman, Barbara. 1998. *Gender Vertigo: American Families in Transition.* New Haven, CT: Yale University Press.

———. 2004. "Gender as a Social Structure: Theory Wrestling with Activism." *Gender & Society* 18:429–50.

Russell, Diana. 1975. *The Politics of Rape.* New York: Stein and Day.

Sampson, Rana. 2002. "Acquaintance Rape of College Students." Problem-Oriented Guides for Police Series, No. 17. Washington, DC: U.S. Department of Justice, Office of Community Oriented Policing Services.

Sanday, Peggy. 1990. *Fraternity Gang Rape: Sex, Brotherhood, and Privilege on Campus.* New York: New York University Press.

———. 1996. "Rape-Prone versus Rape-Free Campus Cultures." *Violence against Women* 2:191–208.

Schwartz, Martin and Walter DeKeseredy. 1997. *Sexual Assault on the College Campus: The Role of Male Peer Support.* Thousand Oaks, CA: Sage Publications.

Stombler, Mindy. 1994. "'Buddies' or 'Slutties': The Collective Reputation of Fraternity Little Sisters." *Gender & Society* 8:297–323.

Stombler, Mindy and Patricia Yancey Martin. 1994. "Bringing Women In, Keeping Women Down: Fraternity 'Little Sister' Organizations." *Journal of Contemporary Ethnography* 23:150–84.

Stombler, Mindy and Irene Padavic. 1997. "Sister Acts: Resisting Men's Domination in Black and White Fraternity Little Sister Programs." *Social Problems* 44:257–75.

Swidler, Ann. 2001. *Talk of Love: How Culture Matters.* Chicago: University of Chicago Press.

Weschsler, Henry and Meichun Kuo. 2003. "Watering Down the Drinks: The Moderating Effect of College Demographics on Alcohol Use of High-Risk Groups." *American Journal of Public Health* 93:1929–33.

KEY CONCEPTS

acquaintance rape   doing gender   victimization

DISCUSSION QUESTIONS

1. Think about the most recent party you attended? Can you determine who was "in control" of the food and drink? Who controlled the guest list? What role, if any, did you have in the way the party was structured?

2. Consider the different perspectives on sexual assault presented in the article (individual determinants; rape culture; and particular contexts for sexual assault). Which do you believe offers the best explanation for sexual assault on college campuses?

18

The Social Organization of Toy Stores

CHRISTINE L. WILLIAMS

Christine Williams wrote a book about her ethnographic research working in toy stores. This article is excerpted from that book and talks about how retail toy stores are structured around race and gender. The social organization of toy stores typically places White women in positions to interact with customers and men of color in the back to work inventory. These patterns of work organization are not observed by the average customer, yet create inequality in the workplace.

Living in a consumer society means that we come into contact with retail workers almost every day. Over 22.5 million people work in this job, composing the largest sector of the service economy (Sandikci and Holt 1998, 305; U.S. Bureau of Labor Statistics 2004). But unless you have "worked retail," you probably know little about the working conditions of the job. At best, retail workers are taken for granted by consumers, noticed only when they aren't doing their job. At worst, they are stereotyped as either dim-witted or haughty, which is how they are often portrayed on television and in the movies.

Retail jobs, like other jobs in the service sector, have grown in number and changed dramatically over the past decades. Service jobs gradually have replaced manufacturing jobs as part of the general deindustrialization of the

SOURCE: Christine L. Williams. 2006. *Inside Toyland: Working, Shopping, and Social Inequality*. Berkeley: University of California Press, pp. 48–91.

U.S. economy. This economic restructuring has resulted in boom times for wealthy American consumers as the prices for many commodities have dropped (a consequence of the movement of production overseas). It has also resulted in an erosion of working conditions for Americans in the bottom half of the economy, including service workers. Retail jobs have become increasingly "flexible," temporary, and part time. Over the past decades, workers in these jobs have experienced a loss of job security and benefits, a diminishment in the power of unions, and a lessening of the value of the minimum wage (McCall 2001). Yet while most retail workers have lost ground, the giant corporations they work for have enjoyed unprecedented prosperity and political clout.

George Ritzer (2002) aptly uses the term *McJobs* to describe the working conditions found in a variety of service industries today. The word is a pun on McDonald's, the fast-food giant that introduced and popularized this labor system. McJobs are not careers; they are designed to discourage long-term commitment. They have short promotion ladders, they provide few opportunities for advancement or increased earnings, and the technical skills they require are not transferable outside the immediate work environment. They target sectors of the labor force that presumably don't "need" money to support themselves or their families: young people looking for "fun jobs" before college; mothers seeking part-time opportunities to fit around their family responsibilities; older, retired people looking for the chance to get out of the house and to socialize. However, this image does not resonate with the increasing numbers of workers in these jobs who are struggling to support themselves and their families (Ehrenreich 2001; Talwar 2002). The marketing of McJobs on television commercials for Wal-Mart and fast-food restaurants obscures the harsh working conditions and low pay that contribute to the impoverished state of the working poor.

In addition to contributing to economic inequality, jobs in the retail industry are structured in ways that enhance inequality by gender and race. Although all retail workers are low paid, white men employed in this industry earn more money than any other group. Overall, about as many men as women work in retail trades, but they are concentrated in different kinds of stores. For example, men make up more than three quarters of workers in retail jobs selling motor vehicles, lumber, and home and auto supplies, while women predominate in apparel, gift, and needlework stores (U.S. Bureau of Labor Statistics 2004).

In both stores where I worked, the gender ratio was about 60:40, with women outnumbering men. I was surprised that so many men worked in these toy stores. In my admittedly limited experience, I associated women with the job of selling toys. But I learned that because of the way that jobs are divided and organized, customers usually don't see the substantial numbers of men who are working there too.

Retail work is also organized by race and ethnicity. Ten percent of all employees in the retail trade industry are African American, and 12 percent are of Hispanic origin, slightly less then their overall representation in the U.S. population. But again, whites, African Americans, and Latinas/os are likely to work in different types of stores. For example, African Americans are under-represented (less than 5 percent) in stores that sell hardware, gardening equipment, and needlework supplies and overrepresented (more than 15 percent) in department stores, variety stores, and shoe stores. Similarly Latina/os are underrepresented (less than 6 percent) in bookstores and gas stations and overrepresented (more than 16 percent) in retail florists and household appliance stores (U.S. Bureau of Labor Statistics 2004).

The two stores where I worked had radically different racial compositions. Sixty percent of the workers at the Toy Warehouse [a "big box" store] were African American, and 60 percent of those at Diamond Toys [a boutique, high-end store] were white. Only three African Americans, all women, worked at Diamond Toys. No black men worked at that store. In contrast, only four white women (including me) worked at the Toy Warehouse....

Sociologists have long recognized the workplace as a central site for the reproduction of social inequalities. Studies of factory work in particular have shown us how race and gender hierarchies are reproduced through the social organization of the work. I argue that the labor process in service industries is equally important for understanding social inequality, even though this sector has not come under the same degree of scrutiny by sociologists. I demonstrate how the working conditions at the two stores perpetuate inequality by class, gender, and race. The jobs are organized in such a way as to benefit some groups of workers and discriminate against others.

The stores where I worked represent a range of working conditions in large retail trade establishments. Although both were affiliated with national chains and both were in the business of selling toys, Diamond Toys was unionized and the Toy Warehouse was not. The union protected workers from some of the most egregious aspects of retail work. But ... the union could not overcome the race, gender, and class inequalities that are reproduced by the social organization of the industry.

STRATIFIED SELLING

Diamond Toys and the Toy Warehouse each employed about seventy workers. As in other large retail establishments, the workers were organized in an elaborate hierarchy. Each store was governed by a regional office, which in turn was overseen by the national corporate headquarters. There was no local autonomy in the layout or the merchandise sold in the stores. Within each store, directors were at the top, followed by managers, supervisors, and associates. Directors and managers were salaried employees; everyone else was hourly. Most directors and managers had a college degree. Candidates for these jobs applied to the regional headquarters and, once hired, were assigned to specific stores. These might not be the stores closest to where they lived. Olive, my manager at the Toy Warehouse, had a two-hour commute each way to work, even though there was a Toy Warehouse within five miles of her home.

The hierarchy of jobs and power within the stores was marked by race and gender. In both stores the directors and assistant directors were white men. Immediately below them were managers, who were a more diverse group, including men and women, whites and Latinas/os, and, at the Toy Warehouse, an African American woman (Olive). There were far more managers at Diamond Toys than at the Toy Warehouse; I met at least ten managers during my time there, versus only two at the Toy Warehouse.

The next layer of the hierarchy under managers were supervisors, who were drawn from the ranks of associates. They were among those who had the most seniority and thus the most knowledge of store procedures, and they had limited authority to do things like void transactions at the registers. All of the supervisors at Diamond Toys were white and most were men, while at the Toy Warehouse supervisors were more racially diverse and most were women. It took me a long time to figure out who the supervisors were at the Toy Warehouse. Many of those I thought were supervisors turned out to be regular employees. They had many of the same responsibilities as supervisors, but, as I came to find out, they were competing with each other for promotion to this position. When I asked why they were acting like supervisors, it was explained to me that the Toy Warehouse wouldn't promote anyone before he or she was proficient at the higher job. This policy justified giving workers more responsibilities without more pay. At Diamond Toys, in contrast, job descriptions were clearer and were enforced.

Associates were the largest group of workers at the stores (sometimes referred to as the staff). They included men and women of all races and ethnic groups and different ages, except at Diamond Toys, where I noted that there were no black men. Despite the apparent diversity among the staff, there was substantial segregation by race and gender in the tasks they were assigned. Employees of toy stores are divided between back-and front-of-house workers. The back-of-house employees and managers work in the storage areas, on the loading docks, and in the assembly rooms. In both stores where I worked, the back-of-house workers were virtually all men. The front-of-house workers, the ones who interacted with customers, included both men and women. But there, too, there was job segregation by gender and race, although, as I will discuss, it was harder to discern and on occasion it broke down.

There were two other jobs in the toy store: security guards and janitors, both of whom were subcontracted workers. Both the Toy Warehouse and Diamond Toys employed plainclothes security guards who watched surveillance monitors in their back offices and roamed the aisles looking for shoplifters. At the Toy Warehouse, the individuals who filled those jobs were mostly African American men and women, while only white men and women were hired for security at Diamond Toys. Finally, all of the cleaners at the two stores were Latinas. They were recent immigrants who didn't speak English.

What accounts for the race and gender segregation of jobs in the toy store? Conventional economic theory argues that job segregation is the product of differences in human capital attainment. According to this view, the marketplace sorts workers into jobs depending on their qualifications and preferences. Because men and women of different racial/ethnic groups possess different skills, aptitudes, and work experiences, they will be (and indeed should be) hired into different jobs. Economists generally see this process as benign, if not beneficial, in a society founded on meritocracy, individual liberty, and freedom of choice (Folbre 2001).

In contrast, when sociologists look at job segregation, they tend to see discrimination and structural inequality (Reskin and Roos 1990). Obtaining the right qualifications for a high-paying job is easier for some groups than others. Differential access to college education is an obvious example: society blocks opportunities for poor people to acquire this human capital asset while smoothing the path for the well-to-do. But the sociological critique of job segregation goes deeper than this. Sociologists argue that the definitions of who is qualified and what it means to be qualified for a job are linked to stereotypes about race and gender. Joan Acker (1990) argues that jobs are "gendered," meaning that qualities culturally associated with men (leadership, physical strength, aggression, goal orientation) are built into the job descriptions of the higher-status and higher-paid occupations in our economy. Qualities associated with women (dexterity, passivity, nurturing orientation) tend to be favored in low-paying jobs. In addition to being gendered, jobs are racialized. Black women have been subjected to a different set of gendered stereotypes than white women. Far from being seen as delicate and passive, they have been perceived as dominant, insubordinate, and aggressive (Collins 2000). Those who make hiring decisions draw upon these kinds of racialized stereotypes of masculinity and femininity when appointing workers to specific jobs. . . .

This process may be exacerbated in interactive service work, where employers carefully pick workers who "look right" for the corporate image they attempt to project to the public. A recent court case against Abercrombie & Fitch illustrates this. A suit was brought against the retailer by Asian Americans and Latinas/os who said they were refused selling jobs because "they didn't project what the company called the A & F look" (Greenhouse 2003). Although the company denied the charge, the suit brings to light the common retail practice of matching employees with the image the company is seeking to cultivate. More egregious examples are found in sexualized service work, as in the case of Hooters restaurants (where only buxom young women are hired), and in theme parks like Disney, notorious for its resistance to hiring African Americans (Loe 1996; Project on Disney 1995).

This process of interpellation was apparent in the toy stores where I worked: managers imagined different kinds of people in each job, who came to see themselves in terms of these stereotypical expectations. . . .

My experience illustrates this process of interpellation and resistance. I was hired to be a cashier at both toy stores. I didn't seek out this job, but this was how both managers who hired me envisioned my potential contribution. Only women were regularly assigned to work as cashiers at the Toy Warehouse, and I noticed that management preferred young or light-skinned women for this job. Some older African American women who wanted to work as cashiers had to struggle to get the assignment. Lazelle, for example, who was about thirty-five, had been asking to be put on register over the two months she had been working there. She had been assigned to be a merchandiser. Merchandisers retrieved items from the storeroom, priced items, and checked prices when the universal product codes (UPCs) were missing. Lazelle finally got her chance at the register the same day that I started. We set up next to each other, and I noticed with a bit of envy how much more competent and confident on the register she was compared to me. (Later she told me she had worked registers at other stores, including fast-food restaurants.) I told her that I had been hoping to get assigned to the job of merchandiser. I liked the idea of being free to walk around the store, engage with customers, and learn more about the toys. I had mentioned to Olive that I wanted that job, but she had made it clear that I was destined for cashiering and the service desk (and later, to my horror, computer accounting). Lazelle looked at me as if I were crazy. Merchandising was generally considered to be the worst job in the store because it was so physically taxing. From her point of view I had been assigned the better job, no doubt because of my race, and it seemed to her that I wanted to throw that advantage away.

The preference for whites in the cashier position reflected the importance of this job in the store's general operations. In discount stores like the Toy Warehouse, customers had few opportunities to interact and consult with salesclerks.... The cashier was the only human being that the customer was guaranteed to contact, giving the role enormous symbolic—and economic—importance for the organization. At the point of sale, transactions could break down if the customers were not treated in accordance with their expectations. The preference for white and light-skinned women as cashiers should be interpreted in this light: in a racist and sexist society, such women are generally believed to be the friendliest and most solicitous group and thus best able to inspire trust and confidence.

Personally, I hated working as a cashier. I thought it was a difficult, stressful, and thankless job. Learning to work a cash register is much like learning to use a new computer software package. Each store seems to use a different operating system. After working at these jobs I started to pay attention to every transaction that I made as a customer in a store, and I have yet to see the same computer system twice. The job looks simple from the outside, but because of the way it is organized cashiers have no discretionary power, making them completely dependent on others if anything out of the ordinary happens.... We couldn't even open up our registers to make change for the gum ball machine. Customers would often treat us like morons because we couldn't resolve these minor and routine situations on our own, but we were given no choice or autonomy.

We were, however, held accountable for everything in the register, which had to match the computer printout record of all transactions. At the Toy Warehouse, we were also responsible for requesting "pulls" (this was done automatically by the security personnel at Diamond Toys). A pull is when a manager removes large sums of cash from the register to protect the money in case of a robbery. We were told to request a pull whenever we accumulated more than $500 cash in our registers. If we didn't, and large sums of money were in the till when we closed out our registers at the end of the shift, we would be given a demerit. Interestingly, we were never given any instruction on what to do to protect ourselves in case of an actual robbery....

A few men were regularly assigned to work as cashiers at the Toy Warehouse, but this happened only in the electronics department. The electronics department was cordoned off from the rest of the store by a metal detector gate intended to curtail theft. All of the men with this regular assignment were Asian American. They had sought out this assignment because they were interested in computers and gaming equipment. Working a register in that section may have been more acceptable to them in part because the section was separated from the main registers and in part because Asian masculinity—as opposed to black or white masculinity—is often defined through technical expertise. My sense was that the stereotypical association of Asian American men with computers made these assignments desirable from management's perspective as well.

Occasionally men were assigned to work the registers outside the electronics department, but this happened only when there were staffing shortages or scheduling problems. Once I came to work to find Deshay, a twenty-five-year-old African American, and Shuresh, a twenty-one-year-old second-generation Indian American, both stationed at the main registers. I flew to the back of the store to clock in so I could take my station next to them, eager to observe them negotiating the demands of "women's work." But the minute I took my station they were relieved of cashiering and told to cash out their registers and return to their regular tasks. When a woman was available, the men didn't have to do the job. . . .

Women also crossed over into the men's jobs, but this happened far less frequently. Management never assigned a woman to work in the back areas to make up for temporary staff shortfalls. At each store, only one woman worked in the back of the house, and both women were African American. At the Toy Warehouse, the only woman who worked in the back was Darlene, whom a coworker once described to me as "very masculine" (but also "really great"). Darlene, who worked in a contracting business on the side, took a lot of pride in her physical strength and stamina. She was also a lesbian, which made her the butt of mean-spirited joking (behind her back) but also probably made this assignment less dissonant in the eyes of management (women in nontraditional jobs are often stereotyped as lesbian). At Diamond Toys, the only woman in the back of the house was eighteen-year-old Chandrika. She started working in the back of the house but asked for and received a transfer to gift wrap. Chandrika, who was one of only three African Americans who worked at Diamond Toys, said she hated working in the storeroom because the men there were racist and "very misogynistic," telling sexist jokes and challenging her competence at the job.

Crossing over is a different experience for men and women. When a job is identified as masculine, men often will erect barriers to women, making them feel out of place and unwanted, which is what happened to Chandrika. In contrast, I never observed women trying to exclude men or marginalize men in "their" jobs. On the contrary, men tried to exclude themselves from "women's work." Job segregation by gender is in large part a product of men's efforts to establish all-male preserves, which help them to prove and to maintain their masculinity (Williams 1989). Management colludes in this insofar as they share similar stereotypes of appropriate task assignments for men and women or perceive the public to embrace such stereotypes. But they also insist on employee "flexibility," the widespread euphemism used to describe their fundamental right to hire, fire, and assign employees at will. At the Toy Warehouse, employees were often threatened that their hours would be cut if they were not "flexible" in terms of their available hours and willingness to perform any job. But in general managers shared men's preferences to avoid register duty unless no one else was available.

How and why a specific job comes to be "gendered" and "racialized," or considered appropriate only for women or for men, or for whites or nonwhites, depends on the specific context (which in the case of these toy stores was shaped—but not determined—by their national marketing strategies) . . . . Thus, in contrast to the Toy Warehouse, Diamond Toys employed both men and women as cashiers, and only two of them were African American (both women). At the Toy Warehouse, most of the registers were lined up in the front of the store near the doors. Diamond Toys was more like a department store with cash registers scattered throughout the different sections. The preference for white workers seemed consistent with the marketing of the store's workers as "the ultimate toy experts." In retail service work, professional expertise is typically associated with whiteness, much as it is in domestic service (Wrigley 1995).

Although both men and women worked the registers, there was gender segregation by the type of toy we sold. Only women were assigned to work in the doll and stuffed animal sections, for example, and only men worked in sporting goods and electronics. Also, only women worked in gift wrap. Some sections, like the book department, were gender neutral, but most were as gender marked as the toys we sold . . . .

The most firmly segregated job in the toy store was the job of cleaner. As I have noted, only Latinas filled these jobs. I never witnessed a man or a woman of different race/ethnicity in them . . . .

HOURS, BENEFITS, AND PAY

When I started this project I thought I had the perfect career to combine with a part-time job in retail. As a college professor, I taught two courses per semester that met six hours per week. I thought I could pick a schedule for twenty hours a week that accommodated those teaching commitments. Wrong. To get a job in retail, workers must be willing to work weekends and to change their schedules from one week to another to meet the staffing needs of the store. This is the meaning of the word *flexible* in retail. It is exactly the kind of schedule that is incompatible with doing anything else. . . .

Workers with seniority can gain some control over their schedules, but it takes years of "flexibility" to attain to this status. Moreover, in my experience, this control was guaranteed only at Diamond Toys, thanks to the union. The senior associates who had worked there more than a year had the same schedule from week to week. This didn't apply to the supervisors, though. They had to be willing to fill in as needed, since there had to be a supervisor on the floor at all times. Occasionally they even had to forgo breaks.

The fact that supervisors had less control over their schedules made the job less desirable, but to sweeten the pot the job paid $1 per hour more than what regular associates earned. From the perspective of at least two women senior associates I talked to, it just wasn't worth it to give up control over their schedules. But this effectively prevented them from rising in the hierarchy and making more money, and it contributed to the gender segregation of jobs.

What did people earn? At Diamond Toys we were instructed during our training session not to discuss our pay with anyone. Doing so was pointed out as an example of "unauthorized disclosure of confidential business information," a "serious willful violation" that could result in immediate discharge. So I wasn't about to ask anyone what he or she made. I made $8.75 per hour. I got a sense that that was about average but somewhat higher than what most new hires made (possibly because of my higher educational credentials).

To put this salary into perspective, a forty-hour, full-time, year-round worker making $8.75 per hour would earn about $17,500 before taxes. This was well above the median income of full-time cashiers in 2001, which was about $15,000 per year, and about average for retail sales workers in general, who earned a median income of $18,000 (U.S. Bureau of Labor Statistics 2003). Of course, most retail workers do not work full time and year round, so their incomes are much lower than this. . . .

The unpredictability of scheduling presented a nightmare for many single mothers at the Toy Warehouse. Schedules were posted on Friday for the following week beginning on Sunday. The two-day notice of scheduling made it especially difficult to arrange child care. (In contrast, schedules came out on Tuesday for the following week at Diamond Toys, another benefit of the union.) Even worse, while I was working at the Toy Warehouse, management reduced everyone's hours, purportedly to make up for revenue shortfalls. We were all asked to fill in a form indicating our "availability" to work, from 6:00 A.M. until 10:00 P.M. This form, which was attached to our paychecks, warned that "associates with the flexible availability will get more hours than those who are limited." Part-timers who limited their availability were hit hard when schedules came out the following week, causing a great deal of anger and bad feelings. Angela, an experienced associate, was scheduled for only four hours, and she was so mad that no one could even talk to her. Some said that they were going to apply for unemployment. . . .

One of the reasons management gave for cutting our hours was that the store had been experiencing major problems with "shrink," the retailer term for theft. It was insinuated that the workers were stealing, but I could never figure out how that could happen. In both stores, very elaborate surveillance systems were set up to monitor employees. Hidden cameras recorded activity throughout the store, including the areas around the emergency exits. Our bags, pockets, and purses were checked every time we left the store. Cashiers were monitored continuously via a back-room computer hooked up to every register. As I noted, the contents of the till had to match exactly with the register report. Being even slightly under was enough to cause a major panic. . . .

TAKE THIS JOB AND SHOVE IT—OR NOT

Over the course of working at the two stores, I witnessed a great deal of employee turnover. I outlasted both of the others who were hired with me at the Toy Warehouse, and by the time I left Diamond Toys I was the third in seniority among the ten associates who worked in my section. On one of my last days there I was given the walkie-talkie, the direct line of communication to the storeroom and the managers' office, which indicated that I was at that time the most senior staff member in the section.

For most employees, retail work is a revolving door. Employers know this and expect and even cultivate it. Most new hires are not expected to last through the three-month probation period. One of my coworkers at the Toy Warehouse told me he rarely talked to the new people since they rarely lasted long. Kevin, a twenty-six-year-old African American supervisor, predicted I wouldn't stay very long because I didn't have the right personality for retail work. I asked him, "What kind of people stay at the store?" He said people who were quiet and didn't stress out easily (my major flaw), and then he whispered, "People who can kiss up to management." I asked him, did he do this? And he said yes, but in a different kind of way, not too obvious. I asked how, and he said that he didn't tell people what he "really" thought if he thought he was being screwed over.

Kevin was one of a handful of hourly workers who saw their jobs at the Toy Warehouse as their lifelong work. He told me that he had "grown up" at the Toy Warehouse and could never imagine leaving. There were also a half-dozen or so associates at Diamond Toys who had worked at the store for more than a year. It is much more understandable why workers might choose to stay there, given the union benefits. An extra incentive for staying was the possibility of moving into management. At least a third of the managers at Diamond Toys started working as regular associates. When they became managers they left the union and earned salaries, starting at about $35,000. In contrast, no associate ever moved into management at the Toy Warehouse, although one of my young coworkers maintained that it *could* happen. Vannie, a twenty-one-year-old first-generation immigrant from the Philippines, told me that she had met someone at another store in the chain (which had since closed down) who knew someone who had worked his way up from janitor to store director. I told her that I thought this seemed unlikely. Janitorial services were subcontracted, and director and management positions required a college degree and involved a completely separate application process through the regional office in another state. The story sounded like a Horatio Alger myth to me. But Vannie said that this man was a great inspiration to her, and I believed her.

Although acquiring a management position at Diamond Toys wasn't unprecedented, few of the long-term associates seemed interested in pursuing one. Some seemed resigned to keeping their associate position with its guaranteed hours, schedule, and benefits. Alyss, for example, had worked two years in the doll department and had a set schedule from 8:00 to 4:30, five days per week. She told me she couldn't believe she had worked at the store that long but that at this point she considered the job pretty easy. The only drawbacks were her dealings with Dorothy (the irritable section manager) and the occasional neurotic high-end Barbie collector. She had no interest in pursuing a supervisory position because that would mean losing her schedule and taking on more work. . . .

Many longtime workers developed quasi-familial bonds at the stores, referring to each other as brother and sister, mother and grandmother. This happened at both stores. There were also real kin networks, most evident at the Toy Warehouse. Some of the older employees brought their children or grandchildren with them to the store during their shift. These kids hung out in the break room or played with the demonstration toys, including the video games that were set up in the electronics department. All the employees helped to keep an eye on them. Those with older children sometimes got them jobs in the store.

I was an outsider to these family and friendship networks, but I was often very touched by the mutual support and caring I witnessed among my coworkers, especially at the Toy Warehouse. For the longtime workers, the store was an extension of their family networks and responsibilities. They used their employee discount cards (10 percent at the Toy Warehouse, 30 percent at Diamond Toys) to buy toys for their kin and quasi-kin. But more important was the social and emotional support they experienced there.

I was struck by this once when I was in the break room at the Toy Warehouse and my coworkers Dwain and Lamonica walked in on a day they were not scheduled to work. They were an African American couple in their early twenties who had been dating for a short time. They had come in to pick up their paychecks and to talk to Selma, an African American woman in her forties who had worked at the store for six years. Dwain was upset with his mother, who was giving him a very hard time, telling him he was worthless and criticizing him for being too dependent. During the discussion Lamonica was sitting on a stack of chairs behind Dwain, sucking on her Jamba Juice. She smiled during the conversation or rolled her eyes, as if to say, "What are you going to do?" But she didn't really participate. Selma, on the other hand, was giving him reassurance and sympathy and moral support. I asked her if she knew his mother and she said no, just what Dwain had told her. I realized that Selma was a mentor for Dwain; he clearly cherished her advice and encouragement.

This is the backstage of stores that most shoppers don't see, but it is critical for understanding why people stay in crummy jobs with low pay. Barbara Ehrenreich (2001) considers this conundrum in her study of low-wage work. She wonders why workers don't leave when other, higher-paying opportunities arise. At the Toy Warehouse, most workers stayed no longer than three months. Those who stayed long term did so because that was where their family was.

WHAT A DIFFERENCE A UNION MAKES

The union at Diamond Toys helped to ameliorate some of the most egregious problems with working retail. It guaranteed hours and schedules for senior associates, mandated longer rest breaks, and provided health benefits, vacation pay, and a career track. I earned 17 percent more at the unionized workplace, which was in line with the national 20 percent wage premium that comes with union membership (McCall 2001, 181). We were always allowed to leave when we were scheduled, whereas in the Toy Warehouse we were kept up to an hour later than scheduled to finish cleaning up the store after closing (a practice that routinely resulted in our being scheduled for fewer breaks than we were lawfully due). I was also impressed by how the managers behaved professionally and respectfully toward the workers. They quickly responded to pages and patiently explained procedures to the new people.

The unfortunate exception was my area manager, Dorothy. She rushed around our section barking orders and shouting insults at us. The first time I met her she told me that my name was unacceptable, as there were already two others in the store named Chris or Christine, so she was going to call me by my middle name instead. (Luckily that didn't last long because I couldn't remember to respond to Louise.) Sometimes when the store was busy she would stand next to me and shout, "Hurry, hurry, hurry." She would roll her eyes and mutter about my stupidity whenever I had to ask her a question or get her help to solve a problem. When I paged her she would pick up the phone and say in an exasperated voice, "What is it *now*, Christine?" It was some consolation that she treated everyone this way. We all found different ways to cope. Carl dealt with it by keeping a happy song in his head, he said. Alyss became depressed and frustrated; she often looked on the verge of tears. Dennis dreamed of leaving retail altogether and becoming a full-time teacher. Chandrika told me that to deal with the abuse she prayed. She also claimed to have filed four formal complaints against Dorothy, but nothing ever came of that. I fantasized about making a principled scene on my last day. (I didn't.)

This is the part of the job that a union can't change. Managers are allowed to harass workers, and there is virtually no recourse unless that harassment targets a worker's race, gender, or other legally protected characteristic (Williams 2003). The other managers and the store directors knew about Dorothy's abusive behavior but oddly seemed to tolerate it. They attributed it to personal and family problems she was experiencing. Rumors abounded about the nature of these problems, but it was impossible to verify them. I desperately wanted to understand why she was so mean, but in the end it probably wouldn't have made a difference. Sadistic bosses are an unfortunate fact of life in many hierarchical work organizations (Gherardi 1995).

The union offered no protection from harassing and abusive customers either. Admittedly this was a bigger problem at the Toy Warehouse than at Diamond Toys, but I don't think that customers were better behaved at Diamond Toys because of the union. For many customers, part of the allure of shopping at Diamond Toys was the educated and solicitous, not to mention white, sales staff. The mixture of class, race, and gender frames the customer-server relationship, just as it does the social organization of retail work. . . .

CONCLUSION

Most sociological research on retail stores looks at them as sites of consumption. But stores are also workplaces. Retail work makes up an increasing proportion of the jobs in our economy. Yet these are "bad" jobs. According to Frank Levy (1998), a "good job" is one that pays enough to support a family and provides benefits, security, and autonomy. In contrast, most jobs in retail pay low wages, offer few benefits, have high turnover, and restrict workers' autonomy.

. . . I have argued that the social organization of work in large toy stores also contributes to class, gender, and race inequalities. The Toy Warehouse, which had a predominately African American staff, paid extremely low wages, offered few benefits, and demanded "flexible" workers who made no scheduling demands. The store was segregated by race and gender, with white men in the director positions and African American women in managerial and supervisory roles. Among the staff, only white and light-skinned women and Asian American men were regularly assigned to cashiering positions, and only men (of all racial/ethnic groups) worked in the back room unloading and assembling the toys. African American men and women filled the positions of security guards, stockers, and gofers.

Because Diamond Toys was unionized, it offered better pay than the Toy Warehouse (but not a "living wage"), and its employees received health care and vacation benefits. Schedules were posted in advance, legally mandated breaks were honored, and career ladder promotions were available. For all of these reasons, a union does make a positive difference for workers. . . .

. . . [T]he hierarchical and functional placement of workers according to managerial stereotypes results in advantages for white men and (to a lesser extent) white women and disadvantages for racial/ethnic minority men and women. These stereotypes are perhaps more deeply entrenched than low wages, based as they are on perceptions of customer preferences. Consumers therefore have a role in pressing for changes in these job assignments. But in my view, the struggle for equal access to "bad jobs" is hardly worth an organized effort. There is little point in demanding equal access to jobs that don't support a family. Similarly, career ladders have to be created before equal opportunities for advancement are demanded. The fight against racism and sexism, then, should be folded into efforts to economically upgrade these jobs. The goal of restructuring jobs in toy stores, and in retail work in general, should be self-sufficiency—and hope—for all workers, regardless of race or gender.

REFERENCES

Acker, Joan. 1990. "Hierarchies, Jobs, Bodies: A Theory of Gendered Organizations." *Gender & Society* 4: 139–58.

Collins, Patricia Hill. 2000. *Black Feminist Thought*. New York: Routledge.

Ehrenreich, Barbara. 2001. *Nickel and Dimed: On (Not) Getting By in America*. New York: Metropolitan Books.

Folbre, Nancy. 2001. *The Invisible Heart: Economics and Family Values*. New York: New Press.

Greenhouse, Steven. 2003. "Abercrombie & Fitch Accused of Discrimination in Hiring." *New York Times*, June 17, A1.

Gherardi, Sylvia. 1995. *Gender, Symbolism and Organizational Culture*. Thousand Oaks, CA: Sage Publications.

Levy, Frank. 1998. *The New Dollars and Dreams: American Incomes and Economic Change*. New York: Russell Sage Foundation.

Loe, Meika. 1996. "Working for Men: At the Intersection of Power, Gender, and Sexuality." *Sociological Inquiry* 66: 399–421.

McCall, Leslie. 2001. *Complex Inequality: Gender, Class and Race in the New Economy*. New York: Routledge.

Project on Disney. 1995. *Inside the Mouse*. Durham, NC: Duke University Press.

Reskin, Barbara, and Patricia Roos. 1990. *Job Queues, Gender Queues*. Philadelphia: Temple University Press.

Ritzer, George. 2002. *McDonaldization: The Reader*. Thousand Oaks, CA: Pine Forge Press.

Sandikci, Ozlem, and Douglas Holt. 1998. "Malling Society: Mall Consumption Practices and the Future of Public Space." In *Servicescapes: The Concept of Place in Contemporary Markets*, ed. John Sherry, 305–36. Lincolnwood, IL: NTC Business Books.

Talwar, Jennifer Parker. 2002. *Fast Food, Fast Track: Immigrants, Big Business, and the American Dream*. Boulder, CO: Westview Press.

U.S. Bureau of Labor Statistics. 2003. "Household Data Annual Averages. 39. Median Weekly Earnings of Full-Time Wage and Salary Workers by Detailed Occupation and Sex." Retrieved March 22, 2005, from www.bls.gov/cps/cpsaat39.pdf.

U.S. Bureau of Labor Statistics. 2004. "Household Data Annual Averages. 18. Employed Persons by Detailed Industry, Sex, Race, and Hispanic or Latino Ethnicity." Retrieved April 6, 2005, from www.bls.gov/cps/cpsaat18.pdf.

Williams, Christine L. 1989. *Gender Differences at Work: Women and Men in Nontraditional Occupations*. Berkeley: University of California Press.

Williams, Christine L. 2003. "Sexual Harassment and Human Rights Law in New Zealand." *Journal of Human Rights* 2 (December) 573–84.

Wrigley, Julia. 1995. *Other People's Children*. New York: Basic Books.

KEY CONCEPTS

| dual labor market | gendered institution | service sector |

DISCUSSION QUESTIONS

1. In your most recent in-store shopping experience, did you notice the different roles of workers in the store? How was your shopping experience influenced by the salespeople working in the store?

2. Take the same analysis presented in this article and apply it to one of your work experiences. How did your gender and race–ethnicity influence the job(s) you were given? Was job segregation by gender and race?

Applying Sociological Knowledge:
An Exercise for Students

Think of an organization with which you are familiar (a college, a work organization, or a religious organization). What are the different groups that make up this organization? Do the different groups that make up the organization have different statuses within the organization? If so, describe each group's status. Is there a hierarchy among these different groups and, if so, how does that affect how they interact with each other?

19

The Functions of Crime

EMILE DURKHEIM

This classic essay, written in 1895 and translated many times since, points to crime as an inevitable part of society. Durkheim's main functionalist thesis that criminal behavior exists in all social settings is still the theoretical basis for many sociological inquiries into crime and deviance.

If there is a fact whose pathological nature appears indisputable, it is crime. All criminologists agree on this score. Although they explain this pathology differently, they nonetheless unanimously acknowledge it. However, the problem needs to be treated less summarily.

. . . Crime is not only observed in most societies of a particular species, but in all societies of all types. There is not one in which criminality does not exist, although it changes in form and the actions which are termed criminal are not everywhere the same. Yet everywhere and always there have been men who have conducted themselves in such a way as to bring down punishment upon their heads. If at least, as societies pass from lower to higher types, the crime rate (the relationship between the annual crime figures and population figures) tended to fall, we might believe that, although still remaining a normal phenomenon, crime tended to lose that character of normality. Yet there is no single ground for believing such a regression to be real. Many facts would rather seem to point to the existence of a movement in the opposite direction. From the beginning of the century statistics provide us with a means of following the progression of criminality. It has everywhere increased, and in France the increase is of the order of 300 percent. Thus there is no phenomenon which represents more incontrovertibly all the symptoms of normality, since it appears to be closely bound up with the conditions of all collective life. To make crime a social illness would be to concede that sickness is not something accidental, but on the contrary derives in certain cases from the fundamental constitution of the living creature. This would be to erase any distinction between the physiological and the pathological. It can certainly happen that crime itself has normal forms; this is what happens, for instance, when it reaches an excessively high level. There is no doubt that this excessiveness is pathological in nature. What is normal is simply that criminality

SOURCE: Emile Durkheim. 1982. *The Rules of Sociological Method.* Steven Lukes (Ed.) Translated by W. D. Halls, New York: The Free Press. A division of Macmillan, pp. 64–75.

exists, provided that for each social type it does not reach or go beyond a certain level which it is perhaps not impossible to fix in conformity with the previous rules.

We are faced with a conclusion which is apparently somewhat paradoxical. Let us make no mistake: to classify crime among the phenomena of normal sociology is not merely to declare that it is an inevitable though regrettable phenomenon arising from the incorrigible wickedness of men; it is to assert that it is a factor in public health, an integrative element in any healthy society. At first sight this result is so surprising that it disconcerted even ourselves for a long time. However, once that first impression of surprise has been overcome it is not difficult to discover reasons to explain this normality and at the same time to confirm it.

In the first place, crime is normal because it is completely impossible for any society entirely free of it to exist.

Crime consists of an action which offends certain collective feelings which are especially strong and clear-cut. In any society, for actions regarded as criminal to cease, the feelings that they offend would need to be found in each individual consciousness without exception and in the degree of strength requisite to counteract the opposing feelings. Even supposing that this condition could effectively be fulfilled, crime would not thereby disappear; it would merely change in form, for the very cause which made the well-springs of criminality to dry up would immediately open up new ones.

Indeed, for the collective feelings, which the penal law of a people at a particular moment in its history protects, to penetrate individual consciousnesses that had hitherto remained closed to them, or to assume greater authority—whereas previously they had not possessed enough—they would have to acquire an intensity greater than they had had up to then. The community as a whole must feel them more keenly, for they cannot draw from any other source the additional force which enables them to bear down upon individuals who formerly were the most refractory. . . .

In order to exhaust all the logically possible hypotheses, it will perhaps be asked why this unanimity should not cover all collective sentiments without exception, and why even the weakest sentiments should not evoke sufficient power to forestall any dissentient voice. The moral conscience of society would be found in its entirety in every individual, endowed with sufficient force to prevent the commission of any act offending against it, whether purely conventional failings or crimes. But such universal and absolute uniformity is utterly impossible, for the immediate physical environment in which each one of us is placed, our hereditary antecedents, the social influences upon which we depend, vary from one individual to another and consequently cause a diversity of consciences. It is impossible for everyone to be alike in this matter, by virtue of the fact that we each have our own organic constitution and occupy different areas in space. This is why, even among lower peoples where individual originality is very little developed, such originality does however exist. Thus, since there cannot be a society in which individuals do not diverge to some extent from the collective type, it is also inevitable that among these deviations some assume a criminal character. What confers upon them this character is not the intrinsic importance of the acts but the importance which the common consciousness ascribes to them. Thus if the latter is stronger and possesses sufficient authority to make these divergences very weak in absolute terms, it will also be more sensitive and exacting. By reacting against the slightest deviations with an energy which it elsewhere employs against those that are more weighty, it endues them with the same gravity and will brand them as criminal.

Thus crime is necessary. It is linked to the basic conditions of social life, but on this very account is useful, for the conditions to which it is bound are themselves indispensable to the normal evolution of morality and law.

Indeed today we can no longer dispute the fact that not only do law and morality vary from one social type to another, but they even change within the same type if the conditions of collective existence are modified. Yet for these transformations to be made possible, the collective sentiments at the basis of morality should not prove unyielding to change, and consequently should be only moderately intense. If they were too strong, they would no longer be malleable. Any arrangement is indeed an obstacle to a new arrangement; this is even more the case the more deep-seated the original arrangement. The more strongly a structure is articulated, the more it resists modification; this is as true for functional as for anatomical patterns. If there were no crimes, this condition would not be fulfilled, for such a hypothesis presumes that collective sentiments would have attained a degree of intensity unparalleled in history. Nothing is good indefinitely and without limits. The authority which the moral consciousness enjoys must not be excessive, for otherwise no one would dare to attack it and it would petrify too easily into an immutable form. For it to evolve, individual originality must be allowed to manifest itself. But so that the originality of the idealist who dreams of transcending his era may display itself, that of the criminal, which falls short of the age, must also be possible. One does not go without the other.

Nor is this all. Beyond this indirect utility, crime itself may play a useful part in this evolution. Not only does it imply that the way to necessary changes remains open, but in certain cases it also directly prepares for these changes. Where crime exists, collective sentiments are not only in the state of plasticity necessary to assume a new form, but sometimes it even contributes to determining beforehand the shape they will take on. Indeed, how often is it only an anticipation of the morality to come, a progression towards what will be! . . . The freedom of thought that we at present enjoy could never have been asserted if the rules that forbade it had not been violated before they were solemnly abrogated. However, at the time the violation was a crime, since it was an offence against sentiments still keenly felt in the average consciousness. Yet this crime was useful since it was the prelude to changes which were daily becoming more necessary . . . .

From this viewpoint the fundamental facts of criminology appear to us in an entirely new light. Contrary to current ideas, the criminal no longer appears as an utterly unsociable creature, a sort of parasitic element, a foreign, unassimilable body introduced into the bosom of society. He plays a normal role in social life. For its part, crime must no longer be conceived of as an evil which cannot be circumscribed closely enough. Far from there being cause for congratulation when it drops too noticeably below the normal level, this apparent progress assuredly coincides with and is linked to some social disturbance. Thus the number of crimes of assault never falls so low as it does in times of scarcity.

Consequently, at the same time, and as a reaction, the theory of punishment is revised, or rather should be revised. If in fact crime is a sickness, punishment is the cure for it and cannot be conceived of otherwise; thus all the discussion aroused revolves round knowing what punishment should be to fulfill its role as a remedy. But if crime is in no way pathological, the object of punishment cannot be to cure it and its true function must be sought elsewhere . . . .

KEY CONCEPTS

collective consciousness   deviance   functionalism
social facts

DISCUSSION QUESTIONS

1. According to Durkheim's theory, criminal behavior exists in all societies. Consider the possibility of a society without the ability to punish criminal behavior (no prisons, no courts). How would individuals respond to crime? What informal social control mechanisms would help to maintain order?

2. How could you use Durkheim's theory as the basis for a research project on deviant behavior? What hypotheses could you test that would challenge or support the functionalist view of crime?

20

The Medicalization of Deviance

PETER CONRAD AND JOSEPH W. SCHNEIDER

This essay outlines the social construction of social deviance. The authors specifically refer to the medical profession as redefining certain deviant behaviors as "illness," rather than as "badness." They argue that the "medicalization of deviance changes the social response to such behavior to one of treatment rather than punishment."

SOURCE: Peter Conrad and Joseph W. Schneider. 1992. *Deviance and Medicalization: From Badness to Sickness*. Philadelphia: Temple University Press, pp. 28–37.

Consider the following situations. A woman rides a horse naked through the streets of Denver claiming to be Lady Godiva and after being apprehended by authorities, is taken to a psychiatric hospital and declared to be suffering from a mental illness. A well-known surgeon in a Southwestern city performs a psycho-surgical operation on a young man who is prone to violent outbursts. An Atlanta attorney, inclined to drinking sprees, is treated at a hospital clinic for his disease, alcoholism. A child in California brought to a pediatric clinic because of his disruptive behavior in school is labeled hyperactive and is prescribed methylphenidate (Ritalin) for his disorder. A chronically overweight Chicago housewife receives a surgical intestinal bypass operation for her problem of obesity. Scientists at a New England medical center work on a million-dollar federal research grant to discover a heroin-blocking agent as a "cure" for heroin addiction. What do these situations have in common? In all instances medical solutions are being sought for a variety of deviant behaviors or conditions. We call this "the medicalization of deviance" and suggest that these examples illustrate how medical definitions of deviant behavior are becoming more prevalent in modern industrial societies like our own. The historical sources of this medicalization, and the development of medical conceptions and controls for deviant behavior, are the central concerns of our analysis.

Medical practitioners and medical treatment in our society are usually viewed as dedicated to healing the sick and giving comfort to the afflicted. No doubt these are important aspects of medicine. In recent years the jurisdiction of the medical profession has expanded and encompasses many problems that formerly were not defined as medical entities. . . . There is much evidence for this general viewpoint—for example, the medicalization of pregnancy and childbirth, contraception, diet, exercise, child development norms—but our concern here is more limited and specific. Our interests focus on the medicalization of deviant behavior: the defining and labeling of deviant behavior as a medical problem, usually an illness and mandating the medical profession to provide some type of treatment for it. Concomitant with such medicalization is the growing use of medicine as an agent of social control, typically as medical intervention. Medical intervention as social control seeks to limit, modify, regulate, isolate, or eliminate deviant behavior with medical means and in the name of health. . . .

Conceptions of deviant behavior change, and agencies mandated to control deviance change also. Historically there have been great transformations in the definition of deviance—from religious to state-legal to medical-scientific. Emile Durkheim (1893/1933) noted in *The Division of Labor in Society* that as societies develop from simple to complex, sanctions for deviance change from repressive to restitutive or, put another way, from punishment to treatment or rehabilitation. Along with the change in sanctions and social control agent there is a corresponding change in definition or conceptualization of deviant behavior. For example, certain "extreme" forms of deviant drinking (what is now called alcoholism) have been defined as sin, moral weakness, crime, and most recently illness. . . . In modern industrial society there has been a substantial growth in the prestige, dominance, and jurisdiction of the medical profession (Freidson, 1970). It is only within the last century that physicians have become highly organized, consistently trained, highly paid, and sophisticated in their therapeutic techniques and abilities. . . . The medical profession dominates the organization of health care and has a virtual monopoly on anything that is defined as medical treatment, especially in terms of what constitutes "illness" and what is appropriate medical intervention. . . . Although Durkheim did not predict this medicalization, perhaps in part because medicine of his time was not the scientific, prestigious, and dominant profession of today, it is clear that medicine is the central restitutive agent in our society.

EXPANSION OF MEDICAL JURISDICTION OVER DEVIANCE

When treatment rather than punishment becomes the preferred sanction for deviance, an increasing amount of behavior is conceptualized in a medical framework as illness. As noted earlier, this is not unexpected, since medicine has always functioned as an agent of social control, especially in attempting to "normalize" illness and return people to their functioning capacity in society. Public health and psychiatry have long been concerned with social behavior and have functioned traditionally as agents of social control (Foucault, 1965; Rosen, 1972). What is significant, however, is the expansion of this sphere where medicine functions in a social control capacity. In the wake of a general humanitarian trend, the success and prestige of modern biomedicine, the technological growth of the 20th century, and the diminution of religion as a viable agent of control, more and more deviant behavior has come into the province of medicine. In short, the particular, dominant designation of deviance has changed; much of what was badness (i.e., sinful or criminal) is now sickness. Although some forms of deviant behavior are more completely medicalized than others (e.g., mental illness), recent research has pointed to a considerable variety of deviance that has been treated within medical jurisdiction: alcoholism, drug addiction, hyperactive children, suicide, obesity, mental retardation, crime, violence, child abuse, and learning problems, as well as several other categories of social deviance. Concomitant with medicalization there has been a change in imputed responsibility for deviance: with badness the deviants were considered responsible for their behavior, with sickness they are not, or at least responsibility is diminished (see Stoll, 1968). The social response to deviance is "therapeutic" rather than punitive. Many have viewed this as "humanitarian and scientific" progress; indeed, it often leads to "humanitarian and scientific" treatment rather than punishment as a response to deviant behavior. . . .

A number of broad social factors underlie the medicalization of deviance. As psychiatric critic Thomas Szasz (1974) observes, there has been a major historical shift in the manner in which we view human conduct:

With the transformation of the religious perspective of man into the scientific, and in particular the psychiatric, which became fully articulated during the nineteenth century, there occurred a radical shift in emphasis away from viewing man as a *responsible agent acting in and on the world* and toward viewing him *as a responsive organism being acted upon* by biological and social "forces." (p. 149)

This is exemplified by the diffusion of Freudian thought, which since the 1920s has had a significant impact on the treatment of deviance, the distribution of stigma, and the incidence of penal sanctions.

Nicholas Kittrie (1971), focusing on decriminalization, contends that the foundation of the therapeutic state can be found in determinist criminology, that it stems from the *parens patriae* power of the state (the state's right to help those who are unable to help themselves), and that it dates its origin with the development of juvenile justice at the turn of the century. He further suggests that criminal law has failed to deal effectively (e.g., in deterrence) with criminals and deviants, encouraging a use of alternative methods of control. Others have pointed out that the strength of formal sanctions is declining because of the increase in geographical mobility and the decrease in strength of traditional status groups (e.g., the family) and that medicalization offers a substitute method for controlling deviance (Pitts, 1968). The success of medicine in areas like infectious disease has led to rising expectations of what medicine can accomplish. In modern technological societies, medicine has followed a technological imperative—that the physician is responsible for doing everything possible for the patient—while neglecting such significant issues as the patient's rights and wishes and the impact of biomedical advances on society (Mechanic, 1973). Increasingly sophisticated medical technology has extended the potential of medicine as social control, especially in terms of psychotechnology (Chorover, 1973). Psychotechnology includes a variety of medical and quasimedical treatments or procedures: psychosurgery, psychoactive medications, genetic engineering, disulfiram (Antabuse), and methadone. Medicine is frequently a pragmatic way of dealing with a problem (Gusfield, 1975). Undoubtedly the increasing acceptance and dominance of a scientific world view and the increase in status and power of the medical profession have contributed significantly to the adoption and public acceptance of medical approaches to handling deviant behavior.

THE MEDICAL MODEL
AND "MORAL NEUTRALITY"

The first "victories" over disease by an emerging biomedicine were in the infectious diseases in which specific causal agents—germs—could be identified. An image was created of disease as caused by physiological difficulties located *within* the human body. This was the medical model. It emphasized the internal and biophysiological environment and deemphasized the external and social psychological environment.

There are numerous definitions of "the medical model." ... We adopt a broad and pragmatic definition: the medical model of deviance locates the source of deviant behavior within the individual, postulating a physiological, constitutional, organic, or, occasionally, psychogenic agent or condition that is assumed to cause the behavioral deviance. The medical model of deviance usually, although not always, mandates intervention by medical personnel with medical means as treatment for the "illness." Alcoholics Anonymous, for example, adopts a rather idiosyncratic version of the medical model—that alcoholism is a chronic disease caused by an "allergy" to alcohol—but actively discourages professional medical intervention. But by and large, adoption of the medical model legitimates and even mandates medical intervention.

The medical model and the associated medical designations are assumed to have a scientific basis and thus are treated as if they were morally neutral (Zola, 1975). They are not considered moral judgments but rational, scientifically verifiable conditions. . . . Medical designations *are* social judgments, and the adoption of a medical model of behavior, a political decision. When such medical designations are applied to deviant behavior, they are related directly and intimately to the moral order of society. In 1851 Samuel Cartwright, a well-known Southern physician, published an article in a prestigious medical journal describing the disease "drapetomania," which only affected slaves and whose major symptom was running away from the plantations of their white masters (Cartwright, 1851). Medical texts during the Victorian era routinely described masturbation as a disease or addiction and prescribed mechanical and surgical treatments for its cure (Comfort, 1967; Englehardt, 1974). Recently many political dissidents in the Soviet Union have been designated mentally ill, with diagnoses such as "paranoia with counterrevolutionary delusions" and "manic reformism," and hospitalized for their opposition to the political order (Conrad, 1977). Although these illustrations may appear to be extreme examples, they highlight the fact that all medical designations of deviance are influenced significantly by the moral order of society and thus cannot be considered morally neutral. . . .

Even after a social definition of deviance becomes accepted or legitimated, it is not evident what particular type of problem it is. Frequently there are intellectual disputes over the causes of the deviant behavior and the appropriate methods of control. These battles about deviance designation (is it sin, crime, or sickness?) and control are battles over turf: Who is the appropriate definer and treater of the deviance? Decisions concerning what is the proper deviance designation and hence the appropriate agent of social control are settled by some type of political conflict.

How one designation rather than another becomes dominant is a central sociological question. In answering this question, sociologists must focus on claims-making activities of the various interest groups involved and examine how one or another attains ownership of a given type of deviance or social problem and thus generates legitimacy for a deviance designation. Seen from this perspective, public facts, even those which wear a "scientific" mantle are treated as products of the groups or organizations that produce or promote them rather than as accurate reflections of "reality." The adoption of one deviance designation or another has consequences beyond settling a dispute about social control turf.

. . .When a particular type of deviance designation is accepted and taken for granted, something akin to a paradigm exists. There have been three major deviance paradigms: deviance as sin, deviance as crime, and deviance as sickness. When one paradigm and its adherents become the ultimate arbiter of "reality" in society, we say a hegemony of definitions exists. In Western societies, and American society in particular, anything proposed in the name of science gains great authority. In modern industrial societies, deviance designations have become increasingly medicalized. We call the change in designations from badness to sickness the medicalization of deviance. . . .

REFERENCES

Cartwright, S. W. Report on the diseases and physical peculiarities of the negro race. *N. O. Med. Surg. J.*, 1851, 7, 691–715.

Chorover, S. "Big Brother and psychotechnology." *Psychol. Today*, 1973, 7, 43–54 (Oct.).

Comfort, A. *The anxiety makers*. London: Thomas Nelson & Sons, 1967.

Conrad, P. Soviet dissidents, ideological deviance, and mental hospitalization. Presented at Midwest Sociological Society Meetings, Minneapolis, 1977.

Durkheim, E. *The division of labor in society*. New York: The Free Press, 1933. (Originally published 1893.)

Englehardt, H. T. Jr. The disease of masturbation: Values and the concept of disease. *Bull. Hist. Med.*, 1974, 48, 234–48 (Summer).

Foucault, M. *Madness and civilization*. New York: Random House, Inc. 1965.

Freidson, E. *Profession of medicine*. New York: Harper & Row Publishers Inc. 1970.

Gusfield, J. R. Categories of ownership and responsibility in social issues: Alcohol abuse and automobile use. *J. Drug Issues*, 1975, 5, 285–303 (Fall).

Kittrie, N. *The right to be different: Deviance and enforced therapy*. Baltimore: Johns Hopkins University Press, 1971.

Mechanic, D. Health and illness in technological societies. *Hastings Center Stud.*, 1973, 1(3), 7–18.

Pitts, J. Social control: The concept. In D. Sills (Ed.) *International Encyclopedia of Social Sciences*. (Vol. 14). New York: Macmillan Publishing Co., Inc. 1968.

Rosen, G. The evolution of social medicine. In H. E. Freeman, S. Levine, and L. Reeder (Eds.) *Handbook of medical sociology* (2nd ed.). Englewood Cliffs, NJ: Prentice-Hall, Inc. 1972.

Stoll, C. S. Images of man and social control. *Soc. Forces*, 1968, 47, 119–127 (Dec.).

Szasz, T. *Ceremonial chemistry*. New York: Anchor Books, 1974.

Zola, I. K. In the name of health and illness: On some socio-political consequences of medical influence. *Soc. Sci. Med.*, 1975, 9, 83–87.

KEY CONCEPTS

medicalization of	social control	deviance

DISCUSSION QUESTIONS

1. Alcoholism is an example of a deviant behavior being medicalized. How has this altered the understanding and treatment of alcoholism? How does the involvement of health professionals in the treatment of alcoholism influence societal reaction to excessive drinking?

2. Some argue that rapists should be castrated. How does this illustrate the transformation of understanding rape as a move "from badness to sickness"? What assumptions guide the suggestion that rapists should be castrated as a way of stopping rape?

21

Explaining Rampage School Shootings

KATHERINE S. NEWMAN, CYBELLE FOX,
WENDY ROTH, AND JAL MEHTA

This piece, written before the Virginia Tech shootings in 2007, summarizes the many school shootings that happened in the late 1990s and into 2000 and 2001. The authors consider the question of whether or not this is a new phenomenon and constitutes an epidemic. They also compare school shootings to other forms of violence, specifically urban gun violence. Finally, they consider the many different explanations for the shooters' behavior.

The 1997–1998 academic year left a bloody trail of multiple-victim homicides in communities that imagined themselves violence free. Rampage school shootings had actually erupted before, but in the late 1990s, a string of six incidents created a sense that an epidemic was under way. On October 1, 1997, sixteen-year-old Luke Woodham of Pearl, Mississippi, killed his mother, came to school, and shot nine students, killing two. One month later, Michael Carneal killed three and wounded five. Fourteen-year-old Joseph Todd shot two students in Stamps, Arkansas, two weeks after Michael's rampage. Mitchell Johnson and Andrew Golden left four students and a teacher dead and wounded ten others. A few weeks later, fourteen-year-old Andrew Wurst of Edinboro, Pennsylvania, killed a teacher and wounded three students at a school dance. The killing season for that year ended on May 21 when fifteen-year-old Kip Kinkel murdered his parents and then went on a shooting spree in his Springfield, Oregon, school cafeteria, killing two students and wounding twenty-five.

The next year brought us "Columbine." The sheer scale of the Littleton, Colorado, rampage was so enormous that this one word will, for years to come, conjure up horrific images of dead and wounded children. Eric Harris, age

SOURCE: Katherine S. Newman, Cybelle Fox, Wendy Roth, and Jal Mehta. 2004. *Rampage: The Social Roots of School Shootings*. New York: Basic Books.

seventeen, and Dylan Klebold, eighteen, invaded the school with an arsenal of guns and bombs, killing twelve students and a teacher, wounding twenty-three others, and finally ending their own lives. One month later, T.J. Solomon injured six students in a school shooting in Conyers, Georgia.

For those who have "been there," each new rampage shooting resurrects terrible memories. Christine Olson, an eighteen-year-old honors student who graduated from Westside High School, had a sister in the middle school at the time of the rampage. She remembered well how the Columbine shooting propelled her own community backward:

> When I heard about Columbine, I spent most of the day crying at school just because it brought back so much. It was just strange how it really affected the community again because we thought we . . . were doing so much better. . . . It took about a year for us to [start healing], and then . . . almost exactly a year after, [Columbine] happened.

Ironically, Michael Carneal felt much the same way. He attempted suicide twice during his incarceration at Northern Kentucky Youth Correctional Center after learning of the events in Littleton. In some way, Michael felt responsible for the shootings that followed his own. "When will it ever end?" he asked his therapist.

The answer, sadly, was no time soon. After a brief respite, the shootings continued. On March 5, 2001, fifteen-year-old Charles Andrew (Andy) Williams killed two students and wounded eleven more as well as a campus security guard, and a student teacher at Santana High School in suburban Santee, California. Two weeks later, eighteen-year-old Jason Hoffman wounded three students and two teachers at Granite Hills High School in nearby El Cajon, California.

To many, it seemed that suddenly, mysteriously, the scourge of deadly youth violence had burst free of poor and minority neighborhoods and came calling in the kinds of comfortable communities that residents believe are perfect places to raise kids. If the shooting had happened at Paducah Tilghman High School, located in the city, residents like Marjorie Eagen would not have been surprised.

> Tilghman is at least 50 percent lower socioeconomic, you've got a greater mix of races at Tilghman. Those are the kids from the projects. . . . That's the city school. You've got the kids that have access to guns, that see drugs sold on a daily basis. . . . If it happened [there], everybody would have said, "Did you hear what happened?" It would have been a big thing, but it wouldn't have been an attorney's son from Heath. People move out to Heath to put their kids in school so they don't go to the city school.

Media pundits weighed in on the causes of deadly school shootings, and academic studies, government commissions, congressional working groups, and presidential summits soon followed—too many, some argued, given how seldom school shootings happen. Indeed, critics such as Orlando Patterson, a sociologist, and Michael Eric Dyson, a professor of African-American studies, argued that the only reason these rare events generated the attention they did was that the shooters and most of the victims were white.[1] Had they been black, the attention would have been minimal and the need to explain the pathology less pronounced.

There are good reasons to dwell on rampage school shootings even though they are rare. They are an unprecedented kind of adolescent violence. We do not understand why they happen and have barely begun to consider their long-term consequences. Models derived from the study of urban violence have minimal value in deciphering the causes of rampage school shootings. Understanding what leads to these attacks does not free anyone of the obligation to think just as hard about the far more common incidence of urban shootings, something social scientists and policymakers have been working on for decades.

A steady stream of explanations came forth when rampage school shootings emerged, ranging from bullying to a southern culture of violence, from mental illness to lack of discipline, from violent media to the availability of guns. Do these theories hold water? We address this question by looking first at whether the phenomenon is new, epidemic, and worthy of a policy response, and then at how school rampages differ from other kinds of mass murders. Finally, we look at the ways in which received wisdom falls short of the kind of explanation we need.

EPIDEMIC OR "SCHOOL HOUSE HYPE"?

In the year following the massacre at Columbine High School, the nation's fifty largest newspapers printed nearly 10,000 stories related to the event and its aftermath, averaging about one story per newspaper every other day. No wonder parents worried about their children. A Gallup poll conducted in August 2000 found that 26 percent of American parents feared for their children's safety at school.[2] Twenty-nine percent of the high school students polled by ABC News in March 2001 said that they saw some risk of an attack at their school, but immediately after the Columbine attack, 40 percent saw some risk.[3]

The intense media coverage and renewed fear generated by the string of shootings around the country led some observers to claim that we are in the midst of an epidemic. This view has not gone unchallenged. Critics, such as the Justice Policy Institute, have argued that the widespread "panic" over school shootings is unjustified. In reports issued in July 1998 and April 2000, the institute reminded the nation that school is still the safest place for a child to be. Even during the seemingly deadly 1998–1999 school year, the chances of dying in school from homicide or suicide were less than one in 2 million. The rate of *out-of-school* homicides alone was about forty times higher.[4]

What worried opponents of the epidemic hypothesis was not just the hype but the responses emerging from policymakers. Resources that might be spent on addressing more deadly problems facing America's youth, especially abuse, neglect, and inner-city youth violence, could be wasted on the statistically minor threat of school shootings. In the rush to take action, critics warned, educators might adopt measures that could backfire, including zero-tolerance policies, profiling, or security measures that induce or aggravate a climate of fear or shunt more troubled youths into the criminal justice system. Finally, they feared that intense media coverage of school shootings would spark more violence as copy-cats swung into action.

Both perspectives, the epidemic view and the hype view, have some merit. Schools are indeed the safest place for our children to be, statistically. Yet if parents and children fear school violence, the schools' primary mission will be profoundly impaired. Parents' concerns cannot be dismissed as irrelevant just because they are not entirely rational.

Moreover, their fears are not unfounded. Attacks of the kind we discuss in this book *did* increase in the 1990s. They are assaults of a very specific kind, and it pays to bear their characteristics in mind. As we define them, rampage school shootings must:

- take place on a school-related public stage before an audience;
- involve multiple victims, some of whom are shot simply for their symbolic significance or at random; and,
- involve one or more shooters who are students or former students of the school.

This definition excludes many kinds of shootings that are cause for worry as well. For example, a student who comes to school looking to shoot a particular antagonist, or the school principal, but does not fire at others would not be counted here. Gang violence, revenge killings following drug deals that go bad—these kinds of mayhem are not included in our definition. Rampage school shootings, then, are a subset of a much larger category of murders or attempted murders and are closer in form to workplace or "postal" attacks than they are to single-victim homicides on or off campus. . . .

Rampage shootings are not unique to the United States. At least five multiple-victim, student-perpetrated incidents of school violence have taken place in other nations since 1975—two in Canada, one in Germany, one in the Netherlands, and one in Kenya.[5] Erfurt, Germany, was the unhappy scene of a sixth deadly attack in April 2002, in which a student who had been expelled from his high school returned in a fury and murdered sixteen people and wounded six. And although no rampage shootings occurred on middle or high school campuses after the September 11 terrorist attacks, several very similar shootings erupted on university campuses.[6]

Curiously, the peak in school shootings came at a time when other trend lines for violence were headed in the opposite direction. Decreases were recorded in adult homicides, youth homicides, drug- and gang-related youth homicide in the inner cities, and nonlethal violence in schools for the same period when school shootings spiked. Of course, it must be remembered that this sharp uptick still represents a small number of deaths and that the incidence of homicide on school campuses remains low.

Media coverage does contribute to a school violence panic and certainly aggravates the difficulties for communities in which rampage episodes have occurred. But the press should also be credited with helping to avert some potential rampages. Articles pointing to the unwillingness of students to come forward when they heard warnings of impending violence trained attention on the need for teenagers to speak out and helped foil some near-miss plots. For example, in November 2001 a plot to shoot up a New Bedford, Massachusetts, high school was uncovered solely because one seventeen-year-old participant came forward.

Unfortunately, relaying threats to adults is not the norm. A poll taken in November 2001 revealed that 61 percent of high school students who knew of someone bringing a gun to school did not report it; 56 percent of those who heard another student make a weapons-related threat said nothing. Even so, between 1999 and 2001, at least seven school shootings were prevented when peers reported the plans to school or law enforcement authorities.[7] The media coverage helped to shift the balance, at least in these cases.

RAMPAGE SCHOOL SHOOTINGS AND INNER-CITY VIOLENCE

Deadly violence among youths is indeed concentrated in poor areas of our nation's inner cities. For example, in 1989 the homicide rate for black males of ages fifteen through nineteen in urban areas was almost thirteen times higher than the national average for that age group.[8] Accordingly, researchers concerned about youth violence have concentrated primarily on explaining urban conditions; the results are both heartening and depressing.

Youth violence increased dramatically in the late 1980s and early 1990s, a period in which rates of crime and violence among other age groups actually went down. The increase was concentrated among, but not limited to, black males. Countless studies have tried to explain this trend. Some of them point to the size of the "at-risk" population, but this appears not to be a very influential factor: violence increased in the 1980s across all age categories, regardless of their size, simultaneously. And we are not likely to "age out" of this problem. Youth violence is not simply "kids killing kids." The majority of people who kill adolescents are adults, and most victims of adolescent killers are adults. Kids and guns do not mix well. The increase in homicides committed by youths came entirely in the form of murders involving firearms.[9]

Given these patterns, social scientists want to know two things: First, *which* youths are violent? Boys are more likely to commit violent acts than girls, and almost all violent offenders first manifest their tendencies between the ages of fourteen and eighteen. Violence peaks in the midteens for black and white males, although the rates for black males rise sharply again in their late twenties.[10] Beyond race, gender, and age, other risk factors for violence among youths include domestic violence and abuse, weak family bonding and ineffective supervision, lack of opportunities for education and employment, peers who engage in or accept violence, drug and alcohol use, gun possession, and individual temperament.[11] Finally, public perceptions notwithstanding, violent youths actually seem to be trying to accomplish goals where conventional means to achieve them are out of reach. They look to elevate their status among peers, cement an identity, acquire material goods or power over others, or find justice or retribution. Defiance of authority and the satisfaction that comes from risk taking and impulsivity also figure among the purposes to which violence is put.[12]

The second basic question social scientists ask about youth violence is why it increased in the late 1980s and early 1990s. Major culprits include the increased availability of guns, the crack cocaine epidemic, and a culture of violence, but none of these factors works by itself. Criminologist Mark Moore has suggested a synthesis that starts with the deteriorating social and economic conditions of inner-city neighborhoods in the late 1970s and early 1980s and adds responses of neighborhood residents that produced steadily increasing violence, fear, and breakdown of social control. As sociologist William Julius Wilson has argued, during the 1970s, joblessness increased in inner-city communities because of the outmigration of the middle class and the movement of blue-collar jobs away from U.S. cities. Economic and social stress contributed to the breakdown of families and other community supports for children and adolescents.[13] In search of support, identity, and meaning in the face of declining opportunity, more youths became involved in gangs, and with the introduction of crack cocaine, gangs turned to this volatile end of the drug market, where violence is the only available mode of protection and social control. Guns are readily available and proliferate. The streets become considerably more dangerous, and an environment develops in which victimization of the weak is commonplace.[14]

Despite the high rates of violent death among inner-city minority youths, relatively little of this violence occurs in city schools. For example, as of 2002, the last gun homicide in New York City's schools occurred in East New York's Jefferson High School on February 26, 1992. The city of Chicago has not had a gun homicide in a school since November 20, 1992.[15] National statistics paint a similar picture. In the nine school years between 1992 and 2001, on average only 8.7 students and slightly less than one school staff member per year were killed in violent attacks in urban schools. Student deaths from violent attacks in cities declined steadily over this period from eighteen in 1992–1993 to four in 2000–2001. . . .

Suburban and rural shootings almost always happen on school grounds, whereas this is rarely the case in the cities. In rural communities, school is one of the few "public stages" where an attention-seeking shooter can create a spectacle.[16] In the city, there are many other (potentially more meaningful) stages available. . . . [T]he school plays a central role in the social life of adults and children in suburban and rural settings. The school itself is a highly symbolic target of the attack.

There are some striking differences between urban youth who become violent and rampage shooters in suburban and rural schools. The former are usually labeled early as "problem kids." They are deemed troubled, and then their actions confirm the label. Murder represents the culmination of escalating problem behaviors. In contrast, school shooters are almost always kids no one would have expected to turn violent. They are often stigmatized as geeks or nerds. They act to *defy* their labels.

Finally, urban shootings "unfold through stages of decision making and escalation ... the culmination of lengthy ongoing disputes," whereas shootings in rural and suburban communities are "mass shootings, often with strangers as the victims if not the targets, and no acute ongoing dispute between victims and offenders."[17] Generally, no one is a stranger in the small, close-knit communities where these shootings have occurred. Nonetheless, rampage shooters often cast a wide net beyond individual targets, and some have no specific targets at all. Michael Carneal was surprised to learn the identity of his victims; Mitchell Johnson and Andrew Golden also were unaware of whom they killed. When urban violence erupts because of ongoing conflicts between individuals or groups that have a "beef" with one another, the identities of the opponents are almost always known. Rampage school shooters plan their attacks well in advance; urban shootings are more spontaneous, erupting at the tail end of a dispute.

Rampage school shootings unfold in places where residents say they trust their neighbors so much that they leave their doors unlocked and regularly watch out for each other. Urban youth violence happens in communities where residents are often afraid to leave their houses for fear of the drug dealers who rule the corners and addicts who rob and steal to support a habit. Sociologist Elijah Anderson describes how drugs and poverty create an environment in which the only way to avoid victimization is to maintain one's reputation and presentation as tough and willing to defend oneself at any cost.[18] When a young man lives with mortal fear, he may react quickly, almost spontaneously, to threats as a form of self-protection. Rampage school shooters are not spontaneous at all. Michael Carneal planned his assault over a period of weeks, as did Mitchell [Johnson] and Andrew [Golden]. . . .

Despite the differences, there are important similarities between the two kinds of violence that should not be ignored. One is the potential link between violence against others and violence against oneself—between homicide and suicide. Some school shooters are suicidal or deeply depressed. Eric Harris and Dylan Klebold killed themselves after the Columbine massacre. T. J. Solomon put a pistol in his mouth, but didn't pull the trigger, after he fired into the lobby of his Conyers, Georgia, high school. Friends of Andrew Wurst, the rampage shooter from Edinboro, Pennsylvania, feared that he might kill himself the night he killed his science teacher and wounded two students at a school dance.

For these youths, a violent attack may be a way to "go down in flames" or to commit suicide by forcing others, especially police, to shoot them. Law enforcement personnel, criminologists, and psychiatrists have a term for this situation: "victim-precipitated homicide" or "suicide by cop." It is an "exit strategy" favored by those who seek a spectacular—and particularly masculine—suicidal end, and it is surprisingly common. One study of shootings involving officers of the Los Angeles County Sheriff's Department found that more than 10 percent of the incidents over a ten-year period met this criteria. . . .

Summing up the evidence, we conclude that there are significant differences between the rampage school shootings, which are mostly committed by youth in rural or suburban settings, and the kind of youth violence associated with inner-city settings. The random nature of victimization, the degree of advance planning, and the locations and public nature of the events all separate rampage shootings from urban shootings. We are looking at a separate genre of youth violence.

COMPARING ADULT AND YOUTH MASS MURDERS

Mass murders by adults bear some resemblance to rampage school shootings. In the twenty years between 1976 and 1995, 483 mass murder incidents took place in the United States (by offenders of any age)—on average about two per month.[19] Although we tend to imagine that the victims of such rampages were killed by strangers, in fact only about 20 percent of mass murders involve victims whom the offender did not know. Almost 40 percent of the victims are family members. Some characteristics of mass murders resemble rampage school shootings. For example, compared with single-victim murders, a larger proportion of mass murders occur in small towns or rural areas (43.3 percent compared with 24.1 percent). Mass murderers are also predominantly white (62.9 percent) and male (94.4 percent), and slightly over 40 percent of such crimes are committed by young adults, aged twenty to twenty-nine. . . .

One form of adult rampage shooting that may be especially similar to school shootings is workplace homicide, particularly in instances in which an employee or former employee attacks his boss or coworkers. Disgruntled employees shoot their supervisors and colleagues depressingly often in the United States, and the number of such incidents has doubled since the 1970s, to about two attacks per month.[20]

Two similarities between workplace attacks and school shootings are noteworthy. First, like school shooters, workplace shooters are arguably attacking not just individuals but the institution itself. School shooters may be angry at the entire social system of the school and the community. The same kind of generalized fury may be present in so-called "postal" rampages.[21] Second, like school shooters, workplace shooters often suffer from painful marginalization. They are the oddballs in the office, the guys that others make fun of behind their backs, the ostracized and harassed. For school shooters, relations with peers at the school may deny them status and security. For workplace shooters, being fired may be the last straw in part because it drops their status to zero, especially for middle-aged men—the typical workplace shooter—for whom unemployment is particularly difficult. Shooting asserts power.[22]

From our point of view, workplace shootings and school rampages have some profound and unfortunate similarities. They represent the tips of similar icebergs, where those who feel ostracized, marginalized, and threatened with emasculation react with murderous violence. Obviously, they are more similar in motivation than in design, but if we want to understand why they happen, it is useful to focus on those commonalities.

POPULAR EXPLANATIONS OF SCHOOL SHOOTINGS: HOW DO THEY FARE?

Academics, government working groups, and think tank researchers have tried to account for the patterns we find in school shootings. Perhaps as many as a dozen explanations have emerged in the process. It is important to consider just how well they do the job of explaining the problem . . . and to identify where they fall short. . . .

Mental Illness

Few school shooters are diagnosed with mental illnesses before their crimes. Yet many are discovered afterward to be mentally ill. Depression and schizophrenia or one of its variants are particularly common. Only a small minority of these children are under treatment. Rampage shooters on campus do share troubling mental conditions with other killers. In a study of 102 adult and juvenile rampage killers conducted by the *New York Times*, "33 of the offenders killed themselves after their crimes. Nine tried or wanted to commit suicide, and four killed themselves later. Nine were killed by police officers or others, perhaps committing what some refer to as 'suicide by cop.'"[23] In more than half of the cases, friends, family, and even the offenders themselves tried to get help or warn others about the impending violence. . . .

We might expect adults who routinely deal with adolescents, such as school personnel, to be able to spot mental illness. It turns out to be exceptionally difficult, largely because problems like clinical depression or schizophrenia may be in their early stages, lacking some of the symptoms that manifest themselves later in life. At the onset of the disease, kids are often aware of how different they are from others and, feeling the stigma that comes with this territory, work hard to conceal their troubles. . . . It is not that teachers don't care about their students. Instead . . . the organization and culture of schools make it extremely difficult to spot troubled adolescents before they lash out.

Given the number of adolescents who are depressed and suicidal, mental illness cannot be viewed as a straightforward predictor of rampage school shootings. According to the National Institute of Mental Health, up to 8.3 percent of adolescents suffer from depression, and in 1996 suicide was the third leading cause of death for people aged fifteen to twenty-four. More than 2 million Americans are affected by schizophrenia, which usually develops in the late teens or early twenties. Ten percent of schizophrenia patients eventually commit suicide.[24] These trends dwarf the number of rampage shootings in schools, and for this reason we must treat the relationship of mental illness to these attacks with caution.

"He Just Snapped"

When we are at a loss to explain something, we look for the most proximate or immediate potential cause. We note that a shooter was just rejected by a girl, punished by a parent or a school official, insulted or assaulted by peers. Timing leads to the assumption that the incident caused the boy to "snap." Our legal system supports this kind of logic—though probably more in popular myth, television, and movies than in reality. We allow defendants to plead "temporary insanity," in which a person is briefly unable to tell right from wrong, given the overpowering influence of his surroundings.

We reject the notion that proximate events explain much about school shootings. They may be the straw that broke the camel's back, but at most they help explain *when* a shooting happens rather than *why*. Events that seem to be precipitators usually turn out on closer inspection not to be. . . .

To put great weight on immediate precipitating events also ignores an important aspect of most school shootings: they are usually planned well in advance. . . .

Family Problems

No explanation for school shootings has received more attention than family problems. The Final Report of the Congressional Bipartisan Working Group on Youth Violence is a good example of the notion that it all begins in the home: "Although there is no single cause for youth violence," the report concludes, "the most common factor is family dysfunction."[25] But what kinds of families are these? Here the picture gets much murkier. A Secret Service study of forty-one school shooters who committed their crimes between 1974 and 2000 found that the perpetrators come from a variety of family backgrounds, ranging "from intact families with numerous ties to the community to foster homes with histories of neglect."[26]

Problems within the family, such as divorce, domestic or sexual abuse, frequent relocations, and fragile family relationships as well as lack of awareness or involvement in children's lives, were all cited by the people we interviewed in Westside and Heath as reasons that youths turn violent. . . .

Diminishing adult and parental authority over children is another variant of this theme and was frequently mentioned as a growing problem by community residents we interviewed. Parents are increasingly losing control of their kids, they say, because they work long hours and have little time to spend with them. Moreover, many people in the community feel that parents have relinquished authority for disciplining their children to the school while the school in turn believes that, increasingly, parents are not supportive of the school's disciplinary decisions.

There can be no doubt that some school shooters have difficult family lives. An FBI review of eighteen school shooting cases concluded that potential warning signs of a school shooter include turbulent parent–child relationships, family acceptance of pathological behavior, access to weapons in the home, lack of closeness or intimacy with family members, a child who "rules the roost," and lack of limits or monitoring of television and the Internet.[27] As the FBI correctly warns, many students fit this profile, so distinguishing between shooters and nonshooters solely on the basis of family characteristics is impossible....

Bullying

Bullying at school is probably the most commonly accepted explanation for school shootings, and for good reason. Shooters do express fury at being excluded, teased, and tormented.

Bullying is a nationwide problem. According to the National Association of School Psychologists, about 160,000 children miss school every day for fear of bullying.[28] A 1998 study of almost 16,000 children in grades six though ten found that 10.6 percent are subjected to this kind of torment at least "sometimes," 13 percent bully others at least "sometimes," and an additional 6.3 percent both are bullied and bully others.[29] Victims of harassment tend to exhibit lower levels of social and emotional adjustment. It is harder for them to make friends, and they are lonely. Bullies get involved in delinquency and substances abuse, and they tend to do poorly in school. Kids who both bully and are bullied seem to be at the greatest risk; they are both victims and aggressors.[30]...

Peer Support

In contrast to adult rampage shootings, youth shootings often are committed by groups or with encouragement from peers.[31] In almost half of the thirty-seven school shooting incidents studied by the Secret Service, "attackers were influenced or encouraged by others."[32] The Columbine and Westside shootings were both done by pairs of students, and law enforcement officials and prosecutors have attempted to link other youths to violent plots.[33]...

The importance of influence by peers can be hard to substantiate, however. Even when researchers have access to all records and interview people who have direct knowledge of the individuals involved, the extent of their participation is not easily assessed. Students who are accused of joining a plot or encouraging a shooter keep quiet about their involvement after the fact. Hence, what we know about peer connections derives from the shooters' own perceptions, which are sometimes clouded by mental illness, rage, and their own motivations, including the desire to underplay the contributions of peers to protect them from criminal prosecution or to claim the "credit" for themselves. In Heath, for example, none of the suspected coconspirators would talk to us because of pending lawsuits.

The degree to which shooters receive support from their peers extends beyond the alleged conspiracies to the reactions of those who are forewarned. In over three-quarters of the shootings examined in the Secret Service study, other students were told of the impending attack.[34] Some of these warnings provided acquaintances the opportunity to encourage the shooters, or at least to convey their approval. In other cases, friends may have tried to discourage the shooters or keep them out of trouble. Andrew Wurst's friends knew he had a gun and was suicidal. They took turns watching him to make sure he did not kill himself.[35]...

Peer support for a school shooting must be seen in the proper context. In no case does a shooter appear to be overwhelmingly influenced or manipulated by his peers. But in the complex mixture of factors that lead to a school shooting, peer support—often misinterpreted by the socially marginal or psychologically unstable—can lead an individual further down the path toward violence. . . .

Changing Communities

We tend to think of small-town life as stable and close—the opposite of the rootlessness that seems to predominate in new suburbs or big cities. Yet change has come to many parts of the United States. Dense social ties have given way to more impersonal relations in rural areas and small towns.[36] Demographic change brought on by migration and population turnover have disrupted tight social ties between families and neighbors. Emile Durkheim, classical social theorist of the nineteenth century, argued that rapid social change produced a condition called "anomie." He worried that it would lead to increases in suicide as the interdependence of "old" societies gave way and eroded the social ties that keep people stable.[37] . . .

Laments about parents being less invested in their children's lives do not always square with reality. Heath and Westside are communities that boast a high level of community connectedness and solidarity. Teachers know the parents of their students from neighborhood, family, church, or other connections. People watch out for one another. And although there are newcomers in town, families are deeply rooted: children follow their parents and often their grandparents into the same schools, generation after generation. Indeed, one of the reasons people in both communities gave us for remaining there was just the depth of the connections they enjoyed with their neighbors and fellow church members.

In fact, we argue . . . that precisely the opposite problem played a role in both school shootings. It was not the weakness of social ties in Westside and Heath that proved their undoing, but the strength of those bonds. Dense, all-encompassing, interconnected networks of friends and family can make the lives of misfits unbearable and actually stifle the flow of information about potential warning signs.

Culture of Violence

The first few school shootings occurred in southern states, leading some observers to wonder whether a "southern culture of violence" was behind the crisis. According to this view, the South has evolved a distinctive culture that legitimates violence as a way of solving disputes or of guarding and gaining social status. The antebellum South was notorious for the use of intimidation, threats, and the deployment of real violence to control former slaves.[38] . . .

Is there a culture of honor in these communities? Andrew Golden may be the closest exemplar. He stood guard over the periphery of his family's property and warned kids off with a BB gun. Andrew did not have trouble with the idea that guns could settle arguments. But there is nothing particularly southern about his perspective. Mitchell Johnson—whose worst moments even before the shooting reveal a belligerent side—did not grow up in the South. His formative years were spent in Minnesota. Michael Carneal—who did grow up in the South—was not considered particularly violent and did not himself resort to violence to settle disputes. . . . Indeed, one of Michael's brewing problems before the shooting was that he was not really able to stand up for himself. He was known as the kid who just "took" physical abuse without confronting or standing up to anyone.

The "southern culture hypothesis" was dealt a series of setbacks as shootings and near-miss plots occurred in California, Pennsylvania, Washington, Colorado, Oregon, Massachusetts, Kansas, Florida, Michigan, and New York. If a particularly southern culture were responsible for school shootings, we would expect to see a much greater proportion of school shootings occurring in the South. We don't.

Gun Availability

It has not escaped the public's attention that school shootings depend on access to guns. Mass murders tend not to happen—in school or anywhere else—when knives are the only weapon available. Scholars and media pundits have been quick to point out that the increase in the number of firearms has coincided with the recent string of school shootings.[39] Since 1970, the number of guns in the United States has doubled, to about 200 million. One might conclude that access to guns is spreading rapidly, but the increase has actually been fueled by people who are already gun owners acquiring additional firearms. The proportion of adults who own guns has stayed relatively constant since 1980 at about 30 percent. This is not a low number; it is the highest proportion of any industrialized country, but it hasn't changed much over the years.[40]

National surveys tell us that about 20 percent of youths have brought some sort of weapon to school. Far more youths than this claim that they could easily get a gun. But the geographic differences are significant. In rural areas and small towns, where most school shootings have taken place, access to firearms is more assured.[41] Our own interviews with adolescents in Heath and Westside match these survey findings. Almost all of the kids in these communities told us that it would be very easy for them to get a gun. . . .

The presence of guns is clearly causally related to the shootings, but it is not clear that their increasing availability accounts for the recent spate of massacres in schools. Hunting communities have always kept guns at the ready, but school shootings began to occur fairly recently.

Violent Media

Americans believe that the country is saturated with bloodthirsty imagery: Video games, television, movies, and music are filled with ugly violence. By all accounts, the situation is getting worse. Exposure to violent media has increased dramatically among our youth over the last decade, pushing media influence forward as a prime explanation for the string of school shootings in the mid- to late 1990s. Indeed, when President Clinton held a presidential summit on youth violence in May 1999, he took great pains to welcome high-level representatives from the entertainment industry. Quite apart from the moral objections people have to these sensational materials, psychologists tell us that violence desensitizes viewers to its consequences. Young people are particularly impressionable or ill-equipped to distinguish between drama and reality.

The available scientific evidence on the contribution violent media makes to violence among youth is inconclusive, although it is suggestive. Exposure to media violence is consistently associated with a variety of antisocial behaviors, from trivial violence against toys to serious criminal violence. Children who are exposed to media violence tend to identify violence as the best solution to a problem or to exhibit hostility and nonviolent but aggressive behavior. Consuming violent media inures viewers to the impact of pain; hence they are less aroused or disturbed—and less likely to intervene—when they witness bloody scenes.[42] . . .

Millions of young people play video games full of fistfights, blazing guns, and body slams. Bodies litter the floor in many of our most popular films. Yet only a minuscule fraction of the consumers become violent. Hence, if there is an effect, children are not all equally susceptible to it. In our own interviews, almost all adolescents scoffed at the idea that they were so easily influenced by television, music, movies, and video games.

Michael Carneal did play violent video games, and he saw the usual run of bloody movies. He had been playing violent video games since he was a child, including "Mortal Kombat" and "Mech Warrior," favorites of millions of American teenagers. . . .

While there is clearly no one-to-one correspondence between exposure to violent video games and behavior, we agree with one of his psychiatrists that "Michael's exposure to media violence can be regarded as a factor which contributed to the attitudes, perceptions, and judgment which led to his violent behavior."[43] There is evidence for this perspective earlier in Michael's life. One of his church group leaders remembered that when hypothetical dilemmas were presented to the youth group he participated in, Michael's solutions often involved "shooting someone with a bazooka." Michael and his friends discussed a number of violent fantasies that were in part based on things they had seen in movies and video games. Although it would be far too simplistic to say that Michael's actions were caused by the movies he saw, violent video games and media provided a template for action and images of masculinity that appealed to a boy who felt weak and socially inadequate. . . .

The Copycat Effect

The timing and clustering of school shootings suggests that the later tragedies took their inspiration from the earlier ones. Not all shootings are sparked by copycatting. The rampage in Pearl, Mississippi, happened not long before Michael Carneal's shooting. Michael took his fateful steps a few months before Mitchell and Andrew fired on Westside. Yet there was never any suggestion that they were particularly aware of the events in these other communities.

Research on imitative violence suggests that media coverage affects the form and method of crimes rather than the amount of crime. This may not be the case for individuals who are suicidal, however. There is some evidence, although it is highly contested, that youth suicides spike after highly publicized suicides, especially by celebrities. It seems inconceivable that otherwise healthy and happy adolescents would shoot up their school because others went on rampages before them. Rather, troubled youths may see a model for a solution to their problems in previous school shootings. . . .

NOTES

1. Orlando Patterson, "When 'They' Are 'Us,'" *New York Times*, April 30, 1999, p. A31; Michael Eric Dyson, "Uglier Than Meets the Eye," *Chicago Sun-Times*, March 13, 2001, p. 25.

2. Sixteen months earlier, the day after the Columbine shooting, 50 percent of parents said they feared for their children's safety at school. Susan Gembrowski, "Life Goes on in a Troubled World: Well-Adjusted Teens Thrive Despite Violent Times," *San Diego Union-Tribune*, November 11, 2001, p. A1.

3. There were 500 respondents in this survey. ABC News polls conducted March 11, 2001, and April 25, 1999. Available online at http:abcnews.go.com/sections/GMA/ GoodMorningAmerica/GMA_School_Violence_POLL.html.

4. Elizabeth Donohue, Vincent Schiraldi, and Jason Ziedenberg, *School House Hype: School Shootings and the Real Risks Kids Face in America* (Washington, DC: Justice Policy Institute, 1998); and *School House Hype: Two Years Later* (San Francisco: Center on Juvenile and Criminal Justice, 2000).

5. National Research Council and the Institute of Medicine. *Deadly Lessons: Understanding Lethal School Violence: Case Studies of School Violence Committee*, Mark H. Moore, Carol V. Petrie, Anthony A. Braga, and Brenda L. McLaughlin, eds. (Washington, DC: National Academies Press, 2003), Table 9-2, p. 295.

6. For example, on January 16, 2002, a student killed three and wounded three at the Appalachian School of Law in West Virginia, and on October 28, 2002, a student killed three and committed suicide at the University of Arizona nursing school. Frances X. Clines, "3 Slain at Law School; Student is Held," *New York Times*, January 17, 2002, p. A18; John M. Broder, "Student Kills 3 Instructors and Himself at U. of Arizona," *New York Times*, October 29, 2002, p. A20.

7. Tatsha Robertson, "Across the Nation, School Attack Plots Pose Legal Challenge," *Boston Globe*, December 16, 2001, pp. A1, A26, lists twelve post-Columbine plots, of which seven appear to have been foiled after tips from peers.

8. Lois A. Fingerhut, Deborah D. Ingam, and Jacob J. Feldman, "Firearm and Non-firearm Homicide Among Persons 15 Through 19 Years of Age: Differences by Level of Urbanization, United States, 1979 Through 1989," *Journal of the American Medical Association* 267, no. 22 (1992): 3048–3053.

9. Philip J. Cook and John H. Laub, "The Unprecedented Epidemic in Youth Violence," in Michael Tonry and Mark H. Moore, eds., *Youth Violence, Crime, and Justice: A Review of Research*, vol. 24 (Chicago: University of Chicago Press, 1998).

10. Delbert Elliott, John Hagan, and Joan McCord, *Youth Violence: Children at Risk* (Washington, DC: American Sociological Association, 1998). Black males are more likely than white males to be arrested for the same behavior.

11. Margaret A. Hamburg, "Youth Violence Is a Public Health Concern," in Delbert S. Elliott, Beatrix Hamburg, and Kirk R. Williams, eds., *Violence in American Schools* (New York: Cambridge University Press, 1998).

12. Jeffrey Fagan and Deanna L. Wilkinson, "Social Contexts and Functions of Adolescent Violence," in *Violence in American Schools*.

13. See William Julius Wilson, *The Truly Disadvantaged* (Chicago: University of Chicago Press, 1987), and William Julius Wilson, *When Work Disappears* (New York: Knopf, 1996).

14. Mark H. Moore, "Youth Violence in America," in Michael Tonry and Mark H. Moore, eds., *Youth Violence, Crime, and Justice*.

15. Mindy Thomson Fullilove et al., "What Did Ian Tell God? School Violence in East New York," and John Hagan, Paul Hirschfield, and Carla Shedd, "Shooting at Tilden High: Causes and Consequences," in *Deadly Lessons*.

16. Diane Wilkinson and Jeffrey Fagan, "What We Know About Gun Use Among Adolescents," *Clinical Child and Family Psychology Review* 4, no. 2 (2001) 109–132.

17. Ibid., p. 128.

18. See Elijah Anderson, *Streetwise* (Chicago: University of Chicago Press, 1990), and Elijah Anderson, *Code of the Street* (New York: Norton, 1999).

19. Criminologists James Alan Fox and Jack Levin define a mass murder as a single event involving four or more homicides. These incidents involved 697 offenders and 2,352 victims. James Alan Fox and Jack Levin, "Multiple Homicide: Patterns of Serial and Mass Murder," in *Crime and Justice: A Review of Research*, vol. 23, edited by Michael Tonry (Chicago: University of Chicago Press, 1998).

20. James Alan Fox and Jack Levin, "Firing Back: The Growing Threat of Workplace Violence," *Annals of the American Academy of Political and Social Sciences* 536 (1994) 16–30.

21. Ibid.

22. Ibid.

23. Ford Fessenden, "They Threaten, Seethe, and Unhinge, Then Kill in Quantity," *New York Times*, April 9, 2000, p. A1.

24. National Institute of Mental Health, *Depression Research Fact Sheet and Schizophrenia Research Fact Sheet* (Bethesda, MD: National Institute of Mental Health, 2000).

25. Representatives Jennifer Dunn and Martin Frost, "Bipartisan Working Group on Youth Violence Final Report," 1999. Available at www.house.gov/dunn/workinggroup/wkg.htm.

26. Vossekuil et al., "Safe School Initiative: An Interim Report on the Prevention of Targeted Violence in Schools," p. 5.

27. Mary Ellen O'Toole, "The School Shooter: A Threat Assessment Perspective," Federal Bureau of Investigation, 2000. Available at http://www.fbi.gov/publications/school/school.pdf.

28. Cited in William Pollack, *Real Boys* (New York: Henry Holt, 1998), p. 343.

29. Boys were most likely to report being teased about looks or speech (19.8 percent),

followed by being hit, slapped, or pushed (17.5 percent), receiving sexual comments or gestures (17.5 percent), and being the subject of rumors (16.7 percent). Nansel et al., "Bullying Behaviors Among U.S. Youth: Prevalence and Association with Psychosocial Adjustment," *Journal of the American Medical Association* 285, no.16 (2001): 2094–2100.

30. Ibid.

31. Ford Fessenden, "How Youngest Killers Differ: Peer Support," *New York Times*, April 9, 2000, p. 29.

32. Vossekuil et al., "Safe School Initiative: An Interim Report on the Prevention of Targeted Violence in Schools," p. 6.

33. Fessenden, "How Youngest Killers Differ."

34. Vossekuil et al., "Safe School Initiative: An Interim Report on the Prevention of Targeted Violence in Schools."

35. William DeJong, Joel C. Epstein, and Thomas E. Hart, "Bad Things Happened in Good Communities: The Rampage Shooting in Edinboro, Pennsylvania, and Its Aftermath," *Deadly Lessons*.

36. Laura Vozzella, "School Killings Cast Shadow on Small Towns," *Fort Worth Star-Telegram*, March 29, 1998, p. 1. See also Mercer L. Sullivan and Rob T. Guerette, "The Copycat Factor: Mental Illness, Guns, and the Shooting Incident at Heritage High School, Rockdale County, Georgia," in *Deadly Lessons*.

37. Emile Durkheim, *Suicide: A Study in Sociology* (New York: Free Press, 1951).

38. On culture of violence, see Fox Butterfield *All God's Children: The Bosket Family and the American Tradition of Violence* (New York Bard:, 1998).

39. For example, see Lewis W. Diuguid,"Guns Are Way of Death in America," *Kansas City Star*, April 4, 1998, p. C1, or Robert L. Kaiser, "Kentucky Killings Shatter a Town's Sense of Innocence," *Fort Worth Star-Telegram*, December 7, 1997, p. 1.

40. Philip J. Cook, Mark H. Moore, and Anthony Braga, "Gun Control" in *Crime: Public Policies for Crime Control*, James Q. Wilson and Joan Petersilia, eds. (Oakland, CA: Institute for Contemporary Studies, 2001).

41. Wilkinson and Fagan, "What We Know About Gun Use Among Adolescents"; Josephson Institute of Ethics, "The Ethics of American Youth: 2000 Report Card," 2001, available at www.josephsoninstitute.org; Joseph F. Sheley and James D. Wright, "High School Youth, Weapons, and Violence: A National Survey," National Institute of Justice Research in Brief, 1998.

42. J. Cantor, "Media Violence," *Journal of Adolescent Health* 27, no. 2 (2000) 30–34.

43. Dewey G. Cornell, "Psychological Evaluation: Michael Adam Carneal," September 3, 1998, p. 16.

KEY CONCEPTS

anomie   deviance   victimization

DISCUSSION QUESTIONS

1. Which of the explanations discussed in this article make most sense to you? Which are more psychological? Which have more sociological implications?

2. How are school shootings similar to other kinds of gun violence? How are they different?

22

The Rich Get Richer and the Poor Get Prison

JEFFREY H. REIMAN

This essay challenges the reader to view the criminal justice system from a radically different angle. Specifically, Jeffrey Reiman argues that the corrections system and broader criminal justice policy in the United States simply provide the illusion of fighting crime. In reality, he argues that criminal justice policies reinforce public fears of crimes committed by the poor. These policies, in turn, help to maintain a "criminal class" of disadvantaged people.

A criminal justice system is a mirror in which a whole society can see the darker outlines of its face. Our ideas of justice and evil take on visible form in it, and thus we see ourselves in deep relief. Step through this looking glass to view the American criminal justice system—and ultimately the whole society it reflects—from a radically different angle of vision.

In particular, entertain the idea that the goal of our criminal justice system is not to eliminate crime or to achieve justice, *but to project to the American public a visible image of the threat of crime as a threat from the poor.* To do this, the justice system must present us with a sizable population of poor criminals. To do that, it must fail in the struggle to eliminate the crimes that poor people commit, or even to reduce their number dramatically. Crime may, of course, occasionally decline, as it has recently—*but largely because of factors other than criminal justice policies.* . . .

SOURCE: Jeffrey H. Reiman. 2005. *The Rich Get Richer and the Poor Get Prison: Ideology, Class, and Criminal Justice*, 7ᵗʰ ed. Boston, MA: Allyn and Bacon, pp. 1–2.

In recent years, we have quadrupled our prison population and, in cities such as New York, allowed the police new freedom to stop and search people they suspect. No one can deny that if you lock up enough people, and allow the police greater and greater power to interfere with the liberty and privacy of citizens, you will eventually prevent some crime that might otherwise have taken place. . . . I shall point out just how costly and inefficient this means of reducing crime is, in money for new prisons, in its destructive effect on inner-city life, in reduced civil liberties, and in increased complaints of police brutality. I don't deny, however, that these costly means do contribute *in some small measure* to reducing crime. Thus, when I say . . . that criminal justice policy is failing, I mean that it is failing to eliminate our high crime rates. We continue to see a large population of poor criminals in our prisons and our courts, while our crime-reduction strategies do not touch on the social causes of crime. Moreover, our citizens remain fearful about criminal victimization, even after the recent declines. . . .

Nearly 30 years ago, I taught a seminar for graduate students titled "The Philosophy of Punishment and Rehabilitation." Many of the students were already working in the field of corrections as probation officers, prison guards, or halfway-house counselors. Together we examined the various philosophical justifications for legal punishment, and then we directed our attention to the actual functioning of our correctional system. For much of the semester, we talked about the myriad inconsistencies and cruelties and the overall irrationality of the system. We discussed the arbitrariness with which offenders are sentenced to prison and the arbitrariness with which they are treated once there. We discussed the lack of privacy and the deprivation of sources of personal identity and dignity, the ever-present physical violence, as well as the lack of meaningful counseling or job training within prison walls. We discussed the harassment of parolees, the inescapability of the "ex-con" stigma, the refusal of society to let a person finish paying his or her "debt to society," and the absence of meaningful noncriminal opportunities for the ex–prisoner. We confronted time and again the bald irrationality of a society that builds prisons to prevent crime knowing full well that they do not, and one that does not seriously try to rid its prisons and postrelease practices of those features that guarantee a high rate of *recidivism*, the return to crime by prison alumni. How could we fail so miserably? We are neither an evil nor a stupid nor an impoverished people. How could we continue to bend our energies and spend our hard-earned tax dollars on cures we know are not working?

Toward the end of the semester, I asked the students to imagine that, instead of designing a criminal justice system to reduce and prevent crime, we designed one that would maintain a stable and visible "class" of criminals. What would it look like? The response was electrifying. Here is a sample of the proposals that emerged in our discussion.

First   It would be helpful to have laws on the books against drug use, prostitution, and gambling—laws that prohibit acts that have no unwilling victim. This would make many people "criminals" for what they regard as normal behavior and would increase their need to engage in *secondary crime* (the drug addict's need to steal to pay for drugs, the prostitute's need for a pimp because police protection is unavailable, and so on).

Second   It would be good to give police, prosecutors, and/or judges broad discretion to decide who got arrested, who got charged, and who got sentenced to prison. This would mean that almost anyone who got as far as prison would know of others who committed the same crime but were not arrested, were not charged, or were not sentenced to prison. This would assure us that a good portion of the prison population would experience their confinement as arbitrary and unjust and thus respond with rage, which would make them more antisocial, rather than respond with remorse, which would make them feel more bound by social norms.

Third   The prison experience should be not only painful but also demeaning. The pain of loss of liberty might deter future crime. But demeaning and emasculating prisoners by placing them in an enforced childhood characterized by no privacy and no control over their time and actions, as well as by the constant threat of rape or assault, is sure to overcome any deterrent effect by weakening whatever capacities a prisoner had for self-control. Indeed, by humiliating and brutalizing prisoners, we can be sure to increase their potential for aggressive violence.

Fourth   Prisoners should neither be trained in a marketable skill nor provided with a job after release. Their prison records should stand as a perpetual stigma to discourage employers from hiring them. Otherwise, they might be tempted *not* to return to crime after release.

Fifth   Ex-offenders' sense that they will always be different from "decent citizens," that they can never finally settle their debt to society, should be reinforced by the following means. They should be deprived for the rest of their lives of rights, such as the right to vote. They should be harassed by police as "likely suspects" and be subject to the whims of parole officers who can at any time threaten to send them back to prison for things no ordinary citizens could be arrested for, such as going out of town, or drinking, or fraternizing with the "wrong people."

And so on.

In short, *when asked to design a system that would maintain and encourage the existence of a stable and visible "class of criminals," we "constructed" the American criminal justice system!* . . .

. . . [T]he practices of the criminal justice system keep before the public the *real* threat of crime and the *distorted* image that crime is primarily the work of the poor. The value of this *to those in positions of power* is that it deflects the discontent and potential hostility of Middle America away from the classes above them and toward the classes below them. If this explanation is hard to swallow, it should be noted in its favor that it not only explains the dismal failure of criminal justice policy to protect us against crime but also explains why the criminal justice system functions in a way that is biased against the poor at every stage from arrest to conviction. Indeed, even at an earlier stage, when crimes are defined in law, the system concentrates primarily on the predatory acts of the poor and tends to exclude or deemphasize the equally or more dangerous predatory acts of those who are well off.

In sum, I will argue that *the criminal justice system fails in the fight against crime while making it look as if crime is the work of the poor.* This conveys the image that the real danger to decent, law-abiding Americans comes from below them, rather than from above them, on the economic ladder. This image sanctifies the status quo with its disparities of wealth, privilege, and opportunity, and thus serves the interests of the rich and powerful in America—the very ones who could change criminal justice policy if they were really unhappy with it.

Therefore, it seems appropriate to ask you to look at criminal justice "through the looking glass." On the one hand, this suggests a reversal of common expectations. Reverse your expectations about criminal justice and entertain the notion that the system's real goal is the very reverse of its announced goal. On the other hand, the figure of the looking glass suggests the prevalence of image over reality. My argument is that the system functions the way it does *because it maintains a particular image of crime: the image that it is a threat from the poor.* Of course, for this image to be believable, there must be a reality to back it up. The system must actually fight crime—or at least some crime—but only enough to keep it from getting out of hand and to keep the struggle against crime vividly and dramatically in the public's view, never enough to substantially reduce or eliminate crime.

I call this outrageous way of looking at criminal justice policy the *Pyrrhic defeat* theory. A "Pyrrhic victory" is a military victory purchased at such a cost in troops and treasure that it amounts to a defeat. The Pyrrhic defeat theory argues that the failure of the criminal justice system yields such benefits to those in positions of power that it amounts to success. . . .

The Pyrrhic defeat theory has several components. Above all, it must provide an explanation of *how* the failure to reduce crime substantially could benefit anyone—anyone other than criminals, that is. . . . I argue there that the failure to reduce crime substantially broadcasts a potent *ideological* message to the American people, a message that benefits and protects the powerful and privileged in our society by legitimating the present social order with its disparities of wealth and privilege, and by diverting public discontent and opposition away from the rich and powerful and onto the poor and powerless.

To provide this benefit, however, not just any failure will do. It is necessary that the failure of the criminal justice system take a particular shape. *It must fail in the fight against crime while making it look as if serious crime and thus the real danger to society are the work of the poor.* The system accomplishes this both by what it does and by what it refuses to do. . . . I argue that the criminal justice system refuses to label and treat as crime a large number of acts of the rich that produce as much or more damage to life and limb as the crimes of the poor. . . . [E]ven among the acts treated as crimes, the criminal justice system is biased from start to finish in a way that guarantees that, *for the same crimes*, members of the lower classes are much more likely than members of the middle and upper classes to be arrested, convicted, and imprisoned—thus providing living "proof" that crime is a threat from the poor. . . .

Our criminal justice system is characterized by beliefs about what is criminal, and beliefs about how to deal with crime, that predate industrial society. Rather than being anyone's conscious plan, the system reflects attitudes so deeply embedded in tradition as to appear natural. To understand why it persists even though it fails to protect us, all that is necessary is to recognize that, on the one hand, those who are the most victimized by crime are not those in positions to make and implement policy. Crime falls more frequently and more harshly on the poor than on the better off. On the other hand, there are enough benefits to the wealthy from the identification of crime with the poor and the system's failure to reduce crime that those with the power to make profound changes in the system feel no compulsion nor see any incentive to make them. In short, the criminal justice system came into existence in an earlier epoch and persists in the present because, even though it is failing—indeed, because of the way it fails—it generates no effective demand for change. When I speak of the criminal justice system as "designed to fail," I mean no more than this. I call this explanation of the existence and persistence of our failing criminal justice system the *historical inertia* explanation. . . .

KEY CONCEPTS

labeling theory   social class   social institution

DISCUSSION QUESTIONS

1. What does Reiman mean in arguing that the current criminal justice system works to maintain a class of criminals? Do you agree or disagree that our corrections system fails to rehabilitate and fails to deter crime?

2. If you had the power to change the corrections system in the United States, what changes would you make to help reduce and prevent crime?

Applying Sociological Knowledge: An Exercise for Students

Become a norm breaker! Think of some norm we have in society that you can go out in public and violate. Make sure it is legal! How do people treat you when you stop doing something that is implicitly expected of you? How does it feel to go against what you feel you should be doing? Was this norm something you thought about doing before or is it something that you did without even thinking? Notice how hard it is to deviate from expected norms.

23

The Communist Manifesto

KARL MARX AND FREDERICH ENGELS

The analysis of the class system under capitalism, as developed by Marx and Engels, continues to influence sociological understanding of the development of capitalism and the structure of the class system. In this classic essay, first published in 1848, Marx and Engels define the class system in terms of the relationships between capitalism, the bourgeoisie, and the proletariat. Their analysis of the growth of capitalism and its influence on other institutions continues to provide a compelling portrait of an economic system based on the pursuit of profit.

BOURGEOIS AND PROLETARIANS

The history of all hitherto existing society is the history of class struggles. . . .

Modern industry has established the world market, for which the discovery of America paved the way. This market has given an immense development to commerce, to navigation, to communication by land. This development has, in its turn, reacted on the extension of industry; and in proportion as industry, commerce, navigation, railways extended, in the same proportion the bourgeoisie developed, increased its capital, and pushed into the background every class handed down from the Middle Ages.

We see, therefore, how the modern bourgeoisie is itself the product of a long course of development, of a series of revolutions in the modes of production and of exchange. . . .

The bourgeoisie has at last, since the establishment of modern industry and of the world market, conquered for itself, in the modern representative state, exclusive political sway. The executive of the modern state is but a committee for managing the common affairs of the whole bourgeoisie.

The bourgeoisie, historically, has played a most revolutionary part.

The bourgeoisie, wherever it has got the upper hand, has put an end to all feudal, patriarchal, idyllic relations. It has pitilessly torn asunder the motley feudal ties that bound man to his "natural superiors," and has left remaining no other nexus

SOURCE: Karl Marx and Frederick Engels. 1998. *Manifesto of the Communist Party.* With introduction by Eric Hobsbawm. New York: Verso, pp. 33–51.

between man and man than naked self-interest, than callous "cash payment." It has drowned the most heavenly ecstasies of religious fervour, of chivalrous enthusiasm, of philistine sentimentalism, in the icy water of egotistical calculation. It has resolved personal worth into exchange value, and in place of the numberless indefeasible chartered freedoms, has set up that single, unconscionable freedom—free trade. In one word, for exploitation, veiled by religious and political illusions, it has substituted naked, shameless, direct, brutal exploitation.

The bourgeoisie has stripped of its halo every occupation hitherto honoured and looked up to with reverent awe. It has converted the physician, the lawyer, the priest, the poet, the man of science, into its paid wage labourers.

The bourgeoisie has torn away from the family its sentimental veil, and has reduced the family relation to a mere money relation. . . .

The need of a constantly expanding market for its products chases the bourgeoisie over the whole surface of the globe. It must nestle everywhere, settle everywhere, establish connections everywhere.

The bourgeoisie has through its exploitation of the world market given a cosmopolitan character to production and consumption in every country. To the great chagrin of reactionists, it has drawn from under the feet of industry the national ground on which it stood. All old-established national industries have been destroyed or are daily being destroyed. They are dislodged by new industries, whose introduction becomes a life and death question for all civilized nations, by industries that no longer work up indigenous raw material, but raw material drawn from the remotest zones; industries whose products are consumed, not only at home, but in every quarter of the globe. In place of the old wants, satisfied by the productions of the country, we find new wants, requiring for their satisfaction the products of distant lands and climes. In place of the old local and national seclusion and self-sufficiency, we have intercourse in every direction, universal interdependence of nations. And as in material, so also in intellectual production. The intellectual creations of individual nations become common property. National one-sidedness and narrow-mindedness become more and more impossible, and from the numerous national and local literatures, there arises a world literature.

The bourgeoisie, by the rapid improvement of all instruments of production, by the immensely facilitated means of communication, draws all, even the most barbarian, nations into civilization. The cheap prices of its commodities are the heavy artillery with which it batters down all Chinese walls, with which it forces the barbarians' intensely obstinate hatred of foreigners to capitulate. It compels all nations, on pain of extinction, to adopt the bourgeois mode of production; it compels them to introduce what it calls civilization into their midst, i.e., to become bourgeois themselves. In one word, it creates a world after its own image.

The bourgeoisie has subjected the country to the rule of the towns. It has created enormous cities, has greatly increased the urban population as compared with the rural, and has thus rescued a considerable part of the population from the idiocy of rural life. Just as it has made the country dependent on the towns, so it has made barbarian and semi-barbarian countries dependent on the civilized ones, nations of peasants on nations of bourgeois, the East on the West.

The bourgeoisie keeps more and more doing away with the scattered state of the population, of the means of production, and of property. It has agglomerated population, centralized means of production, and has concentrated property in a few hands. The necessary consequence of this was political centralization. Independent, or but loosely connected provinces, with separate interests, laws, governments and systems of taxation, became lumped together into one nation, with one government, one code of laws, one national class interest, one frontier and one customs tariff. . . .

The weapons with which the bourgeoisie felled feudalism to the ground are now turned against the bourgeoisie itself.

But not only has the bourgeoisie forged the weapons that bring death to itself; it has also called into existence the men who are to wield those weapons—the modern working class—the proletarians.

In proportion as the bourgeoisie, i.e., capital, is developed, in the same proportion is the proletariat, the modern working class, developed—a class of labourers, who live only so long as they find work, and who find work only so long as their labour increases capital. These labourers, who must sell themselves piecemeal, are a commodity, like every other article of commerce, and are consequently exposed to all the vicissitudes of competition, to all the fluctuations of the market.

Owing to the extensive use of machinery and to division of labour, the work of the proletarians has lost all individual character, and, consequently, all charm for the workman. He becomes an appendage of the machine, and it is only the most simple, most monotonous, and most easily acquired knack, that is required of him. Hence, the cost of production of a workman is restricted, almost entirely, to the means of subsistence that he requires for his maintenance, and for the propagation of his race. But the price of a commodity, and therefore also of labour, is equal to its cost of production. In proportion, therefore, as the repulsiveness of the work increases, the wage decreases. Nay more, in proportion as the use of machinery and division of labour increases, in the same proportion the burden of toil also increases, whether by prolongation of the working hours, by increase of the work exacted in a given time or by increased speed of the machinery, etc.

Modern industry has converted the little workshop of the patriarchal master into the great factory of the industrial capitalist. Masses of labourers, crowded into the factory, are organized like soldiers. As privates of the industrial army they are placed under the command of a perfect hierarchy of officers and sergeants. Not only are they slaves of the bourgeois class, and of the bourgeois state; they are daily and hourly enslaved by the machine, by the overseer, and, above all, by the individual bourgeois manufacturer himself. The more openly this despotism proclaims gain to be its end and aim, the more petty, the more hateful and the more embittering it is.

The less the skill and exertion of strength implied in manual labour, in other words, the more modern industry becomes developed, the more is the labour of men superseded by that of women. Differences of age and sex have no longer any distinctive social validity for the working class. All are instruments of labour, more or less expensive to use, according to their age and sex. . . .

But with the development of industry the proletariat not only increases in number; it becomes concentrated in greater masses, its strength grows, and it feels that strength more. The various interests and conditions of life within the ranks of the proletariat are more and more equalized, in proportion as machinery obliterates all distinctions of labour, and nearly everywhere reduces wages to the same low level. The growing competition among the bourgeois, and the resulting commercial crises, make the wages of the workers ever more fluctuating. The unceasing improvement of machinery, ever more rapidly developing, makes their livelihood more and more precarious; the collisions between individual workmen and individual bourgeois take more and more the character of collisions between two classes. Thereupon the workers begin to form combinations (trade unions) against the bourgeois. . . .

This organization of the proletarians into a class, and consequently into a political party, is continually being upset again by the competition between the workers themselves. But it ever rises up again, stronger, firmer, mightier. It compels legislative recognition of particular interests of the workers, by taking advantage of the divisions among the bourgeoisie itself. . . .

Altogether, collisions between the classes of the old society further, in many ways, the course of development of the proletariat. The bourgeoisie finds itself involved in a constant battle: at first with the aristocracy; later on, with those portions of the bourgeoisie itself, whose interests have become antagonistic to the progress of industry; at all times, with the bourgeoisie of foreign countries. In all these battles it sees itself compelled to appeal to the proletariat, to ask for its help, and thus to drag it into the political arena. The bourgeoisie itself, therefore, supplies the proletariat with its own elements of political and general education, in other words, it furnishes the proletariat with weapons for fighting the bourgeoisie.

Further, as we have already seen, entire sections of the ruling classes are, by the advance of industry, precipitated into the proletariat, or are at least threatened in their conditions of existence. These also supply the proletariat with fresh elements of enlightenment and progress.

Finally, in times when the class struggle nears the decisive hour, the process of dissolution going on within the ruling class, in fact within the whole range of old society, assumes such a violent, glaring character, that a small section of the ruling class cuts itself adrift, and joins the revolutionary class, the class that holds the future in its hands. Just as, therefore, at an earlier period, a section of the nobility went over to the bourgeoisie, so now a portion of the bourgeoisie goes over to the proletariat, and in particular, a portion of the bourgeois ideologists, who have raised themselves to the level of comprehending theoretically the historical movement as a whole.

Of all the classes that stand face to face with the bourgeoisie today, the proletariat alone is a really revolutionary class. The other classes decay and finally disappear in the face of modern industry; the proletariat is its special and essential product. . . .

KEY CONCEPTS

capitalism   proletariats   working class

DISCUSSION QUESTIONS

1. What evidence do you see in contemporary society of Marx and Engels' claim that the need for a constantly expanding market means that capitalism "nestles everywhere"?

2. How do Marx and Engels depict the working class and what evidence do you see of their argument by looking at the contemporary labor market?

24

Shadowy Lines That Still Divide

JANNY SCOTT AND DAVID LEONHARDT

Class inequality is growing in the United States at the same time that the nation is more affluent than many other societies. Here the authors describe contemporary developments in the U.S. class system with an eye toward how Americans perceive, as well as experience, class differences and the possibility for upward movement in the class system.

There was a time when Americans thought they understood class. The upper crust vacationed in Europe and worshiped an Episcopal God. The middle class drove Ford Fairlanes, settled the San Fernando Valley, and enlisted as company men. The working class belonged to the AFL-CIO, voted Democratic, and did not take cruises to the Caribbean.

Today, the country has gone a long way toward an appearance of classlessness. Americans of all sorts are awash in luxuries that would have dazzled their grandparents. Social diversity has erased many of the old markers. It has become harder to read people's status in the clothes they wear, the cars they drive, the votes they cast, the god they worship, the color of their skin. The contours of class have blurred; some say they have disappeared.

But class is still a powerful force in American life. Over the past three decades it has come to play a greater, not lesser, role in important ways. At a time when education matters more than ever, success in school remains linked tightly to class. At a time when the country is increasingly integrated racially, the rich are isolating

SOURCE: Janny Scott and David Leonhardt. 2005. *Class Matters*. New York: Times Books.

themselves more and more. At a time of extraordinary advances in medicine, class differences in health and life span are wide and appear to be widening.

And new research on mobility, the movement of families up and down the economic ladder, shows there is far less of it than economists once thought and less than most people believe. In fact, mobility, which once buoyed the working lives of Americans as it rose in the decades after World War II, has lately flattened out or possibly even declined, many researchers say.

Mobility is the promise that lies at the heart of the American dream. It is supposed to take the sting out of the widening gulf between the have-mores and the have-nots. There are poor and rich in the United States, of course, the argument goes; but as long as one can become the other, as long as there is something close to equality of opportunity, the differences between them do not add up to class barriers. . . .

The trends are broad and seemingly contradictory: the blurring of the landscape of class and the simultaneous hardening of certain class lines; the rise in standards of living while most people remain moored in their relative places.

Even as mobility seems to have stagnated, the ranks of the elite are opening. Today, anyone may have a shot at becoming a United States Supreme Court justice or a CEO, and there are more and more self-made billionaires. Only thirty-seven members of last year's Forbes 400, a list of the richest Americans, inherited their wealth, down from almost two hundred in the mid-1980s.

So it appears that while it is easier for a few high achievers to scale the summits of wealth, for many others it has become harder to move up from one economic class to another. Americans are arguably more likely than they were thirty years ago to end up in the class into which they were born.

A paradox lies at the heart of this new American meritocracy. Merit has replaced the old system of inherited privilege, in which parents to the manner born handed down the manor to their children. But merit, it turns out, is at least partly class-based. Parents with money, education, and connections cultivate in their children the habits that the meritocracy rewards. When their children then succeed, their success is seen as earned.

The scramble to scoop up a house in the best school district, channel a child into the right preschool program, or land the best medical specialist are all part of a quiet contest among social groups that the affluent and educated are winning in a rout. . . .

FAITH IN THE SYSTEM

Most Americans remain upbeat about their prospects for getting ahead. A recent *New York Times* poll on class found that 40 percent of Americans believed that the chance of moving up from one class to another had risen over the last thirty years, a period in which the new research shows that it has not. Thirty-five percent said it had not changed, and only 23 percent said it had dropped.

More Americans than twenty years ago believe it possible to start out poor, work hard, and become rich. They say hard work and a good education are more important to getting ahead than connections or a wealthy background. . . .

Most say their standard of living is better than their parents' and imagine that their children will do better still. Even families making less than $30,000 a year subscribe to the American dream; more than half say they have achieved it or will do so.

But most do not see a level playing field. They say the very rich have too much power, and they favor the idea of class-based affirmative action to help those at the bottom. Even so, most say they oppose the government's taxing the assets a person leaves at death. . . .

THE ATTRIBUTES OF CLASS

One difficulty in talking about class is that the word means different things to different people. Class is rank, it is tribe, it is culture and taste. It is attitudes and assumptions, a source of identity, a system of exclusion. To some, it is just money. It is an accident of birth that can influence the outcome of a life. Some Americans barely notice it; others feel its weight in powerful ways.

At its most basic, class is one way societies sort themselves out. Even societies built on the idea of eliminating class have had stark differences in rank. Classes are groups of people of similar economic and social position; people who, for that reason, may share political attitudes, lifestyles, consumption patterns, cultural interests, and opportunities to get ahead. Put ten people in a room and a pecking order soon emerges.

When societies were simpler, the class landscape was easier to read. Marx divided nineteenth-century societies into just two classes; Max Weber added a few more. As societies grew increasingly complex, the old classes became more heterogeneous. As some sociologists and marketing consultants see it, the commonly accepted big three—the upper, middle, and working classes—have broken down into dozens of microclasses, defined by occupations or lifestyles.

A few sociologists go so far as to say that social complexity has made the concept of class meaningless. Conventional big classes have become so diverse—in income, lifestyle, political views—that they have ceased to be classes at all, said Paul W. Kingston, a professor of sociology at the University of Virginia. To him, American society is a "ladder with lots and lots of rungs." . . .

Many other researchers disagree. "Class awareness and the class language is receding at the very moment that class has reorganized American society," said Michael Hout, a professor of sociology at the University of California, Berkeley. "I find these 'end of class' discussions naïve and ironic, because we are at a time of booming inequality and this massive reorganization of where we live and how we feel, even in the dynamics of our politics. Yet people say, 'Well, the era of class is over.'"

One way to think of a person's position in society is to imagine a hand of cards. Everyone is dealt four cards, one from each suit: education, income, occupation, and wealth, the four commonly used criteria for gauging class. Face cards in a few categories may land a player in the upper middle class. At first, a person's class is his parents' class. Later, he may pick up a new hand of his own; it is likely to resemble that of his parents, but not always.

Bill Clinton traded in a hand of low cards with the help of a college education and a Rhodes scholarship and emerged decades later with four face cards. Bill Gates, who started off squarely in the upper middle class, made a fortune without finishing college, drawing three aces. . . .

One surprising finding about mobility is that it is not higher in the United States than in Britain or France. It is lower here than in Canada and some Scandinavian countries but not as low as in developing countries like Brazil, where escape from poverty is so difficult that the lower class is all but frozen in place.

Those comparisons may seem hard to believe. Britain and France had hereditary nobilities; Britain still has a queen. The founding document of the United States proclaims all men to be created equal. The American economy has also grown more quickly than Europe's in recent decades, leaving an impression of boundless opportunity.

But the United States differs from Europe in ways that can gum up the mobility machine. Because income inequality is greater here, there is a wider disparity between what rich and poor parents can invest in their children. Perhaps as a result, a child's economic background is a better predictor of school performance in the United States than in Denmark, the Netherlands, or France, one study found. . . .

BLURRING THE LANDSCAPE

Why does it appear that class is fading as a force in American life?

For one thing, it is harder to read position in possessions. Factories in China and elsewhere churn out picture-taking cellphones and other luxuries that are now affordable to almost everyone. Federal deregulation has done the same for plane tickets and long-distance phone calls. Banks, more confident about measuring risk, now extend credit to low-income families, so that owning a home or driving a new car is no longer evidence that someone is middle class.

The economic changes making material goods cheaper have forced businesses to seek out new opportunities so that they now market to groups they once ignored. Cruise ships, years ago a symbol of the high life, have become the oceangoing equivalent of the Jersey Shore. BMW produces a cheaper model with the same insignia. Martha Stewart sells chenille jacquard drapery and scallop-embossed ceramic dinnerware at Kmart.

"The level of material comfort in this country is numbing," said Paul Bellew, executive director for market and industry analysis at General Motors. "You can make a case that the upper half lives as well as the upper 5 percent did fifty years ago."

Like consumption patterns, class alignments in politics have become jumbled. In the 1950s, professionals were reliably Republican; today they lean Democratic. Meanwhile, skilled labor has gone from being heavily Democratic to almost evenly split.

People in both parties have attributed the shift to the rise of social issues, like gun control and same-sex marriage, which have tilted many working-class voters rightward and upper-income voters toward the left. But increasing affluence plays an important role, too. When there is not only a chicken, but an organic, free-range chicken, in every pot, the traditional economic appeal to the working class can sound off-key.

Religious affiliation, too, is no longer the reliable class marker it once was. The growing economic power of the South has helped lift evangelical Christians into the middle and upper middle classes, just as earlier generations of Roman Catholics moved up in the mid-twentieth century. It is no longer necessary to switch one's church membership to Episcopal or Presbyterian as proof that one has arrived. . . .

The once tight connection between race and class has weakened, too, as many African-Americans have moved into the middle and upper middle classes. Diversity of all sorts—racial, ethnic, and gender—has complicated the class picture. And high rates of immigration and immigrant success stories seem to hammer home the point: The rules of advancement have changed.

The American elite, too, is more diverse than it was. The number of corporate chief executives who went to Ivy League colleges has dropped over the past fifteen years. There are many more Catholics, Jews, and Mormons in the Senate than there were a generation or two ago. Because of the economic earthquakes of the last few decades, a small but growing number of people have shot to the top . . .

But beneath all that murkiness and flux, some of the same forces have deepened the hidden divisions of class. Globalization and technological change have shuttered factories, killing jobs that were once stepping-stones to the middle class. Now that manual labor can be done in developing countries for two dollars a day, skills and education have become more essential than ever.

This has helped produce the extraordinary jump in income inequality. The after-tax income of the top 1 percent of American households jumped 139 percent, to more than $700,000, from 1979 to 2001, according to the Congressional Budget Office, which adjusted its numbers to account for inflation. The income of the middle fifth rose by just 17 percent, to $43,700, and the income of the poorest fifth rose only 9 percent.

For most workers, the only time in the last three decades when the rise in hourly pay beat inflation was during the speculative bubble of the 1990s. Reduced pensions have made retirement less secure.

Clearly, a degree from a four-year college makes even more difference than it once did. More people are getting those degrees than did a generation ago, but class still plays a big role in determining who does or does not. At 250 of the most selective colleges in the country, the proportion of students from upper-income families has grown, not shrunk. . . .

Class differences in health, too, are widening, recent research shows. Life expectancy has increased overall; but upper-middle-class Americans live longer and in better health than middle-class Americans, who live longer and in better health than those at the bottom.

Class plays an increased role, too, in determining where and with whom affluent Americans live. More than in the past, they tend to live apart from everyone else, cocooned in their exurban châteaus. Researchers who have studied census data from 1980, 1990, and 2000 say the isolation of the affluent has increased.

Family structure, too, differs increasingly along class lines. The educated and affluent are more likely than others to have their children while married. They have fewer children and have them later, when their earning power is high. On average, according to one study, college-educated women have their first child at age thirty, up from twenty-five in the early 1970s. The average age among women who have never gone to college has stayed at about twenty-two.

Those widening differences have left the educated and affluent in a superior position when it comes to investing in their children. . . .

The benefits of the new meritocracy do come at a price. It once seemed that people worked hard and got rich in order to relax, but a new class marker in upper-income families is having at least one parent who works extremely long hours (and often boasts about it). In 1973, one study found, the highest-paid tenth of the country worked fewer hours than the bottom tenth. Today, those at the top work more.

In downtown Manhattan, black cars line up outside Goldman Sachs's headquarters every weeknight around nine. Employees who work that late get a free ride home, and there are plenty of them. Until 1976, a limousine waited at 4:30 p.m. to ferry partners to Grand Central Terminal. But a new management team eliminated the late-afternoon limo to send a message: four thirty is the middle of the workday, not the end.

A RAGS-TO-RICHES FAITH

Will the trends that have reinforced class lines while papering over the distinctions persist?

The economic forces that caused jobs to migrate to low-wage countries are still active. The gaps in pay, education, and health have not become a major political issue. The slicing of society's pie is more unequal than it used to be, but most Americans have a bigger piece than they or their parents once did. They appear to accept the trade-offs.

Faith in mobility, after all, has been consciously woven into the national self-image. Horatio Alger's books have made his name synonymous with rags-to-riches success, but that was not his personal story. He was a second-generation Harvard man, who became a writer only after losing his Unitarian ministry because of allegations of sexual misconduct. Ben Franklin's autobiography was punched up after his death to underscore his rise from obscurity.

The idea of fixed class positions, on the other hand, rubs many the wrong way. Americans have never been comfortable with the notion of a pecking order based on anything other than talent and hard work. Class contradicts their assumptions about the American dream, equal opportunity, and the reasons for their own successes and even failures. Americans, constitutionally optimistic, are disinclined to see themselves as stuck.

Blind optimism has its pitfalls. If opportunity is taken for granted, as something that will be there no matter what, then the country is less likely to do the hard work to make it happen. But defiant optimism has its strengths. Without confidence in the possibility of moving up, there would almost certainly be fewer success stories.

KEY CONCEPTS

class consciousness	consumerism	meritocracy
social class	social mobility	social stratification

DISCUSSION QUESTIONS

1. What contradictions do the authors note between the fact of growing class inequality and the high rates of consumption in the United States?
2. What explains the relative absence of class consciousness in the United States?
3. How does the American dream of social mobility affect people's perceptions of the possibility for upward social mobility?

25

Abandoned Before the Storms

The Glaring Disaster of Gender, Race, and Class Disparities in the Gulf

AVIS A. JONES-DEWEEVER AND HEIDI HARTMANN

Hurricane Katrina is widely perceived as having revealed the reality of race and poverty in the United States. Here the authors document how race and class—as well as gender (and the interaction among the three)—influenced the consequences and aftermath of Hurricanes Katrina and Rita.

SOURCE: Chester Hartmann and Gregory D. Squires, eds. 2006. *There Is No Such Thing as a Natural Disaster: Race, Class, and Hurricane Katrina.* New York: Routledge.

As Hurricanes Katrina and Rita ravaged the Gulf Coast region, the impact of the storms made clear the lingering implications of America's persistent divides. For a time, at least, the outcome of race and class disadvantage in America became crystal clear and garnered long overdue attention. Yet glaring gender disparities throughout the affected areas remained largely unexamined. This article fills that void by exploring how the multiple disadvantages faced by the women of the Gulf both increased their vulnerability in a time of crisis and, unless proactively addressed, remain an impediment to their ability to rebuild their lives long after the storms.

Long before the devastation of the hurricanes, many women within the Gulf region were already living at the bottom. Women in the affected states were the poorest women in the nation, among the least likely to have access to health insurance, and, despite high work participation rates, the most likely to find themselves stuck in a cycle of low-wage work. Most disadvantaged were women of color, who often faced quite limited opportunities and outcomes, especially with respect to employment and earnings, educational attainment, and ultimately, the likelihood of living in poverty. To tease out the plight of the women in the region prior to the devastation that compounded their challenges and to catch a glimpse of the opportunities awaiting them in their relocated homes, we use employment and poverty data from federal government sources as well as indicators developed by the Institute for Women's Policy Research (IWPR) through our *Status of Women and the States* series to uncover the multiple disadvantages faced by women in the city of New Orleans, its broader Metropolitan Statistical Area (MSA), and the MSAs of Biloxi–Gulfport–Pascagoula, Mississippi, and Beaumont–Port Arthur, Texas.

LIVING AT THE BOTTOM

The images splashed across CNN in the wake of Katrina exposed to many the disastrous effects of poverty in America. Yet, poverty had long been a festering problem in the United States, growing for four years in a row and leaving the poor further and further behind. By the time Katrina made it to shore, more than 36 million people across the nation were poor and a large proportion (42.3%) lived in deep poverty, with incomes below 50% of the poverty threshold (U.S. Census Bureau 2005). Even for the "average" poor person, the struggle to make ends meet was intense by historical standards. By 2004, the average income of those living in poverty had dropped further below the poverty line than at any time since that statistic was first recorded back in 1975 (Center on Budget and Policy Priorities 2005). America's poor had truly hit bottom in the modern age.

For most Americans, imagining the struggles of the poor in this country is quite difficult, if not disturbing. Poverty is typically tucked away, either confined to an urban enclave avoided by those who aren't within its boundaries by accident of birth, or dispersed broadly, on a lonely country road far away from neighbors, jobs, and, in many respects, opportunity. In the lives of most Americans, one's only brush with poverty is the occasional discomfort felt due to the outstretched hand of a stranger on the street who claims to be homeless, or the annual newscasts from the local food bank come Thanksgiving or Christmas. In this land of opportunity, poverty is hidden from view, like a messy closet one hopes goes undiscovered by an important houseguest. Although we all know it's there—somewhere—for most, poverty goes unseen and unacknowledged, as does its implications for everyday life, and ultimately, survival—that is, unless you're the one who is poor.

For America's poor, poverty is the one reality that all too often cannot be escaped. Especially among the deeply entrenched urban poor, poverty doesn't go on vacation. It doesn't provide an occasional break. Instead, it remains disturbingly consistent and touches lives in every way imaginable. On any one day in 2004, anywhere between 614,000 to 854,000 households in America had at least one family member who went hungry because there was not enough money to ensure that everyone could eat (Nord, Andrews, and Carlson 2005). Millions more lived amid overcrowded or substandard housing conditions, or couldn't rely on consistent housing at all as the escalating price of keeping a roof over one's head continued to far outpace the value of wages, as it had done for the past 40 years (National Low Income Housing Coalition 2004). For many, being poor meant being trapped in failing school districts, unable to assure a quality education and thus, a way out for their children. It meant being relegated to a place where no one else wants to live, isolated, and disjointed from the rest of society, and having a higher likelihood than others to being exposed to environmental dangers (U.S. Environmental Protection Agency 1993). In New Orleans, that meant being trapped on low ground. In the real lives of real people, poverty is having to decide at the end of the month which is most important—electricity, running water, gas, or any of the other pressing expenses that must be paid. Poverty demands frequent juggling because there is simply not enough to fulfill every need. That is the reality of poverty in America, a reality that has become increasingly difficult to escape. And in an era of restructuring, outsourcing, and low, stagnant wages, poverty, America's dirty little secret, not only lives on, but thrives.

As the prevalence and depth of poverty has grown over the years, the crumbling value of the minimum wage has made finding a way out next to impossible. After adjusting for inflation, the value of the minimum wage is at its second lowest point since 1955, ensuring that working, and working very hard, will not be enough to guarantee an above-poverty-level existence (Bernstein and Shapiro 2005). A single mother of two, for example, working 40 hours per week, 52 weeks a year, earns only $10,712 before taxes—a full $5,000 below the poverty threshold (Boushey 2005). Even with the Earned Income Tax Credit, which maxes out at roughly $4,000, her family income still falls below the poverty line. As a result, she falls squarely within the ranks of the working poor (Shulman 2003; Shipler 2004).

The above-cited example is not atypical. The face of the working poor in America is, more often than not, the face of a woman. It is women who make up the majority (60%) of America's minimum-wage workers and 90% of those who remain stuck in low-wage work throughout their prime earning years (Lovell 2004; Rose and Hartmann 2004). And ironically, those women who lived in the metropolitan areas impacted by Hurricanes Katrina and Rita were especially likely to encounter this fate. The women of the Gulf participated in the labor force at about the same rate as women in the nation overall, and, with the lone exception of those who resided in the Beaumont–Port Arthur, Texas area, women in this region were less likely than women nationwide to work part-time or for only part of the year. Yet, despite their strong attachment to the labor force, in each geographic area examined, their median annual earnings trailed those of women nationally, with the women in the city of New Orleans falling furthest behind (Jones-DeWeever et al. 2006).

Unemployment, too, was especially prevalent in the Gulf region. Prior to Katrina, the city of New Orleans was particularly hard hit, with unemployment rates over 50% higher than that of the nation as a whole. Still, for many, even the advantage of holding a job meant little in terms of ensuring a life free of poverty. In each affected area examined, more than two in five of those who were poor experienced poverty despite being employed, and within the affected Metropolitan Statistical Area (MSA) of Biloxi–Gulfport–Pascagoula in Mississippi, more than half of those who lived below the poverty line (52.7%) worked in 2004, far exceeding the national rate of 45.3%.[1]

Taken together, the evidence paints a picture of a hard-working people. Yet a strong work ethic failed to insulate them from the likelihood of being poor. Overall, poverty in the region was more commonplace than was the case nationwide. And, as is typical throughout this nation and around the world, women were considerably more likely to be poor than men, with women in New Orleans most at risk of poverty in the Gulf region. Fully one quarter (25.9%) of the women who made New Orleans their home lived in poverty, well outpacing the national poverty rate (14.5%) for women (Gault et al. 2005).

For those who faced life beyond their working years, poverty was all too common. Like younger women, older women were much more likely to face poverty in their golden years than their male counterparts. With the exception of Biloxi–Gulfport–Pascagoula, in every one of the four areas examined, the chance that women aged 65 years or older would live in poverty in this region exceeded the national rate and roughly doubled the rate of poverty experienced by elderly men. Also, like their younger counter-parts, those who lived within the city of New Orleans were at greatest risk of poverty, as nearly a quarter (24.3%) of the Crescent City's older women lived below the poverty line (Gault et al. 2005).

Women in this region, then, both young and old, and particularly those who lived in New Orleans, faced a remarkably high likelihood of being poor. Most at risk were those who lived up to the responsibility of raising children on their own. In both the city of New Orleans and its broader Metropolitan Statistical Area, the percentage of female-headed families falling below the federal poverty line exceeded the national average, topping off at roughly 40%. . . . And as Katrina loomed, it was these families who faced the impending crisis with the fewest resources at their disposal to make it out alive and to keep their families together as chaos emerged.

Even though poverty in the Katrina- and Rita-impacted areas was significant and clearly outpaced the nation as a whole, the high level of poverty experienced in the Gulf region is consistent with broader poverty proliferation throughout the South. As compiled and ranked by IWPR's *Status of Women in the States* (2004), all eight states in the South Central Region rank among the bottom third of all 50 states and the District of Columbia for the percent of women living above the poverty line. Mississippi ranks dead last, followed closely by Louisiana (ranked 47th) and Texas (ranked 44th). Kentucky (ranked 36th), the highest-ranked state in the region, still falls significantly below the mid-point for the nation as a whole (Werschkul and Williams 2004).

Not only are women in the Gulf region the most likely to be poor, they are also at high risk of being without health insurance. Texas ranked worst in the nation in terms of women's health insurance coverage, with Louisiana and Mississippi also falling near the bottom, ranking 49th and 43rd, respectively (Werschkul and Williams 2004).

So, while no state in the nation has escaped poverty's reach, and although most recently growth in poverty has been starkest in the Midwest (U.S. Census Bureau 2005), the South remains especially plagued by this problem. This southern legacy not only has a gendered dynamic, but a racial one as well.

EXAMINING THE SIGNIFICANCE OF RACE

Lost on no one during the days immediately following the breach of the levees was the high concentration of African Americans left to fend for themselves in the middle of a ravaged city. But long before the storm, New Orleans, not unlike other urban areas across the nation, was in many ways a city sharply divided by race and class, as well as gender. The women of New Orleans and in the Gulf region generally lived very different lives across the racial divide. For example, despite being home to three Historically Black Universities (Dillard University, Southern University at New Orleans, and Xavier University of Louisiana), college degree attainment among African-American women in the city of New Orleans only slightly outpaced the national rate (18.9% vs. 17.5%). In the broader New Orleans MSA, degree attainment by African-American women fell slightly below the national norm (16.4% vs. 17.5%). White women in the city of New Orleans, however, like women in urban areas more generally, had very high college graduation rates. They greatly exceeded their counterparts nationally in college degree attainment (50.6% vs. 27.8%) and more than doubled the rate of degree attainment achieved by African-American women (50.6% vs. 18.9%). Only within the Beaumont–Port Arthur area, where degree attainment is relatively low for all groups, did the difference in degree attainment rates between African-American and white women drop below ten percentage point. . . .

The educational differences experienced by women across the racial divide most certainly carried over to the labor market. Within the managerial/professional fields, for example, white women's representation more than doubled that of African-American women in the city of New Orleans (65.7% of white women had managerial or professional jobs vs. only 27.2% of African-American women). White women also outpaced African-American women within the broader New Orleans metropolitan area, 45.9% vs. 28.6%, for a difference of more than 15 percentage points, significantly outpacing the national divide of 40% vs. 31%.[2]

Similarly, white women in the city of New Orleans earned a higher proportion of the white male salary (80.2%) than the national average (71.7%) and nearly doubled the proportion of white men's earnings achieved by African-American women (80.2% vs. 43.9%). In fact, across each of the four hurricane-impacted areas examined that had available data, African-American women's earnings were less than half those of white men (43.9%) and well below the national ratio of African-American women's earnings to white men's earnings (62.7%). Although still lagging behind white men in each area examined, in no area did white women fail to exceed three-fifths of the earnings of white men. . . .

Looking statewide, women's earnings were especially low among those who lived in areas that bore the brunt of Hurricane Katrina. In fact, of the 43 states with sample sizes large enough to provide a reliable measure of African-American women's earnings, Louisiana ranked worst in the nation with full-time annual earnings of only $19,400. Mississippi followed closely, ranking next to last with median earnings of $19,900 for African-American women. White women also garnered comparatively low earnings in those states. Although exceeding African-American women's earnings, compared to white women in other states, white women's median earnings in Mississippi ($25,700) and Louisiana ($26,500) put them in the bottom third in the nation and well below the $30,900 median earnings for white women nationally. In contrast, for both groups, women's earnings in Texas well outpaced those of their counterparts in Louisiana and Mississippi and placed them within the top third of earnings nationally. Still, the earnings of African-American women in Texas failed to exceed the average earnings of African-American women nationwide, and the earnings of Hispanic women in Texas fell within the bottom third of all states (Werschkul and Williams 2004).

Labor force participation, too, varied greatly across race and gender and between geographic areas. In the city of New Orleans, for example, nearly three-fourths of white women (73.7%), as compared to approximately two-thirds of African-American women (66.5%), participated in the labor force in 2004. But in the New Orleans MSA, the two groups had nearly equal participation rates of 68.9% for African-American women and 67.8% for white women. The widest variation in labor force participation occurred in the Beaumont–Port Arthur area, where over four-fifths (81.2%) of African-American women participated in the labor force as compared with only 64.3% of white women.[3]

Also consistent with national trends, poverty in the Gulf region is highly correlated with race and gender. While overall women are consistently more likely to be poor than men, both throughout the affected areas as well as the nation as a whole, African-American women are especially vulnerable to poverty. Hardest hit were those who lived within the Biloxi–Gulfport–Pascagoula area and the city of New Orleans, where roughly a third lived below the federal poverty line (32.5% and 32.3%). As was the case for African-American women, white women's poverty was most pronounced in the Biloxi-Gulfport-Pascagoula area, where 13.7% were poor; within New Orleans city and its broader MSA, however, white women's poverty was less than the national average of 10%....

What a sad irony that one of the worst disasters in our nation's history occurred in an area populated by some of the nation's most disadvantaged citizens. Both African-American and white women in Mississippi and Louisiana had many fewer resources at their disposal than their counterparts nationwide. African-American women especially found themselves at the bottom. They were the lowest earners across both race and sex in Mississippi and Louisiana ... and, consequently, were the least likely to be equipped with the necessary resources to take themselves and their families out of harm's way. As a result, they largely stayed to face the storm on their own; and for those who survived, life will never be the same again.

LIFE AFTER THE STORMS

One can only imagine the trauma of losing everything save the clothes on one's back, being separated from family, friends, and community, and suddenly forced to begin life anew in a random, unfamiliar place a long way—geographically, socially, and culturally—from home. This is the challenge facing many of those who have been displaced following the devastation brought by Katrina and Rita. But given the probability that those most likely to relocate were also the ones who faced the most severe challenges at home of poverty, limited educational background, and all the implications of a paycheck-to-paycheck existence, what might the relocated settlers now expect in the way of employment and other opportunities while awaiting the possibility of a newly rebuilt city? Unfortunately, our analysis of the data suggests very little in terms of improved economic opportunities.... [I]n their new homes, many evacuees will still be faced with high poverty rates, particularly among women and female-headed households, and, for African-American women, low median earnings for year-round, full-time work. In some areas, such as Jackson, Mississippi; Baton Rouge, Louisiana; Little Rock, Arkansas; Mobile, Alabama; and Atlanta, Georgia, poverty rates are even higher and/or earnings are even lower than in those environments left behind.

A few areas, though, suggest an environment that provides a more solid foundation for evacuees to rebuild their lives. Places like Charlotte, North Carolina; Richmond–Petersburg, Virginia; and Nashville, Tennessee boast comparatively lower poverty rates for women and female-headed families or have significantly higher earnings for African-American women. These positive conditions offer a glimmer of hope for improved opportunities to build a better and brighter future to those who have chosen to make their relocation permanent and to start the task of building a new life, livelihood, and home in new surroundings and without the social networks that had once defined a community.

CREATING A BETTER FUTURE FOR THE WOMEN OF THE GULF

Life will never be the same for those who lost loved ones, material assets, family mementos, and the one place in the world they called home. The best that can be done for them is to ensure that for all, even the most disadvantaged, we use this crisis as a springboard to creating a future replete with opportunities for a better tomorrow. To fulfill that commitment we must ensure that the particular needs of women are not overlooked in the rebuilding process. Acknowledging and addressing the needs of women ultimately amounts to fulfilling the needs of children, families, and entire communities. The following would be a good start:

> Invest in education and training to increase women's earning potential and foster opportunities to participate in the rebuilding process. . . .
>
> Expand earning opportunities through better wages and fair hiring practices. . . .
>
> Stop the clock on TANF benefits for the victims of the storms. . . .
>
> Respect the needs of women and communities in the planning and rebuilding process, and engage the black middle class. . . .
>
> Make the provision of childcare a priority. . . .
>
> Fulfill the promise of attacking the poverty problem. . . .

CONCLUSION

Ultimately, it is our responsibility to learn from the heartbreak of seeing mothers begging for milk for their dehydrated infants, from seeing children and the elderly stranded on rooftops pleading for help, and seeing the dead left without dignity and the dying suffering for days in their own waste. Lest we forget, these are the images of Katrina that should stay with us forever as a reminder of what can happen when we turn a blind eye to the implications of a society divided. We can and must do better. The women, children, and families of the Gulf deserve nothing less.

REFERENCES

Bernstein, Jared and Isaac Shapiro. 2005. *Unhappy Anniversary: Federal Minimum Wage Remains Unchanged for Eighth Straight Year, Falls to 56-Year Low Relative to the Average Wage.* Washington, D.C.: Center on Budget and Policy Priorities and the Economic Policy Institute.

Boushey, Heather. 2005. *The Structure of Poverty in the U.S.* Paper presented at the Congressional Black Caucus Foundation Summit Poverty, Race and Policy: Strategic Advancement of a Poverty Reduction Agenda.

Caiazza, Amy, April Shaw, and Misha Werschkul. 2004. *Women's Economic Status in the States: Wide Disparities by Race, Ethnicity, and Region.* Washington, D.C.: Institute for Women's Policy Research.

Center on Budget and Policy Priorities. 2005. *Economic Recovery Failed to Benefit Much of the Population in 2004.* Washington, D.C.: Center on Budget and Policy Priorities.

Jones-DeWeever, Avis, Olga Sorokina, Erica Williams, Heidi Hartmann, and Barbara Gault. 2006. *The Women of New Orleans and the Gulf Coast: Multiple Disadvantages and Key Assets for Recovery, Part II: Employment and Earnings, Gender, Race, and Class.* Washington, D.C.: Institute for Women's Policy Research.

Lovell, Vicky. 2004. *No Time to be Sick: Why Everyone Suffers When Workers Don't Have Paid Sick Leave.* Washington, D.C: Institute for Women's Policy Research.

National Low Income Housing Coalition. 2004. *America's Neighbors: The Affordable Housing Crisis and the People it Affects.* Washington, D.C.: National Low Income Housing Coalition.

Nord, Mark, Margaret Andrews, and Steven Carlson. 2005. *Household Food Security in the United States, 2004.* Washington, D.C.: USDA Economic Research Service. http://www.ers.usda.gov/Publications/err11/

Office of the Press Secretary, White House. 2005. *President Discusses Hurricane Relief in Address to the Nation.* http://www.whitehouse.gov/news/releases/2005/09/20050915-8.html

Rose, Stephen J. and Heidi Hartmann. 2004. *Still A Man's Labor Market: The Long-Term Earnings Gap.* Washington, D.C.: Institute for Women's Policy Research.

Shipler, David. 2004. *The Working Poor: Invisible in America.* New York: Knopf.

Shulman, Beth. 2003. *The Betrayal of Work: How Low-Wage Jobs Fail 30 Million Americans.* New York: New Press.

U.S. Census Bureau. 2005. *Income, Poverty, and Health Insurance Coverage in the United States: 2004.* Washington, D.C.: U.S. Government Printing Office.

U.S. Environmental Protection Agency. 1993. *Environmental Equity: Reducing Risk for All Communities. Volume I: Workgroup Report to the Administrator.* Washington, D.C.: U.S. Environmental Protection Agency. http://www.epa.gov/compliance/resources/publications/ej/reducing_risk_ com_voll.pdf

Werschkul, Misha and Erica Williams. 2004. *The Status of Women and the States.* Washington, D.C.: Institute for Women's Policy Research.

NOTES

1. IWPR analysis of the 2004 American Community Survey.
2. Ibid.
3. Ibid.

KEY CONCEPTS

Feminization of poverty   poverty line   unemployment rate

DISCUSSION QUESTIONS

1. How did race, class, and gender each play a role in the devastation and recovery from Hurricanes Katrina and Rita?
2. How does the article show you the interaction of race, class, and gender in producing social and economic inequality?
3. What specific measures could be taken to address the problems of poverty in the United States? How would they help victims of these hurricanes?

26

The Color of the Safety Net

THOMAS M. SHAPIRO

By analyzing racial differences in wealth, Thomas M. Shapiro shows the importance of distinguishing wealth from income and the effects of both on persistent racial inequality. Wealth accumulates over time, thus past practices of racial discrimination continue to influence the social and economic status of African Americans.

SOURCE: Thomas M. Shapiro. 2004. *The Hidden Cost of Being African American: How Wealth Perpetuates Inequality.* New York: Oxford University Press.

Vivian Arrora, 40 years old, is the struggling single mother of a young teenage son, Lamar, and 4-year-old twin girls, Bria and Brittany. Vivian, who is African American, grew up in Watts, which is one of the poorest sections of Los Angeles and where about one in every three families falls below the government's poverty line. Several moves have inched her family away from this poverty-stricken black community toward more middle-class West L.A. She tells me she has been attacked and raped several times. With tenacity and determination, she has bootstrapped her family and, as she says, "branched further west, out of a gang-infested area, a drug area." She dreams of owning a house in "peaceful" and middle-class Culver City.

"After I gave birth to the twins, I was just ready to go to work because just receiving AFDC [welfare] just wasn't the thing to do," Vivian begins, "and all my life I've been receiving AFDC." She took vocational classes at a technical school because she "wanted to learn how to do the computer." She completed the program and earned her certificate, acquiring substantial student loans along the way.

The next step was to find a job. "I was out lookin' for a job, and it seemed like nobody wanted to hire me and I got kind of discouraged, and I just kept lookin', I just kept lookin'." A friend then suggested going to a temporary agency.

> We went to the temp agency on a Wednesday. It was raining, and we just kept on. We kept on going, and the rain didn't stop us.... I went in on a Wednesday, and they called me that Thursday and told me to start work that Monday. And I've been working ever since. And I'm like: Am I really, really ready to go to work? Mentally? But once I started, I just, I've been on a roll ever since.

All this occurred two years before we talked. Vivian worked as a temp for a year and then was hired by the county to work full-time, with some medical benefits, processing adoption papers. She is proud of having worked herself off AFDC, declaring, "I'm worth more than 700 dollars a month. I'm worth more than that!" But this clerical work does not pay much—a tad under $20,000 per year, which is about $500 above the official poverty line for her family of four. She may be a poster girl for welfare reform because she successfully transitioned off welfare, but she has joined the swelling ranks of the working poor.

Vivian's job is very important because it provides skills, habits, stability, and self-worth that she said had not existed before, but she still is very concerned about crime and safety where they live and wants to move into a better place, even own her own home someday. Working hard and boot-strapping her family off welfare has neither lifted her out of poverty nor put the American Dream within her reach. I asked her how she found the neighborhood and apartment where she is living now; her answer reminds us of the fragile and precarious living situations of those without safety nets. She was forced out of her last place with 30 days' notice, and the family just

> landed right here. This is not where I really wanted to be, but I was tired when I was looking because I was working full-time, and by the time I got off it was too late to go look, you know, to be out at night, in there with the

kids, nobody to baby-sit, so I have to come home and cook. It's just me. I don't really like the surroundings. I don't like the traffic over here either. Sometimes when I come home I see a lot of guys, they hang out down here at the corner.

She would like to buy a home in a safer neighborhood for Lamar and the twin girls. It would be the next step up on her mobility ladder because it would solidify her present stability and provide improved services for her family and better schools for her children. She faces serious obstacles. She has lots of debt and ruined credit. She does not seem to have the resources or capabilities to work out of her debt trap, at least not on poverty-level wages. Nonetheless, she is thinking about buying a home through a funding program that requires education, training, and clearing her credit. She wonders how she can find the time to do all this while working full-time, because it would mean finding costly day care for Bria and Brittany.

A modest, even small, amount of assets, together with day care provision, would make a huge difference in securing a better future for this resolute full-time working woman and stabilizing this family's mobility up from poverty. For example, if she had assets put aside, Vivian could acquire job skills and training, and these enhanced skills in turn might well lead to a better-paying job. In the view of mortgage lenders, difficulty in getting out of debt reveals a high credit risk, so if she could get out of debt, she would be in a better position to consider seriously buying a home. Vivian's story gives us glimpses of the kind of life that so many others like her live. Her struggles anchor a starting point regarding some broader asset themes of this [article]. Poverty is not merely the lack of adequate income for daily needs and survival; for the Arrora family it means difficulties around community, housing, crime and safety, debt, environment, child care, and schools. While it is no doubt true that there are some people whom no amount of assets could help, because of handicaps or inclination, given how far she has taken the family already, I firmly believe that Vivian Arrora's family is poised for mobility and self-reliance. Lack of assets holds her back.

Kathryn MacDonald, like Vivian, is in her 40s and earning a salary close to the poverty line. She too is a single mother, but her life struggle tells a very different story. Kathryn works about 30 hours a week as a freelance contractor in publishing, earning approximately $16,000 a year. Her boyfriend left her just before her son, Evan, was born and she has raised him alone. She prefers to work part-time so she can spend part of her day with Evan, who she says has attention deficit disorder. According to her this was a major reason why she moved from New Jersey to St. Louis in 1995.

Kathryn and Evan MacDonald live in Florissant, a traditionally working-class and middle-income community in north St. Louis County. Kathryn worked at a large publishing house in Manhattan before moving. She grew weary of the city's frantic work pace, expensive New Jersey housing, and spending so much time away from her son, so they moved. Now Kathryn does the same work in her St. Louis home that she used to do in a Manhattan skyscraper, matching Library of Congress book subject headings to subject titles for publishers. Freelancing half-time at home allows her to spend much more time with Evan and to watch over his educational and social development. She also enjoys the freedom and autonomy of working at home. Kathryn earns a lot less than when she worked full-time in New York, but she is far happier with her life now, even if her earnings only amount to poverty wages. She likes the community and schools, which are largely white.

Kathryn clearly is pleased that things have worked out so well.

> We're in a good neighborhood. My son can go out to play and I don't have to worry about what he's going to get into or who he is going to be encountering. I don't have to worry about him being abducted. . . . I don't have to worry too much about drive-by shootings. I don't have to worry that something terrible is going to happen to him just because he was out on the street.

Kathryn is especially pleased with Evan's school situation. Evan is smart, just a notch below getting into gifted programs, and the system has special programs for bright kids like him. The school also has understanding and knowledgeable teachers working in small-group settings who can help him overcome his ADD.

Normally, $16,000 does not afford a great deal more than what Kathryn calls "life support," much less the kinds of services and opportunities available in middle-class communities. What makes Kathryn's life so different from Vivian Arrora's? How is she able to live on essentially poverty wages and yet plan for a future that looks to have better prospects? How is she able to live in a place that is safe for herself and Evan? How is Kathryn able to find a school where Evan can thrive? It is not as simple as that one is white and the other is black. The answer is transformative assets.

For one thing, Kathryn is free of debt. Her brother has been sending $100 a month for several years to help her out with Evan's educational and day care expenses. She lost a job several years ago, when Evan was 2, but was able to move in with her father for five years. She has no school loans because her family paid her college bills. Even today, unlike the average American, she does not owe any credit card debt.

But her financial stability goes far beyond just lack of debt, Kathryn explains. She has inherited money from her family.

> I have the proceeds from my father's estate, and also my grandmother. I don't even pretend to understand this—my cousin the lawyer handles all this—but if her estate gets to a certain size, she is liable for more estate taxes, so every so often he has to disburse some of that money.

Kathryn tells us that she has already inherited about $125,000, of which about $90,000 remains, and will inherit another $80,000 when her 94-year-old grandmother passes away. "That could be when I buy a house. That could be what pushes me over the top. With that plus with the mortgage I could get, I could get something decent." She hopes to buy a home with her new boyfriend, who will not be able to contribute much because he pays alimony to a first wife.

When her father died, "the first thing I did was take some money out, and we took a vacation." When another chunk of money came from her grandmother's estate, she and Evan took off to a family wedding in Alaska. She dips into the inheritance every few months as bills mount up, especially when her quarterly estimated income tax is due. She is looking into magnet schools and even private schooling for Evan, in case the local public schools cannot continue to meet his special needs and provide an environment in which he can thrive.

If Kathryn MacDonald did not have assets, one might think of her in an entirely different light, and many questions might arise. For instance: What is she doing to better herself? Why is she not working full-time? Why are her ambitions so low? If she were black, the questions might have a harsher tone, and we can imagine the social condemnation and scorn this single mom might face. Although one might question some of Kathryn's choices, her story is an example of how financial inheritance can provide advantages and a head start in life. Maybe even more important for Kathryn MacDonald, assets supply an anchor for her family's middle-class status and identity that her work and income cannot.

Vivian Arrora's and Kathryn MacDonald's stories provide a concrete starting point for considering how racial inequality is passed from one generation to the next. In many ways they are so alike; yet in many other ways their lives are so different. Vivian's legacy is growing up black in a welfare family in Watts and becoming a single mother herself. She is the first in her family to go to college. The big issues for her are work, debt and bad credit, finding time for the kids, the fear of violence, drugs, and gangs, and figuring out a way to buy a home in a stable and safer community. Hers is a remarkable success story, but her mobility from welfare to working poor may have reached its own limit. Her children go to weak urban schools where getting ahead is a difficult task accomplished only by a few. Lamar, Bria, and Brittany will inherit America's lack of commitment to equal education for all.

Kathryn's situation, if not her accomplishments, is very different. She does not worry about drugs, violence, and gangs, the adequacy of the public schools, or finding time to spend with her child. Her upper-middle-class inheritance includes a debt-free present, a substantial amount of assets, and palpable prospects of inheriting considerably more in the near future. Her inheritance, one could argue, includes class standing that sustains her comfortable and respectable middle-class situation. In looking at these legacies and inheritances, we begin to see that family assets are more than mere money; they also provide a pathway for handing down racial legacies from generation to generation.

Finally, Vivian's greatest dream is to own a home in a safe place with decent schools for her kids. As far as I can tell, this is not likely to happen, unless she actually meets and marries her millionaire—or unless a bold and imaginative policy helps to make her hard work pay off. Kathryn's dream home most likely will become a reality after her grandmother passes away. The lives and opportunities of their children already are being acted out upon different stages, and the gulf between Evan and Lamar is likely to widen further. What the two boys make of their lives from these different starting points will be their own doing, but let us not delude ourselves that Kathryn and Evan and Vivian, Lamar, and the twins share even remotely similar opportunities. . . .

THE ASSET PERSPECTIVE

A core part of my argument is that wealth, as distinct from income, offers the key to understanding racial stratification. Thus a wealth perspective provides a fresh way to examine the "playing field." Indeed, I believe that this perspective challenges a standard part of the American credo—that similar accomplishments result in roughly equal rewards—which needs serious reexamination. First, however, I need to outline this wealth perspective and why I believe it is so important.

By wealth I mean the total value of things families own minus their debts. Income, on the other hand, includes earnings from work, interest and dividends, pensions, and transfer payments. The distinction between wealth and income is significant because one signifies ownership and control of resources and the other represents salary or its replacement. However, the difference between the two is often muddled in the public mind, and only recently have the social sciences begun to treat wealth as an intrinsically important indicator of family well-being that is quite different from income. Another perspective on advantage and disadvantage emerges when wealth is used as an indicator of racial inequality. Wealth represents a more permanent capacity to secure advantages in both the short and long term, and it is transferred across generations. Income data is collected regularly, and vast stores of it exist. In contrast, wealth data has not been collected systematically, and issues such as how to value a home, how to view home equity, whether retirement plans should be counted, and how to value a business make it harder to measure.

Wealth has been a neglected dimension of the social sciences' concern with the economic and social status of Americans in general and racial minorities in particular. We have been much more comfortable describing and analyzing occupational, educational, and income inequality than examining the economic foundation of a capitalist society, "private property." When wealth surveys became available in the mid-1980s, journalists and social scientists began to pay more attention to the issue of wealth. The growing concentration of wealth at the top and the growing racial wealth gap have become important public policy issues that undergird many political debates but, unfortunately, not many policy discussions.

Social scientists typically analyze racial inequality as imbalances in the distribution of power, economic resources, and opportunities. Most research on racial inequality has focused on the economic dimension. This economic component has emphasized jobs and wages. Until very recently, the social sciences and the policy arena neglected the effect of wealth disparity and inheritance on the differing opportunities and well-being of white and black families. We are suggesting that wealth motivates much of what Americans do, grounds their life chances, and provides enduring advantages and disadvantages across generations. Wealth ownership is the single dimension on which whites and blacks are most persistently unequal.

. . . The average American family uses income for food, shelter, clothing, and other necessities. Wealth is different, and I . . . argue that it is used differently than income. Wealth is what families own, a storehouse of resources. Wealth signifies a command over financial resources that when combined with income can produce the opportunity to secure the "good life" in whatever form is needed—education, business, training, justice, health, comfort, and so on. In this sense wealth is a special form of money not usually used to purchase milk and shoes or other life necessities. More often it is used to create opportunities, secure a desired stature and standard of living, or pass class status along to one's children. It is obvious that the positions of two families with the same income but widely different wealth assets are not identical, and it is time for us to take this into account in public policy. . . .

KEY CONCEPTS

discrimination   income   wealth

DISCUSSION QUESTIONS

1. What has happened over time to create the wealth differences by race that Shapiro discusses?
2. Why does Shapiro argue that focusing only on income and workplace discrimination does not capture the complete picture of racial inequality in America?

27

Unmarried with Children

KATHRYN EDIN AND MARIA KEFALAS

In this article, the authors share the story of Jen, a single unmarried mother from Philadelphia. Their research explains why so many young teenage mothers choose not to marry the father of their baby. While many conservative thinkers believe teenage pregnant girls dismiss marriage too readily, Edin and Kefalas uncover the information that these girls may choose not to marry out of respect for the institution of marriage. Their social class position in society limits their options, but not their faith in lasting relationships.

SOURCE: Kathryn Edin and Maria Kefalas. "Unmarried with Children." *Contexts* 4, no. 2 (Spring 2005): 16–22.

Jen Burke, a white tenth-grade dropout who is 17 years old, lives with her stepmother, her sister, and her 16-month-old son in a cramped but tidy row home in Philadelphia's beleaguered Kensington neighborhood. She is broke, on welfare, and struggling to complete her GED. Wouldn't she and her son have been better off if she had finished high school, found a job, and married her son's father first?

In 1950, when Jen's grandmother came of age, only 1 in 20 American children was born to an unmarried mother. Today, that rate is 1 in 3—and they are usually born to those least likely to be able to support a child on their own. In our book, *Promises I Can Keep: Why Poor Women Put Motherhood Before Marriage*, we discuss the lives of 162 white, African American, and Puerto Rican low-income single mothers living in eight destitute neighborhoods across Philadelphia and its poorest industrial suburb, Camden. We spent five years chatting over kitchen tables and on front stoops, giving mothers like Jen the opportunity to speak to the question so many affluent Americans ask about them: Why do they have children while still young and unmarried when they will face such an uphill struggle to support them?

ROMANCE AT LIGHTNING SPEED

Jen started having sex with her 20-year-old boyfriend Rick just before her 15th birthday. A month and a half later, she was pregnant. "I didn't want to get pregnant," she claims. "*He* wanted me to get pregnant." "As soon as he met me, he wanted to have a kid with me," she explains. Though Jen's college-bound suburban peers would be appalled by such a declaration, on the streets of Jen's neighborhood, it is something of a badge of honor. "All those other girls he was with, he didn't want to have a baby with any of them," Jen boasts. "I asked him, 'Why did you choose me to have a kid when you could have a kid with any one of them?' He was like, 'I want to have a kid with *you*.'" Looking back, Jen says she now believes that the reason "he wanted me to have a kid that early is so that I didn't leave him."

In inner-city neighborhoods like Kensington, where childbearing within marriage has become rare, romantic relationships like Jen and Rick's proceed at lightning speed. A young man's avowal, "I want to have a baby by you," is often part of the courtship ritual from the beginning. This is more than idle talk, as their first child is typically conceived within a year from the time a couple begins "kicking it." Yet while poor couples' pillow talk often revolves around dreams of shared children, the news of a pregnancy—the first indelible sign of the huge changes to come—puts these still-new relationships into overdrive. Suddenly, the would-be mother begins to scrutinize her mate as never before, wondering whether he can "get himself together"—find a job, settle down, and become a family man—in time.

Jen began pestering Rick to get a real job instead of picking up day-labor jobs at nearby construction sites. She also wanted him to stop hanging out with his ne'er-do-well friends, who had been getting him into serious trouble for more than a decade. Most of all, she wanted Rick to shed what she calls his "kiddie mentality"—his habit of spending money on alcohol and drugs rather than recognizing his growing financial obligations at home.

Rick did not try to deny paternity, as many would-be fathers do. Nor did he abandon or mistreat Jen, at least intentionally. But Rick, who had been in and out of juvenile detention since he was 8 years old for everything from stealing cars to selling drugs, proved unable to stay away from his unsavory friends. At the beginning of her seventh month of pregnancy, an escapade that began as a drunken lark landed Rick in jail on a carjacking charge. Jen moved back home with her stepmother, applied for welfare, and spent the last two-and-a-half months of her pregnancy without Rick.

Rick sent penitent letters from jail. "I thought he changed by the letters he wrote me. I thought he changed a lot," she says. "He used to tell me that he loved me when he was in jail. . . . It was always gonna be me and him and the baby when he got out." Thus, when Rick's alleged victim failed to appear to testify and he was released just days before Colin's birth, the couple's reunion was a happy one. Often, the magic moment of childbirth calms the troubled waters of such relationships. New parents typically make amends and resolve to stay together for the sake of their child. When surveyed just after a child's birth, eight in ten unmarried parents say they are still together, and most plan to stay together and raise the child.

Promoting marriage among the poor has become the new war on poverty, Bush style. And it is true that the correlation between marital status and child poverty is strong. But poor single mothers already believe in marriage. Jen insists that she will walk down the aisle one day, though she admits it might not be with Rick. And demographers still project that more than seven in ten women who had a child outside of marriage will eventually wed someone. First, though, Jen wants to get a good job, finish school, and get her son out of Kensington.

Most poor, unmarried mothers and fathers readily admit that bearing children while poor and unmarried is not the ideal way to do things. Jen believes the best time to become a mother is "after you're out of school and you got a job, at least, when you're like 21. . . . When you're ready to have kids, you should have everything ready, have your house, have a job, so when that baby comes, the baby can have its own room." Yet given their already limited economic prospects, the poor have little motivation to time their births as precisely as their middle-class counterparts do. The dreams of young people like Jen and Rick center on children at a time of life when their more affluent peers plan for college and careers. Poor girls coming of age in the inner city value children highly, anticipate them eagerly, and believe strongly that they are up to the job of mothering—even in difficult circumstances. Jen, for example, tells us, "People outside the neighborhood, they're like, 'You're 15! You're pregnant?' I'm like, it's not none of their business. I'm gonna be able to take care of my kid. They have nothing to worry about." Jen says she has concluded that "some people . . . are better at having kids at a younger age. . . . I think it's better for some people to have kids younger."

WHEN I BECAME A MOM

When we asked mothers like Jen what their lives would be like if they had not had children, we expected them to express regret over foregone opportunities for school and careers. Instead, most believe their children "saved" them. They describe their lives as spinning out of control before becoming pregnant—struggles with parents and peers, "wild," risky behavior, depression, and school failure. Jen speaks to this poignantly. "I was just real bad. I hung with a real bad crowd. I was doing pills. I was really depressed. . . . I was drinking. That was before I was pregnant." "I think," she reflects, "if I never had a baby or anything, . . . I would still be doing the things I was doing. I would probably still be doing drugs. I'd probably still be drinking." Jen admits that when she first became pregnant, she was angry that she "couldn't be out no more. Couldn't be out with my friends. Couldn't do nothing." Now, though, she says, "I'm glad I have a son . . . because I would still be doing all that stuff."

Children offer poor youth like Jen a compelling sense of purpose. Jen paints a before-and-after picture of her life that was common among the mothers we interviewed. "Before, I didn't have nobody to take care of. I didn't have nothing left to go home for. . . . Now I have my son to take care of I have him to go home for. . . . I don't have to go buy weed or drugs with my money. I could buy my son stuff with my money! . . . I have something to look up to now." Children also are a crucial source of relational intimacy, a self-made community of care. After a nasty fight with Rick, Jen recalls, "I was crying. My son came in the room. He was hugging me. He's 16 months and he was hugging me with his little arms. He was really cute and happy, so I got happy. That's one of the good things. When you're sad, the baby's always gonna be there for you no matter what." Lately she has been thinking a lot about what her life was like back then, before the baby. "I thought about the stuff before I became a mom, what my life was like back then. I used to see pictures of me, and I would hide in every picture. This baby did so much for me. My son did a lot for me. He helped me a lot. I'm thankful that I had my baby."

Around the time of the birth, most unmarried parents claim they plan to get married eventually. Rick did not propose marriage when Jen's first child was born, but when she conceived a second time, at 17, Rick informed his dad, "It's time for me to get married. It's time for me to straighten up. This is the one I wanna be with. I had a baby with her, I'm gonna have another baby with her." Yet despite their intentions, few of these couples actually marry. Indeed, most break up well before their child enters preschool.

I'D LIKE TO GET MARRIED, BUT ...

The sharp decline in marriage in impoverished urban areas has led some to charge that the poor have abandoned the marriage norm. Yet we found few who had given up on the idea of marriage. But like their elite counterparts, disadvantaged women set a high financial bar for marriage. For the poor, marriage has become an elusive goal—one they feel ought to be reserved for those who can support a "white picket fence" lifestyle: a mortgage on a modest row home, a car and some furniture, some savings in the bank, and enough money left over to pay for a "decent" wedding. Jen's views on marriage provide a perfect case in point. "If I was gonna get married, I would want to be married like my Aunt Nancy and my Uncle Pat. They live in the mountains. She has a job. My Uncle Pat is a state trooper; he has lots of money. They live in the [Poconos]. It's real nice out there. Her kids go to Catholic school.... That's the kind of life I would want to have. If I get married, I would have a life like [theirs]." She adds, "And I would wanna have a big wedding, a real nice wedding."

Unlike the women of their mothers' and grandmothers' generations, young women like Jen are not merely content to rely on a man's earnings. Instead, they insist on being economically "set" in their own right before taking marriage vows. This is partly because they want a partnership of equals, and they believe money buys say-so in a relationship. Jen explains, "I'm not gonna just get into marrying him and not have my own house! Not have a job! I still wanna do a lot of things before I get married. He [already] tells me I can't do nothing. I can't go out. What's gonna happen when I marry him? He's gonna say he owns me!"

Economic independence is also insurance against a marriage gone bad. Jen explains, "I want to have everything ready, in case something goes wrong. ... If we got a divorce, that would be my house. I bought that house, he can't kick me out or he can't take my kids from me." "That's what I want in case that ever happens. I know a lot of people that happened to. I don't want it to happen to me." These statements reveal that despite her desire to marry, Rick's role in the family's future is provisional at best. "We get along, but we fight a lot. If he's there, he's there, but if he's not, that's why I want a job ... a job with computers ... so I could afford my kids, could afford the house. ... I don't want to be living off him. I want my kids to be living off me."

Why is Jen, who describes Rick as "the love of my life," so insistent on planning an exit strategy before she is willing to take the vows she firmly believes ought to last "forever"? If love is so sure, why does mistrust seem so palpable and strong? In relationships among poor couples like Jen and Rick, mistrust is often spawned by chronic violence and infidelity, drug and alcohol abuse, criminal activity, and the threat of imprisonment. In these tarnished corners of urban America, the stigma of a failed marriage is far worse than an out-of-wedlock birth. New mothers like Jen feel they must test the relationship over three, four, even five years' time. This is the only way, they believe, to insure that their marriages will last.

Trust has been an enormous issue in Jen's relationship with Rick. "My son was born December 23rd, and [Rick] started cheating on me again ... in March. He started cheating on me with some girl—Amanda. ... Then it was another girl, another girl, another girl after. I didn't wanna believe it. My friends would come up to me and be like, 'Oh yeah, your boyfriend's cheating on you with this person.' I wouldn't believe it. . . . I would see him with them. He used to have hickies. He used to make up some excuse that he was drunk—that was always his excuse for everything." Things finally came to a head when Rick got another girl pregnant. "For a while, I forgave him for everything. Now, I don't forgive him for nothing." Now we begin to understand the source of Jen's hesitancy. "He wants me to marry him, [but] I'm not really sure. . . . If I can't trust him, I can't marry him, 'cause we would get a divorce. If you're gonna get married, you're supposed to be faithful!" she insists. To Jen and her peers, the worst thing that could happen is "to get married just to get divorced."

Given the economic challenges and often perilously low quality of the romantic relationships among unmarried parents, poor women may be right to be cautious about marriage. Five years after we first spoke with her, we met with Jen again. We learned that Jen's second pregnancy ended in a miscarriage. We also learned that Rick was out of the picture—apparently for good. "You know that bar [down the street] It happened in that bar. . . . They were in the bar, and this guy was like badmouthing [Rick's friend] Mikey, talking stuff to him or whatever. So Rick had to go get involved in it and start with this guy. . . . Then he goes outside and fights the guy [and] the guy dies of head trauma. They were all on drugs, they were all drinking, and things just got out of control, and that's what happened. He got fourteen to thirty years."

THESE ARE CARDS I DEALT MYSELF

Jen stuck with Rick for the first two and a half years of his prison sentence, but when another girl's name replaced her own on the visitors' list, Jen decided she was finished with him once and for all. Readers might be asking what Jen ever saw in a man like Rick. But Jen and Rick operate in a partner market where the better-off men go to the better-off women. The only way for someone like Jen to forge a satisfying relationship with a man is to find a diamond in the rough or improve her own economic position so that she can realistically compete for more upwardly mobile partners, which is what Jen is trying to do now. "There's this kid, Donny, he works at my job. He works on C shift. He's a supervisor! He's funny, three years older, and he's not a geek or anything, but he's not a real preppy good boy either. But he's not [a player like Rick] and them. He has a job, you know, so that's good. He doesn't do drugs or anything. And he asked my dad if he could take me out!"

These days, there is a new air of determination, even pride, about Jen. The aimless high school dropout pulls ten-hour shifts entering data at a warehouse distribution center Monday through Thursday. She has held the job for three years, and her aptitude and hard work have earned her a series of raises. Her current salary is higher than anyone in her household commands—$10.25 per hour, and she now gets two weeks of paid vacation, four personal days, 60 hours of sick time, and medical benefits. She has saved up the necessary $400 in tuition for a high school completion program that offers evening and weekend classes. Now all that stands between her and a diploma is a passing grade in mathematics, her least favorite subject. "My plan is to start college in January. [This month] I take my math test . . . so I can get my diploma," she confides.

Jen clearly sees how her life has improved since Rick's dramatic exit from the scene. "That's when I really started [to get better] because I didn't have to worry about what *he* was doing, didn't have to worry about him cheating on me, all this stuff. [It was] then I realized that I had to do what I had to do to take care of my son. . . . When he was there, I think that my whole life revolved around him, you know, so I always messed up somehow because I was so busy worrying about what *he* was doing. Like I would leave the [GED] programs I was in just to go home and see what he was doing. My mind was never concentrating." Now, she says, "a lot of people in my family look up to me now, because all my sisters dropped out from school, you know, nobody went back to school. I went back to school, you know? . . . I went back to school, and I plan to go to college, and a lot of people look up to me for that, you know? So that makes me happy . . . because five years ago nobody looked up to me. I was just like everybody else."

Yet the journey has not been easy. "Being a young mom, being 15, it's hard, hard, hard, you know." She says, "I have no life. . . . I work from 6:30 in the morning until 5:00 at night. I leave here at 5:30 in the morning. I don't get home until about 6:00 at night." Yet she measures her worth as a mother by the fact that she has managed to provide for her son largely on her own. "I don't depend on nobody. I might live with my dad and them, but I don't depend on them, you know." She continues, "There [used to] be days when I'd be so stressed out, like, 'I can't do this!' And I would just cry and cry and cry. . . . Then I look at Colin, and he'll be sleeping, and I'll just look at him and think I don't have no [reason to feel sorry for myself]. The cards I have I've dealt myself so I have to deal with it now. I'm older. I can't change anything. He's my responsibility—he's nobody else's but mine—so I have to deal with that."

Becoming a mother transformed Jen's point of view on just about everything. She says, "I thought hanging on the corner drinking, getting high—I thought that was a good life, and I thought I could live that way for eternity, like sitting out with my friends. But it's not as fun once you have your own kid. . . . I think it changes [you]. I think, 'Would I want Colin to do that? Would I want my son to be like that . . .?' It was fun to me but it's not fun anymore. Half the people I hung with are either . . . Some have died from drug overdoses, some are in jail, and some people are just out there living the same life that they always lived, and they don't look really good. They look really bad." In the end, Jen believes, Colin's birth has brought far more good into her life than bad. "I know I could have waited [to have a child], but in a way I think Colin's the best thing that could have happened to me. . . . So I think I had my son for a purpose because I think Colin changed my life. He *saved* my life, really. My whole life revolves around Colin!"

PROMISES I CAN KEEP

There are unique themes in Jen's story—most fathers are only one or two, not five years older than the mothers of their children, and few fathers have as many glaring problems as Rick—but we heard most of these themes repeatedly in the stories of the 161 other poor, single mothers we came to know. Notably, poor women do not reject marriage; they revere it. Indeed, it is the conviction that marriage is forever that makes them think that divorce is worse than having a baby outside of marriage. Their children, far from being liabilities, provide crucial social-psychological resources—a strong sense of purpose and a profound source of intimacy. Jen and the other mothers we came to know are coming of age in an America that is profoundly unequal—where the gap between rich and poor continues to grow. This economic reality has convinced them that they have little to lose and, perhaps, something to gain by a seemingly "ill-timed" birth.

The lesson one draws from stories like Jen's is quite simple: Until poor young women have more access to jobs that lead to financial independence—until there is reason to hope for the rewarding life pathways that their privileged peers pursue—the poor will continue to have children far sooner than most Americans think they should, while still deferring marriage. Marital standards have risen for all Americans, and the poor want the same things that everyone now wants out of marriage. The poor want to marry too, but they insist on marrying well. This, in their view, is the only way to avoid an almost certain divorce. Like Jen, they are simply not willing to make promises they are not sure they can keep.

KEY CONCEPTS

culture of poverty   urban underclass   welfare system

DISCUSSION QUESTIONS

1. Consider a poor urban tenth grader like Jen and a middle-class suburban tenth grader. How would teen pregnancy be different for these two young girls? What opportunities exist for each?

2. In what ways would marriage help young girls like Jen? What do people who push for marriage believe will change for them? What evidence does this article present that contradicts these ideas?

Applying Sociological Knowledge:
An Exercise for Students

Brand labels in a class-based society communicate our status to others. Make a list of all of clothing labels you can think of and then match each label to a ranking in the class system (working class, middle-class, upper-middle class, etc.). What class images are projected by each? Are there class stereotypes suggested by different labels? Whom do they affect and how? In what ways do these labels reproduce our class identities? Do they do any harm?

28

The Garment Industry in the Restructuring Global Economy

EDNA BONACICH, LUCIE CHENG, NORMA CHINCHILLA,
NORA HAMILTON, AND PAUL ONG

*These authors describe the consequences of the new processes in the global economy,
particularly as they affect different classes of people in different international locales.
The authors note the different interpretations that various scholars have given to the
process of globalization—some seeing its positive effects, others being more critical of the
impact of globalization on women workers, immigrants, and the working class.*

Global integration, a long-standing feature of the world economy, is currently
undergoing a restructuring. Generally, until after World War II, the
advanced industrial countries of western Europe and the United States dominated
the world economy and controlled most of its industrial production. The less-
developed countries tended to concentrate in the production of raw materials.
Since the late 1950s, and accelerating rapidly in the 1980s, however, industrial
production has shifted out of the West, initially to Japan, then to the Asian
NICs (newly industrializing countries—namely, Hong Kong, Taiwan, South
Korea, and Singapore), and now to almost every country of the world. Less-
developed countries are not manufacturing mainly for the domestic market or
following a model of "import substitution"; rather, they are manufacturing
for export, primarily to developed countries, and pursuing a development
strategy of export-led industrialization. What we are witnessing has been
termed by some a "new international division of labor" (Fröbel, Heinrichs,
and Kreye 1980).

The developed countries are faced with the problem of "deindustrialization"
in terms of traditional manufacturing, as their manufacturing base is shifted to
other, less-developed countries (Bluestone and Harrison 1988). At the same time,
they are faced with a massive rise in imports that compete with local industries'
products, moving to displace them. This shift is accompanied by the rise of a new

SOURCE: "The Garment Industry in the Restructuring Global Economy,"
by Edna Bonacich, Lucie Cheng, Norma Chinchilla, Nora Hamilton, and Paul
Ong. As it appears in *Global Production: The Apparel Industry in the Pacific Rim.*
Edited by Edna Bonacich, Lucie Cheng, Norma Chinchilla, Nora Hamilton,
and Paul Ong. Reprinted by permission of Temple University Press © 1994
by Temple University. All rights reserved.

kind of transnational corporation (TNC). Of course, TNCs have existed since the beginning of the European expansion, but they concentrated mainly on the production of agricultural goods and raw materials and, in the postwar period, on manufacturing for the host country market. The new TNCs are global firms that are able to use advanced communications and transportation technology to coordinate manufacturing in multiple locations simultaneously. They engage in "off-shore sourcing" to produce primarily for the home market (Grunwald and Flamm 1985; Sklair 1989).

TNCs sometimes engage in direct foreign investment, but globalized production does not depend on it. They can arrange for production in numerous locations through other, looser connections, such as subcontracting and licensing. In other words, TNCs can set up complex networks of global production without owning or directly controlling their various branches.

The nation-state has increasingly declined as an economic unit, with the result that states are often unable to control the actions of powerful TNCs. The TNCs are supragovernmental actors that make decisions on the basis of profit-making criteria without input from representative governments. Of course, strong states are still able to exercise considerable influence over trade policies and over the policies of the governments of developing countries.

Some scholars have used the concept of "commodity chains" to describe the new spatial arrangements of production (Gereffi and Korzeniewicz 1994). The concept shows how design, production, and distribution are broken down and geographically dispersed, with certain places serving as centers within the chain. Power is differentially allocated along the chain, and countries and firms vie to improve their position in the chain.

Focusing on the geographic aspects of global production also has led to the concept of "global cities" (Sassen 1991). These are coordination centers for the global economy, where planning takes place. They house the corporate headquarters of TNCs, as well as international financial services and a host of related business services. These cities have become the "capitals" of the new global economy.

Another way to view the restructuring is to see it as the proletarianization of most of the world. People who had been engaged primarily in peasant agriculture or in other forms of noncapitalist production are now being incorporated into the industrial labor force. Many of these people are first-generation wage-workers, and a disproportionate number of them are women. These "new" workers sometimes retain ties to noncapitalist sectors and migrate between them and capitalist employment, making their labor cheaper than that of fully prole-tarianized workers. But even if they are not attached to noncapitalist sectors, first-generation workers tend to be especially vulnerable to exploitative conditions. Thus, an important feature of the new globalization is that TNCs are searching the world for the cheapest available labor and are finding it in developing countries.

Countries pursuing export-led industrialization typically follow strategies that encourage the involvement of foreign capital. They offer incentives, including tax holidays and the setting up of export processing zones (EPZs), where the bureaucracy surrounding importing and exporting is curtailed; sometimes they also promise cheap and controllable labor. Countries using this development strategy do not plan to remain the providers of cheap labor for TNCs, however: they hope to move up the production ladder, gaining more economic power and control. They want to shift from labor-intensive manufactures to capital-intensive, high-technology goods. They hope to follow the path of Japan and the Asian NICs and become major economic players in the global economy.

Sometimes participation in global capitalist production is foisted on nations by advanced-industrial countries and/or suprastate organizations such as the World Bank and the International Monetary Fund (IMF), where advanced countries wield a great deal of influence. The United States, in particular, has backed regimes that support globalized production and has pushed for austerity programs that help to make labor cheap. At the same time, developed countries, including the United States,have been affected by the restructured global economy. Accompanying the rise in imports and deindustrialization has been a growth in unemployment and a polarization between the rich and the poor (Harrison and Bluestone 1988). This trend has coincided with increased racial polarization, as people of color have faced a disproportionate impact from these developments.

A rise in immigration from less-developed to more-developed countries has also accompanied globalization. The United States, for example, has experienced large-scale immigration from the Caribbean region and from Asia, two areas pursuing a manufacturing-for-export development strategy. At least part of this immigration is a product of globalization, as people are dislocated by the new economic order and are forced to emigrate for survival (Sassen 1988). Dislocations occur not only because global industries displace local ones (as in the case of agribusiness displacing peasants), but also because austerity programs exacerbate the wage gap between rich and poor countries (making the former ever more desirable). Political refugees, often from countries where the United States has supported repressive regimes, have added to the rise in immigration as well. Finally, some immigration results when people move to service global enterprise as managers, trade representatives, or technicians.

In the advanced countries, the immigration of workers has created a "Third World within." In this case, the newly created proletariat is shifting location. These immigrants play a part in the efforts of the advanced countries to hold on to their industries, by providing a local source of cheap labor to counter the low labor standards in competing countries.

In sum, we are seeing a shakeup of the old world economic order. Some countries have used manufacturing for export as a way to become major economic powers (Appelbaum and Henderson 1992; Gereffi and Wyman 1990). These countries now threaten U.S. dominance. Other countries are trying to pursue this same path, but it is not clear whether they will succeed. Meanwhile, despite the fact that the United States is suffering some negative consequences from the global restructuring, certain U.S.-based TNCs are deeply implicated in the process and benefit from it.

CONTRASTING VIEWS OF RESTRUCTURING

The new globalization receives different interpretations and different evaluations (Gondolf, Marcus, and Dougherty 1986). Some focus on the positive side; they see global production as increasing efficiency by allowing each country to specialize in its strengths. Less-developed countries are able to provide low-cost, unskilled labor while developed countries provide management, technical, and financial resources. Together they are able to maximize the efficient use of resources. The result is that more goods and services are produced more cheaply, to the benefit of all. Consumers, in particular, are seen as the great beneficiaries of globalized production, because of the abundance of low-cost, higher-quality goods from which to choose.

Globalization can be seen as part of the new system of flexible specialization (Piore and Sabel 1984). Consumer markets have become more differentiated, making the old, industrial system of mass production in huge factories obsolete. To be competitive today, a firm must be able to produce small batches of differentiated goods for diverse customers. Globalization contributes to this process by enabling firms to produce a vast range of products in multiple countries simultaneously.

Another aspect of the positive view is to see the entrance of less-developed countries into manufacturing for export as a step toward their industrialization and economic development. Although countries may enter the global economy at a tremendous disadvantage, by participating in exports they are able to accumulate capital and gradually increase their power and wealth. Japan and the Asian NICs have demonstrated the possibilities; now other countries can follow a similar path.

Although workers in the advanced countries may suffer some dislocation by the movement of industry abroad, in the long run they are seen to be beneficiaries of this process. While lower-skilled, more labor-intensive jobs will move to the developing countries, the advanced countries will gain higher-technology jobs, as well as jobs in coordinating and managing the global economy. Thus workers in the advanced countries will be "pushed up" to more middle-class positions, servicing and directing the workers in the rest of the world. Moreover, as other countries develop, their purchasing power will increase, leading to larger markets for the products of the developed countries. Growth in exports means growth in domestic production, and thus growth in domestic employment.

Those who favor globalization also note its inevitability. The economic logic that is propelling global production is immensely powerful. Technology allows globalization, and competition forges it; there is really no stopping the process, so the best one can do is adapt on the most favorable terms possible. Nations feel they must get into the game quickly so as not to be left behind.

A favorable standpoint on globalization is typically coupled with an optimistic view of the effects of immigration. Like new nations entering the global economy, immigrant workers are seen as having to suffer in the short run in order to make advances in the future. Instead of being viewed as exploited, the immigrants are seen as being granted an opportunity—one that they freely choose—to better their life circumstances. They may start off being paid low wages because they lack marketable skills, but with time, they or their children will acquire such skills and will experience upward mobility.

In general, a positive view of globalization is accompanied by a belief in the benefits of markets and free trade. The market, rather than political decision making, should, it is felt, be the arbiter of economic decision making. This favorable and inevitable view of globalization is by far the most predominant approach. It is promoted by the U.S. government, by the TNCs, by many governments in developing countries, and by various international agencies. This position receives considerable support from academics, especially economists, who provide governmental agencies with advice. It is the dominant world policy.

There is, however, a less sanguine interpretation of globalization voiced by U.S. trade unionists and many academics who study development, labor, women, inequality, and social class (Castells and Henderson 1987; Kamel 1990; Kolko 1988; Peet 1987; Ross and Trachte 1990; Sklair 1989). In general, their view is that globalization has a differential class impact: globalization is in the interests of capitalists, especially capitalists connected with TNCs, and of sectors of the capitalist class in developing nations. But the working class in both sets of countries is hurt, especially young women workers, who have become the chief employees of the TNCs (Fernandez-Kelly 1983; Fuentes and Ehrenreich 1983; Mies 1986; Nash and Fernandez-Kelly 1983).

Some argue that globalization is part of a response to a major crisis that has emerged in the advanced capitalist countries. In particular, after the post–World War II boom, the economies of these countries stagnated and profits declined; stagnation was blamed on the advances made by workers under the welfare state. Capital's movement abroad, which was preceded in the United States by regional relocation, is an effort to cut labor costs, weaken unions, and restore profitability. Put generally, globalization can be seen, in part, as an effort to discipline labor.

Globalization enables employers to pit workers from different countries against one another. Regions and nations must compete to attract investment and businesses. Competitors seek to undercut one another by offering the most favorable conditions to capital. Part of what they seek to offer is quality, efficiency, and timeliness, but they also compete in terms of providing the lowest possible labor standards: they promise a low-cost, disciplined, and unorganized work force. Governments pledge to ensure these conditions by engaging in the political repression of workers' movements (Deyo 1989).

The disciplining of the working class that accompanies globalization is not limited to conditions in the workplace. It also involves a cutback in state social programs. For example, in the United States, under the Reagan-Bush administrations, efforts were made to curtail multiple programs protecting workers' standard of living; these tax-based programs were seen as hindering capital accumulation. The argument was made that if these funds were invested by the private sector, everyone would benefit, including workers. This same logic has been imposed on developing countries; they have been granted aid and loans on the condition that they engage in austerity programs that cut back on social spending. The impact of such cutbacks is that workers are less protected from engaging in bargains of desperation when they enter the work force.

This view of globalization is accompanied by a pessimism about the policy of export-led development. Rather than believing that performing assembly for TNCs will lead to development, critics fear that it is another form of dependency, with the advanced capitalist countries and their corporations retaining economic (and political) control over the global economy (Bello and Rosenfeld 1990).

Critics also note a negative side to immigrants' experiences (Mitter 1986; Sassen 1988). They see the immigration of workers as, in part, a product of globalization and TNC activity, as workers in less-developed countries find their means of livelihood disrupted by capitalist penetration. Immigrants are thus not just people seeking a better life for themselves, but often those "forced" into moving because they have lost the means to survive. On arrival in the more-advanced economies, they are faced with forms of coercion, including immigration regulations, racism, and sexism, that keep them an especially disadvantaged work force. Especially coercive is the condition of being an undocumented immigrant. Critics point out that those who favor globalization promote the free movement of commodities and capital, but not the free movement of labor, in the form of open borders. Political restrictions on workers add to the weakening of the working class.

In sum, the critical perspective sees globalization as an effort to strengthen the hand of capital and weaken that of labor. The favorable view argues that the interests of capital and labor are not antagonistic and that everyone benefits from capital accumulation, investment, economic growth, and the creation of jobs. Critics, on the other hand, contend that certain classes benefit at the expense of others, and that, even if workers in poor countries do get jobs, these jobs benefit the capitalists much more than they do the workers, and also hurt the workers in the advanced capitalist countries through deindustrialization.

Where does the truth lie? . . .

To a certain extent, one's point of view depends on geographic location. Generally, Asian countries, especially the NICs, appear to be transforming themselves from dependencies into major actors and competitors in the global economy, leading to an optimism about the effects of globalization. This optimism, however, blots out the suffering and labor repression that is still occurring for some workers in these countries, despite the rise in standard of living for the majority.

On the other hand, the Caribbean region generally faces a harsher reality, in part because the closeness and dominance of the United States pose special problems for these countries. They are more likely to get caught in simple assembly for the TNCs, raising questions about whether manufacturing for export will be transformable into broader economic development. Of course, some in these countries are firm believers in this policy and are pursuing it avidly, but there are clear signs that many workers are severely exploited in the process. . . .

Other confusing issues remain. For example, do women benefit from their movement into the wage sector (proletarianization) as a result of globalization?

A case can be made that working outside the home and earning money gives women new-found power in their relations with men. It can also be argued, however, that these women remain under patriarchal control, but that now, in addition to their fathers and husbands, they are under the control of male bosses. They have double and even triple workloads, as they engage in wage labor, domestic labor, and often industrial homework and other forms of informalized labor.

The two points of view lead to different politics. Those who hold the favorable outlook advocate working for the breakdown of all trade and invest-ment barriers and to pushing rapidly ahead toward global integration. Critics are not trying to stem these forces completely, but rather, are attempting to set conditions on them. For example, globalization should be allowed only if labor and environmental standards are protected in the process. Similarly, the rights of workers to form unions should be safeguarded, so that business cannot wantonly pit groups of workers against one another....

REFERENCES

Appelbaum, Richard P., and Jeffrey Henderson, eds. 1992. *States and Development in the Asian Pacific Rim*. Newbury Park, CA: Sage.

Bello, Walden, and Stephanie Rosenfeld. 1990. *Dragons in distress: Asia's miracle economies in crisis*. San Francisco: Institute for Food and Development Policy.

Castells, Manuel, and Jeffrey Henderson. 1987. "Technoeconomic restructuring, sociopoli-tical processes, and spatial transformation: A global perspective." In *Global restructuring and territorial development*, ed. Jeffrey Henderson and Manuel Castells, 1–17. London: Sage.

Deyo, Frederic C. 1989. *Beneath the Miracle: Labor subordination in the new Asian indus-trialism*. Berkeley: University of California Press.

Fernandez-Kelly, M. Patricia. 1983. *For we are sold, I and my people: Women and industry in Mexico's frontier*. Albany: State University of New York Press.

Fröbel, Folker, Jürgen Heinrichs, and Otto Kreye. 1980. *The new international division of labour: Structural unemployment in industrialised countries and industrialisation in developing countries*. Cambridge: Cambridge University Press.

Fuentes, Annette, and Barbara Ehrenreich. 1983. *Women in the global factory*. Boston: South End Press.

Gereffi, Gary, and Miguel Korzeniewicz, eds. 1994. *Commodity chains and global capitalism*. Westport, CT: Greenwood Press.

Gereffi, Gary, and Donald L. Wyman, eds. 1990. *Manufacturing miracles: paths of indus-trialization in Latin America and East Asia*. Princeton: Princeton University Press.

Gondolf, Edward W., Irwin M. Marcus, and James P. Daugherty. 1986. *The global economy: Divergent perspectives on economic change*. Boulder, CO: Westview Press.

Grunwald, Joseph, and Kenneth Flamm. 1985. *The global factory: Foreign assembly in international trade*. Washington, DC: Brookings Institution.

Harrison, Bennett, and Barry Bluestone, 1988. *The great U-turn: Corporate restructuring and the polarizing of America*. New York: Basic Books.

Kamel, Rachael. 1990. *The global factory: Analysis and action for a new economic era.* Philadelphia: American Friends Service Committee.

Kolko, Joyce. 1988. *Restructuring the world economy.* New York: Pantheon.

Mies, Maria. 1986. *Patriarchy and accumulation on a world scale: Women in the international division of labor.* London: Zed Books.

Mitter, Swasti. 1986. *Common fate, common bond: Women in the global economy.* London: Pluto Press.

Nash, June, and M. Patricia Fernandez-Kelly, eds. 1983. *Women, men, and the international division of labor.* Albany: State University of New York Press.

Peet, Richard, ed. 1987. *International capitalism and industrial restructuring.* Boston: Unwin Hyman.

Piore, Michael J., and Charles F. Sabel. 1984. *The second industrial divide: Possibilities for prosperity.* New York: Basic Books.

Ross, Robert J. S., and Kent C. Trachte. 1990. *Global capitalism: The new leviathan.* Albany: State University of New York Press.

Sassen, Saskia. 1988. *The mobility of labor and capital: A study in international investment and labor flow.* Cambridge: Cambridge University Press.

———. 1991. *The global city: New York, London, Tokyo.* Princeton: Princeton University Press.

Sklair, Leslie. 1989. *Assembling for development: The Maquila industry in Mexico and the United States.* London: Unwin Hyman.

KEY CONCEPTS

commodity chain   economic restructuring   world cities

DISCUSSION QUESTIONS

1. What is a *commodity chain*? What evidence do you see of commodity chains in the wardrobe that you wear?

2. Compare and contrast the two perspectives on the new globalization that the authors describe. How do proponents of the positive and critical views of globalization view immigration?

29

The Nanny Chain

ARLIE RUSSELL HOCHSCHILD

Arlie Hochschild identifies the "nanny chain" as a global system of work in which women workers from poor nations provide the "care work" for more privileged workers in other parts of the world. This pattern of labor is transforming social relations of care worldwide and, according to Hochschild, makes care and love a commodity that is transferred and exchanged in the world market.

Vicky Diaz, a 34-year-old mother of five, was a college-educated school-teacher and travel agent in the Philippines before migrating to the United States to work as a housekeeper for a wealthy Beverly Hills family and as a nanny for their two-year-old son. Her children, Vicky explained to Rhacel Parrenas,

> were saddened by my departure. Even until now my children are trying to convince me to go home. The children were not angry when I left because they were still very young when I left them. My husband could not get angry either because he knew that was the only way I could seriously help him raise our children, so that our children could be sent to school. I send them money every moth.

In her book *Servants of Globalization*, Parrenas, an affiliate of the Center for Working Families at the University of California, Berkeley, tells an important and disquieting story of what she calls the "globalization of mothering." The Beverly Hills family pays "Vicky" (which is the pseudonym Parrenas gave her) $400 a week, and Vicky, in turn, pays her own family's live-in domestic worker back in the Philippines $40 a week. Living like this is not easy on Vicky and her family. "Even though it's paid well, you are sinking in the amount of your work. Even while you are ironing the clothes, they can still call you to the kitchen to wash the plates. It . . . [is] also very depressing. The only thing you can do is give all your love to [the two-year-old American child]. In my absence from my children, the most I could do with my situation is give all my love to that child."

Vicky is part of what we could call a global care chain: a series of personal links between people across the globe based on the paid or unpaid work of

SOURCE: Arlie Russell Hochschild. 2000. "The Nanny Chain." *The American Prospect*, 3 (January 2000), pp. 33–36.

caring. A typical global care chain might work something like this: An older daughter from a poor family in a third world country cares for her siblings (the first link in the chain) while her mother works as a nanny caring for the children of a nanny migrating to a first world country (the second link) who, in turn, cares for the child of a family in a rich country (the final link). Each kind of chain expresses an invisible human ecology of care, one care worker depending on another and so on. A global care chain might start in a poor country and end in a rich one, or it might link rural and urban areas within the same poor country. More complex versions start in one poor country and extend to another slightly less poor country and then link to a rich country.

Global care chains may be proliferating. According to 1994 estimates by the International Organization for Migration, 120 million people migrated—legally or illegally—from one country to another. That's 2 percent of the world's population. How many migrants leave loved ones behind to care for other people's children or elderly parents, we don't know. But we do know that more than half of legal migrants to the United States are women, mostly between ages 25 and 34. And migration experts tell us that the proportion of women among migrants is likely to rise. All of this suggests that the trend toward global care chains will continue.

How are we to understand the impact of globalization on care? If, as globalization continues, more global care chains form, will they be "good" care chains or "bad" ones? Given the entrenched problem of third world poverty— which is one of the starting points for care chains—this is by no means a simple question. But we have yet to fully address it, I believe, because the world is globalizing faster than our minds or hearts are. We live global but still think and feel local.

FREUD IN A GLOBAL ECONOMY

Most writing on globalization focuses on money, markets, and labor flows, while giving scant attention to women, children, and the care of one for the other. Most research on women and development, meanwhile, draws a connection between, say, World Bank loan conditions and the scarcity of food for women and children in the third world, without saying much about resources expended on caregiving. Much of the research on women in the United States and Europe focuses on a chainless, two-person picture of "work-family balance" without considering the child care worker and the emotional ecology of which he or she is a part. Fortunately, in recent years, scholars such as Ernestine Avila, Evelyn Nakano Glenn, Pierette Hondagneu-Sotelo, Mary Romero, and Rhacel Parrenas have produced some fascinating research on domestic workers. Building on this work, we can begin to focus on the first world end of the care chain and begin spelling out some of the implications of the globalization of love.

One difficulty in understanding these implications is that the language of economics does not translate easily into the language of psychology. How are we to understand a "transfer" of feeling from one link in a chain to another? Feeling is not a "resource" that can be crassly taken from one person and given to another. And surely one person can love quite a few people; love is not a resource limited the same way oil or currency supply is. Or is it?

Consider Sigmund Freud's theory of displacement, the idea that emotion can be redirected from one person or object to another. Freud believed that if, for example, Jane loves Dick but Dick is emotionally or literally unavailable, Jane will find a new object (say, John, Dick and Jane's son) onto which to project her original feeling for Dick. While Freud applied the idea of displacement mainly to relations within the nuclear family, the concept can also be applied to relations extending far outside it. For example, immigrant nannies and au pairs often divert feelings originally directed toward their own children toward their young charges in this country. As Sau-ling C. Wong, a researcher at the University of California, Berkeley, has put it, "Time and energy available for mothers are diverted from those who, by kinship or communal ties, are their more rightful recipients."

If it is true that attention, solicitude, and love itself can be "displaced" from one child (let's say Vicky Diaz's son Alfredo, back in the Philippines) onto another child (let's say Tommy, the son of her employers in Beverly Hills), then the important observation to make here is that this displacement is often upward in wealth and power. This, in turn, raises the question of the equitable distribution of care. It makes us wonder, is there—in the realm of love—an analogue to what Marx calls "surplus value," something skimmed off from the poor for the benefit of the rich?

Seen as a thing in itself, Vicky's love for the Beverly Hills toddler is unique, individual, private. But might there not be elements in this love that are borrowed, so to speak, from somewhere and someone else? Is time spent with the first world child in some sense "taken" from a child further down the care chain? Is the Beverly Hills child getting "surplus" love, the way immigrant farm workers give us surplus labor? Are first world countries such as the United States importing maternal love as they have imported copper, zinc, gold, and other ores from third world countries in the past?

This is a startling idea and an unwelcome one, both for Vicky Diaz, who needs the money from a first world job, and for her well-meaning employers, who want someone to give loving care to their child. Each link in the chain feels she is doing the right thing for good reasons—and who is to say she is not?

But there are clearly hidden costs here, costs that tend to get passed down along the chain. One nanny reported such a cost when she described (to Rhacel Parrenas) a return visit to the Philippines: "When I saw my children, I thought, 'Oh children do grow up even without their mother.' I left my youngest when she was only five years old. She was already nine when I saw her again but she still wanted for me to carry her [weeps]. That hurt me because it showed me that my children missed out on a lot."

Sometimes the toll it takes on the domestic worker is overwhelming and suggests that the nanny has not displaced her love onto an employer's child but rather has continued to long intensely for her own child. As one woman told Parrenas, "The first two years I felt like I was going crazy. . . . I would catch myself gazing at nothing, thinking about my child. Every moment, every second of the day, I felt like I was thinking about my baby. My youngest, you have to understand, I left when he was only two months old. . . . You know, whenever I receive a letter from my children, I cannot sleep. I cry. It's good that my job is more demanding at night."

Despite the anguish these separations clearly cause, Filipina women continue to leave for jobs abroad. Since the early 1990s, 55 percent of migrants out of the Philippines have been women; next to electronic manufacturing, their remittances make up the major source of foreign currency in the Philippines. The rate of female emigration has continued to increase and includes college-educated teachers, businesswomen, and secretaries. In Parrenas's study, more than half of the nannies she interviewed had college degrees and most were married mothers in their 30s.

Where are men in this picture? For the most part, men—especially men at the top of the class ladder—leave child-rearing to women. Many of the husbands and fathers of Parrenas's domestic workers had migrated to the Arabian peninsula and other places in search of better wages, relieving other men of "male work" as construction workers and tradesmen, while being replaced themselves at home. Others remained at home, responsible fathers caring or helping to care for their children. But some of the men tyrannized their wives. Indeed, many of the women migrants Parrenas interviewed didn't just leave; they fled. As one migrant maid explained:

> You have to understand that my problems were very heavy before I left the Philippines. My husband was abusive. I couldn't even think about my children, the only thing I could think about was the opportunity to escape my situation. If my husband was not going to kill me, I was probably going to kill him. . . . He always beat me up and my parents wanted me to leave him for a long time. I left my children with my sister. . . . In the plane . . . I felt like a bird whose cage had been locked for many years. . . . I felt free. . . . Deep inside, I felt homesick for my children but I also felt free for being able to escape the most dire problem that was slowly killing me.

Other men abandoned their wives. A former public school teacher back in the Philippines confided to Parrenas: "After three years of marriage, my husband left me for another woman. My husband supported us for just a little over a year. Then the support was stopped. . . . The letters stopped. I have not seen him since." In the absence of government aid, then, migration becomes a way of coping with abandonment.

Sometimes the husband of a female migrant worker is himself a migrant worker who takes turns with his wife migrating. One Filipino man worked in Saudi Arabia for 10 years, coming home for a month each year. When he finally returned home for good, his wife set off to work as a maid in America while he took care of the children. As she explained to Parrenas, "My children were very sad when I left them. My husband told me that when they came back home from the airport, my children could not touch their food and they wanted to cry. My son, whenever he writes me, always draws the head of Fido the dog with tears on the eyes. Whenever he goes to Mass on Sundays, he tells me that he misses me more because he sees his friends with their mothers. Then he comes home and cries."

THE END OF THE CHAIN

Just as global capitalism helps create a third world supply of mothering, it creates a first world demand for it. The past half-century has witnessed a huge rise in the number of women in paid work—from 15 percent of mothers of children aged 6 and under in 1950 to 65 percent today. Indeed, American women now make up 45 percent of the American labor force. Three-quarters of mothers of children 18 and under now work, as do 65 percent of mothers of children 6 and under. In addition, a recent report by the International Labor Organization reveals that the average number of hours of work per week has been rising in this country.

Earlier generations of American working women would rely on grandmothers and other female kin to help look after their children; now the grandmothers and aunts are themselves busy doing paid work outside the home. Statistics show that over the past 30 years a decreasing number of families have relied on relatives to care for their children—and hence are compelled to look for nonfamily care. At the first world end of care chains, working parents are grateful to find a good nanny or child care provider, and they are generally able to pay far more than the nanny could earn in her native country. This is not just a child care problem. Many American families are now relying on immigrant or out-of-home care for their *elderly* relatives. As a Los Angeles elder-care worker, an immigrant, told Parrenas, "Domestics here are able to make a living from the elderly that families abandon." But this often means that nannies cannot take care of their own ailing parents and therefore produce an elder-care version of a child care chain—caring for first world elderly persons while a paid worker cares for their aged mother back in the Philippines.

My own research for two books, *The Second Shift* and *The Time Bind,* sheds some light on the first world end of the chain. Many women have joined the law, academia, medicine, business—but such professions are still organized for men who are free of family responsibilities. The successful career, at least for those who are broadly middle class or above, is still largely built on some key traditional components: doing professional work, competing with fellow professionals, getting credit for work, building a reputation while you're young, hoarding scarce time, and minimizing family obligations by finding someone else to deal with domestic chores. In the past, the professional was a man and the "someone else to deal with [chores]" was a wife. The wife oversaw the family, which—in pre-industrial times, anyway—was supposed to absorb the human vicissitudes of birth, sickness, and death that the workplace discarded. Today, men take on much more of the child care and housework at home, but they still base their identity on demanding careers in the context of which children are beloved impediments; hence, men resist sharing care equally at home. So when parents don't have enough "caring time" between them, they feel forced to look for that care further down the global chain.

The ultimate beneficiaries of these various care changes might actually be large multinational companies, usually based in the United States. In my research on a Fortune 500 manufacturing company I call Amerco, I discovered a disproportionate number of women employed in the human side of the company: public relations, marketing, human resources. In all sectors of the company, women often helped others sort out problems—both personal and professional—at work. It was often the welcoming voice and "soft touch" of women workers that made Amerco seem like a family to other workers. In other words, it appears that these working mothers displace some of their emotional labor from their children to their employer, which holds itself out to the worker as a "family." So, the care in the chain may begin with that which a rural third world mother gives (as a nanny) the urban child she cares for, and it may end with the care a working mother gives her employees as the vice president of publicity at your company.

HOW MUCH IS CARE WORTH?

How are we to respond to the growing number of global care chains? Through what perspective should we view them?

I can think of three vantage points from which to see care chains: that of the primordialist, the sunshine modernist, and (my own) the critical modernist. The primordialist believes that our primary responsibility is to our own family, our own community, our own country. According to this view, if we all tend our own primordial plots, everybody will be fine. There is some logic to this point of view. After all, Freud's concept of displacement rests on the premise that some original first object of love has a primary "right" to that love, and second and third comers don't fully share that right. (For the primordialist—as for most all of us—those first objects are members of one's most immediate family.) But the primordialist is an isolationist, an antiglobalist. To such a person, care chains seem wrong—not because they're unfair to the least-cared-for children at the bottom of the chain, but because they are global. Also, because family care has historically been provided by women, primordialists often believe that women should stay home to provide this care.

The sunshine modernist, on the other hand, believes care chains are just fine, an inevitable part of globalization, which is itself uncritically accepted as good. The idea of displacement is hard for the sunshine modernists to grasp because in their equation—seen mainly in economic terms—the global market will sort out who has proper claims on a nanny's love. As long as the global supply of labor meets the global demand for it, the sunshine modernist believes, everything will be okay. If the primordialist thinks care chains are bad because they're global, the sunshine modernist thinks they're good for the very same reason. In either case, the issue of inequality of access to care disappears.

The critical modernist embraces modernity but with a global sense of ethics. When the critical modernist goes out to buy a pair of Nike shoes, she is concerned to learn how low the wage was and how long the hours were for the third world factory worker making the shoes. The critical modernist applies the same moral concern to care chains: The welfare of the Filipino child back home must be seen as some part, however small, of the total picture. The critical modernist sees globalization as a very mixed blessing, bringing with it new opportunities—such as the nanny's access to good wages—but also new problems, including emotional and psychological costs we have hardly begun to understand.

From the critical modernist perspective, globalization may be increasing inequities not simply in access to money—and those inequities are important enough—but in access to care. The poor maid's child may be getting less motherly care than the first world child. (And for that matter, because of longer hours of work, the first world child may not be getting the ideal quantity of parenting attention for healthy development because too much of it is now displaced onto the employees of Fortune 500 companies.) We needn't lapse into primordialism to sense that something may be amiss in this.

I see no easy solutions to the human costs of global care chains. But here are some initial thoughts. We might, for example, reduce the incentive to migrate by addressing the causes of the migrant's economic desperation and fostering economic growth in the third world. Thus one obvious goal would be to develop the Filipino economy.

But it's not so simple. Immigration scholars have demonstrated that development itself can *encourage* migration because development gives rise to new economic uncertainties that families try to mitigate by seeking employment in the first world. If members of a family are laid off at home, a migrant's monthly remittance can see them through, often by making a capital outlay in a small business or paying for a child's education.

Other solutions might focus on individual links in the care chain. Because some women migrate to flee abusive husbands, a partial solution would be to create local refuges from such husbands. Another would be to alter immigration policy so as to encourage nannies to bring their children with them. Alternatively, employers or even government subsidies could help nannies make regular visits home.

The most fundamental approach to the problem is to raise the value of caring work and to ensure that whoever does it gets more credit and money for it. Otherwise, caring work will be what's left over, the work that's continually passed on down the chain. Sadly, the value ascribed to the labor of raising a child has always been low relative to the value of other kinds of labor, and under the impact of globalization, it has sunk lower still. The low value placed on caring work is due neither to an absence of demand for it (which is always high) nor to the simplicity of the work (successful caregiving is not easy) but rather to the cultural politics underlying this global exchange.

The declining value of child care anywhere in the world can be compared to the declining value of basic food crops relative to manufactured goods on the international market. Though clearly more essential to life, crops such as wheat, rice, or cocoa fetch low and declining prices while the prices of manufactured goods (relative to primary goods) continue to soar in the world market. And just as the low market price of primary produce keeps the third world low in the community of nations, the low market value of care keeps low the status of the women who do it.

One way to solve this problem is to get fathers to contribute more to child care. If fathers worldwide shared child care labor more equitably, care would spread laterally instead of being passed down a social-class ladder, diminishing in value along the way. Culturally, Americans have begun to embrace this idea— but they've yet to put it into practice on a truly large scale [see Richard Weissbourd, "Redefining Dad," *TAP,* December 6, 1999]. This is where norms and policies established in the first world can have perhaps the greatest influence on reducing costs along global care chains.

According to the International Labor Organization, half of the world's women between ages 15 and 64 are working in paid jobs. Between 1960 and 1980, 69 out of 88 countries for which data are available showed a growing proportion of women in paid work (and the rate of increase has skyrocketed since the 1950s in the United States, Scandinavia, and the United Kingdom). If we want developed societies with women doctors, political leaders, teachers, bus drivers, and computer programmers, we will need qualified people to help care for children. And there is no reason why every society cannot enjoy such loving paid child care. It may even remain the case that Vicky Diaz is the best person to provide it. But we would be wise to adopt the perspective of the critical modernist and extend our concern to the potential hidden losers in the care chain. These days, the personal is global.

KEY CONCEPTS

emotional labor	global care chain	theory of displacement

DISCUSSION QUESTIONS

1. What does Hochschild mean by the *globalization of love* and how is this phenomenon linked to the status of women in the United States? In other parts of the world?

2. What different perspectives on the care chain does Hochschild identify? What solutions to the problem does each perspective suggest? What would you recommend?

30

New Commodities, New Consumers

Selling Blackness in a Global Marketplace

PATRICIA HILL COLLINS

Patricia Hill Collins examines how the process of globalization requires ever-expanding consumer markets, thus ensuring greater profits. In this context, she argues, Black men and women become highly commoditized—their bodies used for the interests of a global capitalist economy. She also shows how sexuality in a global marketplace is intertwined with racial and gender inequality.

In the eyes of many Americans, African American youth such as hip hop legend Tupac Shakur constitute a threatening and unwanted population. No longer needed for cheap, unskilled labor in fields and factories, poor and working-class black youth find few job opportunities in the large, urban metropolitan areas where most now reside. Legal and undocumented immigrants now do the dirty work in the hotels, laundries, restaurants and construction sites of a growing service economy. Warehoused in inner city ghettos that now comprise the new unit of racial segregation, poor black youth face declining opportunities and an increasingly punitive social welfare state. Because African American youth possess citizenship rights, social welfare programs legally can no longer operate in racially discriminatory ways. Yet, rather than providing African American youth with educational opportunities, elites chose instead to attack the social welfare state that ensured benefits for everyone. Fiscal conservatives have cut funding for public schools, public housing, public health clinics, and public transportation that would enable poor and working-class black youth to get to burgeoning jobs in the suburbs. Hiding behind a rhetoric of colorblindness, elites claim that these policies lack racial intentionality (Guinier and Torres, 2002; Bonilla-Silva, 2003). Yet when it comes to who is affected by these policies, African American youth constitute a sizable segment of the 'truly disadvantaged' (Wilson, 1987).

...African American youth are often conceptualized as a marginalized, powerless and passive population within macroeconomic policies of globalization. They serve as examples of an economic analysis that only rarely examines intersections of class and race. In contrast, I suggest that because African American youth are in the belly of the beast of the sole remaining world superpower,

SOURCE: Patricia Hill Collins. 2006. "New Commodities, New Consumers: Selling Blackness in a Global Marketplace." *Ethnicities* 6, no. 3: 297–317.

they present an important local location for examining new configurations of social class that is refracted through the lens of race, gender, age and sexuality. Stated differently, because they are centrally located within the United States, Black American youth constitute one important population of social actors who negotiate the contradictions of a racialized globalization as well as the new social class relations that characterize it.

This article asks, what might the placement of poor and working-class African American youth in the global political economy, both as recipients of social outcomes of globalization as well as social agents who respond to those outcomes, tell us about the new racialized class formations of globalization? Conversely, what light might the experiences of poor and working-class African American women and men shed on new global forms of racism? These are very large questions, and I briefly explore them by sketching out a two-part argument. First, I investigate how ideas of consumption, commodification and control situate black youth within a global political economy. I suggest that shifting the focus of class analysis from production to consumption provides a better understanding of black youth. Second, I develop a framework for understanding the commodification of black bodies that ties this process more closely to social class relations. In particular, I use the status of African American youth to explore how the literal and figurative commodification of blackness fosters new strategies of control.

NEW COMMODITIES: ADVANCED CAPITALISM AND BLACK BODY POLITICS

... African American youth are a hot commodity in the contemporary global marketplace and global media. Their images have catalyzed new consumer markets for products and services. The music of hip hop culture, for example, follows its rhythm and blues predecessor as a so-called crossover genre that is very popular with whites and other cultural groups across the globe. Circulated through film, television, and music, news and advertising, mass media constructs and sells a commodified black culture from ideas about class, gender and age. Through a wide array of genres ranging from talk shows to feature length films, television situation comedies to CDs, video rentals to cable television, the images produced and circulated within this area all aim to entertain and amuse a highly segmented consumer market. This market is increasingly global and subject to the contradictions of global marketplace phenomena.

One implication of the significance of consumption for understanding social class relations of black youth concerns the constant need to stimulate consumer markets. Contemporary capitalism relies not just on cutting the costs attached to production, but also on stimulating consumer demand. Just as sustaining relations of production requires a steady supply of people to do the work, sustaining relations of consumption needs ever-expanding consumer markets. Moreover, just as people do not naturally work and must be encouraged or compelled to do so, people do not engage in excess consumption without prompting. In this context, advertising constitutes an important site that creates demand for commodities of all sorts. Marketing and advertising often create demand for things that formerly were not seen as commodities, for example, the rapid growth of the bottled water industry, as well as for intangible entities that seem difficult to commodify. In this regard, the rapid growth of mass media and new informational technologies has catalyzed a demand for black culture as a commodity. . . .

Under this ever-expanding impetus to create new consumer markets, nothing is exempt from commodification and sale, including the pain that African American youth experience with poverty and powerlessness. Nowhere is this more evident than in the contradictions of rap. . . . In this context, rap becomes the only place where black youth have public voice, yet it is a public voice that is commodified and contained by what hip hop producers think will sell. Despite these marketplace limitations, rap remains a potential site of contestation, a place where African American youth can rebel against the police brutality, lack of jobs, and other social issues that confront them (Kelley, 1994). Thus, work on the black culture industry illustrates how images of black culture function to catalyze consumption.

The actual bodies of young African Americans may also be commodified as part of a new black body politics. . . . New forms of commodification within the constant pressure to expand consumer markets catalyze a new black body politics where social class relations rest not solely on exploiting labor power and/or mystifying exploitation through images, but also on the appropriation of bodies themselves. Whereas young black bodies were formerly valued for their labor power, under advanced capitalism, their utility lies elsewhere. . . .

. . . The growth of the punishment industry also illustrates how black male bodies are objectified, commodified and incorporated in service to maintaining prisons as consumer markets. In essence, Black men's commodified bodies become used as raw materials for the growing prison industry. It is very simple—no prisoners, means no jobs for all of the ancillary industries that service this growth industry. Because prisons express little interest in rehabilitating prisoners, they need a steady supply of bodies. The focus is less on appropriating the labor of incarcerated black men (although this does happen) than in finding profitable uses for their bodies while the state absorbs the costs of incarceration. If Kentucky Fried Chicken found chickens in short supply, they would close and their profitability would shrink. The Kentucky Fried Chicken Corporation has little interest in extracting labor from its chickens or in coaxing them to change their ways. Rather, the corporation needs a constant supply of cheap, virtually identical chickens to ensure that their business will remain profitable. In this way, prisons made use of the bodies of unemployed, unskilled young black men, the virtually indistinguishable young black men who populate corners of American cities.

The vast majority of young black people who are incarcerated by the punishment industry are male, yet it is important to remember that disproportionately high numbers of young black women are also incarcerated and thus are subject to this form of commodification. Moreover, young black women may also encounter an additional bodily commodification of their sexuality. The majority of sex workers may be female, obscuring the minority of males who also perform sex work as well as the objectification and commodification of black male bodies within mass media as an important component of the sex work industry. In essence, the bodies and images of young African Americans constitute new commodities that are central to global relations of consumption, not marginalized within them.

NEW CONSUMERS: SEX WORK AND HIP HOP CAPITALISM

Here, I want to take a closer look at this process by exploring how sexuality has grown in importance in the commodification of the bodies and images of black American youth and how this sex work in turn articulates with black agency in responding to advanced capitalism. In essence, black youth are now caught up in a burgeoning sex work industry, one that is far broader than commercial sex work as depicted in the media. Young African Americans participate in the sex work industry, not primarily as commercial workers as is popularly imagined, but rather, as representations of commodified black sexuality as well as potential new consumer markets eager to consume their own images.

Racialized images of pimps and prostitutes may be the commercial sex workers who are most visible in the relations of production, yet the industry itself is much broader. A broader definition of sex work suggests how the sex work industry has been a crucial part of the expansion of consumer markets. The sex work industry encompasses a set of social practices, many of which may not immediately be recognizable as sex work, as well as a constellation of representations that create demand for sexual services, attach value to such services, identify sexual commodities with race, gender and age-specific individuals, and rules that regulate this increasingly important consumer market.

... Sex work is permeating the very fabric of African American communities in ways that resemble how sex work has changed the societies of developing countries. In essence, poor and working-class African American youth increasingly encounter few opportunities for jobs in urban neighborhoods while the mass marketing of sexuality permeates consumer markets. In this sense, their situation resembles that of black youth globally who confront similar pressures in response to globalization. At the same time, the situation of African American youth is unique in that the sexualized images that they encounter are of themselves. In essence, their own bodies often serve as symbols of this sexualized culture, placing African American youth in the peculiar position claiming and rejecting themselves. How might this happen?

... I investigate how reconfiguration of the sex work industry within the United States has shaped the domestic relations within African American communities generally and for poor and working-class African American youth in particular.

Nigeria, the most populous nation state on the African continent, provides an important case for building such an analysis. Reporting on patterns of trafficking in Italy, Eshohe Aghatise describes differential mechanisms used to traffic women from Eastern Europe and Nigeria as well as the differential value placed on women within Italian sex markets. The trafficking of Nigerian women and young girls into Italy for prostitution began in the 1980s in response to Nigeria's economic problems caused by structural adjustment policies of the International Monetary Fund (IMF) (Emeagwali, 1995). As Aghtise points out:

> women and girls started leaving Nigeria for Europe on promises of fantastic well-paying jobs to be obtained in factories, offices, and farms. They arrived in Italy only to find themselves lured into prostitution and sold into sexual slavery to pay off debts, which they were told they incurred in being 'helped' to come to Europe. (Aghatise, 2004: 1129)

Most Nigerian victims of trafficking are illiterate and lacked any exposure to urban life.

The shifting patterns of economic and social change within Nigerian society also contributed to the patterns of trafficking. Traffickers preyed not only on the poverty of Nigerian victims, but also on the breakdown of social and cultural values within Nigerian society, in particular, the disintegration of family structures and a weak social welfare state. For many families, sending female children abroad became a status symbol:

> Subscribing to a consumerist model that is widely publicized on television and in magazines with messages of high living in the West, and in the oral reports of 'been-tos' (a popular name in Nigeria given to those who have been to Western Europe, Canada or the United States), many families believe that it is easy to obtain wealth abroad, and that earning money, in whatever way, will be quick. (Aghatise, 2004: 1132)

Aghatise offers an especially harsh criticism of a society that embraces consumerism and sells its daughters to pay for it:

> The beginning years of Nigeria's economic boom from petrol dollars left the legacy of a people who had acquired a taste for a high standard of living and a consumer society that no longer had the means to satisfy its purchasing habits but was not ready to admit or accept it. (Aghatise, 2004: 1133)

Trafficking of women and girls to Italy demonstrates the fraying social fabric of Nigerian society, especially the ways in which women absorb the pressures placed on families under changing public policies. Most of the women trafficked to Italy are from polygamous families from the Edo ethnic group where wives are in a continuous struggle for a share of the family resources for themselves and their children. Even if men have jobs, their earnings are rarely enough to provide for the needs of the entire family.... The worsening conditions within the Nigerian economy, the weak welfare state, and cultural expectations of women meant that women who were trafficked in the 1990s were mainly much younger girls who set out on a job search to help their families....

... With no jobs for its large youth population, poor Nigerian families learned to look the other way when traffickers commodified and exported its girls and women for the international sex industry and/or when girls saw domestic sex work as their only option. They learned to accommodate a changing set of social norms that pushed young girls toward sex work, for some for reasons of basic survival, yet for others as part of the costs of upward social mobility. Poor and working-class African American girls seemingly confront a similar set of challenges in the context of a different set of circumstances. In this regard, the continuum of sex work from sugar daddies, night brides, floating prostitutes, call girls and trafficked women also applies, yet in a different constellation that reflects the political and economic situation of African Americans as well as cultural values of American society.

Two important features may shape young African American women's participation in the sex work industry. For one, because African American girls are American citizens, they cannot be as easily trafficked as other groups of poor women who lack US citizenship. Girls are typically trafficked into the United States, not out of it. African American girls do enjoy some protections from these forms of exploitation, yet expanding the definition of sex work itself suggests that their patterns of participation have changed. For another, commercial sex work is not always a steady activity, but may occur simultaneously with other forms of income-generating work. In the global context, women sex workers also engage in domestic service, informal commercial trading, market-vending, shining shoes, or office work (Kempadoo and Doezama, 1998: 3). In a similar fashion, African American girls may have multiple sources of income, one of which is sex work. The 'night brides' and 'call girls' of Nigeria may find a domestic counterpart among black American adolescent girls, yet this activity would not be labeled 'sex work', nor would it be seen as prostitution. Restricting the concept of sex work and prostitute to the image of the streetwalker thus obscures the various ways that young black women's bodies and images are commodified and then circulated within the sex work industry....

The consequences of this sex-for-material-goods situation can be tragic. The pressures for young black women to engage in sex work have affected the rapid growth of HIV/AIDS among poor black women in the Mississippi Delta and across the rural South. Between 1990 and 2000, Southern states with large African American populations experienced a dramatic increase in HIV infections among African American women. For example, in Mississippi, 28.5 percent of those reporting new HIV infections in 2000 were black women, up from 13 percent in 1990. In Alabama, the number rose to 31 percent, from 13 percent, whereas in North Carolina, it rose to 27 percent, from 18 percent (Sack, 2001). Most of the women contracted HIV through heterosexual contact, and most found out that they were HIV positive when they became pregnant. The women took risks that may at first seem nonsensical. Yet in the context of their lives there was a sense that because they had so little control over other aspects of their lives, they felt that if God wanted them to get AIDS, then they resigned themselves to getting it.

These examples suggest that many young African American women resign themselves to commodifying their bodies as a necessary source of income. They may not be streetwalkers in the traditional sense, but they also view commodified black sexuality as the commodity of value that they can exchange. These relations also become difficult to disrupt in the context of a powerful mass media that defines and sells images of sexualized black women as one icon of seemingly authentic black culture. Young African American women encounter a set of representations that naturalizes and normalizes social relations of sex work. Whether she sleeps with men for pleasure, drugs, revenge, or money, the sexualized bitch constitutes a modern version of the Jezebel, repackaged for contemporary mass media. In discussing this updated Jezebel image, cultural critic Lisa Jones distinguishes between gold diggers/skeezers, namely, women who screw for status, and crack 'hos', namely, women who screw for a fix (1994: 79). Some women are the 'hos' who trade sexual favors for jobs, money, drugs and other material items. The female hustler, a materialist woman who is willing to sell, rent, or use her sexuality to get whatever she wants constitutes this sexualized variation of the bitch. This image appears with increasing frequency, especially in conjunction with trying to catch an African American man with money. Athletes are targets, and having a baby with an athlete is a way to garner income. Black women who are sex workers, namely, those who engage in phone sex, lap dancing, and prostitution for compensation, also populate this universe of sexualized bitches. The prostitute who hustles without a pimp and who keeps the compensation is a bitch who works for herself.

Black male involvement in the sex work industry may not involve the direct exploitation of black men's bodies as much as the objectification and commodification of sexualized black male images within hip hop culture. The prevalence of representations of black men as pimps speaks to this image of black men as sexual hustlers who use their sexual prowess to exploit women, both black and white. Ushered in by a series of films in the "Blaxploitation" era, the ubiquitous black pimp seems here to stay. Kept alive through HBO produced quasi-documentaries such as *Pimps Up, Hos Down*, African American men feature prominently in mass media. Despite these media constructions, actual pimps see themselves more as businessmen than as sexual predators. For example, the men interviewed in the documentary *American Pimp* all discuss the skills involved in being a successful pimp. One went so far as to claim that only African American men made really good pimps. Thus, the controlling image of the black pimp combines all of the elements of the more generic hustler, namely, engaging in illegal activity, using women for economic gain, and refusing to work.

Representations of black women and men as prostitutes and pimps permeate music videos, film and television. In the context of a powerful global mass media, black men's bodies are increasingly objectified within popular culture in ways that resemble the treatment of all women. Violence and sexuality sell, and associating black men with both is virtually sure to please. Yet the real struggle is less about the content of black male and black female images and more about the treatment of black people's bodies as valuable commodities within advertising and entertainment. Because this new constellation of images participates in commodified global capitalism, in all cases, representations of black people's bodies are tied to structures of profitability. Athletes and criminals alike are profitable, not for the vast majority of African American men, but for the people who own the teams, control the media, provide food, clothing and telephone services to the prisons, and who consume seemingly endless images of pimps, hustlers, rapists, and felons. What is different, however, is how these images of authentic blackness generate additional consumer markets beyond the selling of these specific examples of cultural production. . . .

REFERENCES

Aghatise, E. (2004) 'Trafficking for Prostitution in Italy', *Violence Against Women* 10(10): 1126–55.

Emeagwali, G.T. (1995) *Women Pay the Price: Structural Adjustment in Africa and the Caribbean.* Trenton, N.J.: Africa World Press.

Guinier, L. and G. Torres (2002) *The Miner's Canary: Enlisting Race, Resisting Power, Transforming Democracy.* Cambridge, MA: Harvard University Press.

Jones, Lisa (1994) *Bulletproof Diva: Tales of Race, Sex, and Hair.* New York: Doubleday.

Kelley, R.D.G. (1994) *Race Rebels: Culture, Politics, and the Black Working Class.* New York: Free Press.

KEY CONCEPTS

globalization	sex trafficking	sex work

DISCUSSION QUESTIONS

1. What does Hill Collins mean that Black men and women become a new commodity in the current reality of a global economy?

2. What evidence do you see in popular culture for Hill Collins's argument that Black men and women are sexualized in race and gender specific ways?

3. What parallels does Hill Collins draw regarding the social conditions in Nigeria and in the United States that drives Black women into sex work?

Applying Sociological Knowledge: An Exercise for Students

Take a look at the clothes in your closet. Where are they made? Do some research online into the living environment in some of these countries. Go to the CIA World Fact Book (www.cia.gov). Using the links to these different countries, check under "Economy," "People," and "Government" to answer these questions:

1. What is the life expectancy of people in this nation?

2. What is the infant mortality rate?

3. What percent of people live below the poverty line?

4. What is the unemployment rate?

5. What percentage of household income is held by the highest and lowest income groups?

6. What "transnational issues" (see the bottom of the screen page) does the nation face?

Having answered these questions, what would you now say about global stratification?

31

The Souls of Black Folk

W. E. B. DU BOIS

W. E. B. Du Bois, the first African American Ph. D. from Harvard University, is a classic sociological analyst. In this well-known essay, he develops the idea that African Americans have a "double consciousness"—one that they must develop as a protective strategy to understand how Whites see them. Originally writing this essay in 1903, Du Bois also reflects on the long struggle for Black freedom.

Between me and the other world there is ever an unasked question: unasked by some through feelings of delicacy; by others through the difficulty of rightly framing it. All, nevertheless, flutter round it. They approach me in a half-hesitant sort of way, eye me curiously or compassionately, and then, instead of saying directly, How does it feel to be a problem? they say, I know an excellent colored man in my town; or, I fought at Mechanicsville; or, Do not these Southern outrages make your blood boil? At these I smile, or am interested, or reduce the boiling to a simmer as the occasion may require. To the real question, How does it feel to be a problem? I answer seldom a word. . . .

After the Egyptian and Indian, the Greek and Roman, the Teuton and Mongolian, the Negro is a sort of seventh son, born with a veil, and gifted with second-sight in this American world,—a world which yields him no true self-consciousness, but only lets him see himself through the revelation of the other world. It is a peculiar sensation, this double-consciousness, this sense of always looking at one's self through the eyes of others, of measuring one's soul by the tape of a world that looks on in amused contempt and pity. One ever feels his twoness,—an American, a Negro; two souls, two thoughts, two unreconciled strivings: two warring ideals in one dark body, whose dogged strength alone keeps it from being torn asunder.

The history of the American Negro is the history of this strife—this longing to attain self-conscious manhood, to merge his double self into a better and truer self. In this merging he wishes neither of the older selves to be lost. He would not Africanize America, for America has too much to teach the world and Africa. He would not bleach his Negro soul in a flood of white Americanism, for he knows that Negro blood has a message for the world. He simply wishes to make it possible for a man to be both a Negro and an American, without being cursed

SOURCE: W. E. B. Du Bois. 1989. *The Souls of Black Folk,* edited and with an introduction by Donald B. Gibson. New York: Penguin, pp. 3–12.

and spit upon by his fellows, without having the doors of Opportunity closed roughly in his face.

This, then, is the end of his striving: to be a co-worker in the kingdom of culture, to escape both death and isolation, to husband and use his best powers and his latent genius. These powers of body and mind have in the past been strangely wasted, dispersed, or forgotten. The shadow of a mighty Negro past flits through the tale of Ethíopia the Shadowy and of Egypt the Sphinx. Throughout history, the powers of single black men flash here and there like falling stars, and die sometimes before the world has rightly gauged their brightness. Here in America, in the few days since Emancipation, the black man's turning hither and thither in hesitant and doubtful striving has often made his very strength to lose effectiveness, to seem like absence of power, like weakness. And yet it is not weakness—it is the contradiction of double aims. The double-aimed struggle of the black artisan—on the one hand to escape white contempt for a nation of mere hewers of wood and drawers of water, and on the other hand to plough and nail and dig for a poverty-stricken horde— could only result in making him a poor craftsman, for he had but half a heart in either cause. By the poverty and ignorance of his people, the Negro minister or doctor was tempted toward quackery and demagogy; and by the criticism of the other world, toward ideals that made him ashamed of his lowly tasks. The would-be black *savant* was confronted by the paradox that the knowledge people needed was a twice-told tale to his white neighbors, while the knowledge which would teach the white world was Greek to his own flesh and blood. The innate love of harmony and beauty that set the ruder souls of his people a-dancing and a-singing raised but confusion and doubt in the soul of the black artist; for the beauty revealed to him was the soul-beauty of a race which his larger audience despised, and he could not articulate the message of another people. This waste of double aims, this seeking to satisfy two unreconciled ideals, has wrought sad havoc with the courage and faith and deeds of ten thousand of thousands people,—has sent them often wooing false gods and invoking false means of salvation, and at times has even seemed about to make them ashamed of themselves. . . .

The Nation has not yet found peace from its sins; the freedman has not yet found in freedom his promised land. Whatever of good may have come in these years of change, the shadow of a deep disappointment rests upon the Negro people—a disappointment all the more bitter because the unattained ideal was unbounded save by the simple ignorance of a lowly people. . . .

. . . Merely a concrete test of the underlying principles of the great republic is the Negro Problem, and the spiritual striving of the freedmen's sons is the travail of souls whose burden is almost beyond the measure of their strength, but who bear it in the name of an historic race, in the name of this the land of their fathers' fathers and in the name of human opportunity.

KEY CONCEPTS

caste system   double consciousness

DISCUSSION QUESTIONS

1. What does Du Bois mean by "double consciousness" and how does this affect how African American people see themselves and others?

2. In the contemporary world, what examples do you see that Black people are still defined as "a problem," as Du Bois notes? How does this affect the Black experience?

32

Color-Blind Privilege

The Social and Political Functions of Erasing the Color Line in Post Race America

CHARLES A. GALLAGHER

Charles A. Gallagher discusses the problem of a color-blind approach to race and race relations in this country. By denying race as a structural basis for inequality, we fail to recognize the privilege of Whiteness. With the blurring of racial lines, White college students lack a clear understanding of how the existing social, political, and economic systems advantage or privilege Whites.

INTRODUCTION

An adolescent white male at a bar mitzah wears a FUBU shirt while his white friend preens his tightly set, perfectly braided corn rows. A black model dressed in yachting attire peddles a New England yuppie boating look in Nautica advertisements. It is quite unremarkable to observe white, Asian or African-Americans with dyed purple, blond or red hair. White, black and Asian students decorate their bodies with tattoos of Chinese characters and symbols. In cities and suburbs young adults across the color line wear hip-hop clothing and listen to white rapper Eminem and black rapper Jay-Z. A north Georgia branch of the NAACP installs a white biology professor as its president. The music of Jimi

SOURCE: Gallagher, Charles A. 2003. "Color-Blind Privilege: The Social and Political Functions of Erasing the Color Line in Post Race America." *Race, Gender and Class* (June). Reprinted with permission of the author.

Hendrix is used to sell Apple Computers. Du-Rag kits, complete with bandana headscarf and elastic headband, are on sale for $2.95 at hip-hop clothing stores and family centered theme parks like Six Flags. Salsa has replaced ketchup as the best selling condiment in the United States. Companies as diverse as Polo, McDonalds, Tommy Hilfiger, Walt Disney World, Master Card, Skechers sneakers, IBM, Giorgio Armani and Neosporin antibiotic ointment have each crafted advertisements that show a balanced, multiracial cast of characters interacting and consuming their products in a post-race, color-blind world....

Americans are constantly bombarded by depictions of race relations in the media which suggest that discriminatory racial barriers have been dismantled. Social and cultural indicators suggest that America is on the verge, or has already become, a truly color-blind nation. National polling data indicate that a majority of whites now believe discrimination against racial minorities no longer exists. A majority of whites believe that blacks have "as good a chance as whites" in procuring housing and employment or achieving middle class status while a 1995 survey of white adults found that a majority of whites (58%) believed that African Americans were better off finding jobs than whites.[1] Much of white America now see a level playing field, while a majority of black Americans see a field which is still quite uneven.... The color-blind or race neutral perspective holds that in an environment where institutional racism and discrimination have been replaced by equal opportunity, one's qualifications, not one's color or ethnicity, should be the mechanism by which upward mobility is achieved. Whites and blacks differ significantly, however, in their support for affirmative action, the perceived fairness of the criminal justice system, the ability to acquire the "American Dream," and the extent to which whites have benefited from past discrimination.[2]

This article examines the social and political functions color-blindness serves for whites in the United States. Drawing on interviews and focus groups with whites from around the country I argue that color-blindness maintains white privilege by negating racial inequality. Embracing post-race, color-blind perspective provides whites with a degree of psychological comfort by allowing them to imagine that being white or black or brown has no bearing on an individual's or a group's relative place in the socio-economic hierarchy. My interviews included seventeen focus group and thirty individual interviews with whites around the country. While my sample is not representative of the total white population, I used personal contacts and snowball sampling to purposively locate respondents raised in urban, suburban and rural environments. Twelve of the seventeen focus groups were conducted in a university setting, one in a liberal arts college in the Rocky Mountains and the other at a large urban university in the Northeast. Respondents in these focus groups were selected randomly from the student population. The occupational range for my individual interviews was quite eclectic and included a butcher, construction worker, hair stylist, partner in a prestigious corporate law firm, executive secretary, high school principal, bank president from a small town, retail workers, country lawyer and custodial workers. Twelve of the thirty individual interviews were with respondents who were raised in rural and/or agrarian settings. The remaining respondents lived in suburbs of large cities or in urban areas.

What linked this rather disparate group of white individuals together was their belief that race-based privilege had ended. As a majority of my respondents saw it, color-blindness was now the norm in the United States. The illusion of racial equality implicit in the myth of color-blindness was, for many whites, a form of comfort. This aspect of pleasure took the form of political empowerment ("what about whites' rights") and moral gratification from being liberated from "oppressor" charges ("we are not responsible for the past"). The rosy picture that color-blindness presumes about race relations and the satisfying sense that one is part of a period in American history that is morally superior to the racist days of the past is, quite simply, a less stressful and more pleasurable social place for whites to inhabit.

THE NORM OF COLOR-BLINDNESS

The perception among a majority of white Americans that the socio-economic playing field is now level, along with whites' belief that they have purged themselves of overt racist attitudes and behaviors, has made color-blindness the dominant lens through which whites understand contemporary race relations. Color-blindness allows whites to believe that segregation and discrimination are no longer an issue because it is now illegal for individuals to be denied access to housing, public accommodations or jobs because of their race.... Individuals from any racial background can wear hip-hop clothing, listen to rap music (both purchased at Wal-Mart) and root for their favorite, majority black, professional sports team. Within the context of racial symbols that are bought and sold in the market, color-blindness means that one's race has no bearing on who can ... live in an exclusive neighborhood, attend private schools or own a Rolex.

The passive interaction whites have with people of color through the media creates the impression that little, if any, socio-economic difference exists between the races. Research has found that whites who are exposed to images of upper-middle class African Americans ... believe that blacks have the same socioeconomic opportunities as whites. Highly visible and successful racial minorities like Secretary of State Colin Powell and National Security Advisor Condelleeza Rice are further proof to white America that the state's efforts to enforce and promote racial equality has been accomplished. Reflecting on the extent to which discrimination is an obstacle to socio-economic advancement and the perception of seeing African-Americans in leadership roles, Tom explained:

> If you look at some prominent black people in society today and I don't really see [racial discrimination]. I don't understand how they can keep bringing this problem onto themselves. If they did what society would want them to I don't see that society is making problems for them. I don't see it.

. . . The new color-blind ideology does not, however, ignore race; it acknowledges race while ignoring racial hierarchy by taking racially coded styles and products and reducing these symbols to commodities or experiences which whites and racial minorities can purchase and share. It is through such acts of shared consumption that race becomes nothing more than an innocuous cultural signifier. Large corporations have made American culture more homogeneous through the ubiquitousness of fast food, television, and shopping malls but this trend has also created the illusion that we are all the same through consumption. Most adults eat at national fast food chains like McDonalds, shop at mall anchor stores like Sears and J. C. Penney's and watch major league sports, situation comedies or television drama. Defining race only as cultural symbols that are for sale allows whites to experience and view race as nothing more than a benign cultural marker that has been stripped of all forms of institutional, discriminatory or coercive power. The post-race, color-blind perspective allows whites to imagine that depictions of racial minorities working in high status jobs and consuming the same products, or at least appearing in commercials for products whites desire or consume, is the same as living in a society where color is no longer used to allocate resources or shape group outcomes. By constructing a picture of society where racial harmony is the norm, the color-blind perspective functions to make white privilege invisible while removing from public discussion the need to maintain any social programs that are race-based.

. . . Starting with the deeply held belief that America is now a meritocracy, whites are able to imagine that the socio-economic success they enjoy relative to racial minorities is a function of individual hard work, determination, thrift and investments in education. The color-blind perspective removes from personal thought and public discussion any taint or suggestion of white supremacy or white guilt while legitimating the existing social, political and economic arrangements which privilege whites. This perspective insinuates that class and culture, and not institutional racism, are responsible for social inequality. Color-blindness allows whites to define themselves as politically progressive and racially tolerant as they proclaim their adherence to a belief system that does not see or judge individuals by the "color of their skin." This perspective ignores, as Ruth Frankenberg puts it, how whiteness is a "location of structural advantage societies structured in racial dominance."[3] Frankenberg uses the term "color and power evasiveness" rather than color-blindness to convey how the ability to ignore race by members of the dominant group reflects a position of power and privilege. Color-blindness hides white privilege behind a mask of assumed meritocracy while rendering invisible the institutional arrangements that perpetuate racial inequality. The veneer of equality implied in color-blindness allows whites to present their place in the racialized social structure as one that was earned.

Given this norm of color-blindness it was not surprising that respondents in this study believed that using race to promote group interests was a form of racism.

Joe, a student in his early twenties from a working class background, was quite adamant that the opportunity structure in the United States did not favor one racial group over another.

I mean, I think that the black person of our age has as much opportunity as me, maybe he didn't have the same guidance and that might hurt him. But I mean, he's got the same opportunities that I do to go to school, maybe even more, to get more money. I can't get any aid . . . I think that blacks have the same opportunities as whites nowadays and I think it's old hat.

Not only does Joe believe that young blacks and whites have similar educational experiences and opportunity but it is his contention that blacks are more likely or able to receive money for higher education. The idea that race matters in any way, according to Joe, is anachronistic; it is "old hat" in a color-blind society to blame one's shortcomings on something as irrelevant as race.

Believing and acting as if America is now color-blind allows whites to imagine a society where institutional racism no longer exists and racial barriers to upward mobility have been removed. The use of group identity to challenge the existing racial order by making demands for the amelioration of racial inequities is viewed as racist because such claims violate the belief that we are a nation that recognizes the rights of individuals not rights demanded by groups. Sam, an upper middle class respondent in his 20's, draws on a pre- and post-civil rights framework to explain racial opportunity among his peers:

I guess I can understand in my parents' generation. My parents are older, my dad is almost 60 and my mother is in her mid 50's, ok? But the kids I'm going to school with, the minorities I'm going to school with, I don't think they should use racism as an excuse for not getting a job. Maybe their parents, sure, I mean they were discriminated against. But these kids have every opportunity that I do to do well.

In one generation, as Sam sees it, the color line has been erased. Like Sam's view that opportunity structure is open there is, according to Tara, a reason to celebrate the current state of race relations:

I mean, like you are not the only people that have been persecuted—I mean, yea, you have been, but so has every group. I mean if there's any time to be black in America it's now.

Seeing society as race-neutral serves to decouple past historical practices and social conditions from present day racial inequality as was the case for a number of respondents who pointed out that job discrimination had ended. Michelle was quite direct in her perception that the labor market is now free of discrimination stating that "I don't think people hire and fire because someone is black and white now." Ken also believed that discrimination in hiring did not occur since racial minorities now have legal recourse if discrimination occurs:

I think that pretty much we got past that point as far as jobs. I think people realize that you really can't discriminate that way because you will end up losing . . . because you will have a lawsuit against you.

. . . The logic inherent in the color-blind approach is circular; since race no longer shapes life chances in a color-blind world there is no need to take race into account when discussing differences in outcomes between racial groups. This approach erases America's racial hierarchy by implying that social, economic and political power and mobility are equally shared among all racial groups. Ignoring the extent or ways in which race shapes life chances validates whites' social location in the existing racial hierarchy while legitimating the political and economic arrangements which perpetuate and reproduce racial inequality and privilege. . . .

THE COST OF RACIALIZED PLEASURES

Being able to ignore or being oblivious to the ways in which almost all whites are privileged in a society cleaved on race has a number of implications. Whites derive pleasure in being told that the current system for allocating resources is fair and equitable. Creating and internalizing a color-blind view of race relations reflects how the dominant group is able to use the mass media, immigration stories of upward mobility, rags-to-riches narratives and achievement ideology to make white privilege invisible. Frankenberg argues that whiteness can be "displaced," as is the case with whiteness hiding behind the veil of color-blindness. It can also be made "normative" rather than specifically "racial," as is the case when being white is defined by white respondents as being no different than being black or Asian.[4] Lawrence Bobo and associates have advanced a theory of laissez-faire racism that draws on the color-blind perspective. As whites embrace the equality of opportunity narrative they suggest that

> laissez-faire racism encompasses an ideology that blames blacks them-
> selves for their poorer relative economic standing, seeing it as a function
> of perceived cultural inferiority. The analysis of the bases of laissez-faire
> racism underscores two central components: contemporary stereotypes
> of blacks held by whites, and the denial of societal (structural) respon-
> sibility for the conditions in black communities.[5]

As many of my respondents make clear if the opportunity structure is open ("It doesn't matter what color you are"), there must be something inherently wrong with racial minorities or their culture that explains group level differences.

. . . [T]he form color-blindness takes as the nation's hegemonic political discourse is a variant of laissez-faire racism. Historian David Roediger contends that in order for the Irish to be absorbed into the white race in the mid-nineteenth century "the imperative to define themselves as whites came from the particular 'public and psychological wages' whiteness offered" these new immigrants.[6] There is still a "wage" to whiteness, that element of ascribed status whites automatically receive because of their membership in the dominant group. But within the framework of color-blindness the imperative has switched from whites overtly defining themselves or their interests as white, to one where they claim that color is irrelevant; being white is the same as being black, yellow, brown or red. . . .

My interviews with whites around the country suggest that in this post-race era of color-blind ideology Ellison's keen observations about race relations need modification. The question now is what are we to make of a young white man from the suburbs who listens to hip-hop, wears baggy hip-hop pants, a baseball cap turned sideways, unlaced sneakers and a oversized shirt emblazoned with a famous NBA player who, far from shouting racial epithets, lists a number of racial minorities as his heroes? It is now possible to define oneself as not being racist because of the clothes you wear, the celebrities you like or the music you listen to while believing that blacks or Latinos are disproportionately poor or over-represented in low pay, dead end jobs because they are part of a debased, culturally deficient group. Having a narrative that smooths over the cognitive dissonance and oft time schizophrenic dance that whites must do when they navigate race relations is an invaluable source of pleasure.

NOTES

1. The Gallup Organization, "Black/White Relations in the U.S.," *The Gallup Poll Monthly* (June 10, 1997): 1–5; David Shipler, *A Country of Strangers: Blacks and Whites in America* (New York: Vintage, 1998).

2. David Moore, "Americans' Most Important Sources of Information: Local News," *The Gallup Poll Monthly* (September 1995): 2–5; David Moore and Lydia Saad, "No Immediate Signs that Simpson Trial Intensified Racial Animosity," *The Gallup Poll Monthly* (October 1995): 2–5; Kaiser Foundation, *The Four Americas: Government and Social Policy through the Eyes of America's Multi-Racial and Multi-Ethnic Society* (Menlo Park, CA: Kaiser Family Foundation, 1995).

3. O. Ruth Frankenberg, "The Mirage of an Unmarked Whiteness," in *The Making and Unmaking of Whiteness*, ed. Birget Brander Rasmussen, Eric Klineberg, Irene J. Nexica, and Matt Wray (Durham: Duke University Press, 2001).

4. Ibid., 76.

5. Lawrence Bobo and James R. Kluegel, "Status, Ideology, and Dimensions of Whites' Racial Beliefs and Attitudes: Progress and Stagnation," in *Racial Attitudes in the 1990s: Continuity and Change*, ed. Steven A. Tuch and Jack K. Martin, 95 (Westport, CT: Praeger Publishers, 1997).

6. David Roediger, *The Wages of Whiteness: Race and the Making of the American Working Class*, 137 (New York: Verso Press, 1991).

KEY CONCEPTS

color-blind racism   prejudice   White privilege

DISCUSSION QUESTIONS

1. Summarize Gallagher's argument for why a color-blind attitude is still a privileged attitude. What does color-blindness *not* see when viewing race relations in America?

2. What is the problem with a generation of individuals who do not judge others by the "color of their skin"? Can the individualistic ideology of color-blindness coexist with a society of racist practices?

33

Immigrant Women and Work, Then and Now

NANCY FONER

Nancy Foner compares the experiences of contemporary immigrant women with those who immigrated to the United States closer to the early twentieth century. Although changes in society make the experience different in the two waves of immigration, the experiences of immigrant women in both time periods are shaped by gender relations within their ethnic groups, within their families, and in the social structure of gender in society.

Today's immigrant women enter a society that has undergone remarkable changes since the last great immigrant influx at the turn of the twentieth century. Perhaps most dramatic, is the virtual revolution in women's involvement in the labor force. Whereas in 1900, only 20 percent of women in the nation were in the paid labor force, by 2000, the figure was just over 60 percent. There is a difference in who works, too. A hundred years ago, the vast majority of women workers were young and single. It was generally assumed that work outside the home was temporary for a young girl; when she married, she would move back into the domestic domain. Indeed, in 1900, only 6 percent of the nation's married women were in the labor force.

SOURCE: Nancy Foner, 2005, *In a New Land: A Comparative View of Immigration.* New York: New York University Press.

Today, working daughters have given way to working mothers. Women now enter the labor force later—and they stay. Whether they work for economic need, to maintain or raise their family's living standards, or for personal satisfaction, the fact is that by 2000, seven out of ten women in the United States with children under eighteen worked in the paid labor force, many doing so full-time and year-round.

How have these broad changes in women's participation in the American labor force affected the experiences of immigrant women today as compared to the past? ... In this [article] the stress is on the differences, particularly those shaped by the contrasting structure of work opportunities—and cultural norms and attitudes to women's work—that greeted immigrant women on arrival. Comparing a time when few married immigrant women worked for wages to a period when most do highlights the relationship between migrant women's work and their overall status—and helps us to understand the conditions that lead women to experience gains as well as losses when they come to the United States.

... The analysis of earlier immigrants focuses on eastern European Jews and Italians, who were the vast bulk of the new immigrants in New York City at the turn of the twentieth century. Because no two groups predominate this way today, the discussion of the contemporary period draws on material on a larger number of groups, from Asia, the Caribbean, Latin America, and Europe.

A comparison of migrant women in the two eras reveals some striking differences. Wage work has empowered immigrant wives and mothers in today's New York in ways that were not possible for Jewish and Italian married women of an earlier era, who rarely worked outside the home. Yet, despite this contrast, gender inequalities are still very much with us, and, despite improvements when many women move to New York from abroad, migration has not emancipated the latest arrivals. As feminist scholars have emphasized, simple models that portray migration as leading to female emancipation will not do: migration often leads to losses, as well as gains, for women. Among other things, "traditional" patriarchal codes and practices may continue to have an impact, and women—immigrants as well as the native-born—still experience special burdens and disabilities as members of the "second sex." Indeed, immigrant mothers' continued responsibilities for child care and domestic tasks add new complications for them today when they are more likely to work outside, as well as inside, the home.

JEWISH AND ITALIAN WOMEN THEN

From the beginning, in the move itself, Jewish and Italian women typically followed men—husbands, fiancés, and fathers—who led the way. Women were a minority, too. The Italian migration was, more than anything else, a movement of single men coming to make money and go home. In most years of the peak migration between 1880 and 1910, about 80 percent of Italian immigrants to the United States were male. The Jewish movement was mainly a family affair, but even then men predominated; women made up 43 percent of the migration stream to the United States between 1899 and 1910.

What work did women do in the Old World? In eastern Europe, Jewish women had a central role in economic life. Patriarchy ran deep in Jewish communities—women were excluded from seats of power and positions in the religious sphere—but they were expected to, and did, make important economic contributions to their households.... Women's work, throughout the world of eastern European Jews, was considered necessary and respectable....

Large numbers of Jewish wives worked in business or trade, sometimes helping in a store formally run by their husbands or keeping a store or stall on their own, where they sold food, staples, or household wares. Some women were peddlers who stood in the marketplace or went from house to house selling food they had prepared at home or manufactured goods that were bought in small lots in cities. Jewish wives became tough bargainers who developed a knowledge of the marketplace and a certain worldliness about the society outside their own communities. In the market, women had a better command of local languages spoken by the peasants than did the more learned men, and many developed a reputation for being outspoken and aggressive.

The Jewish community itself provided some jobs for women, for example, rolling and baking matzos at Passover. By the end of the nineteenth century, with the development of factory production in Russia and the movement of many Jews to cities, increasing numbers of unmarried Jewish women were drawn to artisans' shops and small factories, where they made matches, cigarettes, and other goods. When they married, Jewish women rarely took factory jobs that demanded long hours away from home, but many were involved in various kinds of home-based artisanal or outwork production. The sewing machine created new opportunities for doing outwork, and thousands of Jewish married and single female homeworkers made dresses or did other kinds of needlework for contractors who then distributed the garments to stores.

In the Sicilian and southern Italian villages that most Italian immigrant women left behind, married women supervised household chores, organized the making of clothes and food preparation, and managed the family budget. Often, they tended animals and tilled the garden, producing food for family consumption and for sale at the local market. While artisans' wives, who helped out in the shop, worked in the privacy of their homes, peasant women's work took them outside the house as they hauled water, sat together at open streams laundering clothes, or did their chores in the street or courtyard alongside neighbors. Wives in poor families often had no choice but to help as day laborers during harvest periods, picking fruits and nuts, husking almonds, and threshing wheat.

These patterns of work underwent significant change in New York. Although it may be too strong to say, ... that immigration disempowered women who came as wives and mothers, and intensified their subordination, for many Jewish and Italian women, the journey to New York imposed new constraints, and they were forced to lead more sheltered lives than they had in the Old World.

Hardly any Jewish or Italian wives went out to work for wages. The 1905 census recorded only 1 percent of immigrant Russian Jewish households in New York City with wives working outside the home; for Italians the figure, at 6 percent, was not much higher. Marriage, typically around the age of twenty to twenty-two, spelled the end of wage work for the vast majority of Italian and Jewish immigrant women. (This was the general pattern for married white women in American society at the time, with fewer than 5 percent of them in the labor force in 1890 and 1900.) Eventually, some returned to the paid work force in the 1930s and 1940s when their children were grown, but immigrant women who came to New York as married adults often never worked outside the home at all.

Most Italian and Russian Jewish wives and mothers earned money by working at home. In the early years of the immigration, in the 1880s and 1890s, many Jewish women did piecework at home in the needle trades, but by the early twentieth century, the numbers had fallen sharply. By this time, taking care of boarders, virtually indistinguishable from other domestic duties, had become a more attractive alternative—and the main way Jewish wives contributed to the family income. According to the Immigration Commission's 1911-report, as many as 56 percent of New York Russian Jewish families had boarders living with them. Many immigrant wives helped their husbands in "momand-pop" stores and some ran the shops on their own. Minding the store was considered an extension of a woman's proper role as her husband's helpmate; often the family lived above or in back of the store so that wives could run back and forth between the shop counter and the kitchen.

Although many Italian wives added to the family income by taking in boarders, this was a less-frequent practice than among Jews. Homework was more common. By the first decade of the twentieth century, most industrial homeworkers in New York City were Italian. Working in the kitchen or a bedroom, Italian women finished garments or made artificial flowers while raising their children and caring for the house. Women were aware that factory jobs paid better, but the demands of caring for young children and household duties, as well as the widely accepted notion that women should leave the workplace after marriage, usually kept them at home.

In one view, immigrant women's "retirement" to the domestic arena was a blessing. By taking in boarders and doing piecework at home, they contributed much-needed money to the family income at the same time as they reared children and performed time-consuming domestic duties. Cleaning, cooking, and doing the laundry were labor-intensive chores for poor immigrant women who could not afford mechanical conveniences or hired help. The weekly laundry, for example, meant a laborious process of soaking, scrubbing, wringing, rinsing, and drying and ironing clothes. Although women did a tremendous amount of daily housework, they defined their own rhythms.

Unlike the factory, where bosses were in control, women exercised real authority and set the pace in their own households. Apart from nurturing and disciplining children, women managed the family budget. Husbands and unmarried sons usually gave them the larger part of their wages each week; most unmarried daughters handed over their entire paycheck. The role of housewife and mother, moreover, if done well, carried with it respectability and the approval of family and neighbors.

Yet women's housebound existence had a downside as well. By and large, married women's lives were more circumscribed in New York than in the Old World. Immigrant mothers did, of course, socialize with friends and neighbors and go out to shop. The Jewish housewife, as the family member most responsible for decisions about household purchases, presided over a process of acquisition of consumption items. But whereas in eastern Europe, Jewish wives were often the worldly ones, in America their housebound existence made it more difficult to learn the new language and customs. Their husbands picked up English in the workplace; their daughters learned American ways in factory work groups. Many Jewish mothers, however, remained fluent only in Yiddish and felt uncomfortable in new situations outside the Jewish community. They had to depend on their children to learn American customs or, as a few managed to do, attend night school to learn English. Italian women working at home were also more insulated than other family members from the world outside. . . .

Most household chores, as well as industrial homework, were done within the four walls of their tenement apartments. Those from small towns and villages, used to doing chores like laundry in the company of other women, now faced the more lonely and difficult task of washing clothing by themselves inside cramped tenement apartments. . . .

Even as modern plumbing freed women from some of the more rigorous chores they had known in the Old World, the more rigorous standards of cleanliness and new household acquisitions complicated housework. In small eastern European towns and villages, women went to the nearest stream or lake once a month to wash clothes; now the laundry was a weekly task. Another example: mattresses in eastern Europe were generally made of straw, and in cold weather feather bedding was common. In America, beds came with mattresses that required sheets and blankets; these needed washing and airing on a regular basis.

For the Jewish women who had been charged with providing a major portion of the family livelihood in eastern Europe, migration reduced their economic role. In New York, immigrant wives' income-earning activities rarely represented the major contribution to the family economy. Industrial homework or taking in boarders was not as lucrative as work outside the home, and wives were seen as helping out their husbands in family businesses. Married women's earnings in America were now eclipsed by the wages of working daughters in the industrial labor force, who emerged as the main female breadwinners in the Jewish family.

IMMIGRANT WOMEN NOW

Much has changed for the latest arrivals. Women immigrants now outnumber men in virtually all the major groups in New York, and more women come on their own rather than follow in the footsteps of men. Today's immigrant women also include a much higher proportion with professional and middle-class backgrounds. Above all, the world they live in gives women opportunities and benefits unheard of a century ago—and this is particularly evident in the sphere of work.

Today, adult immigrant women are the main female contributors to the family income, while their teenage daughters are generally in school. With the expansion of high schools and colleges over the course of the twentieth century and the raising of the school-leaving age, women (and men) start working later than they used to. Today's immigrant daughters are often eighteen or older when they enter the labor market full-time compared to fourteen or fifteen a century ago. Marriage no longer spells a retreat from paid employment outside the home. Industrial homework, while not entirely a thing of the past, is much rarer than in the era of Italian and Jewish immigrants. Now it is socially accepted, even expected, throughout American society that wives and mothers will go out to work.

At the time of the 1990 census, 60 percent of New York City's working-age foreign-born women (compared to 66 percent of the city's working-age women generally) were in the labor force. At one end, Filipino women, who often came specifically to work in health-care jobs, had a labor force participation rate of more than 85 percent; West Indian women were not far behind, with labor force participation rates in the range of 70 to 80 percent. Dominican women were near the bottom, with 52 percent in the work force. Given the wide variety of groups today, and the diversity of immigrant backgrounds, immigrant women occupy an equally wide range of jobs, from nurses, secretaries, and health technicians to domestics and factory workers.

These new patterns have important consequences. Now that most immigrant women work outside the home, they are able to obtain a kind of independence and power that was beyond the reach of Jewish and Italian wives and mothers a century ago, and that was often beyond their own reach before migration. How much improvement women experience when they migrate depends to a large degree on their role in production and their social status in the home country—and gender roles, norms, and ideologies there—as well as their economic role in New York. What is important here is that for the majority of migrant women, the move to New York has led to gains because they earn a regular wage for the first time, earn a higher wage than in the country of origin, or make a larger contribution to the family economy than previously.

In cases where women did not earn an income or earned only a small supplementary income prior to migration, the gains that come with regular wage work in New York are especially striking. The much-cited case of Dominican women fits this pattern. They left a society where, in 2001, 41 percent of women were in the labor force. Now that so many Dominican immigrant women work for wages—often for the first time—and contribute a larger share of the family income, they have more authority in the household and greater self-esteem. . . .

In New York, Dominican women begin to expect to be copartners in "heading" the household, a change from more patriarchal arrangements in the Dominican Republic. . . . In the Dominican Republic, men generally controlled the household budget, even when wives and daughters put in income on a regular or semiregular basis; in New York, Pessar found that husbands, wives, and working children usually pooled their income in a common fund for shared household expenses. Indeed, she reports that Dominican women are eager to postpone or avoid returning to the Dominican Republic, where social pressures and an unfavorable job market would probably mean their retirement from regular employment and a loss of new-found gains.[1]

Of course, many immigrant women, including some Dominicans, had regular salaries before emigration. Even these women often feel a new kind of independence in New York, because jobs in this country pay more than most could ever earn at home and increase women's contribution to the family economy. This is the experience for many Jamaican women, who come from a society where, at the beginning of the twenty-first century, two out of three women were in the labor force. Many Jamaican women I interviewed who had held white-collar jobs before emigration said they had more financial control and more say in family affairs in New York where their incomes are so much larger.

The sense of empowerment that comes from earning a wage—or a higher wage—and having greater control over what they earn comes out in studies of many different groups. Paid work for Chinese garment workers, according to one report, not only contributes to their families' economic well-being, but also has "created a sense of confidence and self-fulfillment which they may never have experienced in traditional Chinese society." "I do not have to ask my husband for money," one woman said, "I make my own."[2] For many Salvadoran women, the ability to earn wages and decide how they should be used is something new. As one woman explained: "Here [in the U.S.] women work just like the men. I like it a lot because managing my own money I feel independent. I don't have to ask my husband for money but in El Salvador, yes, I would have to. Over there women live dependent on their husbands. You have to walk behind him."[3] Or listen to a Trinidadian woman of East Indian descent: "Now that I have a job I am independent. I stand up here as a man."[4]

The female-first migration pattern involving adult married women that is common in some groups reinforces the effects of wage-earning on women's independence. Many women who have lived and worked in New York without their husbands become more assertive.... One study suggests that Asian men who move to the United States as their wives' dependents often have to subordinate their careers, at least initially, to those of their wives since the women have already established themselves in this country.[5]

Work outside the home in New York brings about another change that women appreciate. Many men now help out more *inside* the home than before they moved to New York. Of course, this is not inevitable. Cultural values in different groups as well as the availability of female relatives to lend a hand influence the kind of household help men provide.... Korean men, staunch supporters of patriarchal family values and norms, generally still expect their wives to serve them and resist performing household chores like cooking, dishwashing, and doing the laundry. Such resistance is more effective when the wife's mother or mother-in-law lives in the household, a not infrequent occurrence in Korean immigrant families. Yet much to their consternation, Korean men in New York with working wives often find themselves helping out with household work more than they did in Korea—and wives often make more demands on them to increase their share.

Research on Latin American and Caribbean groups shows that when wives are involved in productive work outside the home, the organization of labor within it changes. We are not talking about a drastic change in the household division of labor or the emergence of truly egalitarian arrangements. Indeed, Latin American and Caribbean women strongly identify as wives and mothers and like being in charge of the domestic domain. What they want—and what they often get—is more help from men than they were accustomed to back home. Mainly, men oblige because they have little choice.

West Indian men, for example, recognize that there is no alternative to pitching in when their wives work and children (particularly daughters) are not old enough to lend a hand. Working women simply cannot shoulder all the domestic responsibilities expected of them, and they do not have relatives available to help as they did back home. Even if close kin live nearby, they are usually busy with work and their own household chores. Wives' wages are a necessary addition to the family income, and West Indians cannot afford to hire household help in New York. . . . Indeed, West Indian couples with young children often arrange their shifts so that the husband can look after the children while the wife works.

More than behavior changes. As men become accustomed to doing more around the house, their notions of what tasks are appropriate—or expected—often also shift. Research shows that Dominican and Jamaican men and women believe that when both partners have jobs, and daughters are too young to help, husbands should pitch in with such tasks as shopping, dishwashing, and child care. Women tend to view their husband's help as a moral victory; men accept their new duties, however reluctantly.

Although the exigencies of immigrant life—women working outside the home, a lack of available relatives to assist, and an inability to hire help—are mainly responsible for men's greater participation in household tasks, American cultural beliefs and values have an influence too. . . .

In addition to the independence, power, and autonomy that wages bring, there are the intrinsic satisfactions from work itself. Women in professional and managerial positions gain prestige from their positions and often have authority over others on the job. Those in lower-level occupations often get a sense of satisfaction from doing their job well and from the new skills they have learned in New York. And there is the sociability involved. In factories, hospitals, and offices, women make friends and build up a storehouse of experiences that enrich their lives and conversations. Indeed, when women are out of work, they often complain of boredom and isolation. . . . Friendships formed on the job may extend outside the bounds of the workplace as women visit and phone each other, attend parties, and go on shopping jaunts with coworkers.

But it is important not to romanticize or idealize immigrant women's work outside the home as a path to self-fulfillment or economic autonomy. If wage work enables many immigrant women to expand their influence and independence, these gains often come at a price. Wage work brings burdens as well as benefits to immigrant women and may create new sets of demands and pressures both on the job and at home. Moreover, despite changes in women's status in New York, premigration gender role patterns and ideologies do not fade away; they continue to affect the lives of migrant women, often in ways that constrain and limit them. Cultural ideals about gender and spousal relations held at the point of origin, . . . influence the outcome of the changing balance of economic resources in the United States.

Wage work, as immigrant women commonly explain, is not an option but a necessity for their family's welfare. And it typically brings a host of difficulties.

On the job, women's wages are still generally lower than men's—and, for many, perhaps most, immigrant women, too meager to sustain their economic independence from men. In addition, women are limited in their choice of work due to gender divisions in the labor market—often confined to menial, low prestige, and poorly paying jobs. Working in the ethnic economy does not help most women, either. Studies of Chinese, Dominican, and Colombian women in New York who work in businesses owned by their compatriots show that they earn low wages and have minimal benefits and few opportunities for advancement. Sociologist Greta Gilbertson argues that some of the success of immigrant small-business owners and workers in the ethnic enclave is due to the marginal position of immigrant women. The many Korean women who work in family businesses are, essentially, unpaid family workers without an independent source of income. Although many are working outside the home for the first time, they are typically thought of as "helpers" to their husbands; the husband not only legally owns the enterprise but also usually controls the money, hires and fires employees, and represents the business in Korean business associations.[6]

For many immigrant women, working conditions are extremely difficult. Apart from the low wages and long hours, most garment workers have to keep up a furious pace in cramped conditions in noisy, often unsafe, sweatshops; domestic workers often have to deal with humiliating and demeaning treatment from employers. Some women with full-time jobs have more than one position to make ends meet. I know many West Indian women, for example, who care for an elderly person on the weekend to supplement what they earn from a five-day child-care job.

Added to this, of course, are the demands of child-care and burdens of household work. Going outside to earn means that childrearing is more complicated than at the turn of the twentieth century, when married women typically worked at home. Only very affluent immigrants can afford to hire maids or housekeepers, and female relatives, if present in New York, are often busy at work themselves. Occasionally, women can juggle shifts with their husbands so one parent is always around, and sometimes an elderly mother or mother-in-law is on hand to help out. Many working women pay to leave their children with babysitters or, less often, in day-care centers. Child-care constraints are clearly a factor limiting women to low-paid jobs with flexible schedules; they may prevent women from working full-time—or, in some cases, at all. Some women leave their young children behind with relatives in the home country so as to manage work more easily, a common pattern among West Indian live-in household workers. . . .

Immigrant women of all social classes have the major responsibilities for household chores as well as childrearing, so that a grueling day at work is often followed or preceded by hours of cooking, cleaning, and washing. "I'm always working," is how Mrs. Darius, a Haitian nursing home aide with eight children put it. Although her husband, a mechanic, did not help much around the house, Mrs. Darius got assistance from her mother who lived with her. Still, there was a lot to do. "I have to work 24 hours. When I go home, I take a nap, then get up again; sometimes I get up at two in the morning, iron for the children, and go back to sleep."[7]

Korean working wives, according to Pyong Gap Min, suffer from overwork and stress owing to the heavy demands on their time. After doing their work outside the home, they put in, on average, an additional twenty-five hours a week on housework, compared to seven hours contributed by their husbands. Altogether, working wives spend seventy-six hours a week on the job and doing housework—twelve more hours than men do. Although professional husbands help out more around the house than other Korean men, their wives still do the lion's share.[8]

Or take the case of Antonia Duarte, a Dominican mother of three, who put in a seventeen-hour day. At 5:00 a.m., she was up making breakfast and lunch for the family. She woke her three children at 6:00, got them dressed, fed, and ready for school, and then took them to the house of a friend, who cared for the four-year-old and oversaw the older children's departure to and return from school. By 7:15, Antonia was on the subway heading for the lamp factory where she worked from 8:00 a.m. to 4:30 p.m. five days a week. She collected her children a little after 5:00 and began preparing the evening meal when she got home. She did not ask her two oldest children to help—the oldest was a twelve-year-old girl—because, "I'd rather they begin their homework right away, before they get too tired." Her husband demanded a traditional meal of rice, beans, plantains, and meat, which could take as long as two hours to prepare. She and the children ate together at 7:00, but her husband often did not get back from socializing with his friends until later. He expected Antonia to reheat the food and serve it on his arrival. By the time she finished her child-care and other domestic responsibilities, it was 11:30 or 12:00. Like other Dominican women, she explained that if she did not manage the children and household with a high level of competence, her husband would threaten to prohibit her from working.[9]

Women in groups where strong traditional patriarchal codes continue to exert an influence may experience other difficulties. In some better-off Dominican families, wives are pressured by husbands to stay out of the work force altogether as a way to symbolize their household's respectability and elevated economic status. It is still a point of pride for a Latin American man to say that his wife doesn't work; part of making it into the middle class is seeing to it that the women in the household remain at home. In many groups, working women who are now the family's main wage earners may feel a special need to tread carefully in relations with their husbands so as to preserve the appearance of male dominance and to avoid making the men feel inadequate. . . .

Finally, there is the fact that women's increased financial authority and independence—or being more economically successful in New York than their husbands—can lead to greater discord with their spouses. Conflicts often develop when men resent and try to resist women's new demands on them; in some cases, the stresses ultimately lead to marital break-ups. Special problems may develop when men are unemployed or unsuccessful at work and become dependent on women's wage-earning abilities, yet still insist on maintaining the perquisites of male privilege in the household. In extreme cases, the reversal of gender roles can lead to serious physical abuse for women at the hands of their spouses. Indeed, in some instances, increased isolation from relatives in the immigrant situation creates conditions for greater abuse by husbands, who are freer of the informal controls that operated in their home communities, where friends and family would have been more likely to intervene and play a mediatory role.

CONCLUSION: IMMIGRANT WOMEN
IN THE TWO ERAS

Comparing immigrant women today and at the turn of the twentieth century makes clear that women's involvement in the world of work is critical to understanding why moving to New York has been liberating in many ways for so many contemporary immigrants—and why, at least for immigrant mothers and wives, it was more limiting in the past. Jewish and Italian women came to New York at a time when there was a social stigma attached to the wife who worked for wages outside the home; the mother's wage was considered a "final defense against destitution," to be undertaken only on account of severe economic or family emergency.[10] Often, Jewish and Italian immigrant wives found themselves more cloistered in their homes than in the Old World. The work they did to earn money—taking in boarders and industrial homework—did not lead to reallocating household tasks among other household members. Because virtually all of their income-producing activities were done in the home, these activities ended up preserving and intensifying the gender division of housework and child care. The main female wage earners in the family, immigrant daughters, handed over their pay to their mothers, who, as managers of daily financial affairs, used it for running the household.

Now that female wage earners are typically wives and mothers, they have more leverage in the household than working younger daughters once had. Indeed, adult women's employment has begun to transform their family relationships more so than in the earlier generation. Because an immigrant working mother today is often absent from the home for forty or forty-five hours a week, or sometimes longer, someone must fill her place—or at least help out. Often, it is her husband. Women's labor force participation, in other words, frequently increases husbands' participation in household work and leads to changes in the balance of power in immigrant families. Daughters in modern-day families, growing up in an era when female labor participation is the norm and the working mother is commonplace, may go even further in redefining family roles as they enter the labor force for an extended period of their lives.

As the main female wage earners in the family, today's immigrant mothers contribute a larger share of the household income than they did a hundred years ago. Their regular access to wages—and to higher wages—in the United States often gives them greater autonomy and power than they had before migration. Working outside the home also broadens their social horizons and enhances their sense of independence. . . .

If immigrant wives and mothers have come a long way in the past hundred years, it is clear that they are not fully emancipated. Not only do they suffer from gender inequalities that are a feature of American society generally, but important vestiges of premigration gender ideologies and role patterns may place additional constraints on them. . . . Many work in low-status, dead-end positions that pay less than men's jobs. Immigrant working wives in all social classes experience a heavy double burden since the household division of labor remains far from equal. If husbands help out with domestic burdens, they may do so only grudgingly, if at all, and it is women, more than men, who make work choices to accommodate and reflect family and child-care needs. While many, perhaps most, immigrant women feel that the benefits of wage work outweigh the drawbacks, others would, if they could afford it, prefer to remain at home....

A comparison of women in the two eras should not, in short, blind us to the barriers and difficulties immigrant women still face. Improvements in women's status today go hand in hand with the persistence of male privilege. At the same time, the comparison is a powerful reminder that "the New York we have lost," to paraphrase Peter Laslett, was hardly a utopia for women, and that working outside the home, for all its problems, has brought significant benefits to migrant women today.[11]

NOTES

1. Pessar, Patricia. 1995. *A Visa for a Dream*. Boston: Allyn and Bacon.

2. Zhou, Min, and Regina Nordquist. 1994. "Work and Its Place in the Lives of Immigrant Women," *Applied Behavioral Science Review* 2, p. 201.

3. Mahler, Sarah. 1996. "Bringing Gender to a Transnational Focus: Theoretical and Empirical Ideas," Unpublished manuscript.

4. Burgess, Judith, and Meryl Gray. 1981. "Migration and Sex Roles: A Comparison of Black and Indian Trinidadians in New York City." In *Female Immigrants in the United States*, edited by Delores Mortimer and Roy S. Bryce-LaPorte. Washington, DC: Research Institute on Immigration and Ethnic Studies, p. 104.

5. Pessar 1995, p. 60.

6. Gilbertson, Greta. 1995. "Women's Labor and Enclave Employment: The Case of Dominican and Caribbean Women in New York City." *International Migration Review* 19: 657–671; Min, Pyong Gap. 1998. *Changes and Conflicts: Korean Immigrant Families in New York*. Boston: Allyn and Bacon, pp. 45–46.

7. Foner, Nancy. 1994. *The Caregiving Dilemma: Work in An American Nursing Home*. Berkeley: University of California Press, p. 107.

8. Min, Pyong Gap. *Caught in the Middle: Korean Communities in New York and Los Angeles*. Berkeley: University of California Press. 1998.

9. Pessar, Patricia. 1982. "Kinship Relations of Production in the Migration Process: The Case of Dominican Emigration to the United States," Occasional Paper No. 12. New York: Center for Latin American and Caribbean Studies, New York University.

10. Weiner, Lynn. 1985. *From Working Girl to Working Mother: The Female Labor Force in the United States, 1820–1980*. Chapel Hill: University of North Carolina Press, pp. 84–85; Tentler, Leslie Woodcock. 1979. *Wage-Earning Women*. New York: Oxford University Press, pp. 139–142.

11. Laslett, Peter. 1965. *The World We Have Lost*. New York: Scribners.

KEY CONCEPTS

division of labor   ethnic enclaves   immigration

DISCUSSION QUESTIONS

1. What are some of the ways that the experiences of immigrant women near the turn of the twentieth century are different than for immigrant women now? How are they similar?

2. How do gender expectations within different ethnic families shape the experiences of immigrant women?

3. How does the gender division of labor shape the working lives of immigrant women? What similarities and differences are there across the different groups Nancy Foner discusses?

34

Seeing More than Black and White

ELIZABETH MARTINEZ

Racism in the United States is more than simply a matter of Black and White. It involves intense racism about other groups as well, especially Latinos. Elizabeth Martinez advocates a broader framework involving many groups when studying and combating racism.

The racial and ethnic landscape has changed too much in recent years to view it with the same eyes as before. We are looking at a multi-dimensional reality in which race, ethnicity, nationality, culture and immigrant status come together with breathtakingly new results. We are also seeing global changes that have a massive impact on our domestic situation, especially the economy and labor force. For a group of Korean restaurant entrepreneurs to hire Mexican cooks to prepare Chinese dishes for mainly African American customers, as happened in Houston, Texas, has ceased to be unusual.

The ever-changing demographic landscape compels those struggling against racism and for a transformed, non-capitalist society to resolve several strategic questions. Among them: doesn't the exclusively Black-white framework discourage the perception of common interests among people of color and thus sustain

SOURCE: Martinez, Elizabeth. 1998. "Seeing More than Black and White." In *De Colores Means All of Us: Latina Views for a Multi-Colored Century*. Cambridge, MA: South End Press. Reprinted by permission of the South End Press.

White Supremacy? Doesn't the view that only African Americans face serious institutionalized racism isolate them from potential allies? Doesn't the Black-white model encourage people of color to spend too much energy understanding our lives in relation to whiteness, obsessing about what white society will think and do?

That tendency is inevitable in some ways: the locus of power over our lives has long been white (although big shifts have recently taken place in the color of capital, as we see in Japan, Singapore and elsewhere). The oppressed have always survived by becoming experts on the oppressor's ways. But that can become a prison of sorts, a trap of compulsive vigilance. Let us liberate ourselves, then, from the tunnel vision of whiteness and behold the many colors around us! Let us summon the courage to reject outdated ideas and stretch our imaginations into the next century.

For a Latina to urge recognizing a variety of racist models is not, and should not be, yet another round in the Oppression Olympics. We don't need more competition among different social groups for the gold medal of "Most Oppressed." We don't need more comparisons of suffering between women and Blacks, the disabled and the gay, Latino teenagers and white seniors, or whatever. Pursuing some hierarchy of oppression leads us down dead-end streets where we will never find the linkage between different oppressions and how to overcome them. To criticize the exclusively Black-white framework, then, is not some resentful demand by other people of color for equal sympathy, equal funding, equal clout, equal patronage or other questionable crumbs. Above all, it is not a devious way of minimizing the centrality of the African-American experience in any analysis of racism.

The goal in re-examining the Black-white framework is to find an effective strategy for vanquishing an evil that has expanded rather than diminished. Racism has expanded partly as a result of the worldwide economic recession that followed the end of the post-war boom in the early 1970s, with the resulting capitalist restructuring and changes in the international division of labor. Those developments generated feelings of insecurity and a search for scapegoats. In the United States racism has also escalated as whites increasingly fear becoming a weakened, minority population in the next century. The stage is set for decades of ever more vicious divide-and-conquer tactics.

What has been the response from people of color to this ugly White Supremacist agenda? Instead of uniting, based on common experience and needs, we have often closed our doors in a defensive, isolationist mode, each community on its own. A fire of fear and distrust begins to crackle, threatening to consume us all. Building solidarity among people of color is more necessary than ever—but the exclusively Black-white definition of racism makes such solidarity more difficult than ever.

We urgently need twenty-first-century thinking that will move us beyond the Black-white framework without negating its historical role in the construction of U.S. racism. We need a better understanding of how racism developed both similarly and differently for various peoples, according to whether they experienced genocide, enslavement, colonization or some other structure of oppression. At stake is the building of a united anti-racist force strong enough to resist White Supremacist strategies of divide-and-conquer and move forward toward social justice for all....

... African Americans have reason to be uneasy about where they, as a people, will find themselves politically, economically and socially with the rapid numerical growth of other folk of color. The issue is not just possible job loss, a real question that does need to be faced honestly. There is also a feeling that after centuries of fighting for simple recognition as human beings, Blacks will be shoved to the back of history again (like the back of the bus). Whether these fears are real or not, uneasiness exists and can lead to resentment when there's talk about a new model of race relations. So let me repeat: in speaking here of the need to move beyond the bipolar concept, the goal is to clear the way for stronger unity against White Supremacy. The goal is to identify our commonalities of experience and needs so we can build alliances.

The commonalities begin with history, which reveals that again and again peoples of color have had one experience in common: European colonization and/or neo-colonialism with its accompanying exploitation. This is true for all indigenous peoples, including Hawaiians. It is true for all Latino peoples, who were invaded and ruled by Spain or Portugal. It is true for people in Africa, Asia and the Pacific Islands, where European powers became the colonizers. People of color were victimized by colonialism not only externally but also through internalized racism—the "colonized mentality."

Flowing from this shared history are our contemporary commonalities. On the poverty scale, African Americans and Native Americans have always been at the bottom, with Latinos nearby. In 1995 the U.S. Census found that Latinos have the highest poverty rate, 24 percent. Segregation may have been legally abolished in the 1960s, but now the United States is rapidly moving toward resegregation as a result of whites moving to the suburbs. This leaves people of color—especially Blacks and Latinos—with inner cities that lack an adequate tax base and thus have inadequate schools. Not surprisingly, Blacks and Latinos finish college at a far lower rate than whites. In other words, the victims of U.S. social ills come in more than one color. Doesn't that indicate the need for new, inclusive models for fighting racism? Doesn't that speak to the absolutely urgent need for alliances among peoples of color?

With greater solidarity, justice for people of color could be won. And an even bigger prize would be possible: a U.S. society that advances beyond "equality," beyond granting people of color a respect equal to that given to Euro-Americans. Too often "equality" leaves whites still at the center, still embodying the Americanness by which others are judged, still defining the national character.

. . . Innumerable statistics, reports and daily incidents should make it impossible to exclude Latinos and other non–Black populations of color when racism is discussed, but they don't. Police killings, hate crimes by racist individuals and murders with impunity by border officials should make it impossible, but they don't. With chilling regularity, ranch owners compel migrant workers, usually Mexican, to repay the cost of smuggling them into the United States by laboring the rest of their lives for free. The 45 Latino and Thai garment workers locked up in an El Monte, California, factory, working 18 hours a day seven days a week for $299 a month, can also be considered slaves (and one must ask why it took three years for the Immigration and Naturalization Service to act on its own reports about this horror) (*San Francisco Examiner,* August 8, 1995). Abusive treatment of migrant workers can be found all over the United States. In Jackson Hole, Wyoming, for example, police and federal agents rounded up 150 Latino workers in 1997, inked numbers on their arms and hauled them off to jail in patrol cars and a horse trailer full of manure (*Los Angeles Times,* September 6, 1997).

These experiences cannot be attributed to xenophobia, cultural prejudice or some other, less repellent term than racism. Take the case of two small Latino children in San Francisco who were found in 1997 covered from head to toe with flour. They explained they had hoped to make their skin white enough for school. There is no way to understand their action except as the result of fear in the racist climate that accompanied passage of Proposition 187, which denies schooling to the children of undocumented immigrants. Another example: Mexican and Chicana women working at a Nabisco plant in Oxnard, California, were not allowed to take bathroom breaks from the assembly line and were told to wear diapers instead. Can we really imagine white workers being treated that way? (The Nabisco women did file a suit and won, in 1997.)

No "model minority" myth protects Asians and Asian Americans from hate crimes, police brutality, immigrant-bashing, stereotyping and everyday racist prejudice. Scapegoating can even take their lives, as happened with the murder of Vincent Chin in Detroit some years ago. . . .

WHY THE BLACK-WHITE MODEL?

A bipolar model of racism has never been really accurate for the United States. Early in this nation's history, Benjamin Franklin perceived a tri-racial society based on skin color—"the lovely white" (Franklin's words), the Black, and the "tawny," as Ron Takaki tells us in *Iron Cages*. But this concept changed as capital's need for labor intensified in the new nation and came to focus on African slave labor. The "tawny" were decimated or forcibly exiled to distant areas; Mexicans were not yet available to be the main labor force. As enslaved Africans became the crucial labor force for the primitive accumulation of capital, they also served as the foundation for the very idea of whiteness—based on the concept of blackness as inferior.

Three other reasons for the Black-white framework seem obvious: numbers, geography and history. African Americans have long been the largest population of color in the United States; only recently has this begun to change. Also, African Americans have long been found in sizable numbers in most parts of the United States, including major cities, which has not been true of Latinos until recent times. Historically, the Black-white relationship has been entrenched in the nation's collective memory for some 300 years—whereas it is only 150 years since the United States seized half of Mexico and incorporated those lands and their peoples. Slavery and the struggle to end it formed a central theme in this country's only civil war—a prolonged, momentous conflict. Above all, enslaved Africans in the United States and African Americans have created an unmatched heritage of massive, persistent, dramatic and infinitely courageous resistance, with individual leaders of worldwide note.

We also find sociological and psychological explanations of the Black-white model's persistence. From the days of Jefferson onward, Native Americans, Mexicans and later the Asian/Pacific Islanders did not seem as much a threat to racial purity or as capable of arousing white sexual anxieties as did Blacks. A major reason for this must have been Anglo ambiguity about who could be called white. Most of the Mexican *ranchero* elite in California had welcomed the U.S. takeover, and Mexicans were partly European—therefore "semi-civilized"; this allowed Anglos to see them as white, unlike lower-class Mexicans. For years Mexicans were legally white, and even today we hear the ambiguous U.S. Census term "Non-Hispanic Whites."

Like Latinos, Asian Americans have also been officially counted as white in some historical periods. They have been defined as "colored" in others, with "Chinese" being yet another category. Like Mexicans, they were often seen as not really white but not quite Black either. Such ambiguity tended to put Asian Americans along with Latinos outside the prevailing framework of racism.

Blacks, on the other hand, were not defined as white, could rarely become upper-class and maintained an almost constant rebelliousness. Contemporary Black rebellion has been urban: right in the Man's face, scary. Mexicans, by contrast, have lived primarily in rural areas until a few decades ago and "have no Mau-Mau image," as one Black friend said, even when protesting injustice energetically. Only the nineteenth-century resistance heroes labeled "bandits" stirred white fear, and that was along the border, a limited area. Latino stereo-types are mostly silly: snoozing next to a cactus, eating greasy food, always being late and disorganized, rolling big Carmen Miranda eyes, shrugging with self-deprecation "me no speek good eengleesh." In other words, *not serious.* This view may be altered today by stereotypes of the gangbanger, criminal or dirty immigrant, but the prevailing image of Latinos remains that of a debased white, at best.

Among other important reasons for the exclusively Black-white model, sheer ignorance leaps to mind. The oppression and exploitation of Latinos (like Asians) have historical roots unknown to most Americans. People who learn at least a little about Black slavery remain totally ignorant about how the United States seized half of Mexico or how it has colonized Puerto Rico. . . .

One other important reason for the bipolar model of racism is the stubborn self-centeredness of U.S. political culture. It has meant that the nation lacks any global vision other than relations of domination. In particular, the United States refuses to see itself as one among some 20 countries in a hemisphere whose dominant languages are Spanish and Portuguese, not English. It has only a big yawn of contempt or at best indifference for the people, languages and issues of Latin America. It arrogantly took for itself alone the name of half the western hemisphere, America, as was its "Manifest Destiny," of course.

So Mexico may be nice for a vacation and lots of Yankees like tacos, but the political image of Latin America combines incompetence with absurdity, fat corrupt dictators with endless siestas. Similar attitudes extend to Latinos within the United States. My parents, both Spanish teachers, endured decades of being told that students were better off learning French or German. The mass media complain that "people can't relate to Hispanics (or Asians)." It takes mysterious masked rebels, a beautiful young murdered singer or salsa outselling ketchup for the Anglo world to take notice of Latinos. If there weren't a mushrooming, billion-dollar "Hispanic" market to be wooed, the Anglo world might still not know we exist. No wonder that racial paradigm sees only two poles.

The exclusively Black-white framework is also sustained by the "model minority" myth, because it distances Asian Americans from other victims of racism. Portraying Asian Americans as people who work hard, study hard, obey the established order and therefore prosper, the myth in effect admonishes Blacks and Latinos: "See, anyone can make it in this society if you try hard enough. The poverty and prejudice you face are all *your* fault."

The "model" label has been a wedge separating Asian Americans from others of color by denying their commonalities. It creates a sort of racial bourgeoisie, which White Supremacy uses to keep Asian Americans from joining forces with the poor, the homeless and criminalized youth. People then see Asian Americans as a special class of yuppie: young, single, college-educated, on the white-collar track—and they like to shop for fun. Here is a dandy minority group, ready to be used against others.

The stereotype of Asian Americans as whiz kids is also enraging because it hides so many harsh truths about the impoverishment, oppression and racist treatment they experience. Some do come from middle- or upper-class families in Asia, some do attain middle-class or higher status in the U.S., and their community must deal with the reality of class privilege where it exists. But the hidden truths include the poverty of many Asian/Pacific Islander groups, especially women, who often work under intolerable conditions, as in the sweatshops. . . .

THE DEVILS OF DUALISM

Yet another cause of the persistent Black-white conception of racism is dualism, the philosophy that sees all life as consisting of two irreducible elements. Those elements are usually oppositional, like good and evil, mind and body, civilized and savage. Dualism allowed the invaders, colonizers and enslavers of today's United States to rationalize their actions by stratifying supposed opposites along race, color or gender lines. So mind is European, male and rational; body is colored, female and emotional. Dozens of other such pairs can be found, with their clear implications of superior-inferior. In the arena of race, this society's dualism has long maintained that if a person is not totally white (whatever that can mean biologically), he or she must be considered Black. . . .

Racism evolves; our models must also evolve. Today's challenge is to move beyond the Black-white dualism that has served as the foundation of White Supremacy. In taking up this challenge, we have to proceed with both boldness and infinite care. Talking race in these United States is an intellectual minefield; for every observation, one can find three contradictions and four necessary qualifications from five different racial groups. Making your way through that complexity, you have to think: keep your eyes on the prize.

KEY CONCEPTS

colonialism	dualism	race

DISCUSSION QUESTIONS

1. Think of four examples of subtle White-to-Black racism. Now think of four examples of subtle White-to-Latino/Latina racism. What are the similarities and what are the differences?

2. The author advocates going beyond a White-Black paradigm of racism. Describe this paradigm briefly and suggest names for a new racial paradigm.

35

Everyday Race-Making

Navigating Racial Boundaries in Schools

AMANDA E. LEWIS

Amanda Lewis's research investigates how race is socially constructed in everyday life. She studies three different school settings where she analyzes how young children engage in what she calls "everyday race-making."

Sociologists working in the racial formation tradition have made the clear case in recent years for understanding the socially constructed and politically contested nature of race and have detailed macroprocesses of construction and formation (Omi & Winant, 1994). What is left to understand is how these processes work themselves out in and on people's everyday lives. How is it that race is reproduced in day-to-day life? How do we become socialized into the racial schema, knowing what these categories mean for us and for others?

In this article, I attempt to map out one piece of this larger puzzle by examining the practices and processes involved in the construction and reconstruction of racial

SOURCE: Amanda E. Lewis. 2003. "Everyday Race-Making: Navigating Racial Boundaries in School." *American Behavioral Scientist* 47, no. 3 (November): 283–305.

boundaries in daily interactions. Here, I explore what I call "everyday race-making," how the color line is redrawn in daily life. I pay close attention to issues of racial identification (here understood as self-definition) as they do or do not conflict with racial ascription (external racial categorization or assignment), as well as examine the nature of the racial ascription process itself (what factors influence external ascriptions and what implications these categorizations have for treatment). Although in my analysis I recognize the necessity to understand race as "fluid, multiple, relational [and] socially constructed" (Phoenix, 1998, p. 860), I also understand it to be socially "real" and determining of life chances. Moreover, I understand racialization not merely as something that happens to people and institutions but understand institutions and people as racializing agents, as forces in the reproduction and transformation of race.

I draw on ethnographic data from research in three school communities to illustrate some of these processes. Schools are arguably one of the central institutions involved in the drawing and redrawing of racial lines (Forman, 2001a; Hall, 1986a; Olsen, 1997; Perry, 2002; Van Ausdale & Feagin, 2001). Although they clearly do not teach racial identity in the way they teach multiplication or punctuation, schools are settings where people acquire some version "of the rules of racial classification" and of their own racial identity (Omi & Winant, 1994, p. 60). Not only does the actual curriculum teach many racial lessons but schools (and school personnel) serve as a source of racial information, a location for interracial interaction, and a means of both affirming and challenging previous racial attitudes and understandings. Although clearly not the only social institution concerned, schools are involved in framing ideas about race and are at the center of many struggles around racial equity.

BACKGROUND

Since the early 1960s, a considerable amount of historical work has been done mapping out the racialization process in the United States (Almaguer, 1994; Berkhofer, 1978; Dinnerstein, Nichols, & Reimers, 1990; Gossett, 1963; Higham, 1965; Horsman, 1981; Ignatiev, 1995; Jordan, 1968; Montejano, 1987; Roediger, 1991; Rogin, 1996; Takaki, 1987). Although this work has contributed greatly to our understanding of the formation of racial categories, it is unclear whether the processes as they outline them are similar to or distinct from the process of racialization today. To understand the contemporary production and reproduction of racial ideology and racial structures, we must look to the day-to-day events and arenas where ideologies and structures are lived out. As Holt (1995) put it, this calls for more study of the everydayness of race: "It is at this level, I will argue, that race is reproduced via the marking of the racial Other and that racist ideas and practices are naturalized, made self-evident, and thus seemingly beyond audible challenge" (p. 7). . . .

What does it mean to take seriously the idea of the social construction of race? To try and understand how racial categories and racial meanings are concretely made and remade, challenged and changed in daily life? As Omi and Winant (1994) put it, "racialized social structure shapes racial experience and conditions meaning" (p. 59), but it is also true that individual actors are involved everyday in negotiating and contesting racial boundaries and, in the process, reproducing or challenging them. Racialization is an ongoing process that takes place continually at both macro-and microlevels and involves questions of who belongs where, what categories mean, and what effect they have on people's life chances and opportunities.

Thus, race is not merely about the representation of difference; it is about inclusion and exclusion (Hall, 1980; Phoenix, 1998). . . . The process of negotiating what racial categories there are, who belongs to which, and how they are different is a continual one, not only because all of these things change over time through social movements (e.g., recent pushes for multiracial categories) but also because the young must regularly be socialized into the existing racial scheme to learn where they fit and how they sit in relation to others. In many ways, studying these processes of negotiation and socialization today requires a new kind of attention because the manner in which they take place has changed over time. Although historically racial categories were produced through explicit violence and overt policing of racial boundaries (e.g., Black codes, Jim Crow laws, lynching), the social production and reproduction of race must today be accomplished under different conditions. As a number of scholars have outlined, the rules of racial discourse have changed as new colorblind or laissez-faire ideologies have become dominant (Bobo, Kluegel, & Smith, 1997; Crenshaw, 1997; Forman, 2001b; Lewis, 2001). This does not mean that racial categories have lost their salience. What it does mean is that in trying to understand and explain how it is that race is produced and contested in the everyday, it is required that we give our analytic attention to subtle and implicit processes as well as those that are overt and explicit.

DATA, METHODS, AND SETTING

Data for this study were collected in three public elementary schools (two urban and one suburban) in California. Over the course of the year, I spent upward of 35 hours a week in the schools (resulting in more than 1,500 pages of field notes) and conducted 85 formal (along with many more informal) interviews with school personnel, parents, and children. All three schools were drawn from Hillside, a large metropolitan area. Each was selected with several criteria in mind. I sought to find three different kinds of schools: a fairly typical and diverse urban school (West City), a fairly typical and homogeneous suburban school (Foresthills), and a school that structurally and culturally was a bicultural or non-White space (Metro2—a Spanish, dual-language program). . . . All three were small to midsize elementary schools and were neither the best nor the worst schools in their respective districts. . . .

FINDINGS

Racial Boundary Formation

Everyday interactions, the moments in which the social category of race takes shape and are given meaning in social interaction, are the means through which boundaries between groups are created, reproduced, and resisted. One is a member of one group at least in part because one is not a member of another. Systems of social inclusion and exclusion are organized (at least in part) around the resulting racial categories and the boundaries between them. Racialization thus involves the assignment of bodies to racial categories (assigning identities to people and groups) and the association of symbols, attributes, qualities, and other meanings with those categories (which then are understood to belong to those bodies in a primordial or natural way). Racial categorizations are used to decide who is similar and different; opportunities and resources are then distributed along racial lines as people are included in and/or excluded from a range of institutions, activities, or opportunities because of their categorization.

The various racial interactions or negotiations in daily life are skirmishes along the borders between racial categories. As categories are applied in inter-personal interactions, the boundaries between categories are simultaneously created or reinforced. One cannot determine who one is without determining simultaneously who one is not and in some manner, at least metaphorically, drawing a boundary (Spelman, 1988)....These racial boundaries are not fixed. They are processes, the ongoing products of the kind of social interactions described below in which identities are assigned and negotiated, produced and reproduced.

External Racial Ascription

One key aspect of how race operates in daily life is in the negotiation around racial identification. Racial identities have never involved only internal self-identification (someone's choice to identify in a particular way). In fact, historically racial identifications were more a matter of what Brubaker and Cooper (2002) have called "imposed identity"; that is, the imposition of identities on previously nonracialized bodies was central to the initial creation of racial categories. Today, racial boundaries are reproduced and/or challenged in part through the modern extension of these earlier racialization mechanisms—the external ascription of specific racial identities to particular bodies in daily interaction. One example of how this works took place in the Metro 2 schoolyard. In a conversation during morning break, Lily and Kate, two fourth-grade girls, stood talking about the class presentation that morning in which Lily had described her ethnic heritage as "Mexican American and European American." She asked Kate about her own background, to which Kate replied, "just Caucasian." Seeing Benjamin (a biracial/bicultural Colombian and Filipino fifth-grader) sitting nearby eating his morning snack, Lily then turned to him and asked, "What are you?" He looked at the two girls for several moments without replying. Eventually, he responded that he would "rather not say." Trying to be helpful, one of the girls offered, "You're Chinese, right?" When he did not respond to either confirm or deny their suggestion, the girls turned away.

Similar to Benjamin, all adults and children must regularly contend with others' racial ascriptions—external racial identifications that may or may not match up with individuals' own self-identifications. In daily interactions such as the one illustrated, people regularly go through the same process as Lily and Kate—they work with available racial categories and meanings, draw on available cues, and make decisions about who they think someone is and where in the racial schema the person they are observing fits.

In another example, Rodney, an African American fourth-grader at West City, made the following evaluation of his Latino fifth-grade peer Mike.

AUTHOR: What about Mike, what is he?

RODNEY: White.

AUTHOR: He's White?

RODNEY: He White to me.

It is safe to say that Mike would have been quite upset had he heard Rodney's ascription of majority identity to him. Yet in his use of the phrase, "He White to me," Rodney very astutely recognizes the reality that external racial ascriptions in many ways matter as much as one's self-identification, if not more. Thus, although what Mike thinks (how he self-identifies) matters, it does not necessarily trump the identity ascribed to him externally. Rodney is reading the world and mapping those he sees into the schema as he understands it, and as far as he is concerned Mike is White. As Richard Jenkins (1996) states, "Identity is often in the eye of the beholder" (p. 2). Although Rodney's reading may matter less to Mike than those of Mike's teacher, a storeowner, or a future employer, peer judgments of racial categorization and performance are far from unimportant or meaningless.

These examples of Lily, Kate, Benjamin, and Rodney illustrate the kinds of external ascriptions that are a daily part of social interaction. Such interactions not only involve external ascriptions or impositions of identity but also involve negotiations. In another example, a fifth-grader, Malik, wrote a story for Martin Luther King Day saying that he was grateful for King because his efforts allowed him to have White friends such as Julio. After hearing the story read out loud, Julio exploded out of his seat, yelling, "I know you ain't talkin' about me. I ain't White." In this case, a fistfight was only narrowly avoided as Julio very aggressively resisted a mistake in external racial identification.

Self-identifications are always occurring in interaction with an external world that assigns identities. Antonio, a Latino father of a biracial/bicultural (Latino and White) Metro 2 fifth-grader, described his son's self-identification:

I think that Omar identifies himself more with Latinos, in part because he lives with me, and all his life he has lived with me, so, you can't obviate that fact. At the same time, he feels part of the whole American culture. He speaks more English than Spanish, for example. But I don't even think that he has determined what he wants to be, and I think that it's definitely up to him, and what he wants. If he wants to be, to determine himself as White, American or White–Latino or Latino–Latino, or Hispanic, I think that will come later.

Omar did in fact identify as both White and Latino in different moments, but he was externally identified almost exclusively as Latino. He was, in practice, not really able to choose. Omar understood this to some extent. As his father explained, Omar recognized his own status when it came to various political issues. For example, he and his father had participated in marches and protests against Proposition 187, and Omar, despite his own U.S. citizenship, had identified with the group he understood to be collectively targeted and unwanted: Latinos. . . .

Factors Shaping Ascription

What is it that leads a particular racial category to be ascribed to a particular person? How do we know where someone belongs, where to place them? Although skin color or phenotype is clearly one key aspect of external racial identifications, it is not the only factor at play. For example, in a case involving school personnel from Metro 2, Angela discussed her son Héctor's struggle to be recognized at school.

Angela: Well, let's see, well his father is originally from El Salvador and Héctor, Spanish is really his first language, although I've never been able to convince anybody at Metro 2 of that. I think they look at him and they think he's White, basically, and don't want to hear otherwise. . . . I mentioned earlier that he had two kindergarten teachers, one of them didn't even know that he spoke Spanish . . . and they're suppose to assess him and they're like writing on report cards, you know, "Spanish coming along . . ." and to me that is just so annoying and I try to talk to them but it's like they think he's White. Whenever [my ex-husband and I] talk to him about being in two different cultures, Héctor goes "yes, but I'm not White, look at my skin," his skin's like very, very pale but he's saying it's brown.

Here, Angela had a long process of negotiating with teachers around their identification of her son as White. Despite her interventions, teachers did not begin to read him or interact with him differently. Héctor's regular frustration at having his Spanish skills underestimated continued. As Héctor's relationships with his teachers illustrates, language is closely linked to, and often works in interaction with, readings of external physical features (phenotype). Héctor's teachers' mistaken assumptions about his Spanish language proficiency are tightly interwoven with their reading of him as White.

Moreover, this example shows that it is not just color that marks people as different or of color. It is likely true that if Héctor had been darker or more Mestizo-looking, his situation would have been different. But this is by no means an objective or easily measured characteristic. We can see from Angela's discussion of her son's defense of his own self-identification that skin color is a key factor in identification processes, but it is key both in his teachers identifying him as White and in his own claims to be otherwise. That skin color is used both in the teachers' reading of him as White and in his own defense of his "brownness" highlights the often-subjective quality of this category.

Here we see that both phenotypic features and language have complicated roles to play in the process of racial ascription. This is true even for those who do not speak a foreign language. When Ms. Washington, an African American first-grade teacher at West City, reflected on her experiences growing up and those things that had made her feel different, she talked about skin color, hair, racially coded standards of beauty, and also about racially coded ways of talking. She described struggling while growing up to find ways of speaking that would allow her to be successful in school but not leave her feeling like an outsider in her own community. In this way, language is not merely a neutral mode of communication but a way of telling who a person is, where they are from, and whether they are in some way collectively different. Not only is language racialized (certain ways of talking are thought of as Black or White) but language itself racializes (e.g., certain ways of talking can mark one as racially "other"). Language and skin color are not the only such markers of difference.

Aside from just looking or sounding in a particular way, how one behaves also plays a role. Similar to gender, race is at least in part about performance (Butler, 1990; Goffman, 1977). In their article on "doing gender," West and Zimmerman (1987) argue that gender is a routine, methodological, and recurring accomplishment. Rather than being something that is innate or internal to the person, it is something produced in interactions and institutions. Similar to gender, race is the product of social doings. And, similar to gender, doing race is unavoidable in societies such as ours, where racial meaning and categories are relevant and enforced and where racial differences are naturalized. As Holt (1995) points out, "Race yet lives because it is part and parcel of the *means* of living" (p. 12).

For instance, when I asked Anne how her biracial White/Latino son Jorge, a fifth-grader, racially self-identified, she raised issues of cultural performance as key to racial identification. Initially, she signaled her son's Latino-ness by stating, "Well, he's a *Velez*" (their last name). Almost immediately, however, she began to backtrack as she remembered a recent incident. Anne's Latina sister-in-law Aunt Maribel takes Anne's boys and her own children to the amusement park every summer; in the past, Jorge had trouble getting on the roller coasters because of his height. This year, Maribel gelled Jorge's hair back to give him some height. As Anne stated, "It worked, and he got on, and it was fabulous." Yet, as Anne reported, when he looked at himself in the mirror with his new hairdo, Jorge's reaction was to laugh as he said, "'I look like a Latino kid,' and [his aunt] goes, 'Well what do you think you are?'"

In this case, Anne referred first to the importance of surname, "He's a *Velez*," in signifying her son's racial status. In reflecting on recent events, however, she backed off as she recalled how her son, in changing his hairstyle, saw himself as becoming something he was not (or at least he did not look like) before: a Latino kid. This incident highlights the performative nature of race; when it comes to everyday interactions, certain kinds of performances or costumes may determine what one is or how one is seen and categorized as much if not more than blood, ancestry, or phenotype. . . .

Racial Meaning, Power, and Exclusion

In regard to race, the delineation of same and different that transpires during racial ascription is not a neutral or benign process but one imbued with power (Hall, 1980). People experience not merely being identified or labeled but, as boundaries are drawn, being simultaneously included or excluded; these are moments when they are treated in a particular way because someone has identified them as a member of a particular racial group. Racial identifications thus are not merely about thought processes but about action; acts of inclusion and exclusion are part of the racialization process. These acts range from explicit (e.g., racial violence) to subtle (e.g., not being recognized as supervisor) and from acts of exclusion to processes of inclusion (e.g., José being included in a White "we" who are hostile to immigration). One Latina mother, Julia, described what often happens when her Latino-ness is recognized.

> Oh, yes, yes, right off the bat [people] recognize the Latino, don't you think? The Hispanic. Not that one feels bad about it, you know, but some people give us bad looks sometimes or . . . or they do not answer the way they should. I have seen racism in that sense particularly. A certain look sometimes says it all. . . .

Contexts of Racialization

Although racialization processes exist in all settings, they do not operate uniformly across time and space. Thus, local contexts, although operating within a larger racial formation, have some impact on the shape of racial boundaries and on how they operate in everyday life. The way people get categorized varies from setting to setting (e.g., the same person may be read differently in different settings), and the meaning of particular labels (e.g., the meanings associated with Black or Latino) as well as the experiential aspects of group membership (e.g., how Latino-ness or Blackness is experienced) varies from place to place. What it means to be Black in a setting where you are one of many is quite different from what it means when you are one of eight in a school of 600. Context clearly matters in at least three ways: (a) spaces themselves can be racially coded, (b) local contexts and institutions can have both direct and indirect influences on identification processes, and (c) the effects of being categorized can vary by context. . . .

CONCLUSION

Race is at play all the time inside and outside of schools. It is a part of what is happening in our many daily interpersonal interactions. It is one lens through which people read the world around them and make decisions on how to act, react, and interact. Using data from ethnographic research in three school communities, I have begun to outline how racialization processes work. These processes describe the ways that racial identities are assigned to individuals and how racial categories are mapped onto groups. These ascriptive processes work primarily through interpersonal interactions in which we attempt to assess what we know about another person first through the instantaneous reading or interpreting of available clues (e.g., visible cues such as skin color or facial features, auditory cues such as accent, spatial cues such as neighborhood) and second through rereading or reinterpretating initial assumptions as additional information becomes available. These processes operate in a largely relational manner in which some people are determined to be same (or "like me") and others are determined to be different. At all steps, institutional processes and dynamics affect these racial interactions and interpretations. Both racial ascription and racial self-identification are contextual processes influenced by local meaning systems, rules, demographics, relationships, and structures.

Moreover, racial ascriptions are also not solely about deciding what categories individuals belong to but about the mapping of systems of meaning onto individuals. When a person is categorized as Black, White, or Asian, he or she is being linked with a category already imbued with meanings. The sameness and/or difference ascertained is not a neutral measure of differences such as shoe size or earring shape but central clues about who a person is. The moment of identification is also a moment of inclusion or exclusion in which an understanding is not merely formed but in many cases subtly or explicitly acted on. These moments of inclusion or exclusion can take form in how one is treated in a particular context (e.g., the slightly cool treatment of a waitress or the particularly welcoming greeting from a new neighbor) or in concrete material processes of who gets access to what kind of resources (e.g., what mortgage rate a bank officer offers).

Although racial categorization is not merely imposed from above, available racial categories and the meanings associated with them cannot be simply rejected or ignored. As Benjamin's failed efforts to reject identification illustrated, one cannot decide to opt-out altogether. Yet, collective action can alter the content or boundaries of categories (e.g., Black efforts to redefine Blackness—"Black is Beautiful"). In addition, individual interventions are not meaningless; they can complicate understandings. However, what one is able to claim for oneself is clearly limited by context and the available categories therein. Although Tiger Woods created the racial/ethnic category Cablinasian to describe himself, it was not one that was widely (or even narrowly) accepted. His actions did, however, encourage an occasionally more complex reading of his person otherwise (and most often still) read as Black.

The state historically has played an important role in the creation and alteration of categories usually in interaction with the groups involved (Almaguer &

Jung, 1999). For instance, recent efforts on the part of multiracial groups and Asian Pacific Islanders, among others, have led to new census categories that allow for different and/or multiple racial designations. At issue in both individual claims and state actions are not only the borders between categories—whether they exist, what shape they take, whether they are permeable or fixed—but also the content of the categories themselves. What does it mean to be White, Black, Latino, and/or Asian? Who decides what category a person belongs to? Are these categories mutually exclusive?

In practice, these questions have no single answer. What the boundaries are and how they work are not established and universally consistent social facts (Almaguer & Jung, 1999; Davis, 1991; Harris & Sim, 2002), and the content and meaning of any racial category is not consistent across space, culture, or time. This indefiniteness lies at the heart of what it means to talk about race as a social construction. Although the idea that race is a social construction is widely accepted, the reality of race in daily life has received too little attention. This article represents an initial effort at addressing this gap. Most of the evidence I find is of entrenched boundaries with persisting consequences for life experiences. However, by examining daily processes of negotiation and reproduction, we do begin to see possibilities for change. In the everyday making and remaking of racial categories, racial meaning, and racial boundaries, people do resist, borders are crossed, and boundaries are challenged—at times yielding moments of retrenchment and solidification of what is but also at times offering the remote possibility for something new.

REFERENCES

Almaguer, T. (1994). *Racial fault lines*. Berkeley: University of California Press.

Almaguer, T., & Jung, M.-K. (1999). The enduring ambiguities of race in the United States. In J. L. Abu-Lughod (Ed.), *Sociology for the twenty-first century* (pp. 213–239). Chicago: University of Chicago Press.

Berkhofer, R. F. (1978). *The White man's Indian*. New York: Random House.

Bobo, L., Kluegel, J. R., & Smith, R. A. (1997). Laissez faire racism: The crystallization of a "kinder, gentler" anti-Black ideology. In S. Tuch & J. Martin (Eds.), *Racial attitudes in the 1990s: Continuity and change* (pp. 15–42). Westport, CT: Praeger.

Bonilla-Silva, E. (1997). Rethinking racism: Toward a structural interpretation. *American Sociologial Review, 62*, 465–480.

Brubaker, R., & Cooper, F. (2002). Beyond identity. *Theory and Society, 29*, 1–47.

Butler, J. (1990). *Gender trouble*. New York: Routledge.

Crenshaw, K. W. (1997). Color-blind dreams and racial nightmares: Reconfiguring racism in the post-civil rights era. In T. Morrison & C. B. Lacour (Eds.), *Birth of a nation hood*. New York: Pantheon Books.

Davis, F. J. (1991). *Who is Black? One nation's definition*. University Park: The Pennsylvania State University Press.

Dinnerstein, L., Nichols, R. L., & Reimers, D. M. (1990). *Natives and strangers: Blacks, Indians, and immigrants in America.* New York: Oxford University Press.

Forman, T. (2001a). *Social change, social context and White youth's racial attitudes.* Unpublished doctoral dissertation, University of Michigan, Ann Arbor, Department of Sociology.

Forman, T. (2001b). Social determinants of White youth's racial attitudes. *Sociological Studies of Children and Youth,* 8, 173–207.

Goffman, E. (1977). The arrangement between the sexes. *Theory and Society,* 4, 301–333.

Gossett, T. F. (1963). *Race: The history of an idea in America.* Dallas, TX: Southern Methodist University Press.

Hall, S. (1980). Race, articulation and societies structured in dominance. In *Sociological theories: Race and colonialism* (pp. 305–345). Paris: UNESCO.

Harris, D. R., & Sim, J. J. (2002). Who is mixed race? Assessing the complexity of lived race. *American Sociological Review,* 67, 614–627.

Higham, J. (1965). *Strangers in the land.* New York: Atheneum.

Holt, T. C. (1995, February). Marking race, race-making, and the writing of history. *American Historical Review,* 100, 1–20.

Horsman, R. (1981). *Race and manifest destiny.* Cambridge, MA: Harvard University Press.

Ignatiev, N. (1995). *How the Irish became White.* New York: Routledge.

Jenkins, R. (1996). *Social identity.* New York: Routledge.

Jordan, W. D. (1968). *White over Black: American attitudes toward the Negro, 1550–1812.* Baltimore, MD: Penguin.

Lewis, A. (2001). There is no "race" in the schoolyard: Colorblind ideology in an (almost) all White school. *American Educational Research Journal,* 38(4), 781–811.

Montejano, D. (1987). *Anglos and Mexicans in the making of Texas, 1836–1986.* Austin: University of Texas Press.

Olsen, L. (1997). *Made in America.* New York: New Press.

Omi, M., & Winant, H. (1994). *Racial formation in the United States: From the 1960s to the 1990s.* New York: Routledge.

Perry, P. (2002). *Shades of White: White kids and racial identities in high school.* Durham, NC: Duke University Press.

Phoenix, A. (1998). Dealing with difference: The recursive and the new. *Ethnic and Racial Studies, 21,* 859–880.

Roediger, D. R. (1991). *The wages of Whiteness: Race and the making of the American working class.* New York: Verso.

Rogin, M. (1996). *Black face, white noise: Jewish immigrants in the Hollywood melting pot.* Berkeley: University of California Press.

Takaki, R. (1993). *A different mirror.* Boston: Little, Brown.

Telles, E. E., & Murguia, E. (1990). Phenotypic discrimination and income differences among Mexican Americans. *Social Science Quarterly,* 71, 682–696.

Van Ausdale, D., & Feagin, J. (2001). *The first R: How children learn race and racism.* Lanham, MD: Rowman & Littlefield.

West, C., & Zimmerman, D. H. (1987). Doing gender. *Gender & Society,* 1, 125–151.

KEY CONCEPTS

everyday race-making   racial formation   racialization

DISCUSSION QUESTIONS

1. What does Lewis mean by "everyday race-making" and what evidence does she give of this happening?

2. How does power influence the way that racial categories are imposed and experienced?

3. Why does Lewis argue that inclusion and exclusion are part of the process of racial categorization?

Applying Sociological Knowledge:
An Exercise for Students

Consider how racial-ethnic groups are portrayed in the media. Take one type of television show (a sitcom, sports broadcast, news, or drama) and watch two such shows for a specified period of time (say, one hour each). As you watch, systematically observe how different racial-ethnic groups (including Whites) are depicted. You should keep a tally of how many characters from different racial-ethnic groups are shown and how they are portrayed. Then ask yourself how images in the media shape racial stereotypes and group prejudice.

36

The Social Construction of Gender

MARGARET L. ANDERSEN

In this essay, Margaret Andersen outlines the meaning of the "social construction of gender." She discusses the difference between the terms "sex" and "gender" and defines sexuality as it relates to both. After a brief discussion of the cultural basis of gender, the essay outlines the difference between a gender roles conceptualization of gender and the gendered institutions approach.

To understand what sociologists mean by the phrase *the social construction of gender*, watch people when they are with young children. "Oh, he's such a boy!" someone might say as he or she watches a 2-year-old child run around a room or shoot various kinds of play guns. "She's so sweet," someone might say while watching a little girl play with her toys. You can also see the social construction of gender by listening to children themselves or watching them play with each other. Boys are more likely to brag and insult other boys (often in joking ways) than are girls; when conflicts arise during children's play, girls are more likely than boys to take action to diffuse the conflict (McCloskey and Coleman, 1992; Miller, Danaber, and Forbes, 1986).

To see the social construction of gender, try to buy a gender-neutral present for a child—that is, one not specifically designed with either boys or girls in mind. You may be surprised how hard this is, since the aisles in toy stores are highly stereotyped by concepts of what boys and girls do and like. Even products such as diapers, kids' shampoos, and bicycles are gender stereotyped. Diapers for boys are packaged in blue boxes; girls' diapers are packaged in pink. Boys wear diapers with blue borders and little animals on them; girls wear diapers with pink borders with flowers. You can continue your observations by thinking about how we describe children's toys. Girls are said to play with dolls; boys play with action figures!

When sociologists refer to the social construction of gender, they are referring to the many different processes by which the expectations associated with being a boy (and later a man) or being a girl (later a woman) are passed on through society. This process pervades society, and it begins the minute a child is born. The exclamation "It's a boy!" or "It's a girl!" in the delivery room sets a course that from that moment on influences multiple facets of a person's life. Indeed, with the

SOURCE: Margaret L. Andersen. 2003. *Thinking About Women: Sociological Perspectives on Sex and Gender.* Boston, MA: Allyn and Bacon, pp. 19–24. Reprinted with permission.

modern technologies now used during pregnancy, the social construction of gender can begin even before one is born. Parents or grandparents may buy expected children gifts that reflect different images, depending on whether the child will be a boy or a girl. They may choose names that embed gendered meanings or talk about the expected child in ways that are based on different social stereotypes about how boys and girls behave and what they will become. All of these expectations—communicated through parents, peers, the media, schools, religious organizations, and numerous other facets of society—create a concept of what it means to be a "woman" or be a "man." They deeply influence who we become, what others think of us, and the opportunities and choices available to us. The idea of the social construction of gender sees society, not biological sex differences, as the basis for gender identity. To understand this fully, we first need to understand some of the basic concepts associated with the social construction of gender.

SEX, GENDER, AND SEXUALITY

The terms *sex, gender,* and *sexuality* have related, but distinct, meanings within the scholarship on women. Sex refers to the biological identity and is meant to signify the fact that one is either male or female. One's biological sex usually establishes a pattern of gendered expectations, although, . . . biological sex identity is not always the same as gender identity; nor is biological identity always as clear as this definition implies.

Gender is a social, not biological, concept, referring to the entire array of social patterns that we associate with women and men in society. Being "female" and "male" are biological facts; being a woman or a man is a social and cultural process—one that is constructed through the whole array of social, political, economic, and cultural experiences in a given society. Like race and class, gender is a social construct that establishes, in large measure, one's life chances and directs social relations with others. Sociologists typically distinguish sex and gender to emphasize the social and cultural basis of gender, although this distinction is not always so clear as one might imagine, since gender can even construct our concepts of biological sex identity.

Making this picture even more complex, sexuality refers to a whole constellation of sexual behaviors, identities, meaning systems, and institutional practices that constitute sexual experience within society. This is not so simple a concept as it might appear, since sexuality is neither fixed nor unidimensional in the social experience of diverse groups. Furthermore, sexuality is deeply linked to gender relations in society. Here, it is important to understand that sexuality, sex, and gender are intricately linked social and cultural processes that overlap in establishing women's and men's experiences in society.

Fundamental to each of these concepts is understanding the significance of culture. Sociologists and anthropologists define culture as "the set of definitions of reality held in common by people who share a distinctive way of life" (Kluckhohn, 1962:52). Culture is, in essence, a pattern of expectations about what are appropriate behaviors and beliefs for the members of the society; thus, culture provides prescriptions for social behavior. Culture tells us what we ought to do, what we ought to think, who we ought to be, and what we ought to expect of others. . . .

The cultural basis of gender is apparent especially when we look at different cultural contexts. In most Western cultures, people think of *man* and *woman* as dichotomous categories—that is, separate and opposite, with no overlap between the two. Looking at gender from different cultural viewpoints challenges this assumption, however. Many cultures consider there to be three genders, or even more. Consider the Navaho Indians. In traditional Navaho society, the *berdaches* were those who were anatomically normal men but who were defined as a third gender and were considered to be intersexed. Berdaches married other men. The men they married were not themselves considered to be berdaches; they were defined as ordinary men. Nor were the berdaches or the men they married considered to be homosexuals, as they would be judged by contemporary Western culture. . . .

Another good example for understanding the cultural basis of gender is the *hijras* of India (Nanda, 1998). Hijras are a religious community of men in India who are born as males, but they come to think of themselves as neither men nor women. Like berdaches, they are considered a third gender. Hijras dress as women and may marry other men; typically, they live within a communal subculture. An important thing to note is that hijras are not born so; they choose this way of life. As male adolescents, they have their penises and testicles cut off in an elaborate and prolonged cultural ritual—a rite of passage marking the transition to becoming a hijra. . . .

These examples are good illustrations of the cultural basis of gender. Even within contemporary U.S. society, so-called "gender bending" shows how the dichotomous thinking that defines men and women as "either/or" can be transformed. Cross-dressers, transvestites, and transsexuals illustrate how fluid gender can be and, if one is willing to challenge social convention, how easily gender can be altered. The cultural expectations associated with gender, however, are strong, as one may witness by people's reactions to those who deviate from presumed gender roles. . . .

In different ways and for a variety of reasons, all cultures use gender as a primary category of social relations. The differences we observe between men and women can be attributed largely to these cultural patterns.

THE INSTITUTIONAL BASIS OF GENDER

Understanding the cultural basis for gender requires putting gender into a sociological context. From a sociological perspective, gender is systematically structured in social institutions, meaning that it is deeply embedded in the social structure of society. Gender is created, not just within family or interpersonal relationships (although these are important sources of gender relations), but also within the structure of all major social institutions, including schools, religion, the economy, and the state (i.e., government and other organized systems of authority such as the police and the military). These institutions shape and mold the experiences of us all.

Sociologists define institutions as established patterns of behavior with a particular and recognized purpose; institutions include specific participants who share expectations and act in specific roles, with rights and duties attached to them. Institutions define reality for us insofar as they exist as objective entities in our experience. . . .

Understanding gender in an institutional context means that gender is not just an attribute of individuals; instead, institutions themselves are *gendered*. To say that an institution is gendered means that the whole institution is patterned on specific gendered relationships. That is, gender is "present in the processes, practices, images and ideologies, and distribution of power in the various sectors of social life" (Acker, 1992:567). The concept of a gendered institution was introduced by Joan Acker, a feminist sociologist. Acker uses this concept to explain not just that gender expectations are passed to men and women within institutions, but that the institutions themselves are structured along gendered lines. Gendered institutions are the total pattern of gender relations—stereotypical expectations, interpersonal relationships, and men's and women's different placements in social, economic, and political hierarchies. This is what interests sociologists, and it is what they mean by the social structure of gender relations in society.

Conceptualizing gender in this way is somewhat different from the related concept of gender roles. Sociologists use the concept of social roles to refer to culturally prescribed expectations, duties, and rights that define the relationship between a person in a particular position and the other people with whom she or he interacts. For example, to be a mother is a specific social role with a definable set of expectations, rights, and duties. Persons occupy multiple roles in society; we can think of social roles as linking individuals to social structures. It is through social roles that cultural norms are patterned and learned. Gender roles are the expectations for behavior and attitudes that the culture defines as appropriate for women and men.

The concept of gender is broader than the concept of gender roles. *Gender* refers to the complex social, political, economic, and psychological relations between women and men in society. Gender is part of the social structure—in other words, it is institutionalized in society. *Gender roles* are the patterns through which gender relations are expressed, but our understanding of gender in society cannot be reduced to roles and learned expectations.

The distinction between gender as institutionalized and gender roles is perhaps most clear in thinking about analogous cases—specifically, race and class. Race relations in society are seldom, if ever, thought of in terms of "race roles." Likewise, class inequality is not discussed in terms of "class roles." Doing so would make race and class inequality seem like matters of interpersonal interaction. Although race, class, and gender inequalities are experienced within interpersonal interactions, limiting the analysis of race, class, or gender relations to this level of social interaction individualizes more complex systems of inequality; moreover, restricting the analysis of race, class, or gender to social roles hides the power relations that are embedded in race, class, and gender inequality (Lopata and Thorne, 1978).

Understanding the institutional basis of gender also underscores the interrelationships of gender, race, and class, since all three are part of the institutional framework of society. As a social category, gender intersects with class and race; thus, gender is manifested in different ways, depending on one's location in the race and class system. For example, African American women are more likely than White women to reject gender stereotypes for women, although they are more accepting than White women of stereotypical gender roles for children. Although this seems contradictory, it can be explained by understanding that African American women may reject the dominant culture's view while also hoping their children can attain some of the privileges of the dominant group (Dugger, 1988).

Institutional analyses of gender emphasize that gender, like race and class, is a part of the social experience of us all—not just of women. Gender is just as important in the formation of men's experiences as it is in women's (Messner, 1998). From a sociological perspective, class, race, and gender relations are systemically structured in social institutions, meaning that class, race, and gender relations shape the experiences of all. Sociologists do not see gender simply as a psychological attribute, although that is one dimension of gender relations in society. In addition to the psychological significance of gender, gender relations are part of the institutionalized patterns in society. Understanding gender, as well as class and race, is central to the study of any social institution or situation. Understanding gender in terms of social structure indicates that social change is not just a matter of individual will—that if we changed our minds, gender would disappear. Transformation of gender inequality requires change both in consciousness and in social institutions. . . .

REFERENCES

Acker, Joan. 1992. "Gendered Institutions: From Sex Roles to Gendered Institutions." *Contemporary Sociology* 21 (September): 565–569.

Dugger, Karen. 1988. "The Social Location of Black and White Women's Attitudes." *Gender & Society* 2 (December): 425–448.

Kluckhohn, C. 1962. *Culture and Behavior.* New York: Free Press.

Lopata, Helene Z., and Barrie Thorne. 1978. "On the Term 'Sex Roles.' " *Signs* 3 (Spring): 718–721.

McCloskey, Laura A., and Lerita M. Coleman. 1992. "Difference Without Dominance: Children's Talk in Mixed- and Same-Sex Dyads." *Sex Roles* 27 (September): 241–258.

Messner, Michael A. 1998. "The Limits of 'The Male Sex Role': An Analysis of the Men's Liberation and Men's Rights Movements' Discourse." *Gender & Society* 12 (June): 255–276.

Miller, D., D. Danaber, and D. Forbes. 1986. "Sex-related Strategies for Coping with Interpersonal Conflict in Children Five and Seven." *Developmental Psychology* 22: 543–548.

Nanda, Serena. 1998. *Neither Man Nor Woman: The Hijras of India.* Belmont, CA: Wadsworth.

KEY CONCEPTS

gender   gendered institution   gender socialization

DISCUSSION QUESTIONS

1. Walk through a baby store. Can you easily identify products for girls and for boys? Could you easily purchase gender neutral clothing?

2. Consider an occupation that is traditionally men's work or traditionally women's work. What happens when a member of the other gender works in that field? What stereotypes and derogatory assumptions do we make about a woman working in a man's occupation or a man working in a woman's occupation?

37

Ain't I a Beauty Queen?

Black Women, Beauty, and the Politics of Race

MAXINE LEEDS CRAIG

Maxine Leeds Craig shows how ideals of beauty are shaped by the dominant culture, but by examining Black women, she also shows how dominant contructions can be contested. In addition, her article analyzes the dynamic between both gender and race in the social construction of women's identities.

In September 1968, as a panel of beauty experts prepared to select the forty-eighth consecutive white Miss America, two protests were under way. One protest denounced beauty contests. The other *was* a beauty contest. On the boardwalk in front of Atlantic City's Convention Center roughly one hundred women who identified themselves as members of Women's Liberation dumped bras, girdles, and false eyelashes into a trash can. Several blocks away, at the Ritz Carlton Hotel, the National Association for the Advancement of Colored People

SOURCE: Maxine Leeds Craig. 2002. *Ain't I a Beauty Queen? Black Women, Beauty, and the Politics of Race.* New York: Oxford University Press, pp. 3–6.

(NAACP) staged the first Miss Black America pageant as a "positive protest" against the exclusion of black women from the Miss America title. As the day progressed, a counterprotest and a breakaway action added to the commotion that transformed the 1968 Miss America pageant from a cliché-ridden celebration of American beauty ideals into the symbolic beginning point of a new movement. Provoked by the Women's Liberation protest, three spectators, including a former Miss Green Bay, Wisconsin, stepped forward from the boardwalk crowd to form a counterpicket in the contest's defense, while inside the Convention Center Peggy Dobbins, a Women's Liberation protester, hurled a stench bomb from the audience.

The Women's Liberation protest captured the attention if not the sympathies of the national media. The image of unruly women mocking symbols of American beauty was broadcast widely by the media, and as a result the 1968 Miss America contest protest lives in the nation's memory as the action that announced the arrival of a new women's movement. The press erroneously reported that bras had been burned at the protest, providing the "bra-burner" image that, for years, allowed critics to place the women's movement on the incendiary fringe along with flag and draft-card burners.

In the nation's black press, the preponderance of articles published about the events surrounding the Miss America pageant focused on the new Miss Black America contest. They described how the winner, Saundra Williams, dazzled the contest's judges with her performance of an African dance she called the "Fiji." Meeting with reporters after her victory, crown nestled in her Afro hairstyle, she described her parents as middle-class Negroes and proudly listed among her prior accomplishments participation in a protest to integrate segregated businesses in Prince Anne, Maryland. Wearing an Afro, performing an "African" dance created in America, claiming middle-class status, putting her body on the line in civil rights demonstrations, and placing her beauty on display to demonstrate racial pride, Saundra Williams breathed life into the symbols and practices of an emergent black identity.

The women who heaped bras and girdles into the trash can at the Convention Center were awkwardly aware of the NAACP protest contest that was taking place a few blocks away. Many of the white women who were early members of the women's movement had been active participants in the Civil Rights Movement during its racially integrated years and had long-standing commitments to the cause of racial equality. Their Civil Rights Movement participation made them sensitive to racial issues and familiar with African American protest. They denounced the Miss America pageant's racism as well as its sexism, handing out leaflets that listed racism as one of ten points of protest against the pageant and what it represented. Their actions on the day of the protest were steeped in metaphors drawn from the black experience of oppression and borrowed from the protest traditions of the Civil Rights Movement. They spoke of the "enslavement" of all women to the demands of beauty standards. As they picketed they sang, "We shall not be used," an adaption of "We shall not be moved," which was frequently sung at Civil Rights Movement protests. One of the women on the picket line was Robin Morgan, who would later emerge as a leader of the women's movement. The *New York Times* introduced her as a poet who was also a "housewife who uses her maiden name." Commenting on the NAACP protest, Morgan said, "We deplore Miss Black America as much as Miss White America but we understand the black issue involved."[1]

The divergence of the Women's Liberation and NAACP protests is striking. It is indicative of the gulf that existed by 1968 between the largely white Left (including the burgeoning women's movement) and even the more integrationist segments of black movements. The first Miss Black America pageant was a demonstration that existed on the cusp between the Civil Rights and Black Power movements. It celebrated black women by creating an all-black contest while maintaining that its goal was to protest the exclusionary practices of the Miss America pageant. By 1968, the Civil Rights Movement had been shattered and the growing Black Power Movement had been fragmented into groups pursuing widely varied courses. Despite the variety of goals and strategies of competing black political organizations, some commonalities existed. From virtually all positions within African American political movements at that time, regardless of location with respect to strategy for black liberation, there was scant room for a separate women's politics.

A *New York Times* reporter covering both demonstrations focused on Bonnie Allen, one of the few black participants in the Women's Liberation protest. Though picketing at the Miss America contest, she said, "I'm for beauty contests, but then again maybe I'm against them. I think black people have a right to protest."[2] The quote, seemingly full of contradictions, is most probably a reporter's composite. Nonetheless, it expresses the difficulty Allen had as she tried to formulate an opinion about these two protests, neither of which adequately addressed her situation. As a black woman in a movement that was established primarily by white women and ignored, viewed with suspicion, or mocked by many African Americans, Allen suggested an uneasy allegiance to both protests. Her inability to explain her position stemmed from what legal scholar Kimberlé Crenshaw has described as black women's "intersectional disempowerment."[3] Caught in the intersection of domination by race and by gender, Bonnie Allen could stand with the NAACP protest by ignoring its objectification of women or stand with the Women's Liberation protest and feel disloyal for not joining her brothers and sisters in unified protest at the Ritz Carlton Hotel.

The Women's Liberation protesters selected the 1968 Miss America pageant as the target of their action because it epitomized the objectification of women within a male dominated culture. Several studies have analyzed the ways in which beauty ideals serve to reinforce male supremacy. These works demonstrate how beauty standards required all women to judge themselves in relation to unrealistic physical ideals but do not explain how Eurocentric standards placed black women and white women in very different situations vis-à-vis beauty as an ideal. In 1968, black women and white women stood in different locations in relation to the institutions that established and perpetuated national beauty ideals: national beauty contests, media advertising, women's magazines, Hollywood movies, and television programming. White women were objectified in these venues; black women were either excluded from them or included in images that reinforced Eurocentric beauty ideals. Black women had to contest their wholesale definition as non-beauties. In response to the exclusion of black women from dominant representations of beauty, African American women's beauty became part of the symbolic repertoire with which champions of the race sought to assert racial pride.

From the perspective of the achievement-oriented NAACP, the Miss America title was a symbol of success worth fighting for. In the Miss America contest's early days, racial exclusion was written into its by-laws. Long after the racist by-law was rescinded the contest remained, in practice, closed to all but white women. To the extent that the annual spectacle of white women established the reigning definition of beauty, it reinforced cultural codes that placed black women outside of the beauty ideal. In the wider culture, black women had been excluded from all that was celebrated in the crowning of Miss America. Where Miss America represented a contradictory ideal of sexualized beauty and chastity, black women had been portrayed in mainstream depictions as ugly and sexually available, a diametrically opposed set of images.

Exclusion from the dominant beauty ideal did not mean that black women had been spared objectification in either dominant or minority culture. On the contrary, whether facing disparagement of their bodies in dominant cultural images or the mixture of appreciation (where alternative standards prevailed) and ridicule (where dominant standards prevailed) within African American communities, black women were diminished when their general worth was determined by their physical appearance. On that basis, black women potentially shared grievances with the white women at the Miss America protest. But Miss America, the symbolic target of the protest, had never stood for black women. Few black women in 1968 were willing to voice protest against the Miss America pageant for what it supposedly said about women generally.

The colliding protests at the Miss America pageant dramatized the complexity of the situation of black women in relation to images of beauty. In order to comprehend the positions of both Saundra Williams, the first Miss Black America, and Bonnie Allen, an isolated black woman in the Women's Liberation protest, one displaying her beauty as a protest, the other picketing displays of beauty, we must take account of the ways in which African American women have been depicted by the dominant culture. We should also consider the often divergent ways African American women have been depicted in African American cultural representations, the ways African Americans have used images of African American women to challenge cultural domination, and the subjective experiences of black women, whose bodies were the objects of ridicule and celebration. . . .

NOTES

1. Curtis, Carlotte. 1968. "Miss America Pageant Is Picketed by 100 Women," *New York Times*, September 8, 1968, p. 81.

2. Curtis, p. 81.

3. Crenshaw, Kimberle. 1992. "Whose Story Is It, Anyway? Feminist and Antiracist Appropriations of Anita Hill." In *Race-ing Justice, En-gendering Power: Essays on Anita*

Hill, Clarence Thomas, and the Construction of Social Reality, edited by Toni Morrison. New York: Random House, p. 406.

KEY CONCEPTS

gender identity   multiracial feminism   racism

DISCUSSION QUESTIONS

1. What ideals of women's beauty are projected in popular culture (magazines, film, video, etc.)? How are they defined for women in different racial and ethnic groups?
2. What contemporary images of beauty appear in magazines whose audience is mostly African American women?
3. In what ways do ideals of beauty for African American women both represent and contest the dominant culture?

38

A Black Woman Took My Job

MICHAEL KIMMEL

Michael Kimmel examines how changes in women's lives resulting from the feminist movement are affecting men. He also argues that the feminist revolution has the possibility of improving life for men as well as for women.

Over the past three generations, women's lives have been utterly and completely transformed—in politics, the military, the workplace, professions, and education. But during that time, the ideology of masculinity has remained relatively intact.

SOURCE: Michael Kimmel. 2004. "A Black Woman Took My Job." New Internationalist. Copyright *New Internationalist*, www.newint.org.

The notions we have about what it means to be a man remain locked in a pattern set decades ago, when the world looked very different. The single greatest obstacle to women's equality today remains the behaviour and attitudes of men.

In the mid-1970s, [scholars] offered what [they] called the four basic rules of masculinity:

1. No Sissy Stuff. Masculinity is based on the relentless repudiation of the feminine.

2. Be a Big Wheel. Masculinity is measured by the size of your paycheck, and marked by wealth, power and status. As a US bumper sticker put it: "He who has the most toys when he dies, wins."

3. Be a Sturdy Oak. What makes a man a man is that he is reliable in a crisis. And what makes him reliable in a crisis is that he resembles an inanimate object. A rock, a pillar, a tree.

4. Give 'em Hell. Exude an aura of daring and aggression. Take risks; live life on the edge.

The past decade has found men bumping up against the limitations of these traditional definitions, but without much of a sense of direction about where they might look for alternatives. We chafe against the edges of traditional masculinity but seem unable or unwilling to break out of the constraints of those four rules. Hence the defensiveness, the anger, the confusion that is everywhere in evidence.

Let me pair up those four rules of manhood with the four areas of change in women's lives—gender identity, the workplace, the balance of work and family life, the sexual landscape—and suggest some of the issues I believe we are facing around the world today.

First, women made gender visible, but most men do not know they are gendered beings. Courses on gender are still populated mostly by women. Most men don't see that gender is as central to their lives as it is to women's. The privilege of privilege is that its terms are rendered invisible. It is a luxury not to have to think about race, or class, or gender. Only those marginalized by some category understand how powerful that category is when deployed against them. I was reminded of this recently when I went to give a guest lecture for a female colleague at my university. (We teach the same course on alternate semesters, so she always gives a guest lecture for me, and I do one for her.) As I walked into the auditorium, one student looked up at me and said: "Oh, finally, an objective opinion!"

The second area in which women's lives have changed is the workplace. Recall the second rule of manhood: Be a Big Wheel. Most men derive their identity as breadwinners, as family providers. Often, though, the invisibility of masculinity makes it hard to see how gender equality will actually benefit us as men. For example, while we speak of the "feminization of poverty" we rarely "see" its other side—the "masculinization of wealth." Instead of saying that US women, on average, earn [a percentage] of what US men earn, what happens if we say that men are earning $1.30 for every dollar women earn? Now suddenly privilege is visible!

Recently I appeared on a television talk show opposite three "angry white males" who felt they had been the victims of workplace discrimination. The show's title was "A Black Woman Took My Job." In my comments to these men, I invited them to consider what the word "my" meant in that title: that they felt that the jobs were originally "theirs." But by what right is that "his" job? Only by his sense of entitlement, which he now perceives as threatened by the movement towards workplace gender equality.

The economic landscape has changed dramatically and those changes have not necessarily been kind to most men. The great global expansion of the 1990s affected the top 20 percent of the labor force. There are fewer and fewer "big wheels." European countries have traded growth for high unemployment, which will mean that more and more men will feel as though they haven't made the grade, will feel damaged, injured, powerless. These are men who will need to demonstrate their masculinity all over again. And here come women into the workplace in unprecedented numbers. Just when men's economic breadwinner status is threatened, women appear on the scene as easy targets for men's anger— or versions of anger. Sexual harassment, for example, is a way to remind women that they are not yet equals in the workplace, that they really don't belong there.

It is also in our interests as men to begin to find a better balance of work and family life. There's a saying that "no man on his deathbed ever wished he had spent more time at the office." But remember the third rule of manhood: Be a Sturdy Oak. What has traditionally made men reliable in a crisis is also what makes us unavailable emotionally to others. We are increasingly finding that the very things that we thought would make us real men impoverish our relationships with other men and with our children. Fatherhood, friendship, partnership all require emotional resources that have been, traditionally, in short supply among men, resources such as patience, compassion, tenderness, attention to process.

In the US, men become more active fathers by "helping out" or by "pitching in"; they spend "quality time" with their children. But it is not "quality time" that will provide the deep intimate relationships that we say we want, either with our partners or with our children. It's quantity time—putting in those long, hard hours of thankless, unnoticed drudge—that creates the foundation of intimacy. Nurture is doing the unheralded tasks, like holding someone when they are sick, doing the laundry, the ironing, washing the dishes. After all, men are capable of being surgeons and chefs, so we must be able to learn how to sew and to cook.

Finally, let's examine the last rule of manhood: Give 'em Hell. What this says to men is: take risks, live dangerously. And this, of course, impacts most dramatically on our bodies, sex, health, and violence. Masculinity is the chief reason why men do not seek healthcare as often as women. Women perform self-exams, seek preventive screenings, and pay attention to diet, substance abuse, far more often than men. Why? As health researcher Will Courtenay writes: "A man who does gender correctly would be relatively unconcerned about his health and well-being in general. He would see himself as stronger, both physically and emotionally, than most women. He would think of himself as independent, not needing to be nurtured by others."[1] Or, as one Zimbabwean man put it, "real men don't get sick."[2]

Indeed. The ideas that we thought would make us "real men" are the very things that endanger our health. One researcher suggested slapping a warning label on us: Caution: Masculinity May Be Hazardous to Your Health. A study of adolescent males in the US [in the 1990s] found that adherence to traditional masculinity ideology was associated with being suspended from school, drinking, use of street drugs, having a high number of sexual partners, not using condoms, being picked up by the police, forcing someone to have sex.[3]

These gender-conforming behaviors increase boys' risk for HIV, STDs, early death by accident, injury, or homicide. It's no exaggeration to say that the spread of HIV is driven by masculinity. HIV risk reduction requires men to take responsibility by wearing condoms. But in many cultures ignoring the health risks to one's partner, eschewing birth control, and fathering many children are signs of masculine control and power.

Finally, let me turn to what may be the single greatest public health issue of all: violence. In the US, men and boys are responsible for 95 percent of all violent crimes. Every day 12 boys and young men commit suicide—7 times the number of girls. Every day, 18 boys and young men die from homicide—10 times the number of girls. From an early age, boys learn that violence is not only an acceptable form of conflict resolution but one that is admired. Four times more teenage boys than girls think fighting is appropriate when someone cuts into the front of a line. Half of all teenage boys get into a physical fight each year.

Violence has been part of the meaning of manhood, part of the way men have traditionally tested, demonstrated, and proved their manhood. Without another cultural mechanism by which young boys can come to think of themselves as men, they've eagerly embraced violence as a way to become men. It would be a major undertaking to enumerate all the health consequences that result from the equation of violence and masculinity.

And just as women are saying "yes" to their own sexual desires, there's an increased awareness of the problem of rape all over the world, especially of date and acquaintance rape. In one recent US study, 45 percent of all college women said that they had had some form of sexual contact against their will, and a full 25 percent had been pressed or forced to have sexual intercourse against their will. When one psychologist asked male undergraduates if they would commit rape if they were certain they could get away with it, almost 50 percent said they would. Nearly 20 years ago, anthropologist Peggy Reeves Sanday proposed a continuum of propensity to commit rape upon which all societies could be plotted—from "rape-prone" to "rape-free." (The US was ranked as a highly rape-prone society, far more than any country in Europe; Norway and Sweden were among the most rape-free.) Sanday found that the single best predictors of rape-proneness were

1. Whether the woman continued to own property in her own name after marriage, a measure of women's autonomy.
2. Father's involvement in child rearing, a measure of how valued parenting is and how valued women's work is.

So women's economic autonomy is a good predictor of their safety—as is men's participation in child rearing. If men act at home the way we say we want to act, women will be safer.

And the news gets better. A 1996 study of Swedish couples found positive health outcomes for wives, husbands, and children when the married couple adopted a partnership model in work-family balance issues. A recent study in the US found that men who shared housework and child care had better health, were happier in their marriages, reported fewer psychological distress symptoms, and—perhaps most important to them—had more sex! That's right, men who share housework have more sex. What could possibly be more in men's "interests" than that?

Another change that is beginning to erode some of those traditional "masculine" traits is the gradual mainstreaming of gay male culture. One of the surprise hit TV shows of the past year has been "Queer Eye for the Straight Guy." Imagine if, 10 years ago, there'd been a TV show in which five flamboyantly gay men showed up at a straight guy's house to go through his clothing, redo his house, and tell him, basically, that he hasn't a clue about how to be socially acceptable. The success of "Queer Eye" has been the partial collapse of homophobia among straight men. And the cause of that erosion is simple: straight women, who have begun to ask straight men: "Why can't you guys be more like gay guys?"

Rather than resisting the transformation of our lives that gender equality offers, I believe that we should embrace these changes, both because they offer us the possibilities of social and economic equality and because they also offer us the possibilities of richer, fuller, happier lives with our friends, with our lovers, with our partners, and with our children. We, as men, should support gender equality, both at work and at home. Not because it's right and just—although it is those things. But because of what it will do for us, as men.

The feminist transformation of society is a revolution-in-progress. For nearly two centuries, we men have met insecurity by frantically shoring up our privilege or by running away. These strategies have never brought us the security and the peace we have sought. Perhaps now, as men, we can stand with women and embrace the rest of this revolution—embrace it because of our sense of justice and fairness, embrace it for our children, our wives, our partners, and ourselves. Ninety years ago, the American writer Floyd Dell wrote an essay called "Feminism for Men." It's first line was this: "Feminism will make it possible for the first time for men to be free."

NOTES

1. W. H. Courtenay, "College Men's Health: An Overview and a Call to Action," *Journal of American College Health* 46, no. 6 (1998).

2. M. Foreman, ed., *AIDS and Men: Taking Risks or Taking Responsibility* (Zed Books, 1999).

3. J. H. Pleck, F. L. Sonenstein, and L. C. Ku, "Masculinity Ideology: Its Impact on Adolescent Males' Heterosexual Relationships," *Journal of Social Issues* 49, no. 3 (1993): 11–29.

KEY CONCEPTS

feminism   hegemonic masculinity   power

DISCUSSION QUESTIONS

1. How does Kimmel portray the traditional forms of masculinity? How are
    they transformed by the feminist movement?
2. What benefits does Kimmel say will accrue to men because of women's
    liberation?

39

Whose Body Is It, Anyway?

PAMELA FLETCHER

*By reflecting on her own life, Pamela Fletcher provides a powerful and painful narrative
that explores social myths about violence against women. She asks us to imagine a world
in which destructive thinking about violence is transformed by women's empowerment.*

RAPE

I never heard the word while growing up. Or, if I had, I blocked it out because
its meaning was too horrific for my young mind: a stranger, a weapon, a dark
place, blood, pain, even death. But I do remember other people's responses to it,
especially those of women. I specifically remember hearing about Rachel when I
was in high school in the '70s. The story was that she "let" a group of boys "pull
a train" on her in the football field one night. I remember the snickers and the
looks of disgust of both the girls and the boys around campus. It was common
knowledge that nobody with eyes would want to fuck Rachel; she had a face
marred by acne and glasses. But, she had *some* body.

While I am writing this essay, I remember the stark sadness and confusion I felt then. This same sadness returns to me now, but I am no longer confused. Then I wondered how she could "do" so many guys and actually like it. Then I thought maybe, she didn't like it after all, and maybe, just maybe, they made her do it. But the word rape never entered my mind. After all, she knew them, didn't she? There was no weapon, no blood. She survived, didn't she? And, just what was she doing there all by herself, anyway? Now, I know what "pulling a train" is. Now I know they committed a violent crime against her body and her soul. Now I know why she walked around campus with that wounded face, a face that none of us girls wanted to look into because we knew intuitively that we would see a reflection of our own wounded selves. So the other girls did not look into her eyes. They avoided her and talked about her like she was "a bitch in heat." Why else would such a thing have happened to her?

I tried to look into Rachel's eyes because I wanted to know something—what, I didn't know. But she looked down or looked away or laughed like a lunatic, you know, in that eerie, loud, nervous manner that irritated and frightened me because it didn't ring true. Now I wonder if she thought such laughter would mask her pain. It didn't.

PAINFUL SILENCE AND DEEP-SEATED RAGE

I remember another story I heard while I was in college. Larry told me that his close friend, Brenda, let Danny stay over one night in her summer apartment after they had smoked some dope, and he raped her. Larry actually said that word.

"Don't tell anyone," Brenda begged him. "I never should have let him spend the night. I thought he was my friend."

Larry told me not to ever repeat it to anyone else. And, trying to be a loyal girlfriend to him and a loyal friend to Brenda, I didn't say anything. When we saw Danny later at another friend's place, we neither confronted nor ignored him. *We acted as though everything was normal.* I felt agitated and angry. I wondered why Larry didn't say anything to Danny, you know, man-to-man, like: "That shit was not cool, man. Why you go and do somethin' like that to the sista?"

It never occurred to me to say anything to Brenda, because I wasn't supposed to know, or I was supposed to act as though I didn't know, stupid stuff like that. I sat there, disconnected from her, watching her interact with people, Danny among them, acting as though everything was normal.

DENIAL

Since I began writing this essay two months ago, I have had such difficulty thinking about my own related experiences. I hadn't experienced rape. Or, had I? For months, in the hard drive of my subconscious mind, I searched for files that would yield any incidence of sexual violence or sexual terrorism. When certain memories surfaced, I questioned whether those experiences were "real rapes."

I have some very early recollections that challenge me. Max, my first boyfriend, my childhood sweetheart, tried to pressure me into having sex with him when we were in junior high. Two of my friends, who were the girlfriends of his two closest friends, also tried to pressure me because they were already "doing it" for their "men."

"Don't be a baby," they teased. "Everybody's doing it."

But I wouldn't cave in, and I broke up with Max because he wasn't a decent boy.

A year later, when we reached high school, I went crawling back to Max because I "loved" him and couldn't stand his ignoring me. He stopped ignoring me long enough to pin me up against the locker to kiss me roughly and to suck on my neck long and hard until he produced sore, purple bruises, what we called hickies. I had to hide those hideous marks from my parents by wearing turtleneck sweaters. Those hickies marked me as his property and gave his friends the impression that he had "done" me, even though we hadn't gotten that far yet. We still had to work out the logistics.

I hated when he gave me hickies, and I didn't like his exploring my private places as he emotionally and verbally abused me, telling me I wasn't pretty like Susan: "Why can't you look like her?" And I remember saying something like, "Why don't you go be with her, if that's what you want?" He answered me with a piercing "don't-you-ever-talk-to-me-like-that-again" look, and I never asked again. He continued, however, to ask me the same question.

In my heart, I realized that the way he treated me was wrong because I felt violated; I felt separated from my body, as if it did not belong to me. But at sixteen I didn't know how or what to feel, except that I felt confused and desperately wanted to make sense out of what it meant to be a girl trapped inside a woman's body. Yes, I felt trapped, because I understood that we girls had so much to lose now that we could get pregnant. Life sagged with seriousness. Now everybody kept an eye on us: our parents, the churches, the schools, and the boys. Confusion prevailed. While we were encouraged to have a slight interest in boys (lest we turn out "funny") so that ultimately we could be trained to become good wives, we were instructed directly and indirectly to keep a safe distance away from them.

We liked boys and we thought we wanted love, but what we really wanted as youth was to have some fun, some clean, innocent fun until we got married and gave our virtuous selves to our husbands just as our mothers had done. We female children had inherited this lovely vision from our mothers and from fairy tales. Yet now we know that those visions were not so much what our mothers had experienced, but what they wished they had experienced, and what they wanted for us.

We thought "going with a boy" in the early '70s would be romance-filled fun that involved holding hands, stealing kisses, exchanging class rings, and wearing big lettered sweaters. Maybe it was, for some of us. But I know that many of us suffered at the hands of love.

I soon learned in high school that it was normal to be mistreated by our boyfriends. Why else would none of us admit to each other the abuse we tolerated? These boys "loved" us, so we believed that they were entitled to treat us in any way they chose. We believed that somehow we belonged to them, body and soul. Isn't that what many of the songs on the radio said? And we just knew somehow that if we did give in to them, we deserved whatever happened, and if we didn't give in, we still deserved whatever happened. Such abuse was rampant because we became and remained isolated from each other by hoisting our romances above our friendships.

We didn't define what they did to us as rape, molestation, or sexual abuse. We called it love. We called it love if it happened with our boyfriends, and we called other girls whores and sluts if it happened with someone else's boyfriend or boyfriends, as in the case of Rachel and "the train."

We called it love because we had tasted that sweet taste of pain. Weren't they one and the same?

REALIZATION

One sharp slap from Max one day delivered the good sense I had somehow lost when I got to high school. After that point I refused to be his woman, his property. When I left home for college, I left with the keen awareness that I had better take good care of myself. In my involvement with Max, I had allowed a split to occur between my body and my soul, and I had to work on becoming whole again.

I knew that I was growing stronger (though in silent isolation from other young women and through intense struggle) when I was able to successfully resist being seduced (read: molested) by several college classmates and when I successfully fought off the violent advances and the verbal abuse (what I now recognize as an attempted rape) of someone with whom I had once been sexually intimate.

But how does a woman become strong and whole in a society in which women are not permitted (as if we need permission!) to possess ourselves, to own our very bodies? We females often think we are not entitled to ourselves, and many times we give ourselves away for less than a song. The sad truth of the matter is that this is how we have managed to survive in our male-dominated culture. Yet, in the wise words of the late Audre Lorde, "the Master's tools will never dismantle the Master's house." In other words, as long as we remain disconnected from ourselves and each other and dependent on abusive males, we will remain weak, powerless, and fragmented.

Just imagine how different our lives would be today if we were not injured by internalized misogyny and sexism. Imagine how different our lives would be if we would only open our mouths wide and collectively and loudly confront males and *really* hold them accountable for the violent crimes they perpetrate against females. Imagine how our lives would be if all mothers tell their daughters the truth about romantic love and teach them to love themselves as females, to value and claim their bodies, and to protect themselves against violent and disrespectful males.

What if we girls in junior high and high school believed we deserve respect rather than verbal and sexual abuse from our male classmates? What if we girls in my high school had confronted the gang of boys who raped Rachel that night in the football field twenty years ago, rather than perpetuated that cycle of abuse and shame she suffered? What if Larry and I had confronted Danny for raping Brenda that summer night in her apartment? What if Brenda had felt safe enough to tell Larry, me, and the police? What if we females believed ourselves and each other to be as important and deserving of our selfhood as we believe males to be? Just imagine.

Envision a time when we women are connected to ourselves and each other, when we no longer feel the need and desire to conspire with men against each other in order to survive in a misogynist, violent culture. We must alter our destructive thinking about being female so that we can begin to accept, love, and cherish our femaleness. It is the essence of our lives.

Readjusting our lens so we can begin to see ourselves and each other as full, capable, and mighty human beings will take as much work as reconstructing our violent society. Neither job is easy, but the conditions and the tasks go hand in hand. Two ways to begin our own transformation are to become physically active in whatever manner we choose so we can take pleasure in fully connecting to ourselves and in growing physically stronger, and to respect, protect, support, and comfort each other. Once we stop denying that our very lives are endangered, we will soon discover that these steps are not only necessary but viable in empowering ourselves and claiming our right to exist as whole human beings in a peaceful, humane world.

KEY CONCEPTS

domestic violence	rape myths	sexual assault

DISCUSSION QUESTIONS

1. As Fletcher thinks back on high school, how does she see women's dependence on men as leading women to ignore the reality of violence against women?

2. In what ways does Fletcher's account resonate with experiences that you have observed—either in high school or college?

3. What support systems are needed to support victims of sexual assault? What systems are necessary to work with perpetrators of violence?

Applying Sociological Knowledge:
An Exercise for Students

Imagine that you wake up tomorrow morning and are a member of the other gender. Make a list of all the things about yourself that you think would have changed. After making your list, make a note of which characteristics are biological or physical, which are attitudinal, which involve behavior, and which are institutional. Then ask yourself how individual and institutional forms of gender are related.

40

Dynamics of Internet Dating

HELENE M. LAWSON AND KIRA LECK

The Internet has influenced social interaction in many ways. This article specifically addresses the process of dating on the Internet. Through interviews with people who use the Internet to meet people, the authors uncover the importance of companionship, comfort, presentation of self, and trust when developing online relationships. The rules of interaction are modified in this technologically advanced form of dating.

The present research focused on the dynamics of Internet dating, a method of courting used by individuals who meet on the Internet and continue online correspondence in hopes of forming a supportive romantic relationship. It sought to determine why people choose to date online, what aspects of face-to-face relations are reproduced, and the rationales and strategies Internet daters use to negotiate and manage problems of risk accompanying the technology.

A BRIEF HISTORY OF DATING PRACTICES

Although the practices of courting vary from culture to culture and change over time, technologies of communication have historically shaped courtship, making it freer and expanding possibilities. The timeless love letter notwithstanding, courtship interaction in the United States has been limited to supervised situations or contained within the bounds of engagement for marriage. This was especially true during the puritan, colonial, and Victorian eras (Hunt, 1959). Historians believe that freer dating practices, such as meeting privately and face-to-face for romantic interactions at scheduled times and places, emerged among middle-class teenagers in the 1920s. These practices developed alongside new technologies such as telephones, automobiles, and drive-in theaters, which allowed teenagers to become more independent from their parents. . . .

In the 1990s, the Internet became a major vehicle for social encounters. Through the Internet, people can interact over greater distances in a shorter period and at less expense than in the past. Theorists have debated the positive

SOURCE: Helene M. Lawson and Kira Leck. 2006. "Dynamics of Internet Dating." *Social Sciences Computer Review* 24, no. 2 (Summer): 189–208.

and negative effects this technology has on social interactions. Initially, theorists such as Zuboff (1991) believed "the Internet reduced face-to-face interaction" and created an "uncomfortable isolation" (pp. 479–482) for people at work. Conversely, Raney (2000) argued that online communication expands social networks. According to Raney, the Pew Internet and American Life Project found supporting evidence for this view in a study in which "more than half of Internet users reported that e-mail was strengthening their family ties. And Internet users reported far more offline social contact than non-users." (p. G7). . . . Today Internet video and sound communications are commonplace, and photographs, video, and sound clips can all be altered or fabricated entirely. These new technologies allow Internet daters enormous latitude to prepare their presentations of self.

USING THE INTERNET FOR DATING

The Internet is a new social institution that has the ability to connect people who have never met face to face and is thus likely to transform the dating process. Beginning with newsgroups such as Usenet and various bulletin boards that operated under the now-obsolete Gopher system, the Internet facilitated the formation of communities. . . .

We explored the phenomenology of Internet dating, which we defined as the pattern of periodic communication between potential partners using the Internet as a medium. We examined the respondents' concerns over the risk of being deceived, their anxieties about physical appearance, and the hazards of romantic involvement.

METHOD

Participants

Because we needed a sample of respondents who could be tracked over time and whose reliability could be verified, we began to investigate the phenomenon of Internet dating by interviewing people who were personally accessible, such as coworkers, acquaintances, and students. Soon the sample expanded because respondents told us about people they knew who dated online, which resulted in a snowball sample. It was not a uniform sample with respect to such attributes as race and socioeconomic status because it favored a White middle class and was instead a sample dictated by sampling logistics. However, we believe that the phenomenon of Internet courtship is largely a White, middle-class phenomenon. . . . For this reason, we believe our sample to be qualitatively representative: It was composed of 32% students, 24% business and clerical workers, 14% trade workers, and 14% professionals and semiprofessionals. The sample also included unemployed persons, small business owners, and housewives.

Because we were interested in romantic dating relationships that could result in commitment, we did not include people interested only in pornography or online sexual encounters as their primary focus. We defined dating as setting up specific times to mutually disclose personal information with potential romantic partners on an ongoing basis. We did not place any other restrictions on whom we were willing to interview. Consequently, the sample included homosexuals and unhappily married persons. Romance was not neccessarily the goal of online dating, but in our sample, three married persons changed partners as a result of Internet interactions.

Interview Questions

Interviews were open-ended and informal. We asked respondents to (a) describe their experiences with Internet dating, (b) state whether these experiences were positive or negative, (c) state how and why they entered the world of online dating, and (d) state whether they used online dating services or met incidentally through chat rooms, online games, or common interest groups. Respondents were eager to relate their experiences, and many interviews lasted an hour or longer.

Interviews were conducted during lunch in restaurants, at respondents' homes, at the home of the first author, in the university cafeteria, and on walks in various neighborhoods. All respondents had ready access to computers in their homes, dorm rooms, or places of work. We watched while they talked back and forth online. In addition, the first author invited three newly paired couples to her home for dinner. Follow-up data were collected in person, on the phone, by e-mail, and by mail. Interviews were later transcribed and coded by keywords according to concepts that emerged through the dialogue, such as trust, time, risk, and need satisfaction.

We limited the number of respondents to 25 men and 25 women because we wanted to compare gender variables in a balanced sample. The men ranged in age from 18 to 58 with a mean age of 32.6. The women ranged in age from 15 to 48 with a mean age of 33. In all, 17 men and 11 women were single (never married), 7 men and 10 women were divorced, and 1 man and 4 women were married. Two men and one woman were gay. Two women and one man were African American. One man was Indian. Six men and seven women were the parents of young children, and as previously stated, five respondents were married when they began to interact romantically online.

RESULTS

Companionship

Lonely people tend to report being dissatisfied with their relationships and are often cynical, rejecting, bored, and depressed. They also have difficulty making friends, engaging in conversations, getting involved in social activities, and dating

(Chelune, Sultan, & Williams, 1980; W. H. Jones, Hobbs, & Hockenbury, 1982). Their tendency to engage in minimal self-disclosure and be unresponsive to conversational partners often results in poor interactions that are unrewarding for both partners, which leads lonely individuals to feel dissatisfied with their relationships (McAdams, 1989). Both relationship dissatisfaction and difficulty with social behaviors may lead lonely people to seek online relationships.

Regardless of their marital status, respondents of all ages tended to report being lonely. They all talked about needing more communication, emotional support, and companionship. Fred, a 19-year-old student who had never been married, said, "I hate being alone. You want to know someone out there at least cares."

Greta, a 43-year-old, unhappily married mother of a 9-year-old, worked a night shift. Her husband worked during the day, and they both dated others online through chat rooms. Chat rooms often require only token (username) identification. The face presented is largely cloaked, but marital status is usually not hidden. Rather, it is explained:

> I guess the big problem is that my husband works 6 days a week, is gone all day long, and doesn't spend time with me. It is like we are strangers living in the same house. We haven't actually gone out with anyone.

Kelly, a 48-year-old, unhappily married student also blamed her lack of communication with her husband for why she dated online:

> I think I qualify for this interview because I date someone online. In our house there is no communication. That is no way to be. It's two people living in the same house like roommates that have totally different lives. We never talk. That is how my life was before I met George [online]. . . .

Regardless of their marital status, . . . individuals seemed to perceive their social lives as incomplete. This may be a reflection of the separation of family and friends because of current societal structure. Thus, it is not surprising that they were highly motivated to become involved in online relationships with people who were willing to talk, listen, and serve a supportive function. . . .

Comfort After a Life Crisis

. . . [S]everal respondents in the present sample reported seeking comfort after a life crisis, such as the loss of a job, a divorce, or a death in the family. Robin, a 32-year-old, never married woman, said,

> I had suffered such a great loss when my grandpa died. We were very close and he raised me. I guess at that particular point in time in my life I needed someone in my life. One night I was searching for someone to talk with. There is a button you can hit to find a random chat partner. I must have gone through about 10 to 15 different people until his name popped up. I read his details that he provided about himself, and I sent him a message. The first night we talked for about 5 to 6 hours straight, nonstop.

Anna, a 39-year-old, divorced woman, also got online because of her recent divorce: "After my divorce, I cried all the time. My friends were tired of listening to me. I wanted a support group so I went into this chat room." . . .

Our society's lack of support structure for individuals who experience life crises may lead them to seek out comfort from online sources. . . . The online setting allows them to select which aspects of themselves to reveal to their online companions, which lessens the probability of unfavorable judgment that may be leveled by real-life friends and family members.

Control Over Presentation and Environment

The Internet provides a medium for people to present themselves in a way that that they think is flattering. Clark (1998) reports that girls describe themselves as "thinner and taller" and otherwise prettier in Internet communications than they actually are. Because contact is mediated, individuals do not have to expose themselves directly on the Internet. In general, "the surest way for a person to prevent threats to his face is to avoid contacts in which these threats are likely to occur" (Goffman, 1967, p. 15).

Jean, a 35-year-old, never married woman, said if you were heavy, you could get to know someone who might like you instead of having to attract people with your looks before they wanted to know you:

> Many of the women I met from my chat room were way overweight. It's easy to sit at home and talk online, say things, and be appealing. I mean it's safe. It's totally safe if you don't ever plan on ever meeting anyone [face to face]. If later on, you do meet them, maybe they will like you anyway. By that time it's worth the risk. . . .

Reid, a 37-year-old, divorced father with two children, said,

> The Internet is a place where people can take risks without consequences. You can experiment with people you wouldn't normally meet or get involved with. You can grocery shop. There are more people to meet. You can play games for a long time. You can look at so many pictures; it's fun like a candy store.

. . . For people who are shy, anxious, and deficient in social skills, use of the Internet may facilitate social interaction because it requires different skills that are necessary for initiating heterosocial interaction in a face-to-face setting. In one study, college students reported using the Internet to meet people because they found it reduced their anxiety about social interaction (Knox, Daniels, Sturdivant, & Zusman, 2001).

Some respondents of both sexes claimed they found it difficult to talk to strangers in social situations such as parties or even in places such as the school cafeteria or a classroom. Rick, a 32-year-old, never married man, said he liked using the Internet because "I'm shy. That is why I went into a chat room. I can say things online that I can't say in person. I am so quiet. But, I can talk on the telephone too."

Pete, a 22-year-old, never married man, did not trust dating in general, but he liked the Internet better than bars:

> Bars are a meat market, and I feel that everybody there is putting on more of a show than actuality. I mean when you meet them [women] in a bar, it's like they are a different person than in real life. And it's the same thing with the Internet, you know, with a lot of women. So many haven't returned messages, or they just leave you hanging, or they pretend to be someone they are not. I'm too shy, too afraid of getting turned down. It's easier, less painful getting turned down on type than it is in person.

Men and women respondents complained that bars were not a good place to get to know prospective partners. Harry argued that he did not trust the character of bar pickups:

> One thing I found with the bar is that most ladies who go there will say yes and say yes to about anybody given the time of night. Some ladies have propositioned me! Let's just say I don't like being in that situation.

Anna also said, "I don't want to go to bars to meet people. This is a lot safer."

Societal expectations for appearance and behavior can result in individuals who do not fit the norm and perceive themselves as deviants who will not be accepted. Furthermore, they may fear negative reprisals from more mainstream members of society and thus may retreat into an online setting where they feel safer and have control.

Freedom from Commitment and Stereotypic Roles

Clark (1998) found that Internet dating is particularly appealing to teenage girls because it allows them to be aggressive while remaining sheltered. Clark argued that "Internet dating affords teenage girls in particular the opportunity to experiment with and claim power within heterosexual relationships," but she questioned whether the resulting relationships were any more emancipative than those found in the real-life experiences of teenagers. She suggested that "power afforded through self-construction on the Internet does not translate into changed gender roles and expectations in the social world beyond cyberspace." The teenage girls in Clark's study were "not interested in meeting the boys with whom they conversed as they might undermine (their) attractive and aggressive on-line persona" (pp. 160–169).

. . . Traditional gender norms that dictate that women wait for men to ask them out and men be assertive leaders are still common today (Mongeau, Hale, Johnson, & Hillis, 1993; Simmel, 1911). However, some research (e.g., Cooper & Sportolari, 1997) and responses from the interviewees suggest that these norms may not operate online.

Cathryn, a 15-year-old girl, stated.

> I like to play but not really be there. I met this boy and we talked about school and movies, but we didn't meet. We live in different states. I don't know much about him really. He's just fun to talk to. I tease him a lot. Sometimes my friends pretend they are older or even guys instead of girls.

This online interaction is free from commitment.

Five of the respondents, both men and women, talked about freedom from commitment and stereotypic sex roles. Anna said,

> We agreed that there would be no expectations and if we didn't like each other, we'd have a few laughs, go to a baseball game or two, have a few beers, who cares. Since I like to travel, I also felt if the guy was a jerk, I had a credit card and would go to a different hotel and stay in San Diego and have a nice vacation....

Ross, a 40-year-old, divorced father who had custody of his 10-year-old son, said,

> There is such a difference between actually talking to somebody and putting things in print. You can make yourself sound like I could be Joe Big Stud or whatever on the Internet. Then when we met, we'd see if we got along.

Greg, a 21-year-old, never married student, said,

> Every few weeks we'd say "Hey, how's it going?" I told her from day one we'd never know each other's real names, where we lived, or anything about it. She didn't know how old I was or if I was married or single or anything. But we loved talking, and we talked about meeting.

Although many respondents initially wanted freedom from commitment, they liked spending a lot of time online getting to know each other. Often after a period of months, they decided to meet face to face. Some changed their minds about having no commitment and increased their involvement, whereas others concluded that they had too little in common to justify continuing the relationship. Thus, as with traditional dating, online daters seemed to want to get to know their partners better before committing....

Trust, Risk, and Lying Online

Trust may not be important in an interaction when compared to that of opening an opportunity for taking a gamble. Goffman (1967) believed, "Chance lies in the attitude of the individual himself—his creative capacity to redefine the world around him into its decisional potentialities" (p. 201). Goffman saw all forms of action as gambling. Similarly, Simmel (1911) argued that when a person is offered a token of trust, the recipient is expected to respond in kind. When people place online personals ads, those who respond may be perceived as offering a gift; the implication is made that "I trust you enough to treat me well.". . .

The Internet has been described as a "revolutionary social space" (Hardey, 2002, p. 577) in which old rules for social interaction are discarded in favor of new ones that may be better suited to the technology. However, Hardey (2002) found that Internet daters' interactions are often guided by "rituals and norms that protect the self" (p. 577), which was originally suggested by Goffman (1967). The technology of the Internet may present new challenges to building intimacy and avoiding rejection, but the basic motivations for protecting the self remain. New risks inspire new coping strategies to maintain an environment of trust. Such an environment is necessary to maintain the solidarity of society, according to Simmel (1978). Giddens (1990) emphasized a need to establish trust among individuals and observed that the alternative to trust is inaction, which in itself may be risky because if we do not take the risk of interacting, we will not develop a supportive friendship network. He saw relationships as "ties based upon trust, where trust is not pre-given but worked upon, and where the work involved means a mutual process of self-disclosure" (p. 121).

To establish close relationships within the constraints of the Internet, people use creative methods to identify themselves as cool and trustworthy. Emotions, abbreviations, unconventional spellings, and specialized grammar are used to weed out people who do not share others' realities or ways of being (Waskul, 2003).[1] Turkle (1995) observed that through photographs, profiles, and narratives, "people create and cycle through a sometimes surprising range of online identities" (p. 10). . . .

Online, people commonly misrepresent their appearance, making it more flattering (Clark, 1998). One sample of college students reported lying about their age, weight, and marital status (Knox et al., 2001). They may also misrepresent their gender (Danet, 1998; Knox et al., 2001). Misrepresentation in online social interactions seems so natural that few seem to give much thought to what usually could be dismissed as a makeover of one's persona. Given the limited amount of information available to respondents about each other in Internet interactions and their transitory nature, deception is common.

Most respondents said they had been lied to more than once, and some reported surprise when this happened. Robin, the 32-year-old, single woman, wanted to trust people:

> I was raised to believe and trust in people when they tell you things. So it was very hard for me to believe that someone could play on another person's feelings the way he did with me [a previous Internet relationship had not worked, and Robin believes he had not told her the truth about being truly interested in meeting her and being there for her]. But I have accepted the fact that it happened, and I have moved on with my life and met [also online] someone better. The only advice I have for people who are thinking of Internet dating is just be careful. There is a Web site out there where you can have someone's background checked out to see if they are telling you the truth. In the back of my mind I had a feeling he [her previous online date] was lying, but for some reason I didn't want to face the reality of it. . . .

Most men and women in this study took physical and emotional risks to gain trust and were willing to continue seeking online relationships even after others had lied to them. A few teenagers and adults who did not want committed relationships took fewer risks by taking on unrealistic roles, not being open, and postponing face-to-face meetings. Others developed symbolic trust indicators to lessen the consequential risks of interacting.

Indicators of Trust

Berger and Luckmann (1967) believed people decide to trust based on intuitive impressions that we refer to as "trust indicators." This research uncovered the presence of early and late trust indicators as part of early and late negotiating strategies that serve to minimize harm to the self.

The development of trust in an online dating relationship requires not only the assurance that the other means no physical harm but also that the other will treat the online persona with ritual deference. A remark such as, "I did not know you were so large; do you use Photoshop?" would be a devastating blow.[2] This is one of the reasons some Internet daters postpone or evade face-to-face meetings. . . .

Younger respondents were concerned with the hermeneutics of keystrokes and codes. Arlene, a 17-year-old interviewed by the first author, used *LOL* (laughing out loud), *BRB* (be right back), and other abbreviations when chatting. We found younger people used this coded language more frequently than did older individuals. Respondents who were not adept in the use of such codes exposed their lack of grace in social interaction and were weeded out. Participants selected for interactions of usually only a few minutes duration were chosen many times based on one word or the speed of their typing. More mature respondents had different early indicators. Lisa, a 41-year-old, divorced woman, said,

> I don't use chat rooms much anymore. They are filled with a vast bastion of people looking for absolutely nothing. They are "players." They are talking to you while having cybersex with someone else and talking with a third person in another room at the same time. If you get serious, they don't like it. They use romance and dating rooms, sex cams, interest and game rooms, and they chat on the side at the same time. . . .

Chet, a 28-year-old, divorced man, said chat rooms were for mindless, immature people. He used dating services also:

> I look for women who are funny, sarcastic, you know, intellectual, sharp-witted. I can't start a conversation with someone who says she wants to come over and have sex the next day. Or the stereotypic interaction with emphasis on age, hobbies. . . . It's mindless, immature.

Janet, an 18-year-old, female student, said she could tell right away if it was going to work:

> You talk to them. If they answer with one-word sentences . . . if the [online] conversation is really unbalanced, I look and see how much I have said and how much they have said. If I tell them what my field of study is and they don't understand anything at all about it. . . . Most people in chat rooms are uneducated, working class, and just plain dumb. You need to weed them out.

Respondents used indicators contained in e-mailed or posted pictures to help evaluate their potential mates and attempted to determine their age and degree of affluence. Clothing, hairstyle, and projected lifestyle were augured from photographs.

Jessie, a 24-year-old woman, focused on economic status:

> I met this man online in a church chat room. He was from South Africa,
> and he sent me e-mail pictures where he was standing in front of a very
> expensive car. His clothes were expensive-looking, too, and his house was
> like a mansion. He said he was a professional businessman with lots of
> money. He said he wanted to come over here to meet me and my family.
> He had never been in the States before. I told my mother about him.

Other indicators deal with time. Through face-to-face relating, we have
come to expect a certain pattern of flow through which a relationship develops.
This pattern is reflected through the timing of conversation and self-disclosure.
Often on the Internet there is a pressure to disclose much in a short time to
establish trust and kinship quickly. Some respondents dislike this pressure. Julian,
25-year-old salesman, observed,

> Internet people are more desperate; things move fast in weird ways. People
> put pictures up for everyone to see, but you don't know their personal
> mannerisms. Do they smell bad? Have a funny laugh? Do they bite their
> nails? The beginning is different. It [meeting online] sets you off on a weird
> path. You get way too intense too soon. There's like a speed to get to know
> each other. All you have is conversation that becomes exaggerated and
> magnified. It becomes drama. People attach deep meaning and feeling
> prematurely. Feelings get hurt. Self-revelation leads to distortion of the
> picture. One woman I met online said, "I think I am ready for a relation-
> ship now." This scared me. I wanted to just may be have at least one date in
> person and get to know her better before committing to a relationship.

Although this respondent felt it was not a good practice to discuss personal
matters too soon, we observed him doing just that in his second e-mail to a
woman he had just recently met online.

To develop intimacy to create a bond with an online partner, Internet daters felt
pressed to self-disclose as much information as they could in the shortest possible time,
though letting people know one's shortcomings begs rejection. Furthermore, disclos-
ing too much too fast violates social conventions and norms. The woman who told
Julian, "I think I am ready for a relationship now," scared her potential partner away. . . .

Once Internet daters find each other compatible, they move on to the next step of
relationship building. This involves spending more time getting to know one another to
build trust. Basic interpersonal trust is either contractual trust based on social contracts as
in family relationships or trust based on time in relations (Govier, 1992). Most respondents liked the time they spent getting to know each other. They said this time helped
develop trust and intimacy. Robin said it seemed safer to get to know people over time:

> I guess I chose the Internet over meeting someone in a bar or on a blind date
> because to me it felt a little safer. In a bar you are meeting someone and you
> get the impression that they want just a one-night stand and that is it. That is
> not how I was raised. On the Internet you could talk to this person for as
> long as you wanted to before you went ahead and met that person.

Josh, a 56-year-old, never married man, also felt he had developed trust during time spent online:

> I felt I knew her even though we had not met yet. She was not a stranger. We had spoken over the phone and e-mailed over a period of months. I was not afraid at all. It didn't even enter my mind. I didn't have any reason to believe she would be any different in person than she appeared to be. . . .

DISCUSSION

The Internet has opened a new avenue for romantic interaction. In the present study, Internet daters reported being able to reach a larger pool of potential partners and experiencing increased freedom of choice among partners. The Internet also raises new issues of negotiating risk and establishing trust. Respondents said they were willing to take risks to take advantage of the new courting opportunities offered by this new technology. Some risks involved physical danger, and others involved loss of face and possible rejection, though interviewees developed rationales and strategies to deal with these risks to trust that they would have positive experiences.

Dating online modified gendered interactions by allowing women to behave more assertively and men to be more open. It also necessitated the development of new strategies based on keystrokes, codes interpreting online photographs, and reading user profiles to develop trust and confirm compatibility. In Internet interactions, gains and losses are only symbolic, and rejection by an online entity identified only as "suv4" can represent no great material loss. It is this very abstraction that motivates people to use the Internet for dating to avoid stereotyped gender roles and the pain of rejection.

The interrelating of Internet daters also reflects old patterns and problems common to all forms of courtship. Even if they do not find objectification and harassment online, meeting offline often brings objectification or harassment into a formerly nonjudgmental relationship. There is irony in seeking a way out of loneliness through a medium that ensures the insularity of participants and perpetuates gender stereotyping once participants meet.

Several old problems remain in Internet dating. It is easy for people to lie to each other, and appearance issues and shyness do not completely disappear when dating online. Rejection and its emotional pain are ultimately a part of Internet dating as much as of dating that is entirely face to face from the start. The fundamental issues of trust, self-presentation, and compatibility carry over from conventional courtship into its Internet variant.

The need to obtain companionship motivates people to seek out romantic relationships in a variety of ways, and the Internet is merely the latest technological development used by people to assist their romantic goals. Participants in the current study reported reducing their loneliness, obtaining comfort, and finding fun and excitement. These benefits appeared to outweigh the risks.

NOTES

1. Emoticons are small icons bearing emotive faces. These can be inserted into text messages.
2. Adobe Photoshop is a very popular photograph manipulation program that allows users to drastically alter photographs and cinematographic video.

REFERENCES

Berger, P. L., & Luckmann, T. (1967). *The social construction of reality: A treatise in the sociology of knowledge.* New York: Anchor.

Chelune, G. J., Sultan, F. E., & Williams, C. L. (1980). Loneliness, self-disclosure, and interpersonal effectiveness. *Journal of Counseling Psychology, 27,* 462–468.

Clark, L. S. (1998). Dating on the Net: Teens and the rise of "pure" relationships. In S. Jones (Ed.), *Cybersociety 2.0: Revisiting computer-mediated communication and community* (pp. 159–181). Thousand Oaks, CA: Sage.

Cooper, A., & Sportolari, L. (1997). Romance in cyberspace: Understanding online attraction. *Journal of Sex Education and Therapy, 22,* 7–14.

Danet, B. (1998). Text as mask: Gender, play, and performance on the Internet. In S. Jones (Ed), *Cybersociety 2.0: Revisiting computer-mediated communication and community* (pp. 129–157). Thousand Oaks. CA: Sage.

Giddens, A. (1990). *The consequences of modernity.* Stanford, CA: Stanford University Press.

Goffman, E. (1959). *The presentation of self in everyday life.* New York: Doubleday.

Goffman, E. (1967). *Interaction ritual: Essays on face-to-face behavior.* Doubleday.

Govier, T. (1992). Trust, distrust, and feminist theory. *Hypatia, 7,* NI.

Hardey, M. (2002). Life beyond the screen: Embodiment and identify through the Internet. *Sociological Review, 50,* 570–585.

Hunt, M. M. (1959). *The natural history of love.* New York: Knopf.

Jones, W. H., Hobbs, S. A., & Hockenbury, D. (1982). Loneliness and social skills deficits. *Journal of Personality and Social Psychology, 42.* 682–689.

Knox, D., Daniels, V., Sturdivant, L., & Zusman, M. (2001). College student use of the Internet for mate rejection. *College Student Journal, 35,* 158–160.

Mongeau, P. A., Hale, J. L., Johnson, K. L., & Hillis, J. D. (1993). Who's wooing whom? An investigation of female initiated dating. In P. J. Kalbfleisch (Ed.), *Interpersonal communication: Evolving interpersonal relationships* (pp. 51–68). Hillsdale, NJ: Lawrence Erlbaum.

Raney, R. F. (2000, May 11). Study finds Internet of social benefit to users. *New York Times.* p. G7.

Simmel, G. (1911). *Phiosophische kultur: Gesammelte essays* [Philosophical culture: Collected essays] (2nd ed.). Leipzig. Germany: Alfred Kroner.

Turkle, S. (1995). *Life on the screen: Identity in the age of the Internet.* New York: Simon & Schuster.

Waskul, D. (2003). *Self-games and body-play: Personhood in online chat and cybersex*. New York: Peter Lang.

Zuboff, S. (1991). New worlds of computer-meditated work. In J. M. Hepslin (Ed.), *Down to earth sociology: Introductory readings* (6th ed., pp. 476–485). New York: Free Press.

KEY CONCEPTS

gender stereotype   presentation of self   social interaction

DISCUSSION QUESTIONS

1. What assumptions about sexuality are made when interacting in chat rooms and other online communications? What methods do users have for revealing sexual preference?

2. Are the reasons for dating online, as described in this article, different from the reasons for dating face-to-face? Why or why not?

41

Get a Life, Girls

ARIEL LEVY

In this article, Ariel Levy delivers an indictment of the "raunch culture" that encourages women to dress and act in overly sexualized ways. She argues that society has moved from shunning women and girls who express genuine sexuality to a society that encourages the "slutty stereotypes of female sexuality." While some argue that women's expression of sexuality in their dress and in their actions is sexually liberating, Levy argues that it is simply an imitation of sexual power. We are not truly free, simply altering the norm of femininity.

SOURCE: Ariel Levy. 2006. "Get a Life, Girls." *The Spectator* (March 4).

Some version of a sexy, scantily clad temptress has been around through the ages, and there has always been a demand for smut.

But whereas this was once a guilty pleasure on the margins—on the almost entirely male margins—now, strippers, porn stars and Playboy bunnies have gone mainstream, writing bestsellers, starring in reality television shows, living a life we're all encouraged to emulate. Prepubescent girls wear "thong" underpants, their mothers drive off to the gym for pole-dancing classes after lunch.

Last week Anita Roddick, founder of the Body Shop, hit out at what she called "pimp and ho chic." "A lot of people seem to think that it's cool to be a pimp or a ho," she said.

"But it's not cool. The reality is dark, evil, appalling and unregulated. There are thousands of ads, mostly focused on women and young girls, that say you are not attractive, you are not sexy, you are not intelligent unless you look like this. Something has gone very wrong." What Dame Anita called "ho chic" I call "raunch culture," and it's everywhere. Men and women alike have developed a taste for kitschy, slutty stereotypes of female sexuality—we don't even think about it any more, we just expect to see women flashing and stripping and groaning everywhere we look.

Not so long ago the revelation that a woman in the public eye had appeared in any kind of pornography would have destroyed her reputation. Think of Vanessa Williams, crowned the first black Miss America in 1983, and how quickly she was dethroned after her nude photos surfaced in Penthouse. She managed to make a comeback as a singer, but the point is that being exposed as a porn star then was something you needed to come back from. Now, it's the comeback itself.

Paris Hilton was just a normal, blonde New York socialite, an heiress with a taste for table-dancing, before she and her former boyfriend Rick Salomon made a video of themselves having sex. Somehow the footage found its way on to the internet and was distributed worldwide, after which Paris Hilton became one of the most recognizable and marketable female celebrities in the world.

Since the advent of the sex tapes, Hilton has become famous enough to warrant a slew of endorsement deals. There is a Paris Hilton jewelry line (belly-button rings feature prominently), a perfume, and a string of nightclubs called Club Paris set to open in London, New York, Atlanta, Madrid, Miami and Las Vegas. She also has a modelling contract for Guess Jeans, and her book, *Confessions of an Heiress*, was a bestseller.

Her debut CD—the first single is entitled "Screwed"—is about to be released. Paris Hilton isn't some disgraced exile of our society. On the contrary, she has become our mascot, the embodiment of our collective fixations—blondness, hotness, richness, antiintellectualism and exhibitionism.

The rise of raunch culture in the West seems counterintuitive to some. What about conservative values and evangelical Christianity? they wonder. But raunch culture transcends political parties both in America and the UK because the values people vote for are not necessarily the same values they live by. Even if people consider themselves conservative, their political ideals may be a reflection of the way they wish things were, rather than an indication of how they plan to lead their lives.

Raunch culture is not essentially progressive; it is essentially commercial. It isn't about opening our minds to the possibilities and mysteries of sexuality. If we were to acknowledge that sexuality is personal and unique, it would become unwieldy. Making sexiness into something simple and, quantifiable makes it easier to explain and to market. If you remove the human factor from sex and make it about stuff—big fake boobs, bleached blonde hair, long nails, poles, thongs—then you can sell it. Suddenly, sex requires shopping: you need plastic surgery, peroxide, a manicure, a mall.

There is a disconnection between sexiness, or "hotness," and sex itself. As Paris Hilton told *Rolling Stone*, "My boyfriends always tell me I'm not sexual. Sexy, but not sexual." And any 14-year-old who has downloaded her sex tapes can tell you that Hilton looks excited when she is posing for the camera, bored when she is engaged in actual sex. (In one tape, Hilton took a cellphone call during intercourse.) She is the perfect sexual celebrity for this moment, because our interest is in the appearance of sexiness, not the existence of sexual pleasure.

Passion isn't the point. The glossy, overheated thumping of sexuality in our culture is less about "connection" than consumption.

"Hotness" has become our cultural currency, and a lot of people spend a lot of time and a lot of money trying to acquire it. Hotness is not the same thing as beauty, which has been valued throughout history. Hot can mean popular. Hot can mean talked about. . . .

These are the literal job criteria for our role models: strippers and porn stars. These are women whose profession is based on faking lust, imitating actual female sexual pleasure and power. If we're all trying to look like porn stars these days (and we are), we're imitating an imitation of arousal. It's a long way from sexual liberation.

This is not a situation foisted upon women. In the West, in the 21st century, we have opportunities and expectations that our mothers never had. We have attained a degree of hard-won (and still threatened) freedom in our personal lives; we are gradually penetrating the highest levels of the workforce; we get to go to college and play sports and be secretaries of state. But to look around, you'd think that all any of us women want to do is to rip off our clothes and shake our butts in men's faces.

So why do we go in for raunch culture?

Why, when the feminist movement was supposed to have freed us from stereotypes, have we deliberately embraced them again?

The freedom to be sexually provocative or promiscuous is not enough freedom; it is not the only "women's issue" worth paying attention to. And we are not even free in the sexual arena. We have simply adopted a new norm, a new role to play: lusty, busty exhibitionist. There are other choices. It's ironic that we call this "adult" entertainment, when reducing sexuality to implants and polyester underpants is really pretty adolescent.

KEY CONCEPTS

| consumerism | culture | sexual revolution |

DISCUSSION QUESTIONS

1. What articles of clothing do you own or see others wear that remind you of the "raunch culture"? How are these different from the fashion trend of five or ten years ago?

2. How is the culture of overly sexualized dress and behavior limiting for women? Is there harm that comes from this?

42

The Long Goodbye

DIANE VAUGHAN

Sociologists typically explain how social relationships are formed in society, but here sociologist Diane Vaughan twists the sociological imagination to ask how relationships end. As she shows, there are discernible patterns in social behavior that are the steps to "breaking up."

One week before the wedding, the bride-to-be had dreamed that for the ceremony she would be dressed all in white—except for black shoes. She interpreted this as a warning, and on the second day of the honeymoon her worst fears began to come true. A disagreement turned into a heated argument during which the husband threw his wedding ring across the room and then punched his new bride in the nose. Before leaving for the hospital, the woman searched for the ring and put it back on her husband's finger. This was the first of a long string of similar episodes. Six years later, while they were walking on a snowy night in Japan, the same thing happened. The woman spent three hours crawling in the snow looking for the ring.

The woman's search for the ring in the snow symbolizes the importance we place on our relationships. They can become so important that we hang on even when they no longer make us happy. We even come up with plausible reasons for doing so: We believe in commitment, we feel bound by the law, we don't want to hurt the other person. We are afraid there may be no one better out there, we believe in our ability to fix things, we are not quitters. We do it for the children, for our parents, for God.

SOURCE: Vaughan, Diane. 2006. "The Long Goodbye." *Psychology Today* 21 (July): 36–39.

Given all these constraints, it's a wonder couples ever manage to uncouple, but they do. The question is how.

In an attempt to find out, I collected case histories, interviewed counselors and sat in on group sessions for separated and divorced people. For each person, the experience was unique, but I found that no matter what type of couple was involved—married or living together, straight or gay, young or old—there was a discernible pattern in each breakup.

In most cases, one person—the initiator—wants out while the partner wants the relationship to continue. Typically, by the time the partner realizes the relationship is going down the drain, the initiator is already so far gone that efforts to save the relationship are futile.

Partners often report that they were unaware or only remotely aware, even at the time of separation, that the relationship was deteriorating. Initiators tell a different story, reporting months or even years of trying to get the other person's attention.

This apparent breakdown in communication occurs because the closeness of relationships sometimes encourages secrecy rather than openness. In the beginning, each person is so intent on discovering everything about the other person that they develop a sensitivity that allows them to pick up on the smallest cues. They are tuned in to each other and constantly explore and test the nature of the relationship. But once the relationship seems secure, couples often replace the intense energy-consuming monitoring of the early days with a style of communication that suppresses rather than reveals information. Instead of attentively probing, couples confirm that the relationship is healthy by using familiar cues and signals: the words "I love you," holidays with the family, good sex, good times with friends, the routine exchange of "How was your day?"

When a close relationship begins to fall apart, this shorthand method of testing the solidity of the union obscures the changes that are taking place and obstructs the sending and receiving of new information. People may even use these familiar cues and signals to avoid confrontation and to hide their true feelings. Taking this easy way out, however, can have a devastating effect on the relationship. Problems can't be resolved until both parties admit that the relationship is in trouble. And the longer one person's unhappiness remains unacknowledged, the more time he or she has to back off from the relationship.

Dissatisfied partners tend to keep their unhappiness secret at first as they contemplate, brood on and quietly assess the situation. This allows them to continue participating in routine aspects of life with the other person. As unhappiness grows, they begin to reveal it but often in a vague and ill-defined manner.

These early attempts to communicate dissatisfaction usually take the form of complaints aimed at changing the other person or the relationship to better suit the needs of the initiator. He or she may complain, for example, about how the other person spends leisure time. "Why do you watch TV all the time? Why don't you turn it off and do something else?" But television isn't usually the issue. The initiator may be questioning the partner's level of commitment or the appropriateness of a relationship with a person who has nothing better to do than watch the tube.

Such complaints fail to communicate that the relationship is the real problem. At this stage, the initiator may not be able to articulate the deeper reasons for his or her unhappiness. The partner, trying to be cooperative, may even turn off the television, thus eliminating the symptom but not the problem.

Instead of complaining, initiators may use sullenness, anger and a decrease in intimacy to suggest unhappiness. Having always remembered to buy a birthday present, they may bring an inappropriate or a thoughtless gift. The reporting of daily events,usually exchanged at dinner,may get briefer and briefer. When such subtle signals fail, initiators may then turn to interests outside the relationship—a hobby, graduate school, the children, friends. In a healthy relationship, outside interests can add to the excitement and well-being of the situation because they are shared in ways that strengthen ties. In a deteriorating relationship, these activities exclude the partner and widen the already-open gap.

When initiators turn away from the relationship and seek solace in others, they often choose people they believe will be sympathetic to their complaints: a relative, a close friend or work associate, a therapist, a lover. By sharing their unhappiness with others, they create bonds and begin (still unintentionally) to prepare for life without the partner.

As outside bonds grow stronger and more important, the thought of breaking up becomes stronger. Initiators often choose confidants who have gone through a breakup or are in the midst of one—people who can provide relevant information and insights. As initiators gather information, they apply it to their own situation, weighing the costs and benefits of leaving the relationship: Can they make it on their own, will the partner get violent, how will friends react, will the children suffer, is there anyone better out there, what about loneliness?

Even at this advanced stage, initiators are likely to communicate their unhappiness indirectly. They may spend less time with the partner, complain about mundane failings, engage in perfunctory sex. They withhold signals that would be sure to get the partner's attention. Their behavior suggests,"I'm unhappy,"but not,"I'm unhappy,and I'm seeing someone else."

Initiators tend to avoid direct confrontation for several reasons. They may be uncertain about their ability to leave the relationship. Many cannot bear to hurt their partner. Some hide their unhappiness to protect themselves from arguments and possible retaliation. Others want out of the relationship but don't want to lose the partner completely.

The signals become increasingly bold, however, as initiators gradually come to see the relationship as unsavable. As outside interests grow more important, they devote less and less time to restoring the relationship. They express discontent to convince the partner that the relationship is not working, rather than to change or improve things.

But even these intensified signals fail to get attention because they are buried in daily routine. A serious airing of feelings, for example, may be interpreted as a sign of some deeper problem at first but as time passes, be seen as only a momentary upset. Or the signals themselves may become routine: What begins as a break in the pattern becomes the pattern. Arguments, working late at the office or even threats to leave lose their impact with repetition.

Even strong, repeated signals of unhappiness may be so inconsistent with the partner's conception of the relationship that they are denied or reinterpreted rather than confronted. Partners may say, for example, "All relationships have trouble. Ours wouldn't be normal if we didn't." "After a while, all couples lose interest in sex." Or partners may interpret negative signals as nothing more than a temporary aberration. When the pressure of work lets up, when the midlife crisis is over, when the children are all in school, when the parents are told that the liaison is gay—then the couple will be happy again.

Ironically, though they are separated in many ways, both members of the relationship remain full-fledged partners in one endeavor: suppressing the true status of the relationship. Even when signals grow so bold that the partner begins to question the initiator, the questions are asked in gentle ways. Nobody wants to discover that a valued way of life is ending. So the partner, too, avoids direct confrontation and gets to hang on a while longer. But by helping the initiator keep the status of the relationship a secret, the partner is missing a chance to do something to solve the problem.

Only when initiators have created what seems to be a secure niche elsewhere and when the costs of staying in the relationship outweigh the benefits will they reveal all to their partners. Until then, they continue to tell without telling. The partners who are about to be left behind will acknowledge the negative signals and confront the problem directly only when the situation becomes so costly in terms of emotional energy, tension and human dignity that they have more to lose by suppression than by revelation. Until then they will continue to know without knowing. It's this conspiracy of secrecy that makes breaking up so very, very hard to do.

KEY CONCEPTS

dyad   sexual orientation

DISCUSSION QUESTIONS

1. How might social factors like gender, the amount of money each partner makes, the presence of children, and so forth influence the process of breaking up that Vaughan describes?

2. People usually think of relationships as ending because of problems between individuals, but what external social forces exert pressure on couples that might influence the stability of relationships? How have such social forces changed in recent years and does this increase or decrease the stability of coupled relationships?

43

The Impact of Multiple Marginalization

PAULA RUST

The idea of multiple marginalization, as described in this article, refers to individuals who are outside the dominant culture because of their race-ethnicity and their sexuality. The author describes many homosexual and bisexual people who are also Asian, Latino, Black, and other non-White race-ethnicities. For many, their sexual identity is specifically problematic for their culture. Unlike White Americans who face marginalization from the heterosexual community but find support among other bisexuals, bisexual people of color face cultural rejection on many levels.

One's sexuality is affected not only by the sexual norms of one's culture of origin but also by the position of one's culture of origin vis-à-vis the dominant culture of the United States. For individuals who belong to marginalized racial-ethnic, religious, or socioeconomic groups, the effects are numerous. Marginalized groups sometimes adopt the attitudes of the mainstream; other times, they reject these attitudes as foreign or inapplicable. McKeon (1992) notes that both processes shape the sexual attitudes of the white working class. On the one hand, the working class absorbs the homophobic attitudes promoted by the middle- and upper-class controlled media. At the same time, working-class individuals are rarely exposed to "liberal concepts of tolerance" taught in institutions of higher education which help moderate overclass heterosexism. On the other hand, working-class sexual norms are less centered around the middle-class notion of "propriety"—a value that working-class individuals cannot as readily afford. The result is a set of sexual norms that differs in complex ways from those facing middle- and upper-class bisexuals.

In marginalized racial and ethnic groups, racism interacts with cultural monosexism and heterosexism in many ways. In general, the fact of racism strengthens ethnic communities' desires to preserve ethnic values and traditions, because ethnicity is embodied and demonstrated via the preservation of these values and traditions. Tremble, Schneider, and Appathurai (1989) wrote, "After all, one can abandon traditional values in Portugal and still be Portuguese. If they are abandoned in the New World, the result is assimilation" (p. 225). Thus, ethnic

SOURCE: Paula Rust. 1996. "The Impact of Multiple Marginalization." In *Bisexuality: The Psychology and Politics of an Invisible Minority,* edited by Beth A. Firestein. Thousand Oaks, CA: Sage.

minorities might cling even more tenaciously to traditional cultures than Euro-Americans do, because any cultural change reflects not a change in ethnic culture but a loss of ethnic culture. To the extent that ethnic values and traditions restrict sexual expression to heterosexuality, ethnic minority bisexuals will be under particular pressure to deny same-sex feelings in demonstration of ethnic loyalty and pride. Attempts to challenge these values and traditions by coming out as bisexual will be interpreted as a challenge to ethnic culture and identity in general.

Because homosexuality represents assimilation, it is stigmatized as a "white disease" or, at least, a "white phenomenon." Individuals who claim a bisexual, lesbian, or gay identity are accused of buying into white culture and thereby becoming traitors to their own racial or ethnic group. Previous researchers have found the attitude that lesbian or gay identity is a white thing among African-Americans and Hispanics and the attitude that homosexuality is a "Western" behavior among Asian-Americans (Chan 1989; Espin 1987; H. 1989; Icard 1986; Matteson 1994; Morales 1989). In the current study, the association of gayness with whiteness was reported most often by African-American respondents. One African-American woman wrote that "when I came out, it was made clear to me that my being queer was in some sense a betrayal of my 'blackness.' Black women just didn't do 'these' kinds of things. I spent a lot of years thinking that I could not be me and be 'really' black too." Morales (1990) found that Hispanic men choose to identify as bisexual even if they are exclusively homosexual, because they see gay identity as representing "a white gay political movement rather than a sexual orientation or lifestyle" (p. 215). A Mexican woman in the current study wrote that she has "felt like . . . a traitor to my race when I acknowledge my love of women. I have felt like I've bought into the White 'disease' of lesbianism." A Puerto Rican woman reported that in Puerto Rico homosexuality is considered an import from the continental States. Chan (1989) found that Asian-Americans tend to deny the existence of gays within the Asian-American community, Wooden et al. (1983) reported this attitude among Japanese Americans, and Carrier et al. (1992) found denial of the existence of homosexuality among Vietnamese-Americans who considered homosexuality the result of seduction by Anglo-Americans. Tremble et al. (1989) suggested that viewing homosexuality as a white phenomenon might permit ethnic minority families to accept their LesBiGay members, while transferring guilt from themselves to the dominant society.

Ironically, whereas racism can strengthen commitment to ethnic values and traditions, it can also pressure ethnic minorities to conform to mainstream values in an effort to gain acceptance from culturally dominant groups. Because members of ethnic minorities are often perceived by Euro-Americans as representatives of their entire ethnic group, the nonconformist behavior of one individual reflects negatively on the whole ethnic group. For example, African-American respondents reported that homosexuality is considered shameful for the African-American community because it reflects badly on the whole African-American community in the eyes of Euro-Americans. A similar phenomenon exists among lesbians and gays, some of whom chastise their more flamboyant members with "How can you expect heterosexual society to accept us when you act like *that!*?"

As one Black bisexual woman put it, "Homosexuality is frowned upon in the black community more than in the white community. It's as if I'm shaming the community that is crying so hard to be accepted by the white community."

The fact of ethnic oppression also interacts with particular elements of ethnic minority culture in ways that affect bisexuals. Specifically, the emphasis on the family found in many ethnic minority cultures is magnified by ethnic oppression in two ways. First, oppression reinforces the prescription to marry and have children among minorities which, for historical reasons, fear racial genocide (Greene 1994; Icard 1986). Second, the fact of racism makes the support of one's family even more important for ethnic minority individuals. As Morales (1989) put it, the "nuclear and extended family plays a key role and constitutes a symbol of their ethnic roots and the focal point of their ethnic identity" (p. 225). Ethnic minority individuals learn techniques for coping with racism and maintaining a positive ethnic identity from their families and ethnic communities; to lose the support of this family and community would mean losing an important source of strength in the face of the ethnic hostility of mainstream society (Amaguer 1993; Chan 1992; Icard 1986). Thus, ethnic minority bisexuals have more to lose if they are rejected by their families than do Euro-American bisexuals. At the same time, they have less to gain because of the racism of the predominantly Euro-American LesBiGay community. Whereas Euro-American bisexuals who lose the support of their families can count on receiving support from the LesBiGay community instead (albeit limited by the monosexism of that community), ethnic minority bisexuals cannot be assured of this alternative source of support.

Because of fear of rejection within their own racial, ethnic, or class communities, many bisexuals—like lesbians and gay men—remain in the closet among people who share their racial, ethnic, and class backgrounds. For example, an African-American-Chicana "decided to stay in the closet instead of risk isolation and alienation from my communities." Sometimes, individuals who remain closeted in their own racial-ethnic or class communities participate in the mainstream lesbian, gay, and bisexual community, which is primarily a Euro-American middle-class lesbian and gay community. Such individuals have to juggle two lives in two different communities, each of which is a valuable source of support for one aspect of their identity, but neither of which accepts them completely. Among people of their own racial, ethnic, or class background, they are not accepted and often not known as bisexuals, and among Euro-American lesbians and gays, they encounter both monosexism and class and racial prejudice or, at the least, a lack of support and understanding for the particular issues that arise for them because of their race, ethnicity, or class. Simultaneously, like other members of their racial, ethnic, or class community, they have to be familiar enough with mainstream Euro-American heterosexual culture to navigate daily life as a racial or ethnic minority; so they are, in effect, tricultural (Lukes and Land 1990; Matteson 1994; Morales 1989). This situation leads not only to a complex social life but might also promote a fractured sense of self, in which one separates one's sexual identity from one's racial identity from one's American identity and experiences these identities as being in conflict with each other, just as are the communities that support each identity. . . .

A positive integration of one's racial, ethnic, or class identity with one's sexual identity is greatly facilitated by support from others who share an individual's particular constellation of identities. For some, finding kindred spirits is made difficult by demographic and cultural realities. But as more and more people come out, there are inevitably more "out" members of racial and ethnic minorities and among these, more bisexuals. Many respondents described the leap forward in the development of their sexual identities that became possible when they finally discovered a community of bisexuals, lesbians, or gays with a similar racial or ethnic background. A Jewish-Chicana reported that she is "finding more people of my ethnic backgrounds going through the same thing. This is affirming." Similarly, a Chicano is "just now starting to integrate my sexuality and my culture by getting to know other gays/bis of color." An African-American woman reported that "it wasn't until I lived in Washington, D.C., for a number of years and met large numbers of Black lesbians that I was able to resolve this conflict for myself." Many Jewish respondents commented on the fact that there are many Jewish bisexuals, lesbians, and gays, and noted that receiving support from these peers was important in the development and maintenance of their positive sexual identities. One man, when asked to describe the effect of his racial or ethnic cultural heritage on his sexuality, said simply "I'm a Jewish-Agnostic Male-oriented Bisexual. There are lots of us." Some Jewish respondents also commented that being racially white facilitated their acceptance in the mainstream LesBiGay community and permitted them to receive support from this community that was not as available to individuals of other racial and ethnic backgrounds. Of course, it is this same assimilationist attitude that caused the Orthodox Jewish woman quoted earlier to find a lack of support among Jewish LesBiGays.

For individuals who belong to racial or ethnic minorities, the discovery that one is bisexual is a discovery that one is a double or triple minority. It is even more the case for bisexuals than for lesbians and gays, because bisexuals are a political and social minority within the lesbian and gay community. Many racial and ethnic minority individuals experience their coming out as a process of further marginalization from the mainstream that is, as an exacerbation of an already undesirable position. An African-American woman described being bi as "just one other negative thing I have to deal with. My race is one and my gender another." This can inhibit coming out for individuals who are reluctant to take on yet another stigmatized identity. For example, Morales (1990) reported that some Hispanic men limit their coming out, because they do not want to risk experiencing double discrimination in their careers and personal lives. A Black woman in the current study said that she is "unwilling to come too far 'out' as I already have so many strikes against me."

Many respondents found, however, that their experiences as racial or ethnic minorities facilitated their recognition and acceptance of their sexuality. This was most common among Jewish respondents, many of whom explained that their history as an oppressed people sensitized them to other issues of oppression. One man wrote, "The Jewish sense of being an outsider or underdog has spurred my rebelliousness; the emphasis on learning and questioning has helped to open my mind." A woman wrote, "My Jewish ethnicity taught me about oppression and the need to fight it. It gave me the tools to be able to assert that the homophobes (like the anti-Semites) are wrong." Some non-Jewish respondents also found that their experiences as ethnic minorities facilitated their coming out as bisexual. For example, a woman of Mexican, Dutch, and Norwegian descent wrote that her cultural background "has made me less afraid to be different." She was already ethnically different, so she was better prepared to recognize and accept her sexual difference. . . .

Many bisexuals of mixed race or ethnicity feel a comfortable resonance between their mixed heritage and their bisexuality. In a society where both racial-ethnic and sexual categories are highly elaborated, individuals of mixed heritage or who are bisexual find themselves straddling categories that are socially constructed as distinct from one another. The paradox presented by this position was described by a bisexual woman of Native-American, Jewish, and Celtic heritage who wrote, "Because I am of mixed ethnicity, I rotate between feeling 'left out' of every group and feeling 'secretly' qualified for several racial/cultural identities. I notice the same feeling regarding my sexual identity." Other respondents of mixed racial and ethnic backgrounds also saw connections between their ethnic heritage and their bisexuality. For example, an Asian-European woman wrote,

> Being multiracial, multicultural has always made me aware of nonbipolar thinking. I have always been outside people's categories, and so it wasn't such a big leap to come out as bi, after spending years explaining my [racial and cultural] identity rather than attaching a single label [to it].

A Puerto Rican who grew up alternately in Puerto Rico and a northeastern state explained,

> The duality of my cultural upbringing goes hand in hand with the duality of my sexuality. Having the best of both worlds (ethnically speaking—I look white but am Spanish) in my everyday life might have influenced me to seek the best of both worlds in my sexual life— relationships with both a man and a woman.

A Black Lithuanian Irish Scottish woman with light skin, freckles, red curly hair, and a "Black political identity," who is only recognized as Black by other Blacks, wrote, "As with my race, my sex is not to be defined by others or absoluted by myself. It is a spectrum."

However, individuals whose mixed heritages have produced unresolved cultural difficulties sometimes transfer these difficulties to their bisexuality. A "Latino-Anglo" who was raised to be a "regular, middle-class, all-American," and who later became acculturated to Latin culture, wrote,

> Since I am ethnically confused and pass as different from what I am, as I do in sexual orientation also, I spend a lot of time underground. . . .
> I think it has definitely been a major factor in the breakup of two very promising long-term relations.

Similarly, a transgendered bisexual respondent of mixed European, Native-American, and North African heritage believes that the pressures she feels as a transgenderist and a bisexual are closely related to the fact that her parents "felt it necessary to hide a large part of their ethnic and racial heritage," although she did not elaborate on the nature of these pressures.

In contrast to bisexuals from marginalized racial-ethnic, religious, or class backgrounds, middle- or upper-class Protestant Euro-Americans experience relatively few difficulties integrating their sexual identities with their cultural backgrounds and other identities. Euro-American bisexuals might have difficulty developing a positive bisexual identity in a monosexist culture, but unlike Bisexuals of Color, they have no particular problems integrating their sexual identity with their racial identity, because these identities are already integrated in the LesBiGay community. Being Euro-American gives them the luxury of not dealing with racial identity. Not surprisingly, when asked how their racial-ethnic background had affected their sexuality, most Euro-Americans did not mention their race at all. Instead, Euro-Americans tended to attribute their sexual upbringing to the peculiarities of their parents, their religion, their class, or their geographic location within the United States. One woman explained,

> I do not associate my racial-ethnic cultural background and my sexuality. Undoubtedly I would think and feel differently if I were of a different background but I'm not able to identify the effect of my background on my sexuality.

REFERENCES

Almaguer, T. (1993). Chicano men: A cartography of homosexual identity and behavior. In H. Abelove, M.A. Barale, & D.M. Halperin (Eds.), *The lesbian and gay studies reader*. New York: Routledge.

Carrier, J., Nguyen, B., & Su, S. (1992). Vietnamese American sexual behaviors and HIV infection. *Journal of Sex Research, 29*(4), 547–560.

Chan, C.S. (1989). Issues of identity development among Asian American lesbians and gay men. *Journal of Counseling and Development, 68*(1), 16–21.

Chan, C.S. (1992). Cultural considerations in counseling Asian American lesbians and gay men. In S. H. Dworkin & F. Guitérrez (Eds.), *Counseling gay men and leshians* (pp. 115–124). Alexandria, VA: American Association for Counseling and Development.

Espin, O. (1987). Issues of identity in the psychology of Latina lesbians. In Boston Lesbian Psychologies Collective (Eds.), *Lesbian psychologies: Explorations and challenges* (pp. 35–51). Urbana: University of Illinois Press.

Greene, B. (1994). Ethnic-minority lesbians and gay men: Mental health and treatment issues. *Journal of Consulting and Clinical Psychology, 62*(2), 243–251.

H., P. (1989). Asian American lesbians: An emerging voice in the Asian American community. In Asian Women United of California (Eds.), *Making waves: An anthology of writings by and about Asian American women* (pp. 282–290). Boston: Beacon.

Icard, L. (1986). Black gay men and conflicting social identities: Sexual orientation versus racial identity. *Journal of Social and Human Sexuality,* 4(1/2), 83–92.

Lukes, C.A., and Land, H. (1990, March). Biculturality and homosexuality. *Social Work,* 155–161.

Matteson, D.R. (1994). *Bisexual behavior and AIDS risk among some Asian American men.* Unpublished manuscript.

McKeon, E. (1992). To be bisexual and underclass. In E.R. Weise (Ed.), *Closer to home: Bisexuality & feminism* (pp. 27–34), Seattle, WA: Seal.

Morales, E.S. (1989). Ethnic minority families and minority gays and lesbians. *Marriage and Family Review,* 14(3/4), 217–239.

Tremble, B., Schneider, M., & Appathurai, C.(1989). Growing up gay or lesbian in a multicultural context. *Journal of Homosexuality,* 17(1–4), 253–267.

Wooden, W.S., Kawasaki, H., & Mayeda, R. (1983). Lifestyles and identity maintenance among gay Japanese American males. *Alternative Lifestyles,* 5(4), 236–243.

KEY CONCEPTS

bisexuality   heterosexism   heterosexuality

DISCUSSION QUESTIONS

1. What does Rust mean by multiple marginalization? What consequences are there for the individual when marginalized from multiple groups?

2. In what ways is bisexuality or homosexuality problematic for racial-ethnic minority cultures? Why do some cultures see bisexuality as a "White disease"?

Applying Sociological Knowledge:
An Exercise for Students

Sexuality is generally thought to be a private matter, and yet public social norms very much shape and regulate sexuality. For a period of one week, observe and keep track of every comment you hear that seems to enforce certain sexual scripts. What have you heard? What assumptions does everyday talk make about heterosexuality? About gays and lesbians? How does everyday public talk shape social norms about sexuality?

44

Weaving Work and Motherhood

ANITA GAREY

This article examines the connection between family and work, as experienced in the lives of mothers. Garey argues that an "orientation model" of balancing work and family permeates representations of working mothers in popular culture and academic studies. She suggests instead that work and motherhood are no longer separate forces but are woven together in mothers' lives.

I grew up in the 1950s, but my 1950s was not the one I read about, many years later, in sociology texts, where families were nuclear, fathers were the sole breadwinners, and mothers stayed out of the labor force and were called housewives. And because children assume that the world they know represents the way the world is, my 1950s family was my norm. And in my 1950s, mothers were employed.

My grandmother, my mother, my aunts, and the woman down the street all held jobs. They weren't always full-time jobs, or year-round jobs, or day jobs (although my grandmother sometimes held two jobs), and they certainly weren't "careers," but the women I saw around me were employed. Everywhere I went I saw employed women: my elementary-school teachers, the receptionist and the assistant in my dentist's office, the nurse at the clinic, the beautician who did my grandmother's hair, the grocery clerks at the A&P, the "cafeteria ladies" at school, bank tellers, the school secretary, the salesclerks at Woolworth's and at the candy store, the ticket-seller at the movie theater, the saleswomen in the clothing department of Sears, the "Avon ladies" who came to the door, and the voices of the telephone operators. I had no doubt that women, including mothers, *worked;* I would have been surprised to find that they didn't.

What I've discovered since then is that not all work counts *as work,* in discussions of "working mothers." As a child, I had not yet learned that what counts as *real* work is full-time (forty hours or more), day-shift, year-round employment in a defined occupation. I had not yet learned how not to see the employment of large numbers of women, many of them mothers. This disjuncture between my experience of women and employment and what I read about the 1950s family sensitized me to the missing stories in generalizations about families, mothers, and employment. . . .

Since the 1950s, there has been a steady increase in the proportion of women in the formal labor force. The major part of this increase has been the result of

SOURCE: Garey, Anita. *Weaving Work and Motherhood.* Temple University Press, Philadelphia, 1999. Reprinted with permission.

married women entering the labor force, and this has been true across racial-ethnic groups.... Two dramatic changes have drawn significant attention to the topic of "working mothers": the increase in the percentage who are employed of mothers with children under six years of age ... and the increase in the percentage of employed mothers not working in home-based employment....

Although more than two-thirds (70 percent) of all married mothers with children under the age of eighteen are in the labor force, scholarly and popular attention to the topic of working mothers has been for the most part narrowly focused on the small percentage of those mothers employed in managerial or professional positions. While there are a number of excellent studies of women employed in blue-collar or pink-collar jobs, sociological studies of work and family use primarily a dual-career family model that focuses on elite and restrictive careers, despite the fact that this does not reflect the experiences of most women, particularly those of most married women with children.... The general image of the "working mother" in popular magazines or in most scholarly discussions of work and family is of the professional or corporate woman, briefcase in hand. The women I saw around me when I was growing up do not fit this image....

While inroads have been made by women, including women with children, into male-dominated, professional occupations, most employed women are concentrated in particular categories of work. In 1990, more women were employed as secretaries than as any other single occupation.... Fifty-two percent of employed women in 1995 were in nonsupervisory positions in sales, service, and secretarial occupations.... Of all employed women in 1995, for example, only 0.3 percent were physicians, 0.4 percent were lawyers and 0.7 percent were college and university professors.... There is a perception that mothers are employed in managerial and professional positions in far larger numbers than is the case, because the majority of employed mothers are missing from cultural images and social analyses of working mothers....

THE ORIENTATION MODEL OF WORK AND FAMILY

... Although there is great variation by race-ethnicity and class in the work experiences of women in the United States, most discussions of women, work, and family are embedded in a conceptual frame that I have termed the orientation model of work and family.[*] For women in the United States, employment and family have been portrayed

---

[*] Patricia Hill Collins argues that, "in contrast to the cult of true womanhood where work is defined as being in opposition to and incompatible with motherhood, work for Black women has been an important and valued dimension of Afrocentric definitions of Black motherhood" (Collins 1987:5). However, the dominant culture in the United States continues to define employment and motherhood as incompatible. So while that combination may be valued within the African-American community, it is likely to be defined as a problem from outside that community. I am talking here about the conceptual frameworks that shape the images we see in popular culture and about the frameworks used in scholarship and policy-making.

dichotomously—and women are described as being either "work oriented" or "family oriented." These concepts are not similarly linked for men. To be a "family man" not only includes but necessitates, providing economically for one's family . . . , and while some men may be referred to as workaholics, this term is reserved for those perceived to be extreme in the time they give to their employment. For men, employment and family are not portrayed as inevitably detracting from one another. For women, work and family are represented as oppositional arenas that have a zero-sum relationship. In this representation, the more a woman is said to be oriented to her work (employment), the less she is seen as oriented to her family. Regardless of the experience of work and family for individual women or within particular groups of women, *the dominant cultural portrayal of work and family for women in the United States classifies women as either work oriented or family oriented*. . . .

The orientation model of work and family, which has framed most sociological analyses of work and family, fits into a larger ideological framework of separate spheres. The ideology of separate spheres divides the social world into two mutually exclusive areas: the public realm of economic and civic life and the private realm of domestic life. This ideology relegates women to the domestic sphere and men to the public arena. . . . It is not surprising that this ideology emerged in England and the United States in the wake of the industrial revolution, when production moved out of households and into factories. With respect to family life, this meant that men, as fathers, were expected to go outside the home to work in the public world, while women, as mothers, were expected to stay home and be in charge of the domestic world of home and children. . . . Phrases like "breadwinner father" and "stay-at-home mom" epitomize the ideology of separate spheres. But when mothers are employed, the model cannot accommodate them very well; thus employed mothers are described in terms of divided relationships to arenas that are seen as separate and oppositional.

In the same way, conceptualizing women's involvement with employment and family in terms of an orientation obscures the integration and connectedness of that involvement. For example, women who postpone having children until they complete their education, or are established in their occupation, or have saved a certain amount of money are acting in ways connected to *both* family and employment considerations, but they would be categorized as work oriented in an orientation model. Women who work non-day shifts or part-time schedules are also attempting to mesh work and family, but anyone using the orientation model would categorize these women as family oriented. The orientation model of work and family is primarily a behaviorist model that categorizes observable behavior, including what people say, without considering the meaning and context of the behavior to the actors. . . .

"WORKING MOTHERS": MAKING SENSE OF OPPOSITIONAL IMAGES

. . . What does it mean to be a working mother, given that work and family are portrayed as conflicting with each other? How do employed women with children think about their lives and about the ways in which work and motherhood fit into those lives? What does it mean to be a worker with children—and a mother who works?

Basically, a working mother is an employed woman who is also responsible for and in a parental relationship with one or more children. But clearly more is contained in the term "working mother" than this simple definition. We don't, for example, have an equivalent term, "working father," for men—even though most men are employed and most men have or will have children. "Working mother" is a conceptual category meant to encompass the relationship for women between being employed and being a parent, but since that relationship is presented as oppositional, it is a conceptual category that often creates more confusion than clarity.

When I interviewed Danielle, a thirty-six-year-old, Euro-American, full-time clerical worker and the mother of a five-year-old child, I told her that I was interested in what being a working mother meant to women who were employed and who were mothers. Danielle's response illustrates the confusion that can occur in the attempt to get at the meaning of "working mother":

> I don't think that means anything. Not to me. I have a real problem with that phrase, because it implies that women who are at home don't work; it also— men who are parents are never called working fathers [exasperated laugh]. And I guess, from that, men who are parents work and a lot of women who are parents work and that's just, that's what *is*. And it's just this nice phrase that can be attached to one group of those parents—but anyway, I'm trying to think if I would describe myself as a working mother. I don't think I would! I think I'd have to take a few sentences to describe myself. I mean that's just too easy a phrase . . . and it doesn't have a lot of meaning.

This employed mother finds that the term "working mother" doesn't fit her experience; it is a gendered and asymmetrical conceptual category. She wonders what it says about her relationship to work if men are not called working fathers and what it says about the work of mothering if the term "mothers" is qualified by "working" when mothers are employed. She does not, however, find available alternative terms or concepts that simultaneously capture her identity as a mother and an employed person; as she says, "I think I'd have to take a few sentences to describe myself."

Why does Danielle have so much trouble using the term "working mother" or finding terms that would articulate her identity as both a worker and a mother? I suggest that the difficulty stems from the fact that the term "working mother" juxtaposes two words with antithetical cultural images: worker/mother; provider/homemaker; public/private. . . .

. . . The very construction of the term "working mother" points our attention in particular directions: "mother" is a noun, and "working" modifies "mother." We do not say "mothering worker," which conjures up an image of a nurturant employee; nor do we say "mother worker," which sounds more like a job category for nannies. Clearly, when one goes from being a "working woman" to being a "working mother," it is "mother" that, linguistically, stands for the essential self. It is the mother who works, not the worker who has children. It is the mother who must fit into the workplace, not the workplace that must adjust to the needs of workers with children. When workers do adjust their work to their family responsibilities, their actions are interpreted as the actions of their selves *as mothers,* not of their selves *as workers.* It is "mother" that is an identity, "working" that is an activity; it is "mother" that is *being,* "working" that is *doing.*

On the one hand, when we conceptualize mothers as being rather than doing, the work of mothering is hidden. I mean not only the work of maintaining children, although that is certainly a part of the work mothers do, but also all of the ways of acting in relation to one's children that constitute not only the expected norm, but also the actual practice of a great many people. On the other hand, by directing our attention to working as *an activity,* we divert our attention away from what it means to women, including mothers, to *be* workers. Everett Hughes ... notes that "a man's work is one of the more important parts of his social identity, of his self, indeed of his fate, in the one life he has to live" .... Whether Hughes meant men or humankind, it has been shown that work is also an important part of the social identity or being of women, including mothers.... Therefore, instead of the orientation model of work and family, we need a framework that makes sense of both the experiences of employed women with children and the experience of *being* employed woman with children.

WEAVING: AN ALTERNATIVE FRAMEWORK

... I suggest that we use the metaphor of weaving as a way to look at the lives of employed women with children. Weaving is both a process (an activity—to weave something) and a product (an object—a weaving, something constituted from available materials), and I use the image in both senses. As a process, weaving is a conscious, creative act. It requires not only vision and planning, but also the ability to improvise when materials are scarce, to vary color and texture in response to available resources, to change direction in design, and to splice new yarn. As a product, a weaving reveals both grand patterns and minor designs; it reveals the connections between pattern changes and how what has come before is linked to what follows; and it reveals the richness or thinness of the materials used....

... I argue that the metaphor of "weaving" better represents the actions and intentions of employed women with children than the current dominant model of individual orientation that pervades discussions of work and family for women. The conceptual framework of weaving allows us to step back and view the whole, to think of the fabric of a life, the strength of the weave, and the intricacy of design. It reminds us not to get lost in the close examination of one moment or one strand, and to remember that moments and strands are parts of the weave but not the weave itself. Work, family, friendships, reflection, vocation, and recreation are parts of a person's life. They are not, separately and on their own, the life or the person.

NOT A ZERO-SUM GAME

... The orientation model of work and family that underlies much of the discussion and thinking about working mothers misrepresents the intersection of employment and motherhood in the lives of women. Motherhood and employment are not incompatible activities in a zero-sum game. And "mothers" and "paid workers" are not opposed categories. The way that we conceptualize the relationship of employment and motherhood is important because the way that we think about an issue shapes what we do about it—or what we think we should do about it. Sociologists stress the importance of understanding the individual's or the group's "definition of the situation," because those definitions—those answers to the question "What's going on here?"—have concrete implications. . . .

There are consequences to defining the situation of employment and motherhood for women in terms of opposition and orientation—and these consequences are not good for women, men, children, and families. By describing as "family oriented" women who are employed on the night shift or who are employed less than forty hours a week (or who put in less than the sixty or seventy hours a week expected in some professions), the orientation model reinforces standard myths about women's marginal relationship to employment—myths that are belied by women's work histories and their expressed feelings about the place of employment in their lives. Defining women's relationship to employment as marginal has had negative consequences for wages, benefits, job security, and national employment and childcare policies. . . .

By describing as "work oriented" mothers who are in male-dominated professions, mothers who are employed forty hours or more per week, mothers who travel as part of their jobs, or even mothers who admit to liking their jobs, the orientation model reproduces cultural ideas that mothers who care about their families do not work full time, or like their jobs, or want to be employed. The expectation that mothers should immerse themselves, to the exclusion of other activities, in the care and nurturance of their children has consequences. For mothers who are employed, the results are often exhaustion from trying to do everything and guilt from feeling they are never doing enough. For nonemployed mothers, the results include economic vulnerability and feelings of resentment toward "working mothers." For fathers, the assumption that children are primarily the responsibility of mothers often results in assumptions that support their nonparticipation in their children's lives. And for children, the result is the lack of societal responsibility for their care and the associated child-unfriendly environment. . . .

Not only does the categorization of these mothers as work oriented reinforce certain cultural constructions of motherhood, it also reproduces cultural understandings of career or commitment to one's job as all-encompassing. This definition of work has consequences for men's and women's relationship to employment and for the organization of the workplace. If work commitment is understood as inherently conflicting with family responsibilities, then men, as breadwinner-fathers, will not be socially expected to share in family work as fathers, sons, and brothers. And women, as mothers, daughters, and sisters, will not be treated seriously in the workplace. Furthermore, employers can use this definition and standard of commitment to rationalize ever-increasing work loads, speed-ups, and demands on workers. If a commitment to job or career is conceptualized as being in conflict with participation in family life, then there is little reason to expect changes in the workplace that will accommodate *both* work *and* family needs. . . .

KEY CONCEPTS

role conflict   role negotiation   role strain

DISCUSSION QUESTIONS

1. Why does Garey think that the orientation model misrepresents women's experiences of work and motherhood? Why is "weaving" a more appropriate metaphor?

2. Given Garey's argument, what social policies are needed to help women and men weave work and parenthood?

45

The Care Crisis

How Women are Bearing the Burden of a National Emergency

RUTH ROSEN

This piece examines the problem women face in balancing work and the care work that is required of families and households. Rosen argues that the women's movement has not been fully realized, and society needs to address the problem of caring for our

SOURCE: Ruth Rosen. "The Care Crisis: How Women Are Bearing the Burden of a National Emergency." *The Nation*, March 12, 2007, pp. 11–16.

*young and our elderly. The efforts to manage work, home, child care, and
elderly parent care have been relegated to the private sphere and must be moved
to the public and, more importantly, political stage. Only with determination
and grassroots efforts will families be able to enact change that will benefit
all families.*

A baby is born. A child develops a high fever. A spouse breaks a leg. A parent
suffers a stroke. These are the events that throw a working woman's delicate
balance between work and family into chaos.

Although we read endless stories and reports about the problems faced by
working women, we possess inadequate language for what most people view as a
private rather than a political problem. "That's life," we tell each other, instead of
trying to forge common solutions to these dilemmas.

That's exactly what housewives used to say when they felt unhappy and unfulfilled
in the 1950s: "That's life." Although magazines often referred to housewives' unex-
plained depressions, it took Betty Friedan's 1963 bestseller to turn "the problem that has
no name" into a household phrase, "the feminine mystique"—the belief that a woman
should find identity and fulfillment exclusively through her family and home.

The great accomplishment of the modern women's movement was to name
such private experiences—domestic violence, sexual harassment, economic discrimination, date rape—and turn them into public problems that could be
debated, changed by new laws and policies or altered by social customs. That is
how the personal became political.

Although we have shelves full of books that address work/family problems, we
still have not named the burdens that affect most of America's working families.

Call it the care crisis.

For four decades, American women have entered the paid workforce—on men's
terms, not their own—yet we have done precious little as a society to restructure the
workplace or family life. The consequence of this "stalled revolution," a term coined
by sociologist Arlie Hochschild, is a profound "care deficit." A broken healthcare
system, which has left 47 million Americans without health coverage, means this
care crisis is often a matter of life and death. Today the care crisis has replaced the
feminine mystique as women's "problem that has no name." It is the elephant in
the room—at home, at work and in national politics—gigantic but ignored.

Three decades after Congress passed comprehensive childcare legislation in
1971—Nixon vetoed it—childcare has simply dropped off the national agenda.
And in the intervening years, the political atmosphere has only grown more
hostile to the idea of using federal funds to subsidize the lives of working families.

The result? People suffer their private crises alone, without realizing that the
care crisis is a problem of national significance. Many young women agonize
about how to combine work and family but view the question of how to raise
children as a personal dilemma, to which they need to find an individual solution.
Most cannot imagine turning it into a political debate. More than a few young
women have told me that the lack of affordable childcare has made them
reconsider plans to become parents. Annie Tummino, a young feminist active
in New York, put it this way: "I feel terrified of the patchwork situation women

are forced to rely upon. Many young women are deciding not to have children or waiting until they are well established in their careers."

... The obstacles ... are formidable, given that government and businesses—as well as many men—have found it profitable and convenient for women to shoulder the burden of housework and caregiving.

It is as though Americans are trapped in a time warp, still convinced that women should and will care for children, the elderly, homes and communities. But of course they can't, now that most women have entered the workforce. In 1950 less than a fifth of mothers with children under age 6 worked in the labor force. By 2000 two-thirds of these mothers worked in the paid labor market.

Men in dual-income couples have increased their participation in household chores and childcare. But women still manage and organize much of family life, returning home after work to a "second shift," of housework and childcare—often compounded by a "third shift," caring for aging parents.

Conservatives typically blame the care crisis on the women's movement for creating the impossible ideal of "having it all." But it was women's magazines and popular writers, not feminists, who created the myth of the Superwoman. Feminists of the 1960s and '70s knew they couldn't do it alone. In fact, they insisted that men share the housework and child-rearing and that government and business subsidize childcare.

A few decades later, America's working women feel burdened and exhausted, desperate for sleep and leisure, but they have made few collective protests for government funded childcare or family-friendly workplace policies. As American corporations compete for profits through layoffs and outsourcing, most workers hesitate to make waves for fear of losing their jobs.

Single mothers naturally suffer the most from the care crisis. But even families with two working parents face what Hochschild has called a "time bind." Americans' yearly work hours increased by more than three weeks between 1989 and 1996, leaving no time for a balanced life. Parents become overwhelmed and cranky, gulping antacids and sleeping pills, while children feel neglected and volunteerism in community life declines....

For the very wealthy, the care crisis is not so dire. They solve their care deficit by hiring full-time nannies or home-care attendants, often from developing countries, to care for their children or parents. The irony is that even as these immigrant women make it easier for well-off Americans to ease their own care burdens, their long hours of paid caregiving often force them to leave their own children with relatives in other countries. They also suffer from extremely low wages, job insecurity and employer exploitation.

Middle- and working-class families, with fewer resources, try to patch together care for their children and aging parents with relatives and baby sitters. The very poor sometimes gain access to federal or state programs for childcare or eldercare; but women who work in the low-wage service sector, without adequate sick leave, generally lose their jobs when children or parents require urgent attention. As of 2005, 21 million women lived below the poverty line—many of them mothers working in these vulnerable situations.

The care crisis starkly exposes how much of the feminist agenda of gender equality remains woefully unfinished. True, some businesses have taken steps to ease the care burden. Every year, *Working Mother* publishes a list of the 100 most "family friendly" companies. In 2000 the magazine reported that companies that had made "significant improvements in 'quality of life' benefits such as telecommuting, onsite childcare, career training, and flextime" were "saving hundreds of thousands of dollars in recruitment in the long run."

Some universities, law firms and hospitals have also made career adjustments for working mothers, but women's career demands still tend to collide with their most intensive child-rearing years. Many women end up feeling they have failed rather than struggled against a setup designed for a male worker with few family responsibilities.

The fact is, market fundamentalism—the irrational belief that markets solve all problems—has succeeded in dismantling federal regulations and services but has failed to answer the question, Who will care for America's children and elderly?

As a result, this country's family policies lag far behind those of the rest of the world. A just-released study by researchers at Harvard and McGill found that of 173 countries studied, 168 guarantee paid maternal leave—with the United States joining Lesotho and Swaziland among the laggards. At least 145 countries mandate paid sick days for short- or long-term illnesses—but not the United States. One hundred thirty-four countries legislate a maximum length for the workweek, not us.

The media constantly reinforce the conventional wisdom that the care crisis is an individual problem. Books, magazines and newspapers offer American women an endless stream of advice about how to maintain their "balancing act," how to be better organized and more efficient or how to meditate, exercise and pamper themselves to relieve their mounting stress. Missing is the very pragmatic proposal that American society needs new policies that will restructure the workplace and reorganize family life.

Another slew of stories insist that there simply is no problem: Women have gained equality and passed into a postfeminist era. Such claims are hardly new. Ever since 1970 the mainstream media have been pronouncing the death of feminism and reporting that working women have returned home to care for their children. Now such stories describe, based on scraps of anecdotal data, how elite (predominantly white) women are "choosing" to "opt out," ditching their career opportunities in favor of home and children or to care for aging parents. In 2000 Ellen Galinsky, president of the Families and Work Institute in New York, wearily responded to reporters, "I still meet people all the time who believe that the trend has turned, that more women are staying home with their kids, that there are going to be fewer dual-income families. But it's just not true."

Such contentious stories conveniently mask the reality that most women have to work, regardless of their preference. They also obscure the fact that an absence of quality, affordable childcare and flexible working hours, among other family-friendly policies, greatly contributes to women's so-called "choice" to stay at home.

In the past few years, a series of sensational stories have pitted stay-at-home mothers against "working women" in what the media coyly call the "mommy wars." When the *New York Times* ran a story on the controversy, one woman wrote the editor. "The word 'choice' has been used ... as a euphemism for unpaid labor, with no job security, no health or vacation benefits and no retirement plans. No wonder men are not clamoring for this 'choice.' Many jobs in the workplace also involve drudgery, but do not leave one financially dependent on another person."

Most institutions, in fact, have not implemented policies that support family life. As a result, many women do feel compelled to choose between work and family. In Scandinavian countries, where laws provide for generous parental leave and subsidized childcare, women participate in the labor force at far greater rates than here—evidence that "opting out" is, more often than not, the result of a poverty of acceptable options.

American women who do leave their jobs find that they cannot easily re-enter the labor force. The European Union has established that parents who take a leave from work have a right to return to an equivalent job. Not so in the United States. According to a 2005 study by the Wharton Center for Leadership and Change and the Forte Foundation, those who held advanced degrees in law, medicine or education often faced a frosty reception and found themselves shut out of their careers. In her 2005 book *Bait and Switch,* Barbara Ehrenreich describes how difficult it was for her to find employment as a midlevel manager, despite waving an excellent résumé at potential employers. "The prohibition on [résumé] gaps is pretty great," she says. "You have to be getting an education or making money for somebody all along, every minute."

Some legislation passed by Congress has exacerbated the care crisis rather than ameliorated it. Consider the 1996 Welfare Reform Act, which eliminated guaranteed welfare, replaced it with Temporary Assistance to Needy Families (TANF) and set a five-year lifetime limit on benefits. Administered by the states, TANF aimed to reduce the number of mothers on welfare rolls, not to reduce poverty.

TANF was supposed to provide self-sufficiency for poor women. But most states forced recipients into unskilled, low-wage jobs, where they joined the working poor. By 2002 one in ten former welfare recipients in seven Midwestern states had become homeless, even though they were now employed.

TANF also disqualified higher education as a work-related activity, which robbed many poor women of an opportunity for upward mobility. Even as the media celebrate highly educated career women who leave their jobs to become stay-at-home moms, TANF requires single mothers to leave their children somewhere, anywhere, so they can fulfill their workfare requirement and receive benefits. TANF issues vouchers that force women to leave their children with dubious childcare providers or baby sitters they have good reasons not to trust.

Some readers may recall the 1970 Women's Strike for Equality, when up to 50,000 women exuberantly marched down New York's Fifth Avenue to issue three core demands for improving their lives: the right to an abortion, equal pay for equal work and universal childcare. The event received so much media attention that it turned the women's movement into a household word.

A generation later, women activists know how far we are from achieving those goals. Abortion is under serious legal attack, and one-third of American women no longer have access to a provider in the county in which they live. Women still make only 77 percent of what men do for the same job; and after they have a child, they suffer from an additional "mother's wage gap," which shows up in fewer promotions, smaller pensions and lower Social Security benefits. Universal childcare isn't even on the agenda of the Democrats.

Goals proposed in 1970, however unrealized, are no longer sufficient for the new century.... If women really mattered, they ask, how would we change public policy and society? As one writer puts it. "What would the brave new world look like if women could press reboot and rewrite all the rules?"

Though no widely accepted manifesto exists, many advocacy organizations—such as the Institute for Women's Policy Research, the Children's Defense Fund, the National Partnership for Women and Families, Take Care Net and MomsRising—have argued that universal healthcare, paid parental leave, high-quality subsidized on-the-job and community childcare, a living wage, job training and education, flexible work hours and greater opportunities for part-time work, investment in affordable housing and mass transit, and the reinstatement of a progressive tax structure would go a long way toward supporting working mothers and their families....

Confronting the care crisis and reinvigorating the struggle for gender equality should be central to the broad progressive effort to restore belief in the "common good." Although Americans famously root for the underdog, they have shown far less compassion for the poor, the vulnerable and the homeless in recent years. Social conservatives, moreover, have persuaded many Americans that they—and not liberals—are the ones who embody morality, that an activist government is the problem rather than the solution and that good people don't ask for help....

The truth is, we're living with the legacy of an unfinished gender revolution. Real equality for women, who increasingly work outside the home, requires that liberals place the care crisis at the core of their agenda and take back "family values" from the right.... So it's up to us, the millions of Americans who experience the care crisis every day, to take every opportunity—through electoral campaigns and grassroots activism—to turn "the problem that has no name" into a household word.

KEY CONCEPTS

care work   grassroots activism   sandwich generation

DISCUSSION QUESTIONS

1. Can you think of anyone who is sandwiched between child care and elder care? Do they work outside the home? What are the issues facing someone in this situation?

2. What is Rosen's argument for why the women's movement is still needed? What are the "women's rights" issues facing women today?

46

Gay Marriage

STEVEN SEIDMAN

The debate about extending the right to marry to gays and lesbians is summarized in this article. Seidman outlines the key issues presented against gay marriage, including religious and historical arguments. Mostly those who are critical of gay marriages argue that it weakens the institution of marriage even further. He then goes on to offer a rebuttal argument. While he recognizes the religious value placed on heterosexual unions, he argues that America is not a solely Christian country, that the dynamics of family structure have changed, and that gender roles within families have changed. Why, then, can't the gender of those who are married change?

Marriage is not just about love and intimacy between two adults. It is an institution. Marriage is recognized by the state, and those that marry get specific rights and benefits such as the right to spousal support, to bring wrongful death suits, to be listed as social security beneficiaries, to get legally recognized divorces, and to claim full parental rights. In America, only a man and woman can marry. The question raised by the gay marriage debate is whether the gender of a spouse should matter in determining who can marry. And, what role should the state have in regulating marriage and intimate choices?

Gay marriage was not an issue in the 1950s and early 1960s when gays and lesbians began their struggle for tolerance. It initially surfaced in public debate in the 1970s.... [A] more confident and assertive gay movement redefined homosexuality: it was no longer a stigma but was viewed as natural and good. The gay movement approached being gay as a positive basis of identity, community, and a fulfilling lifestyle.

By the mid-1980s, the gay movement sought full, across-the-board legal and social equality. Gays wanted to be equal citizens. They pursued equality at work, in schools, in the military, and in the eyes of the law. They also demanded to be treated respectfully in their families, by mass media, and elsewhere.

SOURCE: Steven Seidman. 2003. *The Social Construction of Sexuality.* New York: Norton, pp. 123–133.

Gay marriage became a key issue in the 1990s as part of the pursuit of social equality. Without the right to marry, gays are second-class citizens. Also, many gays and lesbians were turning their attention to intimate relationships. The generation that came out in the late 1960s and 1970s was now middle-aged; like many straight Americans in their thirties, forties, and fifties, their attention often turned to creating families. Many gays and lesbians were in or sought long-term intimate relationships. It was inevitable that they would turn their political focus toward marriage. Furthermore, AIDS pushed the issue of the legal status of their intimate relationships into the heart of their lives. The AIDS crisis forced many gays and lesbians to address concerns such as health coverage of partners, hospital visitation rights, inheritance, and residency rights. AIDS compelled many gays and lesbians to invest enormous amounts of time, energy, and resources into caring for their partners, which made the legal status of their relationships an urgent concern.

Despite unfriendly social conditions, gays have always formed long-term relationships. For example, scholars have uncovered a long and complicated history of gay relationships in nineteenth century America. Sometimes women passed as men to form straight-seeming relationships; sometimes men or women lived together as housemates but were really lovers; sometimes individuals would marry but still carry on romantic, sometimes life-long, same-sex intimate relationships. By the 1990s, many gay men and lesbians were in long-term relationships; the lives of these couples were emotionally, socially, and financially intertwined in ways that were similar to straight marriages. Indeed, through either adoption or artificial insemination, many gay and lesbian couples were adding children to their families.

Gay marriage has today become a front-line issue for gays and lesbians—and not only in the United States. Gay marriage has become a worldwide issue. Many societies have already enacted laws that permit gays to marry (for example, Denmark, Sweden, Norway, and the Netherlands). In Canada, France, and Germany, laws have been passed that offer some type of state recognition to gay relationships.

Gay marriage might not have gained political traction in the United States were it not for broader changes in the American family. We are all aware that today families come in many sizes and shapes. Although the nuclear family, with a breadwinner husband and a stay-at-home wife, might still be an ideal for many Americans, it describes only a minority of actual households. The reality is a dizzying variety of families—cohabiting couples, one-parent households, combined families, marriages without children, lifelong partners who do not live together, and so on. The variety of intimate arrangements, the reality that almost half of all marriages end in divorce, the uncoupling of motherhood from marriage, and the lessening of the stigma attached to being single have all diminished the cultural authority of marriage. Still, make no mistake; marriage is not just one choice among others. The state supports marriage with a cluster of rights and benefits that no other intimate relationship is given. In addition, the right to marry continues to serve as a symbol of first-class citizenship. Many gays and lesbians want to marry, or at least want the right to marry, in order to become equal, respected citizens.

Changes in the social organization of intimacy have also contributed to raising doubts about the reasonableness of excluding gays from marriage. For most of our parents and grandparents, the organization of marriage was more or less fixed. Men were the breadwinners; their lives were focused on making a living. Women were wives and mothers; their lives were centered on domestic tasks. Within the household, men were expected to do "masculine" domestic activities (mow the lawn, take out the garbage, discipline the children), whereas women were responsible for "feminine" tasks such as cooking, cleaning, and child care. Gender shaped the very texture of intimacy. Men were supposed to initiate and direct sex; women were supposed to go along without showing too much interest or pleasure. Men made the big decisions (where to live or how to spend money); women arranged the social affairs of the couple. Today, many women work, have careers, and pursue interests outside the household; men are expected to perform household and child care tasks. We can perhaps reasonably speak of a somewhat new ideal of intimacy: marriage as a relationship between equals in which decisions are openly discussed and household roles are negotiated. In short, gender is becoming less important in organizing marriage. The gender of those that marry would then seem to matter less.

The battle over gay marriage is being fought on two fronts. A war is being waged in the legal courts and in the court of public opinion between mostly gay and lesbian advocates of gay marriage and straight critics. Public opinion is still on the side of critics, but this is beginning to change. There is also debate within the gay movement. Not all gays and lesbians think that pursuing marital rights is the right goal for a movement that once aspired to ideals of sexual liberation. Gay critics view marriage as inimical to a culture championing sexual variation.

Straight critics of gay marriage make several key points. Heterosexual marriage, they say, has deep roots in history. As far back as we know, marriage has always been between a man and a woman. There is something seemingly natural and right about heterosexual marriage. Moreover, heterosexual marriage is a cornerstone of the Judeo-Christian tradition. As some critics quip, God made Adam and Eve, not Adam and Steve. America has a secular government and was founded on secular principles, but a majority of Americans identify themselves as Christian.

Marriage is also a cornerstone of American society. It provides a stable, positive, moral environment essential for shaping good American citizens. Critics ask, will young people acquire clear gender identities without heterosexual marriage? Boys look to their fathers to learn what it means to be a man; wives provide daughters with a clear notion of what it means to be a woman. Gay marriage would create gender confusion.

Critics also raise another issue: marriage has already been weakened by the ease and frequency of divorce, by the increase in rates of illegitimacy and single-parent families, and by families in which both parents work. Legalizing gay marriage would further undermine the stability and strength of this institution. Even more ominously, permitting gays to marry would open the door for all sorts of people to demand the right to marry—polygamists, children, friends, kin.

These objections express real anxieties on the part of Americans about the fragility of marriage. Such concerns should not quickly be dismissed. Marriage has been, and still is, according to virtually all social researchers, a cherished ideal for most Americans. Marriage is bundled with a number of hopes shared by many Americans—for a home, a family, and a sense of community. Instead of bringing additional change to this already weakened institution, critics say, we should find ways to strengthen it.

Advocates of gay marriage have offered forceful rebuttals to these criticisms. No one denies, they say, that heterosexual marriage has deep roots in history, though recent historical scholarship suggests that same-sex intimacies, including marriage, were not as exceptional as once believed. Setting aside past realities, though, the question that must be addressed is this: should the past or tradition always serve as a guide for the present? Consider that racism and sexism also have deep roots in history and social customs. Most of us agree that tradition should not be followed blindly, but examined in light of contemporary thinking and values. After all, imperfect people shaped past practices and traditions. Possibly the historical prejudice against homosexuality is similar to prejudices against nonwhites and women. The battleground for debating the issue of gay marriage should be the present, not the past.

What about the Judeo-Christian disapproval of homosexuality? With all due respect, gay marriage advocates hold that, like other traditions, religious practices too have been made by ordinary individuals who often shared the prejudices of their time. Christian belief and tradition has changed many times in history. The question for Christians is whether gay marriage can be understood as consistent with the spirit of Christianity. Both pro and con arguments have been made in this regard; I'll leave this debate to the faithful. There is, in any event, a more compelling reason to be cautious about religious objections to gay marriage. America is not a Christian nation. Although Judeo-Christian traditions have shaped America, so too have non-Christian and secular traditions. America is, above all, a secular nation. Our government does not officially recognize or promote any particular religion and does not enact legislation, including laws regulating marriage, that need be aligned to any specific faith.

So, appeals to the past, to religious and secular traditions alone, should not exclude the right of gays to marry. But opponents have advanced additional arguments. Some argue that gays and lesbians are psychologically and morally unable to form stable marriages. This argument cannot be taken seriously because it relies on stereotypes that have been exposed and dismissed by social scientific research. The argument that if you extend marital rights to gays and lesbians then all sorts of people and relationships will clamor for the same rights is also not persuasive, since gays are not challenging the institution of marriage as a consensual relationship between two adults. Gays and lesbians simply want the same right of access as heterosexuals.

What are the positive arguments for gay marriage? The key issue, say advocates, is the meaning of marriage today. Too often, critics of gay marriage assume that marriage has always had the same meaning and social organization. This is not the case. Marriage has changed considerably, even in the short history of the United States. Consider that not too long ago marriage was possible only between adults of the same race. Antimiscegenation laws, initially enacted in Maryland in 1661 and not declared unconstitutional by the Supreme Court until 1967, forbid marriage between whites and nonwhites, including blacks, Asians, and Native Americans. Or, consider that throughout most of American history, marriage was rigidly organized around gender roles. Until the twentieth century, women were the legal property of men; a wife could not own her own property and did not even have the right to her own wages. Marriage was thought to be fundamentally for the purpose of having children. Today, we recognize spouses as equal before the law, and gender roles are not enforced by the state. Many couples marry and remain childless by choice.

Marriage today has various meanings. For some, it is about creating a family; for others, it is about social and financial security. For many of us, marriage is fundamentally about love and forging an intimate life with another person. In this companionate ideal, individuals look to marriage to find a deep emotional, social, even spiritual union. In principle, gender plays less of a role in organizing marriage. Men and women share domestic duties; they attempt to negotiate a life together as equals. Spouses want to be respected and fulfilled as individuals. To be sure, social scientists have documented that gender still plays a considerable role in organizing intimacy. For example, women continue to do the lion's share of domestic tasks, including child care. But, today, individuals can demand that their unique wants and desires, regardless of their gender, be considered in the social organization of intimate relationships.

If marriage is about equal individuals forging an intimate, loving, mutually committed relationship, then the gender of the partners should be irrelevant. What should matter is whether the partners agree to marry, and whether they are caring, committed, respectful, responsible, and willing to communicate openly their respective needs and wants. Permitting gays to marry will not change the institution of marriage but strengthen an ideal of this institution as a relationship of loving intimacy between equals.

I think that some people oppose gay marriage because they are threatened by this egalitarian ideal of marriage. This ideal, after all, devalues the role of gender. Some men may fear a loss of status and power, and some women may fear economic insecurity and the loss or devaluation of their chief identity as full-time wives and mothers. For women and men who are deeply invested in gender roles and identities or in marriages that are fundamentally about having children, gay marriage may be viewed as a threat not because it challenges heterosexual privilege, but because it challenges a very specific and narrow idea of gender and the family. . . .

So, for both moral and practical political reasons, I think it is important to defend the right of gays and lesbians to marry. Although I do not expect an end to state support of marriage anytime soon, I do anticipate the distribution of some marital rights to other intimate arrangements—those relationships that, in terms of their intimate ties, look a lot like marriage, that is, relationships involving long-standing emotional, sexual, social, and economic interdependence between two unrelated adults. These intimate unions merit state recognition. In fact, this is already happening. Many of the rights and benefits of marriage are now claimed by "domestic partnerships"—which are recognized in many cities, states, businesses, unions, and colleges—by "civil unions" in Vermont, and by common-law marriages, cohabitation, and single-parent households. In the short run, the best way to promote intimate diversity in the United States is to expand the range of intimate relationships that are recognized by the state.

KEY CONCEPTS

egalitarian   gender roles   miscegenation laws
sexual politics

DISCUSSION QUESTIONS

1. How does Seidman counter the argument that marriage is based on a long history of heterosexual unions? What points does he make to oppose this?
2. What traditional forms of marriage are no longer accepted? How has marriage changed over the years?

47

Divorce and Remarriage

TERRY ARENDELL

Increasingly common in family experience, divorce and remarriage are producing new family patterns and family experiences. Terry Arendell reviews patterns in divorce and remarriage, including the impact on children, custody arrangements, and economic consequences of divorce. In addition, she reviews some of the social dynamics found in stepfamilies.

SOURCE: Arendell, Terry. "Divorce and Remarriage." In *Contemporary Parenting: Challenges and Issues,* edited by Terry Arendell, 154–95. Thousand Oaks, CA: Sage, 1997. Reprinted with permission.

DEMOGRAPHIC PATTERNS IN MARITAL DISSOLUTION AND REMARRIAGE

The divorce rate more than doubled between the early 1960s and mid–1970s. Despite some fluctuations in the annual divorce rates, more than 1 million marriages still are dissolved each year. If trends continue as anticipated, as many as three in five first marriages will end in legal dissolution, as they have since 1980. Second marriages have a somewhat higher termination rate (Gottman, 1994; Kitson & Holmes, 1992; Martin & Bumpass, 1989).

Most likely to divorce are younger adults in shorter term marriages with dependent children. Indeed, children are involved in approximately two thirds of all divorces (U.S. Bureau of the Census, 1995a), and more than half of all children experience their parents' divorces before they reach 18 years of age (Cherlin & Furstenberg, 1994; Furstenberg & Cherlin, 1991; Martin & Bumpass, 1989). Nearly twice as many black children as white children born to married parents will experience parental divorce if trends persist as expected (Amato & Keith, 1991). Marital separation and dissolution rates among parents in other racial and ethnic groups, which generally have been lower than those among whites and blacks, also are increasing (U.S. Bureau of the Census, 1995b). Children who experience divorce spend an average of 5 years in single–parent homes (Glick & Lin, 1986); even among those whose custodial mothers remarry, about half spend 5 years with their mothers alone (Furstenberg, 1990).

Separation and divorce are not the only transitions in parents' marital status and household arrangements experienced by children. Even though remarriage rates are declining, with only about two-thirds of separated or divorced women and about three-fourths of men likely to remarry compared to three-fourths and four-fifths, respectively, in the 1960s, more than one third of adults currently in first marriages will divorce and remarry before their youngest children reach age 18. Thus a high proportion of children will experience the remarriage of the parent, if not both, and the formation of a stepfamily or stepfamilies. Moreover, many children will experience the dissolution of a stepfamily when a parent and stepparent divorce. About one in six children will experience two divorces of the custodial parent before the child reaches age 18 (Furstenberg & Cherlin, 1991). Additionally, increasing numbers of adults, including those who are custodial parents of minor children, are cohabiting. Whether they eventually will marry remains to be seen (see Cherlin & Furstenberg, 1994).

Approximately 1 in 10 children in 1992 lived with a biological parent and a stepparent, and this proportion is expected to increase. About 15% of all children lived in blended families—homes in which children lived with at least one stepparent, stepsibling, or half-sibling. More children lived with at least one half-brother or half-sister than with a stepparent or with at least one stepsibling (Furukawa, 1994). Because the practice of cohabitation, or sharing domestic life and intimacy without legal marriage, is increasing steadily, the number of children who reside with a custodial parent and her or his adult partner, who presumably functions, at least to some extent, as a stepparent—*a quasi-stepparent*— probably is much higher than the official numbers indicate (Cherlin & Furstenberg, 1994, pp. 363–365). Because the large majority of children whose parents divorce live with their mothers, most residential stepparents and quasi-stepparents are men. Census data for 1991 show that among children in single-mother families (which includes never-married as well as divorced), 20% also lived with an adult male (related or unrelated) present in the household. About 37% of children living with a single father also lived with an adult female (related or unrelated) (Furukawa, 1994, pp. 1–2)....

DIVORCE

Postdivorce Parenting

With respect to family functioning, marital dissolution can be a lengthy process, often underway years before the actual spousal separation occurs. Children show the effects of marital dissension and discontent long before divorce (e.g., Amato & Booth, 1996; Block, Block, & Gjerde, 1986, 1988; Shaw, Emery, & Tuer, 1993). Adjustment to the changes wrought by divorce itself can be a gradual and lengthy process, and many parents and children enter a "crisis" period after the marital separation that can last for several years (e.g., Chase-Lansdale & Hetherington, 1990; Hetherington, 1987, 1988; Morrison & Cherlin, 1995). Maccoby and Mnookin (1992), in the Stanford Custody Project, concluded that

> divorcing parents find it difficult to take the time and trouble required to negotiate with children over task assignments and joint plans. Under these conditions of diminished parenting, children tend to become bored, moody, and restless and to feel misunderstood; these reactions lead to an increase in behaviors that irritate their parents, and mutually coercive cycles ensue. (pp. 204–205)

A related phenomenon is single parents' lesser ability to make control demands on their children. Examining data from the National Survey of Families and Households, Thomson, McLanahan, and Curtin (1992) found that single parents of both sexes seem to be "structurally limited" in their ability to control and make demands on a child without the presence of another adult.

The extent and duration of uneven parenting, however, varies by families, with some family units adapting fairly rapidly to their altered circumstances and arrangements, achieving stable and healthy family functioning rather soon after divorce. Some units take much longer to find an equilibrium. Others have a delayed reaction, functioning well initially and then encountering adjustment difficulties (e.g., Kitson & Holmes, 1992). In addition, "some show intense and enduring deleterious outcomes" (Hetherington, 1993, p. 40). Whatever the pattern, parental functioning usually recovers over time, returning nearly to the level found in intact families (Hetherington, 1988; Hetherington, Cox, & Cox, 1982). That is, most family units formed by divorce establish workable and functional interactional processes (Maccoby & Mnookin, 1992; Wallerstein & Blakeslee, 1989).

One of the first major tasks facing parents in divorce is that of determining children's living arrangements as family members separate into two households. Most custody decisions occur with little discussion between the parents, and relatively few custody allocations are actually litigated. Yet the working out of parenting and parental relationships after divorce, including children's access to and involvement with the nonresidential parent if parenting is not shared, can be complicated and difficult, involving various changes and intraparental conflicts. Of the four relationships between married persons that must be altered in divorce—parental, economic, spousal, and legal (Maccoby & Mnookin, 1992)— the parental divorce is perhaps the most difficult to achieve (Ahrons & Wallisch, 1987, p. 228; see also Bohannon, 1970).

Custody Arrangements for Minor Children

Three residential patterns are available for children in divorce: maternal, paternal, and dual. Primary physical custody, maternal or paternal, is the situation in which children spend more than 10 overnights in a 2-week period with a particular parent (Mnookin, Maccoby, Albiston, & Depner, 1990, pp. 40–41). Dual or shared custody is defined as the situation in which "the children spend at least a third of their time in each household" (Maccoby & Mnookin, 1992, p. 203). Shared custody is unusual. Even in California, where dual custody probably is more common than anywhere else, only about one in six children actually lives in a shared custody situation. And in these circumstances, "more often than not" mothers handle the bulk of the managerial aspects of child rearing (Maccoby & Mnookin, 1992, p. 269).

As has been the case for most of this century, maternal custody is predominant; more than 85% of children whose parents are divorced are in the custody of their mothers (U.S. Bureau of the Census, 1995a). A somewhat higher proportion of offspring actually reside with their mothers because, in legally mandated dual-custody situations, children often spend relatively little time with their fathers (e.g., Maccoby & Mnookin, 1992; Seltzer, 1991; Seltzer & Bianchi, 1988). Overwhelmingly, then, it is mothers who become the primary parents in divorce....

Economic Support of Children

Although the preseparation parenting division of labor persists after divorce with mothers doing most of the parenting, what does change is the economic providing for minor children. Whereas men's earned incomes provide the larger share of the economic resources available to intact married families, divorced custodial women assume most financial responsibilities for their offspring. The overwhelming body of scholarly research and governmental and other policy studies shows that fathers' contributions to the economic support of their children are much reduced after marital dissolution (e.g., Kellan, 1995; Maccoby & Mnookin, 1992) despite many men's claims to the contrary (e.g., Arendell, 1995). Approximately three fourths of divorced mothers have child support agreements, but only about half of those women receive the full amounts ordered in the agreements (Holden & Smock, 1991; Scoon-Rogers & Lester, 1995). One fourth receive no payment whatsoever, and the other one fourth receive irregular payments in amounts less than those ordered. According to the Congressional Research Service, only about $13 billion of the $34 billion in outstanding support orders was collected in 1993 (Kellan, 1995, p. 27). Moreover, child support payments amounted to only about 16% of the incomes of divorced mothers and their children in 1991. The average monthly child support paid by divorced fathers contributing economic support in 1991 was $302, amounting to $3,623 for the year (Scoon-Rogers & Lester, 1995). Fathers' limited or lack of financial contributions to the support of their children not residing with them is not offset by other kinds of assistance (Teachman, 1991, p. 360).

As a group, women's incomes drop more than 30% following divorce. About 40% of divorcing women lose more than half of their family incomes, whereas fewer than 17% of men experience this large a drop (Hoffman & Duncan, 1988). Men, in general, experience an increase in their incomes—an average of 15%—partially because they share less of their incomes with their children (Furstenberg & Cherlin, 1991; Kitson & Holmes, 1992; Maccoby & Mnookin, 1992). For many women, the financial hardships accompanying divorce become the overriding experience, affecting psychological well-being and parenting as well as dictating decisions such as where to live, what type of child care to use, and whether or not to obtain health care (Arendell, 1986; Kurz, 1995).

Children's economic well-being after divorce is directly related to their mothers' economic situations. Those living with single mothers are far more likely to be poor than are children in other living arrangements; families headed by single mothers are nearly six times as likely to be impoverished as are families having both parents present (U.S. Bureau of the Census, 1995a). This is not the experience of children being raised by single fathers because men's wages are higher than women's (Holden & Smock, 1991; Scoon-Rogers & Lester, 1995; Seltzer & Garfinkel, 1990); about one eighth of custodial fathers, compared to nearly two fifths of divorced custodial mothers, are poor (Scoon-Rogers & Lester, 1995). Divorced women and their children do not regain their predivorce standards of living until 5 years after the marital breakups. Women's decisions to remarry often involve economic considerations; the surest route to financial well-being for many women is remarriage, not their employment, even when it is full-time (Furstenberg & Cherlin, 1991; Kitson & Morgan, 1990)....

CHILD OUTCOMES IN DIVORCE

How children fare with divorce is a crucial question, one intimately related to issues of parenting. The arguments vary, with assertions ranging from children being irreparably damaged to children adapting successfully to divorce. Most research evidence suggests that a large majority of children adjust reasonably well to their parents' marital dissolutions....

Some argue that the research findings on the effects of divorce on children are not so clear-cut (see, for review, Bolgar, Sweig-Frank, & Paris, 1995). But even those arguments are tempered when large data sets are the bases of analysis, especially those involving longitudinal studies and not just small, nonrepresentative samples (Amato & Booth, 1996). That a majority of children seem to cope with and adapt well to the change in their parents' marital status is particularly salient because many children enter the divorce phase already disadvantaged by exposure, often of long duration, to parental strife and conflict (Block et al., 1986, 1988; Chase-Lansdale & Hetherington, 1990). Furthermore, as numerous scholars point out, children who experience the dissolution of their parents' marriages may well have to cope with multiple adverse circumstances including family events prior to divorce (e.g., Furstenberg & Teitler, 1994). Allen (1993, p.47), for example, argued that when scholars (and others) uncritically compare divorced families to nondivorced ones, they imply that two-parent intact families inevitably result in positive parenting outcomes. Other events that might be more detrimental than divorce itself, as she notes, are father abandonment; failure to pay child support; neglect; intersection of class, race, and gender with poverty; and women's inequality in traditional families.

Some earlier findings suggested that a child's sex and age mattered in post-divorce adjustment. Hetherington (1993, pp. 48–49), drawing from recent work, concluded that these variables—sex and age—are not pivotal factors in children's divorce responses and adjustments (see also Furstenberg & Teitler, 1994; Garasky, 1995). Sex differences in adverse responses, previously attributed to boys, disappeared in Hetherington's (1993) longitudinal study as children moved into adolescence. Where age mattered, it was for adolescents, all of whom showed somewhat increased problem behaviors. Children with divorced and remarried parents did show more such problem behaviors than did those whose parents remained married. "Adolescence often triggered problems in children from divorced and remarried families who had previously seemed to be coping well" (p. 49). Furstenberg and Teitler (1994) summarized findings pertaining to adolescents:

> The findings indicate that certain effects of divorce are quite persistent even when we consider a wide range of predivorce conditions. Early timing of sexual activity, nonmarital cohabitation, and high school dropout do appear to be more frequent for children from divorced families. (p. 188).

The researchers note that these outcomes may be a result of growing up in single-parent homes or witnessing parents' marital transitions, among other things, not just divorce itself (p. 188).

Also, in contrast to earlier arguments, being reared by same-sex parents appears not to be inherently beneficial to children (Powell & Downey, 1995). And, although it may seem counterintuitive, children's overall well-being in divorce does not seem related to the extent of involvement or quality of parent-child relationship with the noncustodial father (e.g., Amato & Keith, 1991; Bolgar et al., 1995; Furstenberg & Cherlin, 1991). . . .

REMARRIAGE

Research attention to stepparenting has increased dramatically in the past 15 years as the divorce and remarriage rates have escalated and remain high. For instance, Coleman and Ganong (1990, p. 925) noted that there were only a handful of studies published prior to 1980 but more than 200 during the decade of the 1980s. The increased attention has continued into the 1990s (e.g., see Booth & Dunn, 1994).

The circumstances leading to the formation of stepfamilies vary. They especially include the marriages of formerly unmarried teen mothers, widowed parents, and divorced ones. Prior to the early 1970s, the death of a spouse was the principal prior circumstance leading a parent to remarry, not divorce as is now the case. Even just among those formed by divorced parents, stepparent families are diverse in composition. For instance, Dunn and Booth (1994, p. 220) noted that two scholars, Burgoyne and Clark (1982), had identified 26 different types. Children may reside with either a stepmother or a stepfather, although the latter is far more common given the preponderance of mother custody. Or, children may have a nonresidential stepparent. Additionally, children may have stepsiblings and half-siblings with whom they may or may not share residences. More specifically,

> somewhere between two-fifths and half of these children [whose parents remarry] will have a stepsibling, although most will not typically live with him or her. And for more than a quarter, a half-sibling will be born within four years. Thus, about two-thirds of children living in stepfamilies will have either half-siblings or stepsiblings. (Furstenberg, 1990, p. 154)

Depending on their cognitive developmental stage, children construct family relatedness with stepsiblings and half-siblings in various ways, adding to the complexity in understanding family relationships (Bernstein, 1988). . . .

The amount of domestic life and parenting shared with nonresidential family members can range greatly between family units and across time for particular children. Variations among stepfamilies occur, moreover, not only in their configurations but also in their functioning.

The remarriage of a divorced parent and creation of a stepfamily entail numerous disruptions and transitions. Altered by the entry of another adult into the family is the family system established by the custodial parent and children following divorce.

. . . Children, and sometimes the custodial parent, often resist a newcomer's efforts to exert authority and alter the existing family dynamics (Hetherington, 1993; Hetherington et al., 1992). Disruption is not limited to the relationship between the stepparent and stepchildren; it can involve the relationship between the custodial parent and children as well. Conflict within the original unit often increases (Brooks-Gunn, 1994, p. 179). Other problems may include a decline in parental supervision and responsibility as the parent divides her time between a new spouse and her children, shifting alliances between family members, and open tension and disputes between children and stepparent and between children of the original unit (e.g., Brooks-Gunn, 1994, p. 170; Hetherington, 1993; Hetherington et al., 1992). Nor are interpersonal tensions and difficulties limited to the residential unit. They may involve the noncustodial parent, his spouse, or other relatives, such as grandparents, aunts, or uncles. Dealing with the larger family context is an ongoing, lengthy, and demanding process (Mills, 1988; more generally, see Beer, 1988).

Some stepparents respond to children's resistance by becoming more authoritarian and dogmatic. Others, on the other hand, withdraw emotionally and cease their attempts to forge intimate relationships. They move to "exhibiting little warmth, control, or monitoring. [These] stepparents are not necessarily negative, they are just distant" (Brooks-Gunn, 1994, p. 179; Hetherington, 1993). Whatever the strategy assumed by stepparents, it has direct impacts on the home ambiance and parent-child relationships. In turn, these all affect the interactional dynamics between spouses; the effects become circular and interactive.

As with other kinds of family transitions, restabilization often follows the initial disequilibrium experienced by the newly formed stepfamily (Ahrons & Wallisch, 1987; Hetherington, 1993; Hetherington et al., 1992). The successful integration of a stepparent into a family is a gradual process, sometimes taking years (e.g., Papernow, 1988, p. 60). Not all families reach such a level; indeed, a large number of stepfamilies dissolve through divorce long before they ever approach the place of becoming smoothly functioning households. . . .

CONCLUSION

In conclusion, a sizable proportion of American children will experience their parents' divorce. The majority of these children will be parented predominantly by one parent, not by both parents. Many of these children also will experience the formation of a stepparent family when a parent remarries. In many families, both parents will remarry, resulting in situations where children have both a live-in and a live-out stepparent. And numerous children will experience another parental divorce. Children, then, are experiencing multiple transitions in the composition and arrangements of their families. Current evidence indicates that the vast majority of children adjust to these changes successfully. What is most crucial in children's well-being and positive outcomes, according to a growing body of research, is the quality and constancy of the parenting by the primary parent. Experiencing relatively low intraparental and other family conflict is crucial for children's adjustment to changing circumstances and positive development.

REFERENCES

Ahrons, C. R., & Wallisch, L. (1987). Parenting in the binuclear family relationships between biological and stepparents. In K. Pasley & M. Ihinger-Tallman (Eds.), *Remarriage and stepparenting: Current research and theory* (pp. 225–256). New York: Guilford.

Allen, K. R. (1993).The dispassionate discourse of children's adjustment to divorce. *Journal of Marriage and the Family, 55,* 46-50.

Amato, P. R., & Booth, A. (1996). A prospective study of divorce and parent-child relationships. *Journal of Marriage and the Family, 58,* 356-365.

Amato, P. R., & Keith, B. (1991). Parental divorce and the well-being of children: A metaanalysis. *Psychological Bulletin, 11,* 26-46.

Arendell, T. (1986). *Mothers and divorce: Legal, economic, and social dilemmas.* Berkeley: University of California Press.

Arendell, T. (1995). *Fathers and divorce.* Thousand Oaks, CA: Sage.

Beer, W. R. (Ed.) (1988). *Relative strangers: Studies of stepfamilies processes.* Totowa, NJ: Rowman & Littlefield.

Bernstein, A. C. (1988). Unraveling the tangles: Children's understanding of stepfamily kinship. In W. Beer (Ed.), *Relative strangers* (pp. 83–111). Totowa, NJ: Rowman & Littlefield.

Block, J. H., Block, J., & Gjerde, P. F. (1986). Personality of children prior to divorce: A prospective study. *Child Development, 57,* 827-840.

Block, J. H., Block, J., & Gjerde, P. F. (1988). Parental functioning and the home environment in families of divorce: Prospective and concurrent analyses. *Journal of American Academy of Child and Adolescent Psychiatry, 27,* 207-213.

Bohannon, P. (1970). *Divorce and after.* Garden City, NY: Doubleday

Bolgar, R., Sweig-Frank, H., & Paris, J. (1995). Childhood antecedents of interpersonal problems in young adult children of divorce. *Journal of the American Academy of Child and Adolescent Psychiatry, 34(2),* 143-150.

Booth, A., & Dunn, J. (Eds.). (1994). *Stepfamilies: Who benefits? Who does not?* Mahwah, NJ: Lawrence Erlbaum.

Brooks-Gunn, J. (1994). Research on stepparenting families: Integrating disciplinary approaches and informing policy. In A. Booth & J. Dunn (Eds.), *Stepfamilies: Who benefits? Who does not?* (pp. 167–204). Mahwah, NJ: Lawrence Erlbaum.

Burgoyne, J., & Clark, D. (1982). Parenting in stepfamilies. In R. Chester, P. Diggory, & M. Sutherland (Eds.), *Changing patterns of child-bearing and child-rearing* (pp. 133–147). London: Academic Press.

Chase-Lansdale, P. L., & Hetherington, E. M. (1990). The impact of divorce on life-span development: Short and long term effects. In D. Featherman & R. Lerner (Eds.), *Life span development and behavior* (Vol. 10, pp. 105–150). Hillsdale, NJ: Lawrence Erlbaum.

Cherlin, A. J., & Furstenberg, F. F., Jr. (1994). Stepfamilies in the United States: A reconsideration. *Annual Review of Sociology, 20,* 359-381.

Coleman, M. & Ganong, L. H. (1990). Remarriage and stepfamily research in the 1980s: Increased interest in an old family form. *Journal of Marriage and the Family, 52,* 925-940.

Dunn, J., & Booth, A. (1994). Stepfamilies: An overview. In A. Booth & J. Dunn (Eds.), *Stepfamilies: Who benefits? Who does not?* (pp. 217–224). Mahwah, NJ: Lawrence Erlbaum.

Furstenberg, F. F. (1990). Coming of age in a changing family system. In S. Feldman & G. Elliot (Eds.), *At the threshold: The developing adolescent* (pp. 147–170). Cambridge, MA: Harvard University Press.

Furstenberg, F. F., & Cherlin, A. (1991). *Divided families: What happens to children when parents part.* Cambridge, MA: Harvard University Press.

Furstenberg, F., & Teitler, J. O. (1994). Reconsidering the effects of marital disruption: What happens to the children of divorce in early adulthood. *Journal of Family Issues, 15,* 173-190.

Furukawa, S. (1994). The diverse living arrangements of children: Summer 1991. In U.S. Bureau of the Census, *Current Population Reports* (Series P70–38). Washington, DC: Government Printing Office.

Garasky, S. (1995).The effects of family structure on educational attainment: Do the effects vary by the age of the child? *American Journal of Economics and Sociology, 54(1), 89–106.*

Glick, P. C., & Lin, S.-L. (1986). Recent changes in divorce and remarriage. *Journal of Marriage and the Family, 48,* 737-747.

Gottman, J. M. (1994). *What predicts divorce? The relationship between marital processes and marital outcomes.* Hillsdale, NJ: Lawrence Erlbaum.

Hetherington, E. M. (1987). Family relations six years after divorce. In K. Pasley & M. Ihinger-Tallman (Eds.), *Remarriage and stepparenting: Current research and theory* (pp. 185–205). New York: Guilford.

Hetherington, E. M. (1988). Parents, children, and siblings six years after divorce. In R. Hinde & J. Stevenson-Hinde (Eds.), *Relationships within families* (pp. 311–331). Cambridge, UK: Clarendon.

Hetherington, E. M. (1993). An overview of the Virginia Longitudinal Study of Divorce and Remarriage with a focus on early adolescence. *Journal of Family Psychology, 7,* 39-56.

Hetherington, E. M., & Clingempeel, W. G., with Anderson, E., Deal, J., Hagan, M. S., Hollier, A., & Lindner, M. (1992). Coping with marital transitions: A family systems perspective. *Monographs of the Society for Research in Child Development, 57*(2-3, Serial No. 227), 1-14.

Hetherington, E., Cox, M., & Cox, R. (1982). Effects of parents and children. In M. Lamb (Ed.), *Nontraditional families: Parenting and child development* (pp. 233–288). Hillsdale, NJ: Lawrence Erlbaum.

Hoffman, S. D., & Duncan, G. D. (1988). What are the consequences of divorce? *Demography, 23,* 641-645.

Holden, K., & Smock, P. J. (1991). The economic costs of marital dissolution: Why do women bear a disproportionate cost? *Annual Review of Sociology, 17,* 51-78.

Kellan, S. (1995). Child custody and support. *Congressional Quarterly Researcher, 5(2),* 25-48.

Kitson, G. C., with Holmes, W. M. (1992). *Portrait of divorce: Adjustment to marital breakdown.* New York: Guilford.

Kitson, G., & Morgan, L. (1990). The multiple consequences of divorce: A decade review. *Journal of Marriage and the Family, 52,* 913-924.

Kurz, D. (1995). *For better or for worse: Mothers confront divorce.* New York: Routledge.

Maccoby, E. E., & Mnookin, R. H. (1992). *Dividing the child: Social and legal dilemmas of custody.* Cambridge, MA: Harvard University Press.

Martin, T. C., & Bumpass, L. L. (1989). Recent trends in marital disruption. *Demography, 26(1),* 37-51.

Mills, D. M. (1988). Stepfamilies in context. In W. Beer (Ed.), *Relative strangers* (pp. 1–29).Totowa, NJ: Rowman & Littlefield.

Mnookin, R., Maccoby, E. E., Albiston, C. R., & Depner, C. E. (1990). Private ordering revisited: What custodial arrangements are parents negotiating? In S. Sugarman & H.H. Kay (Eds.), *Divorce reform at the crossroads* (pp. 37–74). New Haven, CT: Yale University Press.

Morrison, D. R., & Cherlin, F. J. (1995). The divorce process and young children's wellbeing: A perspective analysis. *Journal of Marriage and the Family, 57,* 800-812.

Papernow, P. L. (1988). Stepparent role development: From outsider to intimate. In W. Beer (Ed.), *Relative strangers* (pp. 54–82). Totowa, NJ: Rowman & Littlefield.

Powell, B., & Downey, D. B. (1995, August). *Well-being of adolescents in single-parent households: The case of the same-sex hypothesis.* Washington, DC: Paper presented at the annual meeting of the American Sociological Association. Washington, DC.

Scoon-Rogers, L., & Lester, G. H. (1995). Child support for custodial mothers and fathers: 1991. In U.S. Bureau of the Census, *Current population reports* (Series P60–187).Washington, DC: Government Printing Office.

Seltzer, J. A. (1991). Relationships between fathers and children who live apart: The father's role after separation. *Journal of Marriage and the Family, 53,* 79-101.

Seltzer, J. A., & Bianchi, S. M. (1988). Children's contact with absent parents. *Journal of Marriage and the Family, 50,* 663-677.

Seltzer, J. A. & Garfinkel, I. (1990). Inequality in divorce settlements: An investigation of property settlements and child support awards. *Social Science Research, 19,* 82-111.

Shaw, D. S., Emery, R. E., & Tuer, M. D. (1993). Parental functioning and children's adjustment in families of divorce: A prospective study. *Journal of Abnormal Child Psychology, 21,* 119-134.

Teachman, J. (1991). Contributions to children by divorced fathers. *Social Problems, 38,* 358-371.

Thomson, E., McLanahan, S. S., & Curtin, R. B. (1992). Family structure, gender, and parental socialization. *Journal of Marriage and the Family, 54,* 368-378.

U.S. Bureau of the Census. (1995a). Child support for custodial mothers and fathers: 1991. In *Current population reports* (Series P60–187).Washington, DC: Government Printing Office.

U.S. Bureau of the Census. (1995b). *Statistical abstract of the United States, 1994.* Washington, DC: Government Printing Office.

Wallerstein, J., & Blakeslee, S. (1989). *Second chances: Men, women, and children a decade after divorce.* New York: Ticknor & Fields.

KEY CONCEPTS

divorce rate   marriage rate

DISCUSSION QUESTIONS

1. What are the factors that research has identified as affecting children's wellbeing after divorce? Are there others that you would add?

2. Remarriage produces disruptions and transitions in family life that affect all family members. What are some of the processes that emerge in the creation of stepfamilies and how are different family members affected by these changes?

48

The Protestant Ethic and the Spirit of Capitalism

MAX WEBER

Max Weber's classic analysis of the Protestant ethic and the spirit of capitalism shows how cultural belief systems, such as a religious ethic, can support the development of specific economic institutions. His multidimensional analysis shows how capitalism became morally defined as something more than pursuing monetary interests and, instead, has been culturally defined as a moral calling because of its consistency with Protestant values.

The impulse to acquisition, pursuit of gain, of money, of the greatest possible amount of money, has in itself nothing to do with capitalism. This impulse exists and has existed among waiters, physicians, coachmen, artists, prostitutes, dishonest officials, soldiers, nobles, crusaders, gamblers, and beggars. One may say that it has been common to all sorts and conditions of men at all times and in all countries of the earth, wherever the objective possibility of it is or has been given. It should be taught in the kindergarten of cultural history that this naïve idea of capitalism must be given up once and for all. Unlimited greed for gain is not in the least identical with capitalism, and is still less its spirit. Capitalism may even be identical with the restraint, or at least a rational tempering, of this irrational

SOURCE: Weber, Max. *The Protestant Ethic and the Spirit of Capitalism,* translated by Talcott Parsons, 17–27, 44–83, 157–83. New York: Scribner, 1958. Reprinted with permission.

impulse. But capitalism is identical with the pursuit of profit, and forever renewed profit, by means of continuous, rational, capitalistic enterprise. . . .

If any inner relationship between certain expressions of the old Protestant spirit and modern capitalistic culture is to be found, we must attempt to find it, for better or worse, not in its alleged more or less materialistic or at least anti-ascetic joy of living, but in its purely religious characteristics. . . .

In the title of this study is used the somewhat pretentious phrase, the *spirit* of capitalism. What is to be understood by it? The attempt to give anything like a definition of it brings out certain difficulties which are in the very nature of this type of investigation.

If any object can be found to which this term can be applied with any understandable meaning, it can only be an historical individual, i.e. a complex of elements associated in historical reality which we unite into a conceptual whole from the standpoint of their cultural significance. . . .

"Remember, that *time* is money. He that can earn ten shillings a day by his labour, and goes abroad, or sits idle, one half of that day, though he spends but sixpence during his diversion or idleness, ought not to reckon *that* the only expense; he has really spent, or rather thrown away, five shillings besides.

"Remember, that *credit* is money. If a man lets his money lie in my hands after it is due, he gives me the interest, or so much as I can make of it during that time. This amounts to a considerable sum where a man has good and large credit, and makes good use of it. . . .

"The most trifling actions that affect a man's credit are to be regarded. The sound of your hammer at five in the morning, or eight at night, heard by a creditor, makes him easy six months longer; but if he sees you at a billiard-table, or hears your voice at a tavern, when you should be at work, he sends for his money the next day; demands it, before he can receive it, in a lump." . . .

Truly what is here preached is not simply a means of making one's way in the world, but a peculiar ethic. The infraction of its rules is treated not as foolishness but as forgetfulness of duty. That is the essence of the matter. It is not mere business astuteness, that sort of thing is common enough, it is an ethos. *This* is the quality which interests us.

When Jacob Fugger, in speaking to a business associate who had retired and who wanted to persuade him to do the same, since he had made enough money and should let others have a chance, rejected that as pusillanimity and answered that "he (Fugger) thought otherwise, he wanted to make money as long as he could," the spirit of his statement is evidently quite different from that of Franklin.[1] What in the former case was an expression of commercial daring and a personal inclination morally neutral, in the latter takes on the character of an ethically coloured maxim for the conduct of life. The concept spirit of capitalism is here used in this specific sense, it is the spirit of modern capitalism. For that we are here dealing only with Western European and American capitalism is obvious from the way in which the problem was stated. Capitalism existed in China, India, Babylon, in the classic world, and in the Middle Ages. But in all these cases, as we shall see, this particular ethos was lacking. . . .

And in truth this peculiar idea, so familiar to us today, but in reality so little a matter of course, of one's duty in a calling, is what is most characteristic of the social ethic of capitalistic culture, and is in a sense the fundamental basis of it. It is an obligation which the individual is supposed to feel and does feel towards the content of his professional activity, no matter in what it consists, in particular no matter whether it appears on the surface as a utilization of his personal powers, or only of his material possessions (as capital).

Rationalism is an historical concept which covers a whole world of different things. It will be our task to find out whose intellectual child the particular concrete form of rational thought was, from which the idea of a calling and the devotion to labour in the calling has grown, which is, as we have seen, so irrational from the standpoint of purely eudæmonistic self-interest, but which has been and still is one of the most characteristic elements of our capitalistic culture. We are here particularly interested in the origin of precisely the irrational element which lies in this, as in every conception of a calling. . . .

. . . Like the meaning of the word, the idea is new, a product of the Reformation. This may be assumed as generally known. It is true that certain suggestions of the positive valuation of routine activity in the world, which is contained in this conception of the calling, had already existed in the Middle Ages, and even in late Hellenistic antiquity. We shall speak of that later. But at least one thing was unquestionably new: the valuation of the fulfillment of duty in worldly affairs as the highest form which the moral activity of the individual could assume. This it was which inevitably gave every-day worldly activity a religious significance, and which first created the conception of a calling in this sense. . . . Late Scholasticism, is, from a capitalistic viewpoint, definitely backward. Especially, of course, the doctrine of the sterility of money which Anthony of Florence had already refuted.

. . . For, above all, the consequences of the conception of the calling in the religious sense for worldly conduct were susceptible to quite different interpretations. The effect of the Reformation as such was only that, as compared with the Catholic attitude, the moral emphasis on and the religious sanction of, organized worldly labour in a calling was mightily increased. . . .

The real moral objection is to relaxation in the security of possession, the enjoyment of wealth with the consequence of idleness and the temptations of the flesh, above all of distraction from the pursuit of a righteous life. In fact, it is only because possession involves this danger of relaxation that it is objectionable at all. For the saints' everlasting rest is in the next world; on earth man must, to be certain of his state of grace, "do the works of him who sent him, as long as it is yet day." Not leisure and enjoyment, but only activity serves to increase the glory of God, according to the definite manifestations of His will.

Waste of time is thus the first and in principle the deadliest of sins. The span of human life is infinitely short and precious to make sure of one's own election. Loss of time through sociability, idle talk, luxury, even more sleep than is necessary for health, six to at most eight hours, is worthy of absolute moral condemnation. It does not yet hold, with Franklin, that time is money, but the proposition is true in a certain spiritual sense. It is infinitely valuable because every hour lost is lost to labour for the glory of God. Thus inactive contemplation is also valueless, or even directly reprehensible if it is at the expense of one's daily work. . . .

It is true that the usefulness of a calling, and thus its favour in the sight of God, is measured primarily in moral terms, and thus in terms of the importance of the goods produced in it for the community. But a further, and, above all, in practice the most important, criterion is found in private profitableness. For if that God, whose hand the Puritan sees in all the occurrences of life, shows one of His elect a chance of profit, he must do it with a purpose. Hence the faithful Christian must follow the call by taking advantage of the opportunity. "If God shows you a way in which you may lawfully get more than in another way (without wrong to your soul or to any other), if you refuse this, and choose the less gainful way, you cross one of the ends of your calling, and you refuse to be God's steward, and to accept His gifts and use them for Him when He requireth it: you may labour to be rich for God, though not for the flesh and sin."

Wealth is thus bad ethically only in so far as it is a temptation to idleness and sinful enjoyment of life, and its acquisition is bad only when it is with the purpose of later living merrily and without care. But as a performance of duty in a calling it is not only morally permissible, but actually enjoined. . . .

Let us now try to clarify the points in which the Puritan idea of the calling and the premium it placed upon ascetic conduct was bound directly to influence the development of a capitalistic way of life. As we have seen, this asceticism turned with all its force against one thing: the spontaneous enjoyment of life and all it had to offer. . . .

On the side of the production of private wealth, asceticism condemned both dishonesty and impulsive avarice. What was condemned as covetousness, Mammonism, etc., was the pursuit of riches for their own sake. For wealth in itself was a temptation. But here asceticism was the power "which ever seeks the good but ever creates evil"; what was evil in its sense was possession and its temptations. For, in conformity with the Old Testament and in analogy to the ethical valuation of good works, asceticism looked upon the pursuit of wealth as an end in itself as highly reprehensible; but the attainment of it as a fruit of labour in a calling was a sign of God's blessing. And even more important: the religious valuation of restless, continuous, systematic work in a worldly calling, as the highest means to asceticism, and at the same time the surest and most evident proof of rebirth and genuine faith, must have been the most powerful conceivable lever for the expansion of that attitude toward life which we have here called the spirit of capitalism.

When the limitation of consumption is combined with this release of acquisitive activity, the inevitable practical result is obvious: accumulation of capital through ascetic compulsion to save. The restraints which were imposed upon the consumption of wealth naturally served to increase it by making possible the productive investment of capital. . . .

One of the fundamental elements of the spirit of modern capitalism, and not only of that but of all modern culture: rational conduct on the basis of the idea of the calling, was born—that is what this discussion has sought to demonstrate—from the spirit of Christian asceticism. . . .

The Puritan wanted to work in a calling; we are forced to do so. For when asceticism was carried out of monastic cells into everyday life, and began to dominate worldly morality, it did its part in building the tremendous cosmos of the modern economic order. This order is now bound to the technical and economic conditions of machine production which today determine the lives of all the individuals who are born into this mechanism, not only those directly concerned with economic acquisition, with irresistible force....

Since asceticism undertook to remodel the world and to work out its ideals in the world, material goods have gained an increasing and finally an inexorable power over the lives of men as at no previous period in history. Today the spirit of religious asceticism—whether finally, who knows? has escaped from the cage. But victorious capitalism, since it rests on mechanical foundations, needs its support no longer. The rosy blush of its laughing heir, the Enlightenment, seems also to be irretrievably fading, and the idea of duty in one's calling prowls about in our lives like the ghost of dead religious beliefs. Where the fulfillment of the calling cannot directly be related to the highest spiritual and cultural values, or when, on the other hand, it need not be felt simply as economic compulsion, the individual generally abandons the attempt to justify it at all. In the field of its highest development, in the United States, the pursuit of wealth, stripped of its religious and ethical meaning, tends to become associated with purely mundane passions, which often actually give it the character of sport.

No one knows who will live in this cage in the future, or whether at the end of this tremendous development entirely new prophets will arise, or there will be a great rebirth of old ideas and ideals, or, if neither, mechanized petrification, embellished with a sort of convulsive self-importance. For of the last stage of this cultural development, it might well be truly said: "Specialists without spirit, sensualists without heart; this nullity imagines that it has attained a level of civilization never before achieved." . . .

The modern man is in general, even with the best will, unable to give religious ideas a significance for culture and national character which they deserve. But it is, of course, not my aim to substitute for a one-sided materialistic an equally one-sided spiritualistic causal interpretation of culture and of history. Each is equally possible, but each, if it does not serve as the preparation, but as the conclusion of an investigation, accomplishes equally little in the interest of historical truth.

NOTES

1. The quotations are attributed to Benjamin Franklin.

KEY CONCEPTS

capitalism   Protestant ethic

DISCUSSION QUESTIONS

1. Weber is known for developing a multidimensional view of human society. What role does he see the Protestant ethic as playing in the development of capitalism?

2. Weber's analysis sees western capitalists as not pursuing money just for the sake of money, but because of the moral calling invoked by the Protestant ethic. Given the place of consumerism in contemporary society, how do you think Weber might modify his argument were he writing now? In other words, are there still remnants of the Protestant ethic in our beliefs about stratification? If so, how do they fit with contemporary capitalist values?

49

America and the Challenges of Religious Diversity

ROBERT WUTHNOW

Robert Wuthnow points out that Americans generally think of the United States as a Christian nation, but that idea is being challenged by the increasing diversity of religious practice in the United States. His article, based on an extensive survey of Americans' religious beliefs and practices, examines the challenges that religious diversity poses for national identity.

Questions about racial and ethnic differences and questions about the impact of immigration have attracted extraordinary interest in recent years. Questions about religion and its cultural effects are just as important. They involve beliefs and convictions, assumptions about good and evil, individual and group identities, and concerns about how to live together. These questions were not resolved during the nation's formative era. And they certainly have not faded away. The growing presence of American Muslims, Hindus, Buddhists, and other new immigrant groups makes these questions more pressing than ever. The United States has a strong

SOURCE: Robert Wuthnow. 2005. *America and the Challenges of Religious Diversity*. Princeton, NJ: Princeton University Press.

tradition respecting the rights of diverse religious communities. But American culture is also a product of its distinctive Christian heritage. This heritage exists in tension with the nation's religious and cultural diversity.

The tension between America's Christian heritage and its religious and cultural diversity became evident in the days following the September 11, 2001, attacks on the Pentagon and World Trade Center. In a speech to Congress, President Bush declared to Muslims, "We respect your faith." Yet Bush had also said that only Christians have a place in heaven. How did he reconcile these views? What did he mean by "respect"?

An apparent inconsistency in political rhetoric like this would hardly merit attention were it not for the fact that it points to something much deeper. American identity is an odd mixture of religious particularism and cultural pluralism. Although it is not an established religion, Christianity is the nation's majority religion, and its leaders and followers have often claimed it had special, if not unique, access to divine truth. Yet the reality of religious pluralism, including beliefs and practices different from those embraced by Christianity, has also had a profound impact on American culture. These strands of our national identity are not just contradictory or conflicting impulses. They are inextricably bound together in ways that feed our collective imagination and evoke questions about who we are.

Through a large number of in-depth interviews, data from a new national survey, and published materials about the past and present, I examine the terms in which the relationship between America's Christian heritage and its growing religious diversity is being debated. I emphasize the perceptions of ordinary Americans as well as those of community leaders and the languages in which these perceptions are framed. I argue that interpretations of religious diversity have been, and continue to be, a profound aspect of our national identity.

It has become popular among social observers to argue that American religion is so thoroughly composed of private beliefs and idiosyncratic practices that belief and practice ultimately do not matter. People pick and choose in whatever way helps them to get ahead (or, at least, to get along). Their beliefs are so shallow that inconsistencies make no difference. Some observers also argue that Americans can hold fundamentally incommensurate beliefs in their personal lives, but live amicably in public. This is a recent litany in the literature on pluralism. Let religious subgroups believe whatever they want to, the argument goes, but count on laws and norms of civic decorum to maintain social order. In this view, religion and civic life function without mutual influence. Pluralism is culturally uncomplicated.

The evidence I present here suggests that these views are wrong. I show that pluralism and religious practices are intertwined. How people think about pluralism is influenced by their religious convictions. And religious convictions are influenced by their experiences with pluralism. This means that cultural interpretations of religious questions matter. They matter, not so much as formal expressions of what theologians or religious organizations teach, but in the way that Michael Polanyi described the *tacit* knowledge in which all human behavior is inscribed. Tacit knowledge matters because we prefer to live in a world, even if it is a world of our own construction, that make sense, rather than in a world without sense. Understood this way, it makes a difference how people think about questions of God, death, salvation, heaven, good and evil, other religions, and the teachings of their own tradition. It certainly matters to the many Muslims, Hindus, Buddhists, and practitioners of other non-Western religions who now make up a growing minority of the U.S. population. It also matters to Americans who claim to be Christians or Jews, or who are self-styled spiritual shoppers. They may sometimes deny that it does. But when they confront religious diversity, and when they think about what it means to be religious or spiritual in a diverse society, they articulate tacit assumptions about what it means to be human and what it means to be an American.

Religious identities matter to the collective life of society as well as to the personal lives of individuals. Religious identities are among the ways in which cultural assumptions about what is right and good, or better and best, are organized. Americans believe they are a special people with a distinctive mission to fulfill in the world. This belief is associated historically with our understanding of religion. To say that we are a Christian nation has been a normative statement as well as a descriptive one. Christian values and practices occupied a special place in our thinking. To say that some people were Christian implied that others were not. Our moral universe included assumptions about Jews, Muslims, Hindus, Buddhists, and practitioners of Native American religions. They, too, had duties to fulfill, roles in the cultural drama to perform. Religious diversity was inscribed in the moral order.

Another popular approach to religion among social scientists is to deal with it as if it were purely an expression of something else, such as class, race, gender, and region, or to explain its trends and patterns with reference to demography, organizations, leadership styles, and theories about rational choice. These reductionistic approaches give social scientists an excuse to avoid the *content* of religion. What people believe, or say they believe, and the language in which they make sense of their beliefs and practices are somehow, in the view of these scholars, too marginal, too normative, or too difficult to measure for any self-respecting social scientist to tackle. This is the point at which narrow definitions of disciplinary boundaries get in the way of knowledge.

I choose to emphasize what people think and the cultural idioms in which they express their thoughts. This is how people make sense of their beliefs and practices. It is how they negotiate meaning when faced with multiple religious teachings and traditions. If people were guided only by demography or social position, there would be no need to know what they say or think. But there is a well-established tradition in the social sciences (counting Max Weber and George Herbert Mead among those who observe it) that says that the meaning-making activity of humans is crucial to our understanding of society. Making sense of religious diversity is one of these meaning-making activities.

Still, listening to what people say would be of little value if their views merely echoed the writings of theologians and social philosophers. If ordinary people were guided by these writings, one would want to spend the time one has for scholarly reflection understanding these tomes and writing commentaries about them. Worthy as that may be, it does not provide much of a picture of the society in which we actually live.

When rank-and-file Americans talk about religious diversity, they disclose an implicit cultural text composed of narrative fragments from personal experience, from conversations with friends and neighbors, from the media, from books and magazines, and in many instances from ruminations about questions raised in Sunday school, a high school youth group, a course in comparative religions, or a visit to another country. It is possible to identify themes and variations in this subterranean text. Some people find ways to embrace religious diversity as fully as possible. Others assert loyalty to the tradition in which they were raised (or are presently involved), but acknowledge the validity of other traditions. A substantial number of Americans adamantly reject the truth of religions other than their own. In each of these orientations, people articulate a bricolage of ideas that both reflects and subverts public images of cultural diversity. Patterns of avoidance minimizing considered engagement among religious traditions are evident. And these avoidances illuminate the behavior of religious organizations and their leaders.

I do not argue that the present encounter with religious diversity is entirely new or without precedent. My argument is rather that America and American Christianity have always existed in a world of religious differences and with some awareness of these differences. I further argue, however, that this awareness is probably greater among rank-and-file Americans now than in the past because of mass communications, immigration, and our nation's role in the global economy. . . .

Whereas historical treatments of religious diversity in America have typically emphasized the divisions *within* Christianity (especially those separating Protestants and Catholics, Protestant denominations, and the various sects), my emphasis here is on the encounter between American Christians and other major religious traditions, such as Islam, Buddhism, and Hinduism. I understand that the tensions between Protestants and Catholics, or even between rival branches of Presbyterianism, were sometimes as fierce as anything evident currently between Christians and non-Christians. I am nevertheless interested in the fact that Christians have always had to formulate arguments about people who were clearly outside the Christian tradition by virtue of belonging to other major religious traditions. I am interested in how these arguments played into our national identity historically and how they are being revised at present.

My aim is not to encourage readers to conclude that religions are interchangeable. Nor do I believe the best way to live in a pluralistic society is to combine bits and pieces of several religions. I do insist that the growing religious diversity of our society poses a *significant cultural challenge*. The fact that Muslims, Hindus, and Buddhists are now a significant presence in the United States raises fundamental questions about our historic identity as a Christian nation. This new reality requires is to rethink our national identity and to face difficult choices about how pluralistic we are willing to be. It requires people of all religions, as well as scholars and community leaders, to take notice. If a person's best friend in elementary school belonged to a different religion, and if this person takes religion seriously, he or she will surely think about his or her faith differently than would have been the case if everyone in school belonged to the same religion. If one's neighbors and coworkers hold beliefs vastly different from one's own, this too will evoke a response. We can try to understand and become more aware of these influences, and thus make more informed choices about how we respond, rather than letting circumstances dictate our responses.

. . . News coverage from around the world includes images of religious leaders, adherents, and their places of worship. The nation's expansive economic and military activities render these images more newsworthy than they would have been in the past. Apart from media, exposure to the world's religions comes increasingly through first-hand encounters. During the last third of the twentieth century, approximately twenty-two million immigrants came to the United States. Like the surge of immigration that occurred between 1890 and 1920, most of these immigrants came from countries in which Christians are the dominant religion. Yet, in contrast to that earlier period, the recent immigration included millions of people from countries in which Christians are only a small minority. Thus, in little more than a generation, the United States has witnessed an unprecedented increase in the diversity of major religious traditions represented among its population. More Americans belong to religions outside of the Christian tradition than ever before. The new immigrants include large numbers of Muslims, Hindus, Buddhists, and followers of other traditions and spiritual practices. Their presence greatly increases the likelihood of personal interaction across these religious lines.

Recent immigrants and their descendants generally do not live isolated from other Americans in homogeneous enclaves. They frequently work in middle-class occupations and live in the same neighborhoods as other Americans do. Their mosques, temples, and meditation centers are often located in close proximity to churches and synagogues. The typical American, therefore, can more readily encounter people of other religions as neighbors, friends, and coworkers.

Diversity is always challenging, whether it is manifest in language differences or in modes of dress, eating, and socializing. Seeing people with different habits and lifestyles makes it harder to practice our own unreflectively. When religion is involved, these challenges are multiplied. Religious differences are instantiated [represented] in dress, food, holidays, and family rituals; they also reflect historic teachings and deeply held patterns of belief and practice. These beliefs and practices may be personal and private, but they cannot easily be divorced from questions about truth and morality. Believing that one's faith is correct and behaving in ways that reflect this belief may well be different in the presence of diversity than in its absence.

How have we responded to the religious diversity that increasingly characterizes our neighborhoods, schools, and places of work? Has it sunk into our awareness that the temple or mosque down the street is not just another church? Does it matter that our coworkers have radically different ideas of the sacred than we do? Or do we perceive these ideas as so different from our own? Are our views of America affected by having neighbors whose beliefs and lifestyles may run counter to our own? Does it bother us to read about hate crimes directed at Muslims or Hindus?

Historic interpretations of Christian teachings encourage Christians to practice the acceptance and love exhibited by Mother Teresa. Stories about Jesus' willingness to violate social boundaries separating Jews and Gentiles exemplify how Christianity may encourage openness to racial, ethnic, and cultural diversity.

Yet Christianity has also taught that only by accepting Jesus as their savior can believers overcome sinfulness and gain divine redemption. According to some interpretations of this teaching, the followers of other religions must convert to Christianity if they are to know God.

Throughout America's history, our sense of who we are has been profoundly influenced by our religious beliefs and practices. Christianity's claim to be the unique representative of divine truth has been one of these influences. We have thought of ourselves as a chosen people, a city on a hill, and a new Israel. We have considered ourselves defenders of the faith, a God-fearing people, and a Christian nation. At present, we remain one of the most religiously committed of all nations, at least if religious commitment is measured in numbers professing belief in God and attending services at houses of worship. Our identity is still marked by this fact. Many Americans take for granted that we are a Christian society, even if they implicitly make a place in this notion for Jews and unbelievers. Others take pride in our national accomplishments, our democratic traditions, and our extensive voluntary associations, assuming that these reflect Christian values.

If our understanding of what it means to be American reflects our religious heritage, our collective identity is also influenced by how we think about religious diversity. Until recently, we were able to think of ourselves as a Christian civilization, divided by the historic cleavages separating Protestants from Catholics and, among Protestants, Methodists from Baptists, Presbyterians from Episcopalians, Congregationalists from Quakers, and so on. We were a diverse nation because of the national origins from which the various denominational groups had come and because of racial, ethnic, and regional divisions in which religious disunity was embedded. We took pride in this diversity. It seemed like a mark of distinction.

We clearly do have a long history of religious diversity. This history has affected our laws, encouraging us to avoid governmental intrusion in religious affairs that might lead to an establishment of one tradition in favor of others. And it has taught us a kind of civic decorum that discourages blatant expressions of racist, ethnocentric, and nativist ideas. Yet it will not do, now in the face of new diversity, simply to rewrite our nation's history as a story of diversity and pluralism.

The reality of large numbers of Americans who are Muslims, Buddhists, Zoroastrians, Sikhs, Hindus, and followers of other non-Western religions poses a new challenge to American self-understandings. When Christian leaders and their followers think about it, they will have more trouble knowing what exactly to think about their neighbors who belong to these other religions than they ever did simply thinking about the differences between Methodists and Baptists or Protestants and Catholics. That is, *if* they stop to think about it.

But the truth is, we know very little at this point about how ordinary Americans are responding to religious diversity. And, for that matter, we know little more about how religious leaders are dealing with diversity. We do know, for example, that religious leaders occasionally form interfaith alliances that include representatives of the world's major religious traditions, and we know that other leaders are sometimes quoted in newspapers as saying that the followers of a particular religion other than their own are condemned to hell. Such headlines, however, seldom tell us much about how things are going in local communities or what people really believe and think. . . .

. . . How well are we managing to face the new challenges of religious and cultural diversity? Are we merely managing in the sense of making do, muddling our way by avoiding the issues whenever possible and responding superficially whenever we must? Or are we managing better than that? Are we taking advantage of the opportunities that diversity provides and moving toward a more mature pluralism than we have known in the past?

These are, in my view, among the most serious questions we currently face as a nation. In our public discourse about religion we seem to be a society of schizophrenics. On the one hand, we say casually that we are tolerant and have respect for people whose religious traditions happen to be different from our own. On the other hand, we continue to speak as if our nation is (or should be) a Christian nation, founded on Christian principles, and characterized by public references to the trappings of this tradition. That kind of schizophrenia encourages behavior that no well-meaning people would want if they stopped to think about it for very long. It allows the most open-minded among us to get by without taking religion very seriously at all. It permits religious hate crimes to occur without much public attention or outcry. The members of new minority religions experience little in the way of genuine understanding. The churchgoing majority seldom hear anything to shake up their comforting convictions. The situation is rife with misunderstanding and, as such, holds little to prevent outbreaks of religious conflict and bigotry. It is little wonder that many Americans retreat into their private worlds whenever spirituality is mentioned. It is just easier to do that than to confront the hard questions about religious truth and our national identity.

KEY CONCEPTS

pluralism   religiosity   secular

DISCUSSION QUESTIONS

1. How has immigration changed the religious composition of the U.S. population?

2. How does Wuthnow explain the seeming contradiction between a national identity that respects religious diversity while at the same time being centered in a Christian tradition?

3. What are some of the new questions that Wuthnow identifies as stemming from increased religious diversity in the United States?

50

Religion in the Lives of American Adolescents

MARK REGNERUS, CHRISTIAN SMITH, AND MELISSA FRITSCH

This research study documents the influence of religion in the lives of young Americans. In the study, derived from a large, national survey, Mark Regnerus and his colleagues show the influence of religion on a variety of social outcomes. Here they examine the social factors that are related to religious behavior and beliefs among adolescents.

Contrary to some popular images, religion plays a significant role in the lives of many adolescents in the United States, according to a number of surveys and public opinion polls (see Smith et al. 2002, 2003). For instance, in the early 1990s the Gallup organization reported that some 76 percent of adolescents (ages 13–17) believed in a personal God and that 74 percent prayed at least occasionally (Gallup and Bezilla 1992). Data from the Monitoring the Future project suggest that the overall level of religiousness among U.S. adolescents is relatively high (Donahue and Benson 1995). According to those data, the percentage of high school seniors attending religious services weekly dropped from around 40 percent (1976–81) to 31 percent by 1991 but has remained stable throughout the 1990s. Additionally, nearly 30 percent of 12th graders indicated that religion was a "very important" part of their lives, a figure that has held steady since the inception of the project (Johnston, Bachman and O'Malley 1999).

Data from Monitoring the Future and other sources also suggest an important developmental component of adolescent religious involvement. In short, the frequency of attendance tends to decline between eighth and 12th grades (Potvin, Hoge and Nelson 1976; Benson, Donahue and Erickson 1989; Roehlkepartain and Benson 1993). In 1997, about 44 percent of eighth graders reported attending religious services weekly, as compared with 38 percent of 10th graders and 31 percent of 12th graders. This drop in attendance might reflect growing autonomy as teens mature. Among 12th graders, regular religious service attendance might be likely to result as much from individual volition as from influences of intergenerational transmission and parental socialization. At the same time, there are few age differences in religious salience. Moreover, a study of

SOURCE: Mark Regnerus, Christian Smith, and Melissa Fritsch. 2003. *Religion in the Lives of American Adolescents: A Literature Review.* Chapel Hill, NC: National Study of Youth and Religion.

adolescents in Iowa found that while frequency of religious service attendance dipped during the high school years, levels of participation in other religious activities tended to rise over the same period (King, Elder and Whitbeck 1997).

Among adolescents, as among adults, religious involvement is patterned by gender and race. On average, girls are consistently "more religious" than boys—i.e., more likely to attend services weekly and to report that religion is "very important" to them—by several percentage points. Blacks are more likely than whites to attend religious services regularly (40 percent vs. 29 percent) and vastly more likely to indicate that religion has high importance in their lives (55 percent vs. 24 percent) (Johnston, Bachman and O'Malley 1999). Although much less is known (especially in comparative perspective) about the religious involvement of Latino and Asian-American adolescents, some data indicate that they, too, report greater involvement and commitment than non-Hispanic white youth (Benson 1993).

RELIGIOUS INVOLVEMENT

The sources behind the development of religious involvement in youth are several, though parents easily constitute the strongest influence. Some scholars go so far as to suggest that "religiosity, like class, is inherited" (Myers 1996: 858). Parent-child transmission of religiosity and religious identity is indeed quite powerful. But it's not inevitable. On the whole, mainline Protestant parents are having greater difficulty retaining their children within the mainline Protestant fold than are evangelical Protestant parents (Smith 1998). Religious socialization also is more likely to occur in families characterized by considerable warmth and closeness (Ozorak 1989). Mothers are generally thought to be more influential than fathers in the development of religiousness in adolescent children (Benson, Masters and Larson 1997; Bao et al. 1999).

Studies nearly universally find adolescent girls to be more religious—both privately and publicly—than boys. Ozorak (1989) suggests, in keeping with results from several national datasets, that polarization in religiosity occurs during adolescence. That is, the decreases (in religiosity) of somewhat or moderately religious youth mask the increases of the (fewer) very religious. In her study of 390 adolescents, she found parents' religious affiliation and practices related to all aspects of religiosity during early and middle adolescence, though much less so among older adolescents. Cohesive families, she found, curb this diminishing influence somewhat. Erickson's (1992) analysis of Search Institute data on adolescents found parents' religious influence on religious behavior to be minimal for boys but quite robust for girls.

Overall, his analysis suggests that "adolescent religious development is triggered by home religious habits and religious education, while the (direct) influence of both parents and peers is less important than previously suggested" (1992: 146). Gunnoe and Moore's (2002) research using data from three waves of the National Survey of Children showed that subsequent young adult religiosity was best predicted by peers' religious service attendance patterns during high school, ethnicity (African-American) and gender (female). Maternal religiosity and living in the South were also related to religiosity, but parenting style was not.

Scott Myers' (1996) longitudinal analysis of parents and—12 years later—their adult children revealed that while one's religiosity is "determined largely by the religiosity of one's parents," it also is fostered among families where parents enjoy marital happiness, display moderate strictness, support and show affection toward their children and in households where the husband is employed and the wife is not. Perkins' (1987) study of college youth revealed that 69 percent of students with two highly religious parents reported a strong personal faith themselves, compared to only 39 percent of students with only one devout parent. A novel study of teenage twin girls (1,687 pairs) and their parents provides unusual opportunity to distinguish environmental or socialized development of religiosity from that which is "inherited" (Heath et al. 1999). In it they show evidence suggesting that black girls display considerably higher "heritability" of religious involvement and religious values in contrast to white girls or girls of other races or ethnicities. While still underdeveloped in its ramifications, this suggests that there may be a genetic component to the transmission of religiosity. Why it is more powerfully observed among blacks remains unknown, though a potential genetic trait is in keeping with their genetic tendency toward earlier physical maturation.

Most studies have focused less on parental influences on youth religiosity and more on characteristics of the youth themselves that are conducive to religiosity. Particularly since public religiosity appears to decline during adolescence, what factors stimulate or impede that from happening are of central interest. Dudley's (1999) longitudinal research on Seventh-Day Adventist youth revealed predictable factors related to adolescents' subsequent maintenance of regular attendance patterns—namely, intent to remain a faithful attender, parents' attendance patterns and integration into church day school. In an earlier study, Dudley and Laurent (1988) concluded that the quality of relationship with pastors and parents, as well as opportunities for their own religious involvement, self-concept and the influence of peer groups and mass media each played a role in explaining alienation from religion in a sample of 390 youth.

Youthful attraction to cults is surprisingly understudied, due perhaps to its relative infrequency (despite occasional hype suggesting otherwise). In describing the personality profile of a youth susceptible to cult overtures, both Hunter (1998) and Parker (1985) list identity confusion, alienation from family members, weak social and religious ties and feelings of helplessness or powerlessness (external locus of control). Cult members are generally more apt to come from upper middle class homes exhibiting democratic parental authority structures. Typically, relationships with family decline precipitously prior to membership. A structured sense of belonging and an escape from perceived "normlessness" attract many recruits. Hunter suggests that 18- to 23-year-olds are most at risk for successful cult recruitment. Youth who are converted rapidly to more mainstream religions or religious traditions often are influenced by social pressures as well. Social bonds made possible through religion—whether traditional or "cult-like"—are attractive. Choosing a strong religious identity also is stimulated frequently by role models (Parker 1985).

Most research about conversion focuses less on gradual and developmental conversion types and more on sudden forms, though the former types are much more common in most religious traditions. An historical study of adolescence and revivals in antebellum Boston documented evidence suggesting revivalists particularly targeted adolescents and fashioned their methods to provoke emotions that were popularly associated with youth (Schwartz 1974). Countering anti-revivalists' fears of unleashing pent-up inhibitions, the revivalists argued that youthful emotions are valid and constituted fertile soil in which true religious sentiments could root. The spiritual autobiographies of both distant past and present day tend to describe periods of youthful "wildness, corruption and indiscipline before the onset of (religious) conviction," as Thompson (1984: 140) depicts adolescent culture in colonial Massachusetts.

Religious "doubting" among adolescents was the focus of Kooistra and Pargament's (1999) study of 267 Catholic and Dutch Calvinist (Reformed) school students. First, religious doubting was common—78 percent indicated currently having doubts. Catholic school students displayed considerably more doubt than the Reformed students, who were higher on several religiosity counts. Among the latter, religious doubting was associated with adverse life events, conflictual family patterns and emotional distress.

King, Elder and Whitbeck (1997) assess religious involvement among rural Iowa and inner-city Philadelphia youth. The Philadelphia youth were much less likely to be involved than the Iowa adolescents: 5 percent of the former attended religious services more than once per week, compared to 29 percent of Iowans. None of the Philadelphia sample mentioned youth group participation; 20 percent of the Iowa sample did so. Nevertheless, the Iowa sample did exhibit common developmental traits of religiosity, such as declining average levels of attendance across adolescence. Girls were more involved than boys. More private forms of religiosity remained stable, as did overall participation in organized religious activities. The most consistently and intensely religious youth were distinguished by several factors: farm residence, self-identified as "born again" and having and identifying with religious parents. . . .

REFERENCES

Bao, Wan Ning, Les B. Whitebeck, Danny R. Hoyt, and Rand D. Conger. 1999. "Perceived Parental Acceptance as a Moderator of Religious Transmission among Adolescent Boys and Girls." *Journal of Marriage and the Family* 61:362–374.

Benson, Peter L. 1993. "Troubled Journey: A Portrait of 6th-12th Grade Youth." Search Institute, Minneapolis, MN.

Benson, Peter L., Michael J. Donahue, and Joseph A. Erickson. 1989. "Adolescence and Religion: A Review of the Literature from 1970 to 1986." In *Research in the Social Scientific Study of Religion: A Research Annual*, vol. 1, edited by M. L. Lynn and D. O. Moberg, 153–181. Greenwich, CT: JAI Press, Inc.

Benson, Peter L., Kevin S. Masters, and David B. Larson. 1997. "Religious Influences on Child and Adolescent Development." In *Handbook of Child and Adolescent Psychiatry, Vol. 4, Varieties of Development*, edited by N. E. Alessi, 206–219. New York: John Wiley and Sons, Inc.

Dudley, Roger L. 1999. "Youth Religious Commitment over Time: A Longitudinal Study of Retention." *Review of Religious Research* 41:110–121.

Dudley, Roger L. and C. Robert Laurent. 1989. "Alienation from Religion in Church-Related Adolescents." *Sociological Analysis* 49:408–420.

Erickson, Joseph A. 1992. "Adolescent Religious Development and Commitment: A Structural Equation Model of the Role of Family, Peer Group, and Educational Influences." *Journal for the Scientific Study of Religion* 31:131–152.

Gallup, George and Robert Bezilla. 1992. *The Religious Life of Young Americans*. Princeton, NJ: George H. Gallup International Institute.

Gunnoe, Marjorie Lindner and Kristin A. Moore. (2002). "Predictors of Religiosity Among Youth Aged 17–22: A Longitudinal Study of the National Survey of Children." *Journal for the Scientific Study of Religion* 41:613–622.

Heath, Andrew C., Pamela A. Madden, Julia D. Grant, Tara L. McLaughlin, Alexandre A. Todorov, and Kathleen K. Bucholz. 1999. "Resiliency Factors Protecting against Teenage Alcohol Use and Smoking: Influences of Religion, Religious Involvement and Values, and Ethnicity in the Missouri Adolescent Female Twin Study." *Twin Research* 2:145–55.

Hunter, Eagan. 1998. "Adolescent Attraction to Cults." *Adolescence* 33:709–714.

Iannaccone, Laurence. 1994. "Why Strict Churches Are Strong." *American Journal of Sociology* 99:1180–1212.

Johnston, Lloyd D., Jerald G. Bachman, and Patrick M. O'Malley. 1999. *Monitoring the Future: Questionnaire Responses from the Nation's High School Seniors*. Ann Arbor: Institute for Social Research.

King, Valarie, Glen H. Elder, Jr., and Les B. Whitbeck. 1997. "Religious Involvement among Rural Youth: An Ecological and Life-Course Perspective." *Journal of Research on Adolescence* 7:431–456.

Kooistra, William P. and Kenneth I. Pargament. 1999. "Religious Doubting in Parochial School Adolescents." *Journal of Psychology and Theology* 27:33–42.

Myers, Scott M. 1996. "An Interactive Model of Religiosity Inheritance: The Importance of Family Context." *American Sociological Review* 61:858–866.

Ozorak, Elizabeth W. 1989. "Social and Cognitive Influences on the Development of Religious Beliefs and Commitment in Adolescence." *Journal for the Scientific Study of Religion* 28:448–463.

Parker, Mitchell S. 1985. "Identity and the Development of Religious Thinking." *New Directions for Child Development* 30:43–60.

———. 1987. "Parental Religion and Alcohol Use Problems as Intergenerational Predictors of Problem Drinking among College Youth." *Journal for the Scientific Study of Religion* 26:340–357.

Potvin, Raymond H., Dean R. Hoge, and Hart M. Nelsen. 1976. *Religion and American Youth, with Emphasis on Catholic Adolescents and Young Adults*. Washington: Office of Research Policy and Program Development, U.S. Catholic Conference.

Roehlkepartain, Eugene C. and Peter L. Benson. 1993. *Youth in Protestant Churches*. Minneapolis, MN: Search Institute.

Schwartz, Hillel. 1974. "Adolescence and Revivals in Ante-Bellum Boston." *Journal of Religious History* 8:144–158.

Smith, Christian. 2003. "Theorizing Religious Effects among American Adolescents." *Journal for the Scientific Study of Religion* 42 (1):17–30.

Smith, Christian, Melinda Lundquist Denton, Robert Faris, and Mark Regnerus. 2002. "Mapping American Adolescent Religious Participation." *Journal for the Scientific Study of Religion*. 41 (4):397–612.

Smith, Christian, Robert Faris, Melinda Lundquist Denton, and Mark Regnerus. 2003. "Mapping American Adolescent Subjective Religiosity and Attitudes of Alienation Toward Religion: A Research Report." *Sociology of Religion* 64 (1):111–113.

Smith, Tom W. 1998. "A Review of Church Attendance Measures." *American Sociological Review* 63:131–136.

Thompson, Roger. 1984. "Adolescent Culture in Colonial Massachusetts." *Journal of Family History* 9:127–144.

KEY CONCEPTS

conversion	religiosity	religious socialization

DISCUSSION QUESTIONS

1. What are some of the major social factors that influence youth's religious involvement?

2. How is religion among young people related to that of their parents?

51

The Truth about Boys and Girls

SARA MEAD

There has recently been much national attention to an alleged crisis in boys' education. Based on a review of research on boys' and girls' educational achievement, Sara Mead addresses the question of whether there is a crisis in the education of boys.

SOURCE: Sara Mead. 2006. "The Truth about Boys and Girls." *Education Sector*. www.educationsector.org.

If you've been paying attention to the education news lately, you know that American boys are in crisis. After decades spent worrying about how schools "shortchange girls,"[1] the eyes of the nation's education commentariat are now fixed on how they shortchange boys. In 2006 alone, a *Newsweek* cover story, a major *New Republic* article, a long article in *Esquire*, a "Today" show segment, and numerous op-eds have informed the public that boys are falling behind girls in elementary and secondary school and are increasingly outnumbered on college campuses. A young man in Massachusetts filed a civil rights complaint with the U.S. Department of Education, arguing that his high school's homework and community service requirements discriminate against boys.[2] A growth industry of experts is advising educators and policymakers how to make schools more "boy friendly" in an effort to reverse this slide.

It's a compelling story that seizes public attention with its "man bites dog" characteristics. It touches on Americans' deepest insecurities, ambivalences, and fears about changing gender roles and the "battle of the sexes." It troubles not only parents of boys, who fear their sons are falling behind, but also parents of girls, who fear boys' academic deficits will undermine their daughters' chances of finding suitable mates.

But the truth is far different from what these accounts suggest. The real story is not bad news about boys doing worse; it's good news about girls doing better.

In fact, with a few exceptions, American boys are scoring higher and achieving more than they ever have before. But girls have just improved their performance on some measures even faster. As a result, girls have narrowed or even closed some academic gaps that previously favored boys, while other long-standing gaps that favored girls have widened, leading to the belief that boys are falling behind.

There's no doubt that some groups of boys—particularly Hispanic and black boys and boys from low-income homes—are in real trouble. But the predominant issues for them are race and class, not gender. Closing racial and economic gaps would help poor and minority boys more than closing gender gaps, and focusing on gender gaps may distract attention from the bigger problems facing these youngsters.

The hysteria about boys is partly a matter of perspective. While most of society has finally embraced the idea of equality for women, the idea that women might actually surpass men in some areas (even as they remain behind in others) seems hard for many people to swallow. Thus, boys are routinely characterized as "falling behind" even as they improve in absolute terms.

In addition, a dizzying array of so-called experts have seized on the boy crisis as a way to draw attention to their pet educational, cultural, or ideological issues. Some say that contemporary classrooms are too structured, suppressing boys' energetic natures and tendency to physical expression; others contend that boys need more structure and discipline in school. Some blame "misguided feminism" for boys' difficulties, while others argue that "myths" of masculinity have a crippling impact on boys.[3] Many of these theories have superficially plausible rationales that make them appealing to some parents, educators, and policy-makers. But the evidence suggests that many of these ideas come up short.

Unfortunately, the current boy crisis hype and the debate around it are based more on hopes and fears than on evidence. This debate benefits neither boys nor girls, while distracting attention from more serious educational problems—such as large racial and economic achievement gaps—and practical ways to help both boys and girls succeed in school.

A NEW CRISIS?

The Boy Crisis. At every level of education, they're falling behind. What to do?

NEWSWEEK COVER HEADLINE, JAN. 30, 2006

Newsweek is not the only media outlet publishing stories that suggest boys' academic accomplishments and life opportunities are declining. But it's not true. Neither the facts reported in these articles nor data from other sources support the notion that boys' academic performance is falling. In fact, overall academic achievement and attainment for boys is higher than it has ever been.

Long-Term Trends

Looking at student achievement and how it has changed over time can be complicated. Most test scores have little meaning themselves; what matters is what scores tell us about how a group of students is doing relative to something else: an established definition of what students need to know, how this group of students performed in the past, or how other groups of students are performing. Further, most of the tests used to assess student achievement are relatively new, and others have changed over time, leaving relatively few constant measures.

The National Assessment of Educational Progress (NAEP), commonly known as "The Nation's Report Card," is a widely respected test conducted by the U.S. Department of Education using a large, representative national sample of American students. NAEP is the only way to measure national trends in boys' and girls' academic achievements over long periods of time.[4] There are two NAEP tests. The "main NAEP" has tracked U.S. students' performance in reading, math, and other academic subjects since the early 1990s. It tests students in grades four, eight, and 12. The "long-term trend NAEP" has tracked student performance since the early 1970s. It tests students at ages 9, 13, and 17.

Reading

The most recent main NAEP assessment in reading, administered in 2005, does not support the notion that boys' academic achievement is falling. In fact, fourth-grade boys did better than they had done in both the previous NAEP reading assessment, administered in 2003, and the earliest comparable assessment, administered in 1992. Scores for both fourth-and eighth-grade boys have gone up and down over the past decade, but results suggest that the reading skills of fourth- and eighth-grade boys have improved since 1992.[5]

The picture is less clear for older boys. The 2003 and 2005 NAEP assessments included only fourth- and eighth-graders, so the most recent main NAEP data for 12th-graders dates back to 2002. On that assessment, 12th-grade boys did worse than they had in both the previous assessment, administered in 1998, and the first comparable assessment, administered in 1992. At the 12th-grade level, boys' achievement in reading does appear to have fallen during the 1990s and early 2000s.[6]

Even if younger boys have improved their achievement over the past decade, however, this could represent a decline if boys' achievement had risen rapidly in previous decades. Some commentators have asserted that the boy crisis has its roots in the mid- or early-1980s. But long-term NAEP data simply does not support these claims. In fact, 9-year-old boys did better on the most recent long-term reading NAEP, in 2004, than they have at any time since the test was first administered in 1971. Nine-year-old boys' performance rose in the 1970s, declined in the 1980s, and has been rising since the early 1990s.

Like the main NAEP, the results for older boys on the long-term NAEP are more mixed. Thirteen-year-old boys have improved their performance slightly compared with 1971, but for the most part their performance over the past 30 years has been flat. Seventeen-year-old boys are doing about the same as they did in the early 1970s, but their performance has been declining since the late 1980s.[7]

The main NAEP also shows that white boys score significantly better than black and Hispanic boys in reading at all grade levels. These differences far outweigh all changes in the overall performance of boys over time. For example, the difference between white and black boys on the fourth-grade NAEP in reading in 2005 was 10 times as great as the improvement for all boys on the same test since 1992.

And while academic performance for minority boys is often shockingly low, it's not getting worse. The average fourth-grade NAEP reading scores of black boys improved more from 1995 to 2005 than those of white and Hispanic boys or girls of any race.

Math

The picture for boys in math is less complicated. Boys of all ages and races are scoring as high—or higher—in math than ever before. From 1990 through 2005, boys in grades four and eight improved their performance steadily on the main NAEP, and they scored significantly better on the 2005 NAEP than in any previous year. Twelfth-graders have not taken the main NAEP in math since 2000. That year, 12th-grade boys did better than they had in 1990 and 1992, but worse than they had in 1996.[8]

Both 9- and 13-year-old boys improved gradually on the long-term NAEP since the 1980s (9-year-old boys' math performance did not improve in the 1970s). Seventeen-year-old boys' performance declined through the 1970s, rose in the 1980s, and remained relatively steady during the late 1990s and early 2000s.[9] As in reading, white boys score much better on the main NAEP in math than do black and Hispanic boys, but all three groups of boys are improving their math performance in the elementary and middle school grades.[10] . . .

Overall Long-Term Trends

A consistent trend emerges across these subjects: There have been no dramatic changes in the performance of boys in recent years, no evidence to indicate a boy crisis. Elementary-school-age boys are improving their performance; middle school boys are either improving their performance or showing little change, depending on the subject; and high school boys' achievement is declining in most subjects (although it may be improving in math). These trends seem to be consistent across all racial subgroups of boys, despite the fact that white boys perform much better on these tests than do black and Hispanic boys.

Evidence of a decline in the performance of older boys is undoubtedly troubling. But the question to address is whether this is a problem for older boys or for older students generally. That can be best answered by looking at the flip side of the gender equation: achievement for girls.

The Difference Between Boys and Girls

To the extent that tales of declining boy performance are grounded in real data, they're usually framed as a decline relative to girls. That's because, as described above, boy performance is generally staying the same or increasing in absolute terms.

But even relative to girls, the NAEP data for boys paints a complex picture. On the one hand, girls outperform boys in reading at all three grade levels assessed on the main NAEP. Gaps between girls and boys are smaller in fourth grade and get larger in eighth and 12th grades. Girls also outperform boys in writing at all grade levels.

In math, boys outperform girls at all grade levels, but only by a very small amount. Boys also outperform girls—again, very slightly—in science and by a slightly larger margin in geography. There are no significant gaps between male and female achievement on the NAEP in U.S. history. In general, girls outperform boys in reading and writing by greater margins than boys outperform girls in math, science, and geography.

But this is nothing new. Girls have scored better than boys in reading for as long as the long-term NAEP has been administered. And younger boys are actually catching up: The gap between boys and girls at age 9 has narrowed significantly since 1971—from 13 points to five points—even as both genders have significantly improved. Boy-girl gaps at age 13 haven't changed much since 1971—and neither has boys' or girls' achievement.

At age 17, gaps between boys and girls in reading are also not that much different from what they were in 1971, but they are significantly bigger than they were in the late 1980s, before achievement for both genders—and particularly boys—began to decline.

The picture in math is even murkier. On the first long-term NAEP assessment in 1973, 9- and 13-year-old girls actually scored better than boys in math, and they continued to do so throughout the 1970s. But as 9- and 13-year-olds of both genders improved their achievement in math during the 1980s and 1990s, boys *pulled ahead* of girls, opening up a small gender gap in math achievement that now favors boys. It's telling that even though younger boys are now doing better than girls on the long-term NAEP in math, when they once lagged behind, no one is talking about the emergence of a new "girl crisis" in elementary- and middle-school math.

Seventeen-year-old boys have always scored better than girls on the long-term NAEP in math, but boys' scores declined slightly more than girls' scores in the 1970s, and girls' scores have risen slightly more than those of boys since. As a result, older boys' advantage over girls in math has narrowed.

Overall, there has been no radical or recent decline in boys' performance relative to girls. Nor is there a clear overall trend—boys score higher in some areas, girls in others.

The fact that achievement for older students is stagnant or declining for both boys and girls, to about the same degree, points to another important element of the boy crisis. The problem is most likely not that high schools need to be fixed to meet the needs of boys, but rather that they need to be fixed to meet the needs of *all* students, male and female. The need to accurately parse the influence of gender and other student categories is also acutely apparent when we examine the issues of race and income.

We Should Be Worried About Some Subgroups of Boys

There are groups of boys for whom "crisis" is not too strong a term. When racial and economic gaps combine with gender achievement gaps in reading, the result is disturbingly low achievement for poor, black, and Hispanic boys.

But the gaps between students of different races and classes are much larger than those for students of different genders—anywhere from two to five times as big, depending on the grade. The only exception is among 12th-grade boys, where the achievement gap between white girls and white boys in reading is the same size as the gap between white and black boys in reading and is larger than the gap between white and Hispanic boys. Overall, though, poor, black, and Hispanic boys would benefit far more from closing racial and economic achievement gaps than they would from closing gender gaps. While the gender gap picture is mixed, the racial gap picture is, unfortunately, clear across a wide range of academic subjects.

In addition to disadvantaged and minority boys, there are also reasons to be concerned about the substantial percentage of boys who have been diagnosed with disabilities. Boys make up two-thirds of students in special education—including 80 percent of those diagnosed with emotional disturbances or autism—and boys are two and a half times as likely as girls to be diagnosed with attention deficit hyperactivity disorder (ADHD).[11] The number of boys diagnosed with disabilities or ADHD has exploded in the past 30 years, presenting a challenge for schools and causing concern for parents. But the reasons for this growth are complicated, a mix of educational, social, and biological factors. Evidence suggests that school and family factors—such as poor reading instruction, increased awareness of and testing for disabilities, or over-diagnosis—may play a role in the increased rates of boys diagnosed with learning disabilities or emotional disturbance. But boys also have a higher incidence of organic disabilities, such as autism and orthopedic impairments, for which scientists don't currently have a completely satisfactory explanation. Further, while girls are less likely than boys to be diagnosed with most disabilities, the number of girls with disabilities has also grown rapidly in recent decades, meaning that this is not just a boy issue.

Moving Up and Moving On

Beyond achievement, there's the issue of attainment—student success in moving forward along the education pathway and ultimately earning credentials and degrees. There are undeniably some troubling numbers for boys in this area. But as with achievement, the attainment data does not show that boys are doing worse.

Elementary-school-age boys are more likely than girls to be held back a grade. In 1999, 8.3 percent of boys ages 5–12 had been held back at least one grade, compared with 5.2 percent of girls. However, the percentage of boys retained a grade has declined since 1996, while the percentage of girls retained has stayed the same.[12]

Mirroring the trends in achievement noted above, racial and economic differences in grade retention are as great as or greater than gender differences. For example, white boys are more likely than white girls to be retained a grade, but about equally likely as black and Hispanic girls. Black and Hispanic boys are much more likely to be held back than either white boys or girls from any racial group. Similarly, both boys and girls from low-income homes are much more likely to be held back, while boys from high-income homes are less likely to be held back than are girls from either low-or moderate-income families.[13]

Boys are also much more likely than girls to be suspended or expelled from school. According to the U.S. Department of Justice, 24 percent of girls have been suspended from school at least once by age 17, but so have fully 42 percent of boys.[14] This is undeniably cause for concern.

Boys are also more likely than girls to drop out of high school. Research by the Manhattan Institute found that only about 65 percent of boys who start high school graduate four years later, compared with 72 percent of girls. This gender gap cuts across all racial and ethnic groups, but it is the smallest for white and Asian students and much larger for black and Hispanic students. Still, the gaps between graduation rates for white and black or Hispanic students are much greater than gaps between rates for boys and girls of any race.[15] These statistics, particularly those for black and Hispanic males, are deeply troubling. There is some good news, though, because both men and women are slightly more likely to graduate from high school today than they were 30 years ago.[16]

Aspirations and Preparation

There is also some evidence that girls who graduate from high school have higher aspirations and better preparation for postsecondary education than boys do. For example, a University of Michigan study found that 62 percent of female high school seniors plan to graduate from a four-year-college, compared with 51 percent of male students.[17] Girls are also more likely than boys to have taken a variety of college-preparatory classes, including geometry, algebra II, chemistry, advanced biology, and foreign languages, although boys are more likely to have taken physics.

But this is another case where boys are actually improving, just not as fast as girls. . . .

The Allegedly Disappearing Big Man on Campus

Forty-two men for every 58 women go to college now, undergrad and grad. That means 1 in 4 female students can't find a male peer to date.

ESQUIRE, JULY 2006

Women now significantly outnumber men on college campuses, a phenomenon familiar enough to any sorority sister seeking a date to the next formal.

RICHARD WHITMIRE, *THE NEW REPUBLIC*, JAN. 23, 2006

To hear commentators tell it, college campuses are becoming all-female enclaves, suffering from a kind of creeping Wellesleyfication. But . . . men are enrolling in college in greater numbers than ever before and at historically high rates.

This is undeniably good news for the nation, as more and more future workers will need college credentials to compete in the global economy. Why, then, all the anxiety? Because . . . women are increasing college enrollment at an even faster rate.

Of men graduating from high school in spring 2001, 60 percent enrolled in college in the following fall, compared with 64 percent of women. The gap is smaller among those enrolling in four-year institutions: 41 percent of men, compared with 43 percent of women.

Women are, however, more likely to graduate from college once they get there. Sixty-six percent of women who enrolled in college as freshmen seeking a bachelor's degree during the 1995–96 school year had completed a bachelor's degree by 2001, compared with 59 percent of men.[18] As with high school graduation rates, this appears to be the area in which gender-focused concerns are most justified, with men less likely to stay in school and earn a degree.

Because men are less likely to go to college and more likely to drop out, the share of college students who are men has declined. From 1970 to 2001, men's share of college enrollment fell from 58 to 44 percent, while women's share blossomed from 42 to 56 percent. And fully 57 percent of bachelor's degrees in 2001 were awarded to women.[19]

But these numbers don't necessarily indicate an emerging crisis. Like many other trends in gender and education, they're nothing new. In fact, nearly two-thirds of the increase in women's share of college enrollment occurred more than two decades ago, between 1970 and 1980.

Overall trends, moreover, can be misleading. Women are overrepresented among both nontraditional students—older students going back to college after working or having a family—and students at two-year colleges. Among students enrolled in four-year colleges right out of high school, or traditional college students, the percentages of men and women are closer—and the dating situation is not as dire as Whitmire and *Esquire* suggest.

More important, even as their share of enrollment on college campuses declines, young men are actually more likely to attend and graduate from college than they were in the 1970s and 1980s. The share of men 25 to 29 who hold a bachelor's degree has also increased, to 22 percent—a rate significantly higher than that for older cohorts of men.[20] But the number of women enrolling in and graduating from college has increased much more rapidly during the same time period. The proportion of women enrolling in college after high school gradua- tion, for example, increased nearly 50 percent between the early 1970s and 2001, and nearly 25 percent of women ages 25 to 29 now hold bachelor's degrees.

While it's possible to debate whether men's college attendance is increasing fast enough to keep up with economic changes, it's simply inaccurate to imply that men are disappearing from college campuses or that they are doing worse than they were 10 or 20 years ago. Men's higher-education attainment is not declining; it's increasing, albeit at a slower rate than that of women.

In addition, while women have outstripped men in undergraduate enroll- ment, women still earn fewer than half of first professional degrees, such as law, medicine, and dentistry, and doctorates. Women do earn more master's degrees than men, but female graduate students are heavily concentrated in several traditionally female fields, most notably education and psychology.[21]

Outcomes of Education

With women attending and graduating from college at higher rates than men, we might expect young women, on average, to be earning more than men. But the reality is the opposite.

Female college degrees are disproportionately in relatively low-paying occu- pations like teaching. As a result, women ages 25–34 who have earned a bachelor's degree make barely more money than men of the same age who went to college but didn't get a bachelor's degree.[22] Further, recent female college graduates earn less than their male counterparts, even after controlling for choice of field.[23]

In other words, the undeniable success of more women graduating from high school, going to college, and finishing college ultimately results in women remaining behind men economically—just by not as much as before. Far from surging ahead of men, women are still working to catch up.

THE SOURCE OF THE BOY CRISIS:
A KNOWLEDGE DEFICIT AND A SURPLUS
OF OPPORTUNISM

It's clear that some gender differences in education are real, and there are some groups of disadvantaged boys in desperate need of help. But it's also clear that boys' overall educational achievement and attainment are not in decline—in fact, they have never been better. What accounts for the recent hysteria?

It's partly an issue of simple novelty. The contours of disadvantage in education and society at large have been clear for a long time—low-income, minority, and female people consistently fall short of their affluent, white, and male peers. The idea that historically privileged boys could be at risk, that boys could be shortchanged, has simply proved too deliciously counterintuitive and "newsworthy" for newspaper and magazine editors to resist.

The so-called boy crisis also feeds on a lack of solid information. Although there are a host of statistics about how boys and girls perform in school, we actually know very little about why these differences exist or how important they are. There are many things—including biological, developmental, cultural, and educational factors—that affect how boys and girls do in school. But untangling these different influences is incredibly difficult. Research on the causes of gender differences is hobbled by the twin demons of educational research: lack of data and the difficulty of drawing causal connections among multiple, complex influences. Nor do we know what these differences mean for boys' and girls' future economic and other opportunities.

Yet this hasn't stopped a plethora of so-called experts—from pediatricians and philosophers to researchers and op-ed columnists—from weighing in with their views on the causes and likely effects of educational gender gaps. In fact, the lack of solid research evidence confirming or debunking any particular hypothesis has created fertile ground for all sorts of people to seize on the boy crisis to draw attention to their pet educational, cultural or ideological issues.

The problem, we are told, is that the structured traditional classroom doesn't accommodate boys' energetic nature and need for free motion—or it's that today's schools don't provide enough structure or discipline. It's that feminists have demonized typical boy behavior and focused educational resources on girls—or it's the "box" boys are placed in by our patriarchal society. It's that our schools' focus on collaborative learning fails to stimulate boys' natural competitiveness—or it's that the competitive pressures of standardized testing are pushing out the kind of relevant, hands-on work on which boys thrive.

The boy crisis offers a perfect opportunity for those seeking an excuse to advance ideological and educational agendas. Americans' continued ambivalence about evolving gender roles guarantees that stories of "boys in crisis" will capture public attention. The research base is internally contradictory, making it easy to find superficial support for a wide variety of explanations but difficult for the media and the public to evaluate the quality of evidence cited. Yet there is not sufficient evidence—or the right kind of evidence—available to draw firm conclusions. As a result, there is a sort of free market for theories about why boys are underperforming girls in school, with parents, educators, media, and the public choosing to give credence to the explanations that are the best marketed and that most appeal to their pre-existing preferences.

Unfortunately, this dynamic is not conducive to a thoughtful public debate about how boys and girls are doing in school or how to improve their performance. . . .

NOTES

1. *How Schools Shortchange Girls* (Washington, D.C.: American Association of University Women, 1992). http://www.aauw.org/research/girls_education/hssg.cfm.

2. Adrienne Mand Lewin, "Can Boys Really Not Sit Still in School?" ABCnews.com, Jan. 26, 2006.

3. See Christina Hoff Sommers, *The War Against Boys* (New York: Simon & Schuster, 2000) and William Pollack, *Real Boys: Protecting Our Sons from the Myths of Boyhood* (New York: Random House, 1998).

4. Individual states also administer their own assessments, but none can be used to gauge long-term trends as well as the NAEP, nor do they offer the advantage of a national sample. Overall, differences between boys' and girls' performance on NAEP assessments match those found on state assessments—girls tend to do better than boys on tests of English/language arts, and boys tend to do slightly better in math.

5. Marianne Perie, Wendy S. Grigg, and Patricia L. Donahue, *The Nation's Report Card: Reading 2005* (Washington, D.C.: U.S. Department of Education, Institute of Education Sciences, National Center for Education Statistics 2005). http://nces. ed.gov/nationsreportcard/reading/.

6. *Ibid.*

7. M. Perie, R. Moran, and A.D. Lutkus, *NAEP 2004 Trends in Academic Progress: Three Decades of Student Achievement in Reading and Mathematics* (Washington, D.C.: U.S. Department of Education, Institute of Education Sciences, National Center for Education Sciences, National Center for Education Statistics, 2005). //nces.ed.gov/ nationsreportcard/ltt/results2004/.

8. Marianne Perie, Wendy S. Grigg, and Gloria S. Dion, *The Nation's Report Card: Math 2005* (Washington, D.C.: U.S. Department of Education, National Center for Education Statistics, 2005).

9. The main NAEP and long-term trend NAEP are different assessments, so it is not necessarily inconsistent that older boys' performance rose on the main NAEP from 1992 to 1996 while it did not rise on the long-term trend NAEP.

10. *2004 Trends in Academic Progress*, op. cit.; author analysis using NAEP data explorer. //nces.ed.gov/nationsreportcard/nde/.

11. Office of Special Education Programs, *25th Annual Report to Congress* (Washington, D.C.: U.S. Department of Education, 2003). http://www.ed.gov/about/offices/list/ osers/osep/research.html; Centers for Disease Control and Prevention, "Mental Health in the United States: Prevalence of and Diagnosis and Medication Treatment for Attention Deficit/Hyperactivity Disorder—United States, 2003," *MMWR*

Weekly, Sept. 2, 2005. http://www.cdc.gov/mmwr/preview/mmwrhtml/mm5434a2.htm.

12. Catherine Freeman, *Trends in Educational Equity of Girls and Women* (Washington, D.C.: U.S. Department of Education, National Center for Education Statistics, 2004). http://nces.ed.gov/pubsearch/pubsinfo.asp?pubid=2005016.

13. *Trends in Educational Equity of Girls and Women, op.cit.*

14. Office of Juvenile Justice and Delinquency Prevention, *Juvenile Offenders and Victims: 2006 National Report* (Washington, D.C.: U.S. Department of Justice, Office of Justice Programs, 2006). http://ojjdp.ncjrs.org/ojstatbb/nr2006/index.html.

15. Jay P. Greene and Marcus Winters, *Leaving Boys Behind: Public High School Graduation Rates* (Manhattan Institute, April 2006). http://www.manhattan-institute.org/html/cr_48.htm#05; research published by the Urban Institute reaches similar findings. See Christopher B. Swanson, *Who Graduates? Who Doesn't? A Statistical Portrait of Public High School Graduation 2001* (Washington, D.C.: The Urban Institute, 2003). http://www.urban.org/publications/410934.html.

16. *Trends in Educational Equity of Girls and Women: 2004, op.cit.*

17. As cited in *Trends in Educational Equity of Girls and Women*: 2004, op.cit. and Richard Whitmire, "Boy Trouble," *The New Republic*, Jan. 23, 2006.

18. *Trends in Educational Equity of Girls and Women: 2004, op.cit.*

19. Katharin Peter and Laura Horn, *Gender Differences in Participation and Completion of Undergraduate Education and How They Have Changed Over Time* (Washington, D.C.: U.S. Department of Education, Institute of Education Sciences, National Center for Education Statistics, February 2005). http://nces.ed.gov/pubsearch/pubsinfo.asp?pubid=2005169.

20. *Trends in Educational Equity of Girls and Women: 2004, op.cit.*

21. *Ibid.*

22. *Highlights of Women's Earnings in 2004* (Washington, D.C.: U.S. Department of Labor, U.S. Bureau of Labor Statistics, September 2005).

23. *Trends in Educational Equity of Girls and Women: 2004, op.cit.*

KEY CONCEPTS

educational attainment   standardized ability test   tracking

DISCUSSION QUESTIONS

1. Based on the research evidence, is there a crisis in boys' educational achievement?

2. What evidence is there of a gender gap in educational achievement?

3. In Mead's analysis, what explains what she calls the "national hysteria" about boys' education?

52

Dishonoring the Dead

JONATHAN KOZOL

The U.S. Supreme Court decision Brown v. Board of Education *(1954) declared racial segregation to be unconstitutional. Yet, in recent years, racial segregation has increased in the nation's schools. In this article Jonathan Kozol discusses the racial isolation of many of the nation's minority students.*

Many Americans I meet who live far from our major cities and who have no first-hand knowledge of realities in urban public schools seem to have a rather vague and general impression that the great extremes of racial isolation they recall as matters of grave national significance some 35 or 40 years ago have gradually, but steadily, diminished in more recent years. The truth, unhappily, is that the trend, for well over a decade now, has been precisely the reverse. Schools that were already deeply segregated 25 or 30 years ago, like most of the schools I visit in the Bronx, are no less segregated now, while thousands of other schools that had been integrated either voluntarily or by the force of law have since been rapidly resegregating both in northern districts and in broad expanses of the South.

"At the beginning of the twenty-first century," according to Professor Gary Orfield and his colleagues at the Civil Rights Project at Harvard University, "American public schools are now 12 years into the process of continuous resegregation. The desegregation of black students, which increased continuously from the 1950s to the late 1980s, has now receded to levels not seen in three decades. . . . During the 1990s, the proportion of black students in majority white schools has decreased . . . to a level lower than in any year since 1968. . . . Almost three fourths of black and Latino students attend schools that are predominantly minority," and more than two million, including more than a quarter of black students in the Northeast and Midwest, "attend schools which we call apartheid schools" in which 99 to 100 percent of students are nonwhite. The four most segregated states for black students, according to the Civil Rights Project, are New York, Michigan, Illinois, and California. In California and New York, only one black student in seven goes to a predominantly white school.

SOURCE: Jonathan Kozol, 2006. *The Shame of the Nation: The Restoration of Apartheid Schooling in America.* New York: Three Rivers Press.

During the past 25 years, the Harvard study notes, "there has been no significant leadership towards the goal of creating a successfully integrated society built on integrated schools and neighborhoods." The last constructive act by Congress was the 1972 enactment of a federal program to provide financial aid to districts undertaking efforts at desegregation, which, however, was "repealed by the Reagan administration in 1981." The Supreme Court "began limiting desegregation in key ways in 1974"—and actively dismantling existing integration programs in 1991.

"Desegregation did not fail. In spite of a very brief period of serious enforcement . . ., the desegregation era was a period in which minority high school graduates increased sharply and the racial test score gaps narrowed substantially until they began to widen again in the 1990s. . . . In the two largest educational innovations of the past two decades— standards-based reform and school choice—the issue of racial segregation and its consequences has been ignored."

"To give up on integration, while aware of its benefits," write Orfield and his former Harvard colleague Susan Eaton, "requires us to consciously and deliberately accept segregation, while aware of its harms. . . . Segregation, rarely discussed, scarcely even acknowledged by elected officials and school leaders"— an "exercise in denial," they observe, "reminiscent of the South" before the integration era—"is incompatible with the healthy functioning of a multiracial generation."

Racial isolation and the concentrated poverty of children in a public school go hand in hand, moreover, as the Harvard project notes. Only 15 percent of the intensely segregated white schools in the nation have student populations in which more than half are poor enough to be receiving free meals or reduced price meals. "By contrast, a staggering 86 percent of intensely segregated black and Latino schools" have student enrollments in which more than half are poor by the same standards. A segregated inner-city school is "almost six times as likely" to be a school of concentrated poverty as is a school that has an overwhelmingly white population.

"So deep is our resistance to acknowledging what is taking place," Professor Orfield notes, that when a district that has been desegregated in preceding decades now abandons integrated education, "the actual word 'segregation' hardly ever comes up. Proposals for racially separate schools are usually promoted as new educational improvement plans or efforts to increase parental involvement. . . . In the new era of 'separate but equal,' segregation has somehow come to be viewed as a type of school reform"—"something progressive and new," he writes—rather than as what it is: an unconceded throwback to the status quo of 1954. But no matter by what new name segregated education may be known, whether it be "neighborhood schools, community schools, targeted schools, priority schools," or whatever other currently accepted term, "segregation is not new . . . and neither is the idea of making separate schools equal. It is one of the oldest and extensively tried ideas in U.S. educational history" and one, writes Orfield, that has "never had a systematic effect in a century of trials."

Perhaps most damaging to any effort to address this subject openly is the refusal of most of the major arbiters of culture in our northern cities to confront or even clearly name an obvious reality they would have castigated with a passionate determination in another section of the nation 50 years before and which, moreover, they still castigate today in retrospective writings that assign it to a comfortably distant and allegedly concluded era of the past. There is, indeed, a seemingly agreed-upon convention in much of the media today not even to use an accurate descriptor such as "racial segregation" in a narrative description of a segregated school. Linguistic sweeteners, semantic somersaults, and surrogate vocabularies are repeatedly employed. Schools in which as few as three or four percent of students may be white or Southeast Asian or of Middle Eastern origin, for instance—and where *every other child* in the building is black or Hispanic—are referred to, in a commonly misleading usage, as "diverse." Visitors to schools like these discover quickly the eviscerated meaning of the word, which is no longer a descriptor but a euphemism for a plainer word that has apparently become unspeakable.

School systems themselves repeatedly employ this euphemism in descriptions of the composition of their student populations. In a school I visited in fall 2004 in Kansas City, Missouri, for example, a document distributed to visitors reports that the school's curriculum "addresses the needs of children from diverse backgrounds." But as I went from class to class I did not encounter any children who were white or Asian—or Hispanic, for that matter—and when I later was provided with the demographics of the school, I learned that 99.6 percent of students there were African-American. In a similar document, the school board of another district, this one in New York State, referred to "the diversity" of its student population and "the rich variations of ethnic backgrounds. . . ." But when I looked at the racial numbers that the district had reported to the state, I learned that there were 2,800 black and Hispanic children in the system, one Asian child, and three whites. Words, in these cases, cease to have real meaning; or, rather, they mean the opposite of what they say.

One of the most disheartening experiences for those who grew up in the years when Martin Luther King and Thurgood Marshall were alive is to visit public schools today that bear their names, or names of other honored leaders of the integration struggles that produced the temporary progress that took place in the three decades after *Brown,* and to find how many of these schools are bastions of contemporary segregation. It is even more disheartening when schools like these are not in segregated neighborhoods but in racially mixed areas in which the integration of a public school would seem to be most natural and where, indeed, it takes a conscious effort on the part of parents or of school officials in these districts to *avoid* the integration option that is often right at their front door.

In a Seattle neighborhood, for instance, where approximately half the families were Caucasian, 95 percent of students at the Thurgood Marshall Elementary School were black, Hispanic, Native American, or of Asian origin. An African-American teacher at the school told me of seeing clusters of white parents and their children on the corner of a street close to the school each morning waiting for a bus that took the children to a school in which she believed that the enrollment was predominantly white. She did not speak of the white families waiting for the bus to take their children to another public school with bitterness, but wistfully.

"At Thurgood Marshall," according to a big wall-poster in the lobby of the school, "the dream is alive." But school assignment practices and federal court decisions that have countermanded long-established policies that previously fostered integration in Seattle's schools make the realization of the dream identified with Justice Marshall all but unattainable today.

"Thurgood Marshall must be turning over in his grave," one of the teachers at the school had told the principal, as he reported this to me. The principal, understandably, believed he had no choice but to reject the teacher's observation out of hand. "No, sister," he had told the teacher. "If Justice Marshall was still roamin' nowadays and saw what's goin' on here in this school, he would say 'Hallelujah' and 'Amen!'" Legal scholars may demur at this, but he had a school to run and he could not allow the ironies of names, or history, to undermine the passionate resolve he brought to winning victories for children in the only terms he was allowed.

In the course of two visits to the school, I had a chance to talk with a number of teachers and to spend time in their classrooms. In one class, a teacher had posted a brief summation of the *Brown* decision on the wall; but it was in an inconspicuous corner of the room and, with that one exception, I could find no references to Marshall's struggle against racial segregation in the building.

When I asked a group of fifth grade boys who Thurgood Marshall was and what he did to have deserved to have a school named after him, most of the boys had no idea at all. One said that he used to run "a summer camp." Another said he was "a manager"—I had no chance to ask him what he meant by this, or how he'd gotten this impression. Of the three who knew that he had been a lawyer, only one, and only after several questions on my part, replied that he had "tried to change what was unfair"—and, after a moment's hesitation, "wanted to let black kids go to the same schools that white kids did." He said he was "pretty sure" that this school was not segregated because, in one of the other classrooms on the same floor, there were two white children.

There is a bit of painful humor that I've heard from black schoolteachers who grew up during the era of the integration movement and have subsequently seen its goals abandoned and its early victories reversed. "If you want to see a *really* segregated school in the United States today, start by looking for a school that's named for Martin Luther King or Rosa Parks." . . .

KEY CONCEPTS

| apartheid | hypersegregation | segregation |

DISCUSSION QUESTIONS

1. How are housing segregation and school segregation linked?
2. What factors does Kozol identify as related to school segregation?
3. Why do you think there is now little effort to integrate schools across the nation?

53

Charter Schools and the Public Good

LINDA A. RENZULLI AND VINCENT J. ROSCIGNO

The charter school movement has arisen in part because of national concerns that public schools are failing. Here Linda Renzulli and Vincent Roscigno evaluate whether charter schools are successful in promoting student achievement. They also assess the extent to which charter schools are held accountable for meeting educational standards.

According to the U.S Department of Education, "No Child Left Behind is designed to change the culture of America's schools by closing the achievement gap, offering more flexibility, giving parents more options, and teaching students based on what works." Charter schools—a recent innovation in U.S. education—are one of the most visible developments aimed at meeting these goals. Although they preceded the 2002 No Child Left Behind (NCLB) Act, charter schools are now supported politically and financially through NCLB. Charter schools are public schools set up and administered outside the traditional bureaucratic constraints of local school boards, with the goal of creating choice, autonomy, and accountability.

Unlike regular public schools, charter schools are developed and managed by individuals, groups of parents, community members, teachers, or education-management organizations. In exchange for their independence from most state and local regulations (except those related to health, safety, and nondiscrimination), they must uphold their contracts with the local or state school board or risk being closed. Each provides its own guidelines for establishing rules and procedures, including curriculum, subject to evaluation by the state in which it resides.

SOURCE: Linda A. Renzulli and Vincent J. Roscigno. "Charter Schools and the Public Good." *Contexts* 6, no. 1 (Winter 2007): 31–36. © 2007 by the American Sociological Association.

Charter schools are among the most rapidly growing educational institutions in the United States today. No charter schools existed before 1990, but such schools are now operating in 40 states and the District of Columbia. According to the Center for Educational Reform, 3,977 charter schools are now educating more than a million students.

Charter schools have received bipartisan support and media accolades. This, however, is surprising. The true academic value of the educational choices that charter schools provide to students, as well as their broader implications for the traditional system of public education, are simply unknown—a fact that became obvious in November 2004, when voters in the state of Washington rejected—for the third time—legislation allowing the creation of charter schools. Driven by an alliance of parents, teachers, and teacher unions against sponsorship by powerful figures such as Bill Gates, this rejection went squarely against a decade-long trend. Reflecting on Washington's rejection, a state Democrat told the *New York Times*, "Charter schools will never have a future here now until there is conclusive evidence, nationwide, that these schools really work. Until the issue of student achievement gets resolved, I'd not even attempt to start over again in the Legislature."

THE RATIONALE

Most justifications for charter schools argue that the traditional system of public schooling is ineffective and that the introduction of competition and choice can resolve any deficiencies. The leading rationale is that accountability standards (for educational outcomes and student progress), choice (in curriculum, structure, and discipline), and autonomy (for teachers and parents) will generate higher levels of student achievement. The result will be high-quality schools for all children, particularly those from poor and minority backgrounds, and higher levels of student achievement.

While wealthy families have always been able to send their children to private schools, other Americans have historically had fewer, if any, options. Proponents suggest that charter schools can address such inequality by allowing all families, regardless of wealth, to take advantage of these new public educational options. Opponents contend that charter schools cannot fix broader educational problems; if anything, they become instruments of segregation, deplete public school systems of their resources, and undermine the public good.

Given the rationales for charter schools before and after the NCLB Act, it is surprising how few assessments have been made of charter school functioning, impacts on achievement, or the implications of choice for school systems. Only a handful of studies have attempted to evaluate systematically the claims of charter school effectiveness, and few of these have used national data. The various justifications for charter schools—including the desire to increase achievement in the public school system—warrant attention, as do concrete research and evidence on whether such schools work. The debate, however, involves more than simply how to enhance student achievement. It also involves educational competition and accountability, individual choice and, most fundamentally, education's role in fostering the "public good."

IS THERE PROOF IN THE PUDDING?

Do students in charter schools do better than they would in traditional public schools? Unfortunately, the jury is still out, and the evidence is mixed. Profiles in the *New Yorker, Forbes, Time,* and *Newsweek,* for example, highlight the successes of individual charter schools in the inner cities of Washington, DC and New York, not to mention anecdotal examples offered by high-profile advocates like John Walton and Bill Gates. While anecdotes and single examples suggest that charter schools may work, they hardly constitute proof or even systematic evidence that they always do. In fact, broader empirical studies using representative and national data suggest that many charter schools have failed.

One noteworthy study, released by the American Federation of Teachers (AFT) in 2004, reports that charter schools are not providing a better education than traditional public schools. Moreover, they are not boosting student achievement. Using fourth- and eighth-grade test scores from the National Assessment of Educational Progress across all states with charter schools, the report finds that charter-school students perform *less* well, on average, in math and reading than their traditional-school counterparts. There appear to be no significant differences among eighth-graders and no discernable difference in black–white achievement gaps across school type.

Because the results reported in the AFT study—which have received considerable media attention—do not incorporate basic demographic, regional, or school characteristics simultaneously, they can only relate average differences across charter schools and public schools. But this ignores the huge effects of family background, above and beyond school environment. Without accounting for the background attributes of students themselves, not to mention other factors such as the race and social-class composition of the student body, estimates of the differences between charter schools and traditional public schools are overstated.

In response to the 2004 AFT report, economists Carolyn Hoxby and Jonah Rockoff compared charter schools to surrounding public schools. Their results contradict many of the AFT's findings. They examined students who applied to but did not attend charter schools because they lost lotteries for spots. Hoxby and Rockoff found that, compared to their lotteried-out fellow applicants, students who attended charter schools in Chicago scored higher in both math and reading. This is true especially in the early elementary grades compared to nearby public schools with similar racial compositions. Their work and that of others also shows that older charter schools perform better than newly formed ones—perhaps suggesting that school stability and effectiveness require time to take hold. Important weaknesses nevertheless remain in the research design. For example,

Hoxby and Rockoff conducted their study in a single city—Chicago—and thus it does not represent the effect of charter schools in general.

As with the research conducted by the AFT, we should interpret selective case studies and school-level comparisons with caution. Individual student background is an important force in shaping student achievement, yet it rarely receives attention in this research or in the charter school achievement debate more generally. The positive influence of charter schools, where it is found, could easily be a function of more advantaged student populations drawn from families with significant educational resources at home. We know from prior research that parents of such children are more likely to understand schooling options and are motivated to ensure their children's academic success. Since family background, parental investments, and parental educational involvement typically trump school effects in student achievement, it is likely that positive charter school effects are simply spurious.

More recently, a report by the National Center for Education Statistics (NCES), using sophisticated models, appropriate demographic controls, and a national sample, has concurred with the AFT report—charter schools are not producing children who score better on standardized achievement tests. The NCES report showed that average achievement in math and reading in public schools and in charter schools that were linked to a school district did not differ statistically. Charter schools not associated with a public school district, however, scored significantly *less* well than their public school counterparts.

Nevertheless, neither side of the debate has shown conclusively, through rigorous, replicated, and representative research, whether charter schools boost student achievement. The NCES report mentioned above has, in our opinion, done the best job of examining the achievement issue and has shown that charter schools are not doing better than traditional public schools when it comes to improving achievement.

Clearly, in the case of charter schools, the legislative cart has been put before the empirical horse. Perhaps this is because the debate is about more than achievement. Charter school debates and legislation are rooted in more fundamental disagreements over competition, individualism, and, most fundamentally, education's role in the public good. This reflects an important and significant shift in the cultural evaluation of public education in the United States, at the crux of which is the application of a market-based economic model, complete with accompanying ideas of "competition" and "individualism."

COMPETITION AND ACCOUNTABILITY

To whom are charter schools accountable? Some say their clients, namely, the public. Others say the system, namely, their authorizers. If charter schools are accountable to the public, then competition between schools should ensure academic achievement and bureaucratic prudence. If charter schools are accountable to the system, policies and procedures should ensure academic achievement and bureaucratic prudence. In either case, the assumption is that charter schools will close when they are not successful. The successful application of these criteria, however, requires clear-cut standards, oversight, and accountability—which are currently lacking, according to many scholars. Indeed, despite the rhetoric of their advocates and legislators, charter schools are seldom held accountable in the market or by the political structures that create them.

In a "market" view of accountability, competition will ultimately breed excellence by "weeding out" ineffectual organizations. Through "ripple effects," all schools will be forced to improve their standards. Much like business organizations, schools that face competition will survive only by becoming more efficient and producing a better overall product (higher levels of achievement) than their private and public school counterparts.

Social scientists, including the authors of this article, question this simplistic, if intuitively appealing, application of neoliberal business principles to the complex nature of the educational system, children's learning, and parental choice for schools. If competition were leading to accountability, we would see parents pulling their children out of unsuccessful charter schools. But research shows that this seldom happens. Indeed, parents, particularly those with resources, typically choose schools for reasons of religion, culture, and social similarity rather than academic quality.

Nor are charter schools accountable to bureaucrats. Even though charter schools are not outperforming traditional public schools, relatively few (10 percent nationally) have actually been closed by their authorizers over the last decade. Although we might interpret a 10 percent closure rate as evidence of academic accountability at work, this would be misleading. Financial rather than academic issues are the principal reasons cited for these closures. By all indications, charter schools are not being held accountable to academic standards, either by their authorizers or by market forces.

In addition to measuring accountability through student performance, charter schools should also be held to standards of financial and educational quality. Here, some charter schools are faltering. From California to New York and Ohio, newspaper editorials question fiscal oversight. There are extreme cases such as the California Charter Academy, a publicly financed but privately run chain of 60 charter schools. Despite a budget of $100 million dollars, this chain became insolvent in August 2004, leaving thousands of children without a school to attend.

More direct accountability issues include educational quality and annual reports to state legislatures; here, charter-school performance is poor or mixed. In Ohio, where nearly 60,000 students now attend charter schools, approximately one-quarter of these schools are not following the state's mandate to report school-level test score results, and only 45 percent of the teachers at the state's 250 charter schools hold full teaching certification. Oversight is further complicated by the creation of "online" charter schools, which serve 16,000 of Ohio's public school students.

It is ironic that many charter schools are not held to the very standards of competition, quality, and accountability that legislators and advocates used to justify them in the first place. Perhaps this is why Fredrick Hess, a charter school researcher, recently referred to accountability as applied to charter schools as little more than a "toothless threat."

INDIVIDUALISM OR INEQUALITY?

The most obvious goal of education is student achievement. Public education in the United States, however, has also set itself several other goals that are not reducible to achievement or opportunity at an individual level but are important culturally and socially. Public education has traditionally managed diversity and integration, created common standards for the socialization of the next generation, and ensured some equality of opportunity and potential for meritocracy in the society at large. The focus of the charter school debate on achievement—rooted in purely economic rationales of competition and individual opportunism—has ignored these broader concerns.

Individual choice in the market is a key component of neoliberal and "free-market" theory—a freedom many Americans cherish. Therefore, it makes sense that parents might support choice in public schooling. Theoretically, school choice provides them market power to seek the best product for their children, to weigh alternatives, and to make changes in their child's interest. But this power is only available to informed consumers, so that educational institutions and policies that provide choice may be reinforcing the historical disadvantages faced by racial and ethnic minorities and the poor.

We might expect that students from advantaged class backgrounds whose parents are knowledgeable about educational options would be more likely to enroll in charter schools. White parents might also see charter schools as an educational escape route from integrated public schools that avoids the financial burden of private schooling. On the other hand, the justification for charter schools is often framed in terms of an "educational fix" for poor, minority-concentrated districts in urban areas. Here, charter schools may appear to be a better opportunity for aggrieved parents whose children are attending poorly funded, dilapidated public schools

National research, at first glance, offers encouraging evidence that charter schools are providing choices to those who previously had few options: 52 percent of those enrolled in charter schools are nonwhite compared to 41 percent of those in traditional public schools. These figures, however, tell us little about the local concentrations of whites and nonwhites in charter schools, or how the racial composition and distribution of charter schools compares to the racial composition and distribution of local, traditional public schools.

African-American students attend charter and noncharter schools in about the same proportion, yet a closer look at individual charter schools within districts reveals that they are often segregated. In Florida, for instance, charter schools are 82 percent white, whereas traditional public schools are only 51 percent white. Similar patterns are found across Arizona school districts, where charter school enrollment is 20 percent more white than traditional schools. Amy Stuart Wells's recent research finds similar tendencies toward segregation among Latinos, who are underrepresented in California's charter schools. Linda Renzulli and Lorraine Evans's national analysis of racial composition within districts containing charter schools shows that charter school formation often results in greater levels of segregation in schools between whites and nonwhites. This is not to suggest that minority populations do not make use of charter schools. But, when they do, they do so in segregated contexts.

Historically, racial integration has been a key cause of white flight and it remains a key factor in the racial composition of charter schools and other schools of choice. Decades of research on school segregation have taught us that when public school districts become integrated, through either court mandates or simple population change, white parents may seek alternative schools for their children. Current research suggests that a similar trend exists with charter schools, which provide a public-school option for white flight without the drawbacks of moving (such as job changes and longer commutes). While those from less-privileged and minority backgrounds have charter schools at their disposal, the realities of poor urban districts and contemporary patterns of racial residential segregation may mean that the "choice" is between a racially and economically segregated charter school or an equally segregated traditional school, as Renzulli's research has shown. Individualism in the form of educational choice, although perhaps intuitively appealing, in reality may be magnifying some at the very inequalities that public education has been attempting to overcome since the *Brown v. Board of Education* decision in 1954.

Regarding equality of opportunity and its implications for the American ideal of meritocracy, there is also reason for concern. Opponents have pointed to the dilution of district resources where charter schools have emerged, especially as funds are diverted to charter schools. Advocates, in contrast, argue that charter schools have insufficient resources. More research on the funding consequences of charter school creation is clearly warranted. Why, within a system of public education, should some students receive more than others? And what of those left behind, particularly students from disadvantaged backgrounds whose parents may not be aware of their options? Although evidence on the funding question is sparse, research on public schools generally and charter school attendance specifically suggests that U.S. public education may be gravitating again toward a system of separate, but not equal, education.

THE FUTURE OF PUBLIC EDUCATION

Variation across charter schools prevents easy evaluation of their academic success or social consequences for public education. Case studies can point to a good school or a bad one. National studies can provide statistical averages and comparisons, yet they may be unable to reveal the best and worst effects of charter schools. Neither type of research has yet fully accounted for the influence of family background and school demographic composition. Although conclusions about charter school effectiveness or failure remain questionable, the most rigorous national analyses to date suggest that charter schools are doing no better than traditional public schools.

Certainly some charter schools are improving the educational quality and experience of some children. KIPP (Knowledge Is Power Program) schools, for example, are doing remarkable things for the students lucky enough to attend them. But for every KIPP school (of which there are only 45, and not all are charter schools), there are many more charter schools that do not provide the same educational opportunity to students, have closed their doors in the middle of the school year, and, in effect, isolate students from their peers of other races and social classes. Does this mean that we should prevent KIPP, for example, from educating students through the charter school option? Maybe. Or perhaps we should develop better program evaluations—of what works and what does not—and implement them as guideposts. To the dismay of some policymakers and "competition" advocates, however, such standardized evaluation and accountability would undercut significant charter school variations if not the very nature of the charter school innovation itself.

Student achievement is only part of the puzzle when it comes to the charter school debate; we need to consider social integration and equality as well. These broader issues, although neglected, warrant as much attention as potential effects on achievement. We suspect that such concerns, although seldom explicit, probably underlie the often contentious charter school and school choice debate itself. We believe it is time to question the logics pertaining to competition, choice, and accountability. Moreover, we should all scrutizine the existing empirical evidence, not to mention educational policy not firmly rooted in empirical reality and research. As Karl Alexander eloquently noted in his presidential address to the Southern Sociological Society, "The charter school movement, with its 'let 1,000 flowers bloom' philosophy, is certain to yield an occasional prize-winning rose. But is either of these approaches [to school choice] likely to prove a reliable guide for broad-based, systemic reform—the kind of reform that will carry the great mass of our children closer to where we want them to be? I hardly think so." Neither do we.

KEY CONCEPTS

charter school   educational attainment   No Child Left Behind

DISCUSSION QUESTIONS

1. What factors, other than the school itself, do Renzulli and Roscigno identify as affecting student achievement?

2. What do the authors mean, in the context of schooling, when they argue that "choice" is a racially and economically based concept?

54

The Service Society and the Changing Experience of Work

CAMERON LYNNE MACDONALD AND CARMEN SIRIANNI

The U.S. economy has changed from being based primarily on manufacturing to being based on service industries. The transition to more "service work" has changed the character of workplace control. The service economy is embedded in systems of race, gender, and class stratification that are revealed in patterns of employment and perceptions of who is most fit for particular jobs.

We live and work in a service society. Employment in the service sector currently accounts for 79 percent of nonagricultural jobs in the United States (U.S. Department of Labor 1994: 83). More important, 90 percent of the new jobs projected to be created by the year 2000 will be in service occupations, while the number of goods-producing jobs is projected to decline (Kutscher 1987: 5). Since the mid-nineteenth century the U.S. economy has been gradually transformed from an agriculture-based economy to a manufacturing-based economy to a service-based economy. Near the turn of the century, employment distribution among the three major economic sectors was equally divided at roughly one-third each. Since then, agriculture's labor market share has declined rapidly, now accounting for only about 3 percent of U.S. jobs, while the service sector provides over 70 percent and the goods-producing sector about 25 percent.

The decrease in proportion of manufacturing jobs occurred not because U.S. corporations manufacture fewer goods, but primarily because they use fewer workers to make the goods they produce (Albrecht and Zemke 1990). They use fewer workers due to increasing levels of automation and the exportation of manufacturing functions to low-wage job markets overseas. In addition, the feminization of the work force has created a self-fulfilling cycle in which the entrance of more women into the work force has led to increased demand for

SOURCE: Macdonald, Cameron Lynne, and Carmen Sirianni, eds. 1996. "The Service Society and the Changing Experience of Work." In *Working in the Service Society*, 1–24. Philadelphia: Temple University Press. Reprinted with permission.

those consumer services once provided gratis by housewives (cleaning, cooking, child care, etc.), which in turn has produced more service jobs that are predominantly filled by women.

Still, these trends fail to account fully for the dominance of service work in the U.S. economy, since companies outside of the service sector also contain service occupations. For example, 13.2 percent of the employees in the manufacturing sector work in service occupations such as clerical work, customer service, telemarketing, and transportation (Kutscher 1987). Further, manufacturing and technical occupations are comprised increasingly of service components as U.S. firms adopt Total Quality Management (TQM) and other customer-focused strategies to generate a competitive edge in the global economy. When production efficiency and quality are maximized, the critical variable in the struggle for economic dominance is the quality of interactions with customers. As one business school professor remarks, "Sooner or later, new technology becomes available to everyone. Customer-oriented employees are a lot harder to copy or buy" (Schlesinger and Heskett 1991: 81). So whether one believes that U.S. manufacturing is going to Mexico, to automation, or to the dogs, it is clear that the United States is increasingly becoming a service society and that service work is here to stay.

What do we mean when we speak of "service work"? By definition, a service is intangible; it is produced and consumed simultaneously, and the customer generally participates in its production (Packham 1992).... Service work includes jobs in which face-to-face or voice-to-voice interaction is a fundamental element of the work. "Interactive service work" (Leidner 1993) generally requires some form of what Arlie Hochschild (1983) has termed "emotional labor," meaning the conscious manipulation of the workers' self-presentation either to display feeling states and/or to create feeling states in others. In addition, the guidelines, or "feeling rules," for this emotional labor are created by management and conveyed to the worker as a critical aspect of the job....

Much managerial and professional work also entails emotional labor. For example, doctors are expected to display an appropriate "bedside manner," lawyers are expert actors in and out of the courtroom, and managers, at the most fundamental level, try to instill feeling states and thus promote action in others. However, there remains a critical distinction between white-collar work and work in the emotional proletariat: in management and in the professions, guidelines for emotional labor are generated collegially and, to a great extent, are self-supervised. In front-line service jobs, workers are given very explicit instructions concerning what to say and how to act, and both consumers and managers watch to ensure that these instructions are carried out. However, one could argue that even those in higher ranking positions increasingly experience the kinds of monitoring of their interactive labor encountered by those lower on the occupational ladder, be it by customers, supervisors, or employees....

Given the rising dominance of service occupations in the labor force, what are the special difficulties and opportunities that workers encounter in a service society? A key problem seems to be how to inhabit the job. In the past there was a clear distinction between *careers,* which required a level of personalization, emotion management, authenticity in interaction, and general integration of personal and workplace identities, and *jobs,* which required the active engagement of the body and parts of the mind while the spirit and soul of the worker might be elsewhere. Workers in service occupations are asked to inhabit jobs in ways that were formerly limited to managers and professionals alone. They are required to bring some level of personal identity and self-expression into their work, even if it is only at the level of basic interactions, and even if the job itself is only temporary. The assembly-line worker could openly hate his job, despise his supervisor, and even dislike his co-workers, and while this might be an unpleasant state of affairs, if he completed his assigned tasks efficiently, his attitude was his own problem. For the service worker, inhabiting the job means, at the very least, pretending to like it, and, at most, actually bringing his whole self into the job, liking it, and genuinely caring about the people with whom he interacts.

This demand has several implications: who will be asked to fill what jobs, how they are expected to perform, and how they will respond to those demands. Because personal interaction is a primary component of all service occupations, managers continually strive to find ways to oversee and control those interactions, and worker responses to these attempts vary along a continuum from enthusiastic compliance to outright refusal. Hiring, control of the work process, and the stresses of bringing one's emotions to work are all shaped by the characteristics of the worker and the nature of the work. The self-presentation and other personal characteristics of the worker make up the work process and the work product, and are increasingly the domain of management-worker struggles (see Leidner 1993). In addition, because much of the labor itself is invisible, contests over control of the labor process are often more implicit than explicit....

There are three trends emanating from the rising dominance of service work. First, the need to supervise the production of an intangible, good service, has given rise to particularly invasive forms of workplace control and has led managers to attempt to oversee areas of workers' personal and psychic lives that have heretofore been considered off-limits. Second, the fact that workers' personal characteristics are so firmly linked to their "suitability" for certain service occupations continues to lead to increasing levels of stratification within the service *labor* force. Finally, ... how [do] workers respond to these and other aspects of working in the service society, and how might they build autonomy and dignity into their work, ensuring that service work does not equal servitude? ...

GENDER, RACE, AND STRATIFICATION IN THE SERVICE SECTOR

Service industries tend to produce two kinds of jobs: large numbers of low-skill, low-pay jobs and a smaller number of high-skill, high-income jobs, with very few jobs that could be classified in the middle. As Joel Nelson (1994) notes, "Service workers are more likely than manufacturing workers to have lower incomes, fewer opportunities for full-time employment, and greater inequality in earnings" (p. 240). A typical example of this kind of highly stratified work force can be found in fast food industries. These firms tend to operate with a small core of managers and administrators and a large, predominantly part-time work force who possess few skills and therefore are considered expendable (Woody 1989).

As a result, service jobs fall into two broad categories: those likely to be production-line jobs and those likely to be empowered jobs. This distinction not only refers to the level of responsibility and autonomy expected of workers, but also to wages, benefits, job security, and potential for advancement. While empowered service jobs are associated with full-time work, decent wages and benefits, and internal job ladders, production-line jobs offer none of these. Some researchers have described the distinction between empowered and production-line service jobs as one between "core" and "periphery" jobs in the service economy (Hirschhorn 1988; Walsh 1990; Wood 1989). . . .

The core/periphery distinction may be a misleading characterization of functions in service industries, however. In many firms contingent workers perform functions essential to the operation of the firm and can comprise up to two-thirds of a firm's labor force while "core" workers perform nonessential functions (Walsh 1990). For example, a majority of key functions in industries such as hospitality, food service, and retail sales are performed by workers who, based on their level of benefits, pay, and job security, would be considered periphery workers. In low-skill service positions, job tenure has no relation to output or productivity. Therefore, employers can rely on contingent workers to provide high-quality service at low costs. As T. J. Walsh (1990: 527) points out, it is therefore likely that the poor compensation afforded these workers is due not to their productivity level but to the perceptions of their needs, level of commitment, and availability.

Given the proliferation of service jobs in the United States, key questions for labor analysts are what kinds of jobs are service industries producing, and who is likely to fill them? At the high end, service industries demand educated workers who can rapidly adapt to changing economic conditions. This means that employers may demand a college degree or better for occupations that formerly required only a high school diploma,

> even though many of the job-holders' activities have not changed,
> or appear relatively simple, because they want workers to be more
> responsive to the general situation in which they are working and
> the broader purposes of their work. (Hirschhorn 1988: 35)

In addition, core workers are frequently expected to take on more responsibilities, work longer hours, and intensify their output.

At the low end of the spectrum of service occupations are periphery workers who are frequently classified as part-time, temporary, contract, or contingent. These flexible-use workers act as a safety valve for service firms, allowing managers to redeploy labor costs in response to market conditions. In addition, they allow managers to minimize overhead because they rarely qualify for benefits and generally receive low wages. In 1988, 86 percent of all part-time workers worked in service sector industries, and this trend has continued since (Tilly 1992: p. 30). Labor analysts argue that the bulk of the expansion in part-time and contingent work is due to the expansion of the service sector, which has always used shift and part-time workers as cost-control mechanisms. . . .

Service sector expansion has also sparked the rapid growth of the temporary help industry. Over the past decade, the number of temporary workers has tripled (Kilborn 1995). As the chairman of Manpower, Inc., the largest single employer in the United States, remarked, "The U.S. is going from just-in-time manufacturing to just-in-time employment. The employer tells us, 'I want them delivered exactly when I want them, as many as I need, and when I don't need them, I don't want them here'" (quoted in Castro 1993: 44). Like part-time workers, temporary workers carry no "overhead" costs in terms of taxes and benefits, and they are on call as needed. Since they experience little workplace continuity, they are less likely than full-time, continuous employees to organize or advocate changes in working conditions.

Women, youth, and minorities comprise the bulk of the part-time and contingent work force in the service sector. For example, Karen Brodkin Sacks (1990) has noted that the health care industry is "so stratified by race and gender that the uniforms worn to distinguish the jobs and statuses of health care workers are largely redundant" (p. 188). Patients respond to the signals implicitly transmitted via gender and race and act accordingly, offering deference to some workers and expecting it from others. Likewise in domestic service, race and gender determine who gets which jobs. White American and European women are most likely to be hired for domestic jobs defined primarily as child care, while women of color predominate in those defined as house cleaning (Rollins 1985), regardless of what the actual allocation of work might be. The same kinds of stratification can be found in secretarial work, food service, hotels, and sales occupations (Hochschild 1983; Leidner 1991). In all of these occupational groups, demographic characteristics of the worker determine the job title and thus other factors such as status, pay, benefits, and degree of autonomy on the job.

As a result, the shift to services has had a differential impact on various sectors of the labor market, increasing stratification between the well employed and the underemployed. The service sector work force is highly feminized, especially at the bottom. Within personal and business services, for example, Bette Woody (1989) found that "men are concentrated in high-ticket, high commission sales jobs, and women in retail and food service" (p. 57). And although the decline in manufacturing forced male workers to move into service industries, Jon Lorence (1992: 150) notes that within the service sector, occupations are highly stratified by gender. From 1950 to 1990, 60 percent of all new service sector employment and 74 percent of all new low-skill jobs were filled by women. . . .

Service sector employment is equally stratified by race. For example, Woody (1989) finds that the shift to services has affected black women in two important ways. First, it has meant higher rates of unemployment and underemployment for black men, which increased pressure on black women to be the primary bread-winners for their families and contributed to the overall reduction in black family income. Second, although the increase in service sector jobs has meant greater opportunity for black women and has allowed them to move out of domestic service into the formal economy, as Evelyn Nakano Glenn (1996) also notes, they have remained at the lowest rung of the service employment ladder. Unlike some white women who "moved up" to male-intensive occupations with the shift to services, black women "moved over" to sectors traditionally employing white women (Woody 1989: 54). These traditionally male occupations are not only low-security, low-pay jobs, but they lack internal career ladders. As Ruth Needleman and Anne Nelson (1988) note, "There is no progression from nurse to physician, from secretary to manager" (p. 297).

Service work differs most radically from manufacturing, construction, or agricultural work in the relationship between worker characteristics and the job. Even though discrimination in hiring, differential treatment, differential pay, and other forms of stratification exist in all labor markets, service occupations are the only ones in which the producer in some sense equals the product. In no other area of wage labor are the personal characteristics of the workers so strongly associated with the nature of work. Because at least part of the job in all service occupations is to "manufacture social relations" (Filby 1992; 37), traits such as gender, race, age, and sexuality serve a signaling function, indicating to the customer/employer important cues about the tone of the interaction. Women are expected to be more nurturing and empathetic than men and to tolerate more offensive behavior from customers (Hochschild 1983; Leidner 1991; Pierce 1996, Sutton and Rafaeli 1989). Similarly, both women and men of color are expected to be deferential and to take on more demeaning tasks (Rollins 1985; Woody 1989). In addition, a given task may be viewed as more or less demeaning depending on who is doing it.

These occupations are so stratified that worker characteristics such as race and gender determine not only who is considered desirable or even eligible to fill certain jobs, but also who will want to fill certain jobs and how the job itself is performed. Worker characteristics shape what is expected of a worker by management and customers, how that worker adapts to the job, and what aspects of the job he or she will resist or embrace. The strategies workers use to adapt to the demands of service jobs are likely to differ according to gender and other characteristics. Women are more likely to embrace the emotional demands (e.g., nurturing, care giving) of certain types of service jobs because these demands generally fit their notion of gender-appropriate behavior. As Pierce notes, heterosexual men tend to resist these demands because they find "feminine" emotional labor demeaning; in response they either reframe the nature of the job to emphasize traditional masculine qualities or distance themselves by providing service by rote, making it clear that they are acting under duress.

All of these interconnections between worker, work, and product result in tendencies toward very specific types of labor market stratification. A long and heated debate has raged concerning the ultimate impact of deindustrialization on the structure of the labor market. On one side are those who argue that a shift to a service-based economy will produce skill upgrading and a leveling of job hierarchies as information and communications technologies reshape the labor market (see, for example, Bell 1973). Others take a more pessimistic view, arguing that the shift to services will give rise to two trends: "towards polarization and towards the proliferation of low-wage jobs" (Bluestone and Harrison 1988: 126). In a sense, both positions are correct but for different segments of the labor market.

Overall, the transition to a service-based economy will likely mean a more stratified work force in which more part-time and contingent jobs are filled predominantly by women, minorities, and workers without college degrees. In jobs lacking internal career ladders, these workers have little chance for upward mobility. Contingent workers are also less likely to organize successfully due to their tenuous attachment to specific employers and to the labor force in general. At the opposite end of the service sector occupational spectrum are highly educated managerial, professional, and para-professional workers, who will have equally weak attachments to employers, but who, due to their highly marketable skills, will move with relative ease from one well-paying job to another. Given this divided and economically segregated work force, what are the opportunities for workers to advocate, collectively or individually, for greater security, better working conditions, and a voice in shaping their work? . . .

REFERENCES

Albrecht, Karl, and Ron Zemke. 1990. *Service America!: Doing Business in the New Economy.* New York: Warner Books.

Bell, Daniel. 1973. *The Coming of Post-Industrial Society: A Venture in Social Forecasting.* New York: Basic Books.

Bluestone, Barry, and Bennett Harrison. 1988. *The Great U-Turn: Corporate Restructuring and the Polarizing of America.* New York: Basic Books.

Castro, Janice. 1993. "Disposable Workers." *Time Magazine,* March 29, pp. 43–57.

Filby, M. P. 1992, " 'The Figures, the Personality, and the Bums': Service Work and Sexuality." *Work, Employment, and Society* 6 (March) 23–42.

Glenn, Evelyn Nakano. 1996. "From Servitude to Service Work: Historical Continuities in the Racial Division of Paid Reproductive Labor." In Cameron Lynne Macdonald and Carmen Sirianni, eds., *Working in the Service Society,* pp. 115–156. Philadelphia: Temple University Press.

Hirschhorn, Larry. 1988. "The Post-Industrial Economy: Labor, Skills, and the New Mode of Production." *The Service Industries Journal* 8: 19–38.

Hochschild, Arlie Russell. 1983. *The Managed Heart: Commercialization of Human Feeling.* Berkeley: University of California Press.

Kilborn, Peter T. 1995. "In New Work World, Employers Call All the Shots: Job Insecurity, a Special Report." *New York Times,* July 3, p. A1.

Kutscher, Ronald E. 1987. "Projections 2000: Overview and Implications of the Projections to 2000." *Monthly Labor Review* (September): 3–9.

Leidner, Robin. 1993. *Fast Food, Fast Talk: Service Work and the Routinization of Everyday Life*. Berkeley: University of California Press.

Leidner, Robin. 1991. "Serving Hamburgers and Selling Insurance: Gender, Work, and Identity in Interactive Service Jobs." *Gender & Society* 5: 154–77.

Lorence, Jon. 1992. "Service Sector Growth and Metropolitan Occupational Sex Segregation." *Work and Occupations* 19: 128–56.

Needleman, Ruth, and Anne Nelson. 1988. "Policy Implications: The Worth of Women's Work." In Anne Starham, Eleanor M. Miller, and Hans O. Mauksch, eds., *The Worth of Women's Work: A Qualitative Synthesis* (pp. 293–308). Albany: SUNY Press.

Nelson, Joel I. 1994. "Work and Benefits: The Multiple Problems of Service Sector Employment." *Social Problems* 41: 240–55.

Packham, John. 1992. "The Organization of Work on the Service-Sector Shop Floor." Unpublished paper.

Pierce, Jennifer L. 1996. *Gender Trials: Emotional Lives in Contemporary Law Firms*. Berkeley: University of California Press.

Rollins, Judith. 1985. *Between Women: Domestics and Their Employers*. Philadelphia: Temple University Press.

Sacks, Karen Brodkin. 1990. "Does It Pay to Care?" In Emily K. Abel and Margaret K. Nelson, eds., *Circles of Care: Work and Identity in Women's Lives* (pp. 188–206). Albany: SUNY Press.

Schlesinger, Leonard, and James Heskett. 1991. "The Service-Driven Service Company." *Harvard Business Review* (September–October): 71–81.

Sutton, Robert I., and Anat Rafaeli. 1989. "The Expression of Emotion in Organizational Life." *Research in Organizational Behavior* 11: 1–42.

Tilly, Chris. 1992. "Short Hours, Short Shrift: The Causes and Consequences of Part-Time Employment." In Virginia L. Du Rivage, ed., *New Policies for the Part-Time and Contingent Work-Force* (pp. 15–43). Armonk, NY: M. E. Sharpe, Economic Policy Institute Series.

U.S. Department of Labor, Bureau of Labor Statistics. 1994. *Monthly Labor Review* (July): 74–83.

Walsh, T. J. 1990. "Flexible Labor Utilization in the Private Service Sector." *Work, Employment, and Society* 4: 517–30.

Wood, Stephen, ed. 1989. *The Transformation of Work? Skill, Flexibility, and the Labor Process*. London: Unwin Hyman.

Woody, Bette. 1989. "Black Women in the Emerging Services Economy." *Sex Roles* 21: 45–67.

KEY CONCEPTS

division of labor   downsizing   service sector

DISCUSSION QUESTIONS

1. How has the transition to a service-based economy changed employment patterns in the contemporary economy? What implications does this have for the relationship between management and labor?

2. How do race and gender stratification influence the perceptions of some workers as suitable for particular forms of service work? What evidence have you witnessed of this phenomenon in your experiences in a service-based society?

55

The Time Divide

JERRY A. JACOBS AND KATHLEEN GERSON

Jerry A. Jacobs and Kathleen Gerson examine the time conflicts that families feel as they try to integrate the demands of work and the demands of family. By studying the different demands of work and family, Jacobs and Gerson assess how time is being shaped by contemporary societal forms.

The late twentieth century witnessed dramatic changes in the ways Americans organize their work and family lives. As men's earnings stagnated and women became increasingly committed to work outside the home, the breadwinner-homemaker household that predominated during the middle of the twentieth century gave way to a diverse range of work and family arrangements. Today, as a new century begins, we face a greatly altered family landscape, in which dual-income and single-parent families far outnumber the once ascendant two-parent, one-earner household.

These fundamental alterations in family arrangements and the gender composition of the labor force have created new challenges for the American workplace and new dilemmas for American families, most of whom cannot count on a full-time, unpaid caretaker to attend to children and the household. Women and men alike must now juggle the competing demands of work and family, often without institutional assistance or road maps to guide their way.

SOURCE: Jerry A. Jacobs and Kathleen Gerson. 2005. *The Time Divide: Work, Family, and Gender Inequality.* Cambridge, MA: Harvard University Press.

The rise of conflicts between the social organization of work and the demands of family life has raised a host of questions about the changing nature of work and family life and its consequences for workers and workplaces. Have workers' ties to paid work, and especially the time they devote to it, changed substantially, and if so, how? In what way does the structure of jobs and work settings shape the options and dilemmas facing American workers and their families? How do women and men experience these conflicts and dilemmas, and what are the implications for their families? And what organizational strategies and national policies can provide satisfying resolutions to the conflicts facing American workers and families?

These questions lie at the center of growing debates over the current state and future prospects for family and work in America. Central to this debate is the issue of time. Like money, time is a valuable resource that constantly provokes questions about how it should be allocated and spent. Unlike money, however, the overall supply of time cannot be expanded. There can only be twenty-four hours in a day, seven days in a week, and fifty-two weeks in a year. When time squeezes arise, it is not possible to create more time. It is thus not surprising that the starting point for understanding work and family change centers on the issue of time. . . .

Are Americans working more than ever, and, if so, why and with what consequences?

Many American families find their lives increasingly rushed. The tempo may vary from steady to hectic to frantic, but a large and growing group perceives that life moves at a faster pace than it did for previous generations. Are these perceptions accurate? In this article, we address the thicket of competing claims about whether or not daily life has become more hurried and less leisurely, and, if so, whether overwork is the main cause of this dilemma. Is the shortage of "time for life" real, or is it an ambiguous or isolated social trend that has become exaggerated by a lack of historical perspective and a persuasive concern for social criticism?

In important ways, these questions—and the prevailing answers—are too simple. No one trend can adequately portray the complicated changes taking place in the American labor force and among American households. A more complete account would recognize the diversity among workers and their families. In fact, while a large segment of the labor force *is* working longer and harder than ever, another group of workers is confronting the problem of finding enough work. It is true that Americans are increasingly torn between commitments to work and to family life. Less apparent, however, are the ways in which the American labor force is diverging, as some workers face increasing demands on their time at work and others struggle to find enough work to meet their own and their families' needs. Once this diversity, and indeed divergence, is acknowledged, the competing claims about whether Americans are working more or enjoying more leisure can be resolved.

For a growing group of American families, feelings of overwork are real and well founded. Workers in some households are putting in a great deal of time at work. And many who are not working especially long days are nevertheless facing the challenges of managing a two-earner or a single-parent household. The lack of time for family life is not simply a matter of questionable choices made by some individuals, but instead reflects the way choices are shaped by our economy and the structure of our work organizations. The experience of feeling squeezed for time reflects fundamental and enduring changes in the nature and composition of American society. From this perspective, the scarcity of time for the tasks of daily life is not just a personal problem, but a public issue of great importance....

TRENDS IN THE WORKWEEK SINCE 1970

There is little doubt that workers are facing new and unprecedented challenges in meeting the demands of family and work. The transformation of family structure and household composition in the closing decades of the twentieth century have clearly created a significant time bind for many workers. This time squeeze is both less general and more varied than overarching arguments about national trends have implied. To understand the dynamics of change and its diverse consequences, we need to examine the full range of situations among a varied group of workers....

Although the average American workweek has not changed dramatically over the past several decades, a growing group of Americans are clearly, and strongly, pressed for time. These workers include employees who are putting in especially long days at work each week, often against their desire, and people in dual-earner and single-parent families who cannot rely on a support system anchored by a nonemployed member. The intransigence in the structure of work and the rise of highly demanding jobs, especially at the upper levels of the occupational hierarchy, present dilemmas and problems for many workers. And, increasingly, women and men alike face challenging work without the traditional, unpaid spouses once taken for granted by husbands in upwardly mobile careers and highly demanding jobs. Yet employers have not readily responded to these changed realities, assuming that devoted workers have—or should have—unpaid partners who are home full-time to take care of the many domestic tasks on which not only family life but successful careers and secure communities depend. In the context of these dramatic social shifts in Americans' private lives, it is no surprise that many Americans feel that they are squeezed for time and working more than ever.

A "time bind" has clearly emerged in the contemporary United States. It is, however, rooted as much in the changing nature of family life and women's commitments as in the expansion of working time for individuals. Working parents in dual-earner and single-parent households have always faced a time bind, and the principal change over the past thirty years has been a marked growth in the number of people living in these family situations. The time squeeze created by spending more hours at paid work is, moreover, not universal, but rather is concentrated among professionals and managers, who are especially likely to shape the terms of public discussion and debate. Although there are also less affluent workers who put in substantial overtime or who work at two (or more) jobs, this group represents a smaller proportion of blue-collar workers than do overworked Americans who are professionals and managers. These trends suggest that while it is important to address the new challenges and insecurities facing American workers and their families, we should be careful to move beyond generalizations to focus on the variety of dilemmas workers are encountering.

KEY CONCEPTS

labor force participation rate   time bind   working families

DISCUSSION QUESTIONS

1. How has women's labor force participation influenced what families feel as the "time bind"?
2. How is the time bind influenced by gender roles in society?
3. What can employers do to reduce some of the stresses caused by combining work and family?

56

Nickel-and-Dimed: On (Not) Getting By in America

BARBARA EHRENREICH

Barbara Ehrenreich, a journalist and faculty member, "posed" for several weeks as an unskilled worker to find out how people survive doing low-wage work. Her account, later published in the best-selling book, Nickel and Dimed, *is a compelling portrait of the conditions of work for millions of people in the United States.*

At the beginning of June 1998 I leave behind everything that normally soothes the ego and sustains the body—home, career, companion, reputation,

SOURCE: Ehrenreich, Barbara. "Nickel-and-Dimed: On (Not) Getting By in America." *Harper's Magazine* 298 (January 1999): 37 ff. Reprinted with permission.

ATM card—for a plunge into the low-wage workforce. There, I become another, occupationally much diminished "Barbara Ehrenreich"—depicted on job-application forms as a divorced homemaker whose sole work experience consists of housekeeping in a few private homes. I am terrified, at the beginning, of being unmasked for what I am: a middle-class journalist setting out to explore the world that welfare mothers are entering, at the rate of approximately 50,000 a month, as welfare reform kicks in. Happily, though, my fears turn out to be entirely unwarranted: during a month of poverty and toil, my name goes unnoticed and for the most part unuttered. In this parallel universe where my father never got out of the mines and I never got through college, I am "baby," "honey," "blondie," and, most commonly, "girl."

My first task is to find a place to live. I figure that if I can earn $7 an hour—which, from the want ads, seems doable—I can afford to spend $500 on rent, or maybe, with severe economies, $600. In the Key West area, where I live, this pretty much confines me to flophouses and trailer homes—like the one, a pleasing fifteen-minute drive from town, that has no air-conditioning, no screens, no fans, no television, and, by way of diversion, only the challenge of evading the landlord's Doberman pinscher. The big problem with this place, though, is the rent, which at $675 a month is well beyond my reach. All right, Key West is expensive. But so is New York City, or the Bay Area, or Jackson Hole, or Telluride, or Boston, or any other place where tourists and the wealthy compete for living space with the people who clean their toilets and fry their hash browns. Still, it is a shock to realize that "trailer trash" has become, for me, a demographic category to aspire to.

So I decide to make the common trade-off between affordability and convenience, and go for a $500-a-month efficiency thirty miles up a two-lane highway from the employment opportunities of Key West, meaning forty-five minutes if there's no road construction and I don't get caught behind some sun-dazed Canadian tourists. I hate the drive, along a roadside studded with white crosses commemorating the more effective head-on collisions, but it's a sweet little place—a cabin, more or less, set in the swampy back yard of the converted mobile home where my landlord, an affable TV repairman, lives with his bartender girlfriend. Anthropologically speaking, a bustling trailer park would be preferable, but here I have a gleaming white floor and a firm mattress, and the few resident bugs are easily vanquished.

Besides, I am not doing this for the anthropology. My aim is nothing so mistily subjective as to "experience poverty" or find out how it "really feels" to be a long-term low-wage worker. I've had enough unchosen encounters with poverty and the world of low-wage work to know it's not a place you want to visit for touristic purposes; it just smells too much like fear. And with all my real-life assets—bank account, IRA, health insurance, multiroom home—waiting indulgently in the background, I am, of course, thoroughly insulated from the terrors that afflict the genuinely poor.

No, this is a purely objective, scientific sort of mission. The humanitarian rationale for welfare reform—as opposed to the more punitive and stingy impulses that may actually have motivated it—is that work will lift poor women out of poverty while simultaneously inflating their self-esteem and hence their future value in the labor market. Thus, whatever the hassles involved in finding child care, transportation, etc., the transition from welfare to work will end happily, in greater prosperity for all. Now there are many problems with this comforting prediction, such as the fact that the economy will inevitably undergo a downturn, eliminating many jobs. Even without a downturn, the influx of a million former welfare recipients into the low-wage labor market could depress wages by as much as 11.9 percent, according to the Economic Policy Institute (EPI) in Washington, D.C.

But is it really possible to make a living on the kinds of jobs currently available to unskilled people? Mathematically, the answer is no, as can be shown by taking $6 to $7 an hour, perhaps subtracting a dollar or two an hour for child care, multiplying by 160 hours a month, and comparing the result to the prevailing rents. According to the National Coalition for the Homeless, for example, in 1998 it took, on average nationwide, an hourly wage of $8.89 to afford a one-bedroom apartment, and the Preamble Center for Public Policy estimates that the odds against a typical welfare recipient's landing a job at such a "living wage" are about 97 to 1. If these numbers are right, low-wage work is not a solution to poverty and possibly not even to homelessness.

It may seem excessive to put this proposition to an experimental test. As certain family members keep unhelpfully reminding me, the viability of low-wage work could be tested, after a fashion, without ever leaving my study. I could just pay myself $7 an hour for eight hours a day, charge myself for room and board, and total up the numbers after a month. Why leave the people and work that I love? But I am an experimental scientist by training. In that business, you don't just sit at a desk and theorize; you plunge into the everyday chaos of nature, where surprises lurk in the most mundane measurements. Maybe, when I got into it, I would discover some hidden economies in the world of the low-wage worker. After all, if 30 percent of the workforce toils for less than $8 an hour, according to the EPI, they may have found some tricks as yet unknown to me. Maybe—who knows?—would even be able to detect in myself the bracing psychological effects of getting out of the house, as promised by the welfare wonks at places like the Heritage Foundation. Or, on the other hand, maybe there would be unexpected costs—physical, mental, or financial—to throw off all my calculations. Ideally, I should do this with two small children in tow, that being the welfare average, but mine are grown and no one is willing to lend me theirs for a month-long vacation in penury. So this is not the perfect experiment, just a test of the best possible case: an unencumbered woman, smart and even strong, attempting to live more or less off the land.

On the morning of my first full day of job searching, I take a red pen to the want ads, which are auspiciously numerous. Everyone in Key West's booming "hospitality industry" seems to be looking for someone like me—trainable, flexible, and with suitably humble expectations as to pay. I know I possess certain traits that might be advantageous—I'm white and, I like to think, well-spoken and poised—but I decide on two rules: One, I cannot use any skills derived from my education or usual work—not that there are a lot of want ads for satirical essayists anyway. Two, I have to take the best-paid job that is offered me and of course do my best to hold it; no Marxist rants or sneaking off to read novels in the ladies' room. In addition, I rule out various occupations for one reason or another: Hotel front-desk clerk, for example, which to my surprise is regarded as unskilled and pays around $7 an hour, gets eliminated because it involves standing in one spot for eight hours a day. Waitressing is similarly something I'd like to avoid, because I remember it leaving me bone tired when I was eighteen, and I'm decades of varicosities and back pain beyond that now. Telemarketing, one of the first refuges of the suddenly indigent, can be dismissed on grounds of personality. This leaves certain supermarket jobs, such as deli clerk, or house-keeping in Key West's thousands of hotel and guest rooms. Housekeeping is especially appealing, for reasons both atavistic and practical: it's what my mother did before I came along, and it can't be too different from what I've been doing part-time, in my own home, all my life.

So I put on what I take to be a respectful-looking outfit of ironed Bermuda shorts and scooped-neck T-shirt and set out for a tour of the local hotels and supermarkets. Best Western, Econo Lodge, and HoJo's all let me fill out applica-tion forms, and these are, to my relief, interested in little more than whether I am a legal resident of the United States and have committed any felonies. My next stop is Winn-Dixie, the supermarket, which turns out to have a particularly onerous application process, featuring a fifteen-minute "interview" by computer since, apparently, no human on the premises is deemed capable of representing the corporate point of view. I am conducted to a large room decorated with posters illustrating how to look "professional" (it helps to be white and, if female, permed) and warning of the slick promises that union organizers might try to tempt me with. The interview is multiple choice: Do I have anything, such as child-care problems, that might make it hard for me to get to work on time? Do I think safety on the job is the responsibility of management? Then, popping up cunningly out of the blue: How many dollars' worth of stolen goods have I purchased in the last year? Would I turn in a fellow employee if I caught him stealing? Finally, "Are you an honest person?"

Apparently, I ace the interview, because I am told that all I have to do is show up in some doctor's office tomorrow for a urine test. This seems to be a fairly general rule: if you want to stack Cheerio boxes or vacuum hotel rooms in chemically fascist America, you have to be willing to squat down and pee in front of some health worker (who has no doubt had to do the same thing herself). The wages Winn-Dixie is offering—$6 and a couple of dimes to start with—are not enough, I decide, to compensate for this indignity.

I lunch at Wendy's, where $4.99 gets you unlimited refills at the Mexican part of the Superbar, a comforting surfeit of refried beans and "cheese sauce." A teenage employee, seeing me studying the want ads, kindly offers me an applica-tion form, which I fill out, though here, too, the pay is just $6 and change an hour. Then it's off for a round of the locally owned inns and guest-houses. At "The Palms," let's call it, a bouncy manager actually takes me around to see the rooms and meet the existing housekeepers, who, I note with satisfaction, look pretty much like me—faded ex-hippie types in shorts with long hair pulled back in braids. Mostly, though, no one speaks to me or even looks at me except to proffer an application form. At my last stop, a palatial B&B, I wait twenty minutes to meet "Max," only to be told that there are no jobs now but there should be one soon, since "nobody lasts more than a couple weeks." (Because none of the people I talked to knew I was a reporter, I have changed their names to protect their privacy and, in some cases perhaps, their jobs.)

Three days go by like this, and, to my chagrin, no one out of the approximately twenty places I've applied calls me for an interview. I had been vain enough to worry about coming across as too educated for the jobs I sought, but no one even seems interested in finding out how overqualified I am. Only later will I realize that the want ads are not a reliable measure of the actual jobs available at any particular time. They are, as I should have guessed from Max's comment, the employers' insurance policy against the relentless turnover of the low-wage work-force. Most of the big hotels run ads almost continually, just to build a supply of applicants to replace the current workers as they drift away or are fired, so finding a job is just a matter of being at the right place at the right time and flexible enough to take whatever is being offered that day. This finally happens to me at a one of the big discount hotel chains, where I go, as usual, for housekeeping and am sent, instead, to try out as a waitress at the attached "family restaurant," a dismal spot with a counter and about thirty tables that looks out on a parking garage and features such tempting fare as "Polish [sic] sausage and BBQ sauce" on 95-degree days. Phillip, the dapper young West Indian who introduces himself as the manager, interviews me with about as much enthusiasm as if he were a clerk processing me for Medicare, the principal questions being what shifts can I work and when can I start. I mutter something about being woefully out of practice as a waitress, but he's already on to the uniform: I'm to show up tomorrow wearing black slacks and black shoes; he'll provide the rust-colored polo shirt with HEARTHSIDE embroidered on it, though I might want to wear my own shirt to get to work, ha ha. At the word "tomorrow," something between fear and indignation rises in my chest. I want to say, "Thank you for your time, sir, but this is just an experiment, you know, not my actual life."

So begins my career at the Hearthside, I shall call it, one small profit center within a global discount hotel chain, where for two weeks I work from 2:00 till 10:00 P.M. for $2.43 an hour plus tips. In some futile bid for gentility, the management has barred employees from using the front door, so my first day I enter through the kitchen, where a red-faced man with shoulder-length blond hair is throwing frozen steaks against the wall and yelling, "Fuck this shit!" "That's just Jack," explains Gail, the wiry middle-aged waitress who is assigned to train me. "He's on the rag again"—a condition occasioned, in this instance, by the fact that the cook on the morning shift had forgotten to thaw out the steaks. For the next eight hours, I run after the agile Gail, absorbing bits of instruction along with fragments of personal tragedy. All food must be trayed, and the reason she's so tired today is that she woke up in a cold sweat thinking of her boyfriend, who killed himself recently in an upstate prison. No refills on lemonade. And the reason he was in prison is that a few DUIs caught up with him, that's all, could have happened to anyone. Carry the creamers to the table in a monkey bowl, never in your hand. And after he was gone she spent several months living in her truck, peeing in a plastic pee bottle and reading by candlelight at night, but you can't live in a truck in the summer, since you need to have the windows down, which means anything can get in, from mosquitoes on up.

At least Gail puts to rest any fears I had of appearing overqualified. From the first day on, I find that of all the things I have left behind, such as home and identity, what I miss the most is competence. Not that I have ever felt utterly competent in the writing business, in which one day's success augers nothing at all for the next. But in my writing life I at least have some notion of procedure: do the research, make the outline, rough out a draft, etc. As a server, though, I am beset by requests like bees: more iced tea here, ketchup over there, a to-go box for table fourteen, and where are the high chairs, anyway? Of the twenty-seven tables, up to six are usually mine at any time, though on slow afternoons or if Gail is off, I sometimes have the whole place to myself. There is the touch-screen computer-ordering system to master, which is, I suppose, meant to minimize server-cook contact, but in practice requires constant verbal fine-tuning: "That's gravy on the mashed, okay? None on the meatloaf," and so forth—while the cook scowls as if I were inventing these refinements just to torment him. Plus, something I had forgotten in the years since I was eighteen: about a third of a server's job is "side work" that's invisible to customers—sweeping, scrubbing, slicing, refilling, and restocking. If it isn't all done, every little bit of it, you're going to face the 6:00 P.M. dinner rush defenseless and probably go down in flames. I screw up dozens of times at the beginning, sustained in my shame entirely by Gail's support—"It's okay, baby, everyone does that sometime"— because, to my total surprise and despite the scientific detachment I am doing my best to maintain, I care.

The whole thing would be a lot easier if I could just skate through it as Lily Tomlin in one of her waitress skits, but I was raised by the absurd Booker T. Washingtonian precept that says: If you're going to do something, do it well. In fact, "well" isn't good enough by half. Do it better than anyone has ever done it before. Or so said my father, who must have known what he was talking about because he managed to pull himself, and us with him, up from the mile-deep copper mines of Butte to the leafy suburbs of the Northeast, ascending from boilermakers to martinis before booze beat out ambition. As in most endeavors I have encountered in my life, doing it "better than anyone" is not a reasonable goal. Still, when I wake up at 4:00 A.M. in my own cold sweat, I am not thinking about the writing deadlines I'm neglecting; I'm thinking about the table whose order I screwed up so that one of the boys didn't get his kiddie meal until the rest of the family had moved on to their Key Lime pies. That's the other powerful motivation I hadn't expected—the customers, or "patients," as I can't help thinking of them on account of the mysterious vulnerability that seems to have left them temporarily unable to feed themselves. After a few days at the Hearth-side, I feel the service ethic kick in like a shot of oxytocin, the nurturance hormone. The plurality of my customers are hard-working locals—truck drivers, construction workers, even housekeepers from the attached hotel—and I want them to have the closest to a "fine dining" experience that the grubby circumstances will allow. No "you guys" for me; everyone over twelve is "sir" or "ma'am." I ply them with iced tea and coffee refills; I return, mid-meal, to inquire how everything is; I doll up their salads with chopped raw mushrooms, summer squash slices, or whatever bits of produce I can find that have survived their sojourn in the cold-storage room mold-free. . . .

Ten days into it, this is beginning to look like a livable lifestyle. I like Gail, who is "looking at fifty" but moves so fast she can alight in one place and then another without apparently being anywhere between them. I clown around with Lionel, the teenage Haitian busboy, and catch a few fragments of conversation with Joan, the svelte fortyish hostess and militant feminist who is the only one of us who dares to tell Jack to shut the fuck up. I even warm up to Jack when, on a slow night and to make up for a particularly unwarranted attack on my abilities, or so I imagine, he tells me about his glory days as a young man at "coronary school"—or do you say "culinary"?—in Brooklyn, where he dated a knock-out Puerto Rican chick and learned everything there is to know about food. I finish up at 10:00 or 10:30, depending on how much side work I've been able to get done during the shift, and cruise home to the tapes I snatched up at random when I left my real home—Marianne Faithfull, Tracy Chapman, Enigma, King Sunny Ade, the Violent Femmes—just drained enough for the music to set my cranium resonating but hardly dead. Midnight snack is Wheat Thins and Monterey Jack, accompanied by cheap white wine on ice and whatever AMC has to offer. To bed by 1:30 or 2:00, up at 9:00 or 10:00, read for an hour while my uniform whirls around in the landlord's washing machine, and then it's another eight hours spent following Mao's central instruction, as laid out in the Little Red Book, which was: Serve the people.

I could drift along like this, in some dreamy proletarian idyll, except for two things. One is management. If I have kept this subject on the margins thus far it is because I still flinch to think that I spent all those weeks under the surveillance of men (and later women) whose job it was to monitor my behavior for signs of sloth, theft, drug abuse, or worse. Not that managers and especially "assistant managers" in low-wage settings like this are exactly the class enemy. In the restaurant business, they are mostly former cooks or servers, still capable of pinch-hitting in the kitchen or on the floor, just as in hotels they are likely to be former clerks, and paid a salary of only about $400 a week. But everyone knows they have crossed over to the other side, which is, crudely put, corporate as opposed to human. Cooks want to prepare tasty meals; servers want to serve them graciously; but managers are there for only one reason—to make sure that money is made for some theoretical entity that exists far away in Chicago or New York, if a corporation can be said to have a physical existence at all. Reflecting on her career, Gail tells me ruefully that she had sworn, years ago, never to work for a corporation again. "They don't cut you no slack. You give and you give, and they take."

Managers can sit—for hours at a time if they want—but it's their job to see that no one else ever does, even when there's nothing to do, and this is why, for servers, slow times can be as exhausting as rushes. You start dragging out each little chore, because if the manager on duty catches you in an idle moment, he will give you something far nastier to do. So I wipe, I clean, I consolidate ketchup bottles and recheck the cheesecake supply, even tour the tables to make sure the customer evaluation forms are all standing perkily in their places—wondering all the time how many calories I burn in these strictly theatrical exercises. When, on a particularly dead afternoon, Stu finds me glancing at a USA Today a customer has left behind, he assigns me to vacuum the entire floor with the broken vacuum cleaner that has a handle only two feet long, and the only way to do that without incurring orthopedic damage is to proceed from spot to spot on your knees.

On my first Friday at the Hearthside there is a "mandatory meeting for all restaurant employees," which I attend, eager for insight into our overall marketing strategy and the niche (your basic Ohio cuisine with a tropical twist?) we aim to inhabit. But there is no "we" at this meeting. Phillip, our top manager except for an occasional "consultant" sent out by corporate headquarters, opens it with a sneer: "The break room—it's disgusting. Butts in the ashtrays, newspapers lying around, crumbs." This windowless little room, which also houses the time clock for the entire hotel, is where we stash our bags and civilian clothes and take our half-hour meal breaks. But a break room is not a right, he tells us. It can be taken away. We should also know that the lockers in the break room and whatever is in them can be searched at any time. Then comes gossip; there has been gossip; gossip (which seems to mean employees talking among themselves) must stop. Off-duty employees are henceforth barred from eating at the restaurant, because "other servers gather around them and gossip." When Phillip has exhausted his agenda of rebukes, Joan complains about the condition of the ladies' room and I throw in my two bits about the vacuum cleaner. But I don't see any backup coming from my fellow servers, each of whom has subsided into her own personal funk; Gail, my role model, stares sorrowfully at a point six inches from her nose. The meeting ends when Andy, one of the cooks, gets up, muttering about breaking up his day off for this almighty bullshit.

Just four days later we are suddenly summoned into the kitchen at 3:30 P.M., even though there are live tables on the floor. We all—about ten of us—stand around Phillip, who announces grimly that there has been a report of some "drug activity" on the night shift and that, as a result, we are now to be a "drugfree" workplace, meaning that all new hires will be tested, as will possibly current employees on a random basis. I am glad that this part of the kitchen is so dark, because I find myself blushing as hard as if I had been caught toking up in the ladies' room myself: I haven't been treated this way—lined up in the corridor, threatened with locker searches, peppered with carelessly aimed accusations— since junior high school. Back on the floor, Joan cracks, "Next they'll be telling us we can't have sex on the job." When I ask Stu what happened to inspire the crackdown, he just mutters about "management decisions" and takes the opportunity to upbraid Gail and me for being too generous with the rolls. From now on there's to be only one per customer, and it goes out with the dinner, not with the salad. He's also been riding the cooks, prompting Andy to come out of the kitchen and observe—with the serenity of a man whose customary implement is a butcher knife—that "Stu has a death wish today."

Later in the evening, the gossip crystallizes around the theory that Stu is himself the drug culprit, that he uses the restaurant phone to order up marijuana and sends one of the late servers out to fetch it for him. The server was caught, and she may have ratted Stu out or at least said enough to cast some suspicion on him, thus accounting for his pissy behavior. Who knows? Lionel, the busboy, entertains us for the rest of the shift by standing just behind Stu's back and sucking deliriously on an imaginary joint.

The other problem, in addition to the less-than-nurturing management style, is that this job shows no sign of being financially viable. You might imagine, from a comfortable distance, that people who live, year in and year out, on $6 to $10 an hour have discovered some survival stratagems unknown to the middle class. But no. It's not hard to get my co-workers to talk about their living situations, because housing, in almost every case, is the principal source of disruption in their lives, the first thing they fill you in on when they arrive for their shifts. After a week, I have compiled the following survey:

- Gail is sharing a room in a well-known down-town flophouse for which she and a roommate pay about $250 a week. Her roommate, a male friend, has begun hitting on her, driving her nuts, but the rent would be impossible alone.

- Claude, the Haitian cook, is desperate to get out of the two-room apartment he shares with his girlfriend and two other, unrelated, people. As far as I can determine, the other Haitian men (most of whom only speak Creole) live in similarly crowded situations.

- Annette, a twenty-year-old server who is six months pregnant and has been abandoned by her boyfriend, lives with her mother, a postal clerk.

- Marianne and her boyfriend are paying $170 a week for a one-person trailer.

- Jack, who is, at $10 an hour, the wealthiest of us, lives in the trailer he owns, paying only the $400-a-month lot fee.

- The other white cook, Andy, lives on his dry-docked boat, which, as far as I can tell from his loving descriptions, can't be more than twenty feet long. He offers to take me out on it, once it's repaired, but the offer comes with inquiries as to my marital status, so I do not follow up on it.

- Tina and her husband are paying $60 a night for a double room in a Days Inn. This is because they have no car and the Days Inn is within walking distance of the Hearthside. When Marianne, one of the breakfast servers, is tossed out of her trailer for subletting (which is against the trailer-park rules), she leaves her boyfriend and moves in with Tina and her husband.

- Joan, who had fooled me with her numerous and tasteful outfits (hostesses wear their own clothes), lives in a van she parks behind a shopping center at night and showers in Tina's motel room. The clothes are from thrift shops.

It strikes me, in my middle-class solipsism, that there is gross improvidence in some of these arrangements. When Gail and I are wrapping silverware in napkins—the only task for which we are permitted to sit—she tells me she is thinking of escaping from her roommate by moving into the Days Inn herself. I am astounded: How can she even think of paying between $40 and $60 a day?

But if I was afraid of sounding like a social worker, I come out just sounding like a fool. She squints at me in disbelief, "And where am I supposed to get a month's rent and a month's deposit for an apartment?" I'd been feeling pretty smug about my $500 efficiency, but of course it was made possible only by the $1,300 I had allotted myself for start-up costs when I began my low-wage life: $1,000 for the first month's rent and deposit, $100 for initial groceries and cash in my pocket, $200 stuffed away for emergencies. In poverty, as in certain propositions in physics, starting conditions are everything.

There are no secret economies that nourish the poor; on the contrary, there are a host of special costs. If you can't put up the two months' rent you need to secure an apartment, you end up paying through the nose for a room by the week. If you have only a room, with a hot plate at best, you can't save by cooking up huge lentil stews that can be frozen for the week ahead. You eat fast food, or the hot dogs and styrofoam cups of soup that can be microwaved in a convenience store. If you have no money for health insurance—and the Hearthside's niggardly plan kicks in only after three months—you go without routine care or prescription drugs and end up paying the price. Gail, for example, was fine until she ran out of money for estrogen pills. She is supposed to be on the company plan by now, but they claim to have lost her application form and need to begin the paperwork all over again. So she spends $9 per migraine pill to control the headaches she wouldn't have, she insists, if her estrogen supplements were covered. Similarly, Marianne's boyfriend lost his job as a roofer because he missed so much time after getting a cut on his foot for which he couldn't afford the prescribed antibiotic.

My own situation, when I sit down to assess it after two weeks of work, would not be much better if this were my actual life. The seductive thing about waitressing is that you don't have to wait for payday to feel a few bills in your pocket, and my tips usually cover meals and gas, plus something left over to stuff into the kitchen drawer I use as a bank. But as the tourist business slows in the summer heat, I sometimes leave work with only $20 in tips (the gross is higher, but servers share about 15 percent of their tips with the busboys and bartenders). With wages included, this amounts to about the minimum wage of $5.15 an hour. Although the sum in the drawer is piling up, at the present rate of accumulation it will be more than a hundred dollars short of my rent when the end of the month comes around. Nor can I see any expenses to cut. True, I haven't gone the lentil-stew route yet, but that's because I don't have a large cooking pot, pot holders, or a ladle to stir with (which cost about $30 at Kmart, less at thrift stores), not to mention onions, carrots, and the indispensable bay leaf. I do make my lunch almost every day—usually some slow-burning, high-protein combo like frozen chicken patties with melted cheese on top and canned pinto beans on the side. Dinner is at the Hearthside, which offers its employees a choice of BLT, fish sandwich, or hamburger for only $2. The burger lasts longest, especially if it's heaped with gut-puckering jalapenos, but by midnight my stomach is growling again.

So unless I want to start using my car as a residence, I have to find a second, or alternative, job. I call all the hotels where I filled out housekeeping applications weeks ago—the Hyatt, Holiday Inn, Econo Lodge, Hojo's, Best Western, plus a half dozen or so locally run guesthouses. Nothing. Then I start making the rounds again, wasting whole mornings waiting for some assistant manager to show up, even dipping into places so creepy that the front-desk clerk greets you from behind bulletproof glass and sells pints of liquor over the counter. But either someone has exposed my real-life housekeeping habits—which are, shall we say, mellow—or I am at the wrong end of some infallible ethnic equation: most, but by no means all, of the working housekeepers I see on my job searches are African Americans, Spanish-speaking, or immigrants from the Central European post-Communist world, whereas servers are almost invariably white and monolingually English-speaking. When I finally get a positive response, I have been identified once again as server material. Jerry's, which is part of a well-known national family restaurant chain and physically attached here to another budget hotel chain, is ready to use me at once. The prospect is both exciting and terrifying, because, with about the same number of tables and counter seats, Jerry's attracts three or four times the volume of customers as the gloomy old Hearthside. . . .

I start out with the beautiful, heroic idea of handling the two jobs at once, and for two days I almost do it: the breakfast/lunch shift at Jerry's, which goes till 2:00, arriving at the Hearthside at 2:10, and attempting to hold out until 10:00. In the ten minutes between jobs, I pick up a spicy chicken sandwich at the Wendy's drive-through window, gobble it down in the car, and change from khaki slacks to black, from Hawaiian to rust polo. There is a problem, though. When during the 3:00 to 4:00 P.M. dead time I finally sit down to wrap silver, my flesh seems to bond to the seat. I try to refuel with a purloined cup of soup, as I've seen Gail and Joan do dozens of times, but a manager catches me and hisses "No eating!" though there's not a customer around to be offended by the sight of food making contact with a server's lips. So I tell Gail I'm going to quit, and she hugs me and says she might just follow me to Jerry's herself.

But the chances of this are minuscule. She has left the flophouse and her annoying roommate and is back to living in her beat-up old truck. But guess what? she reports to me excitedly later that evening: Phillip has given her permission to park overnight in the hotel parking lot, as long as she keeps out of sight, and the parking lot should be totally safe, since it's patrolled by a hotel security guard! With the Hearthside offering benefits like that, how could anyone think of leaving?

Gail would have triumphed at Jerry's, I'm sure, but for me it's a crash course in exhaustion management. Years ago, the kindly fry cook who trained me to waitress at a Los Angeles truck stop used to say: Never make an unnecessary trip; if you don't have to walk fast, walk slow; if you don't have to walk, stand. But at Jerry's the effort of distinguishing necessary from unnecessary and urgent from whenever would itself be too much of an energy drain. The only thing to do is to treat each shift as a one-time-only emergency: you've got fifty starving people out there, lying scattered on the battlefield, so get out there and feed them! Forget that you will have to do this again tomorrow, forget that you will have to be alert enough to dodge the drunks on the drive home tonight—just burn, burn, burn! Ideally, at some point you enter what servers call "a rhythm" and psychologists term a "flow state," in which signals pass from the sense organs directly to the muscles, bypassing the cerebral cortex, and a Zen-like emptiness sets in. A male server from the Hearthside's morning shift tells me about the time he "pulled a triple"—three shifts in a row, all the way around the clock—and then got off and had a drink and met this girl, and maybe he shouldn't tell me this, but they had sex right then and there, and it was like, beautiful.

But there's another capacity of the neuromuscular system, which is pain. I start tossing back drugstore-brand ibuprofen pills as if they were vitamin C, four before each shift, because an old mouse-related repetitive-stress injury in my upper back has come back to full-spasm strength, thanks to the tray carrying. In my ordinary life, this level of disability might justify a day of ice packs and stretching. Here I comfort myself with the Aleve commercial in which the cute blue-collar guy asks: If you quit after working four hours, what would your boss say? And the not-so-cute blue-collar guy, who's lugging a metal beam on his back, answers: He'd fire me, that's what. But fortunately, the commercial tells us, we workers can exert the same kind of authority over our painkillers that our bosses exert over us. If Tylenol doesn't want to work for more than four hours, you just fire its ass and switch to Aleve.

True, I take occasional breaks from this life, going home now and then to catch up on e-mail and for conjugal visits (though I am careful to "pay" for anything I eat there), seeing The Truman Show with friends and letting them buy my ticket. And I still have those what-am-I-doing-here moments at work, when I get so homesick for the printed word that I obsessively reread the six-page menu. But as the days go by, my old life is beginning to look exceedingly strange. The e-mails and phone messages addressed to my former self come from a distant race of people with exotic concerns and far too much time on their hands. The neighborly market I used to cruise for produce now looks forbiddingly like a Manhattan yuppie emporium. And when I sit down one morning in my real home to pay bills from my past life, I am dazzled at the two- and three-figure sums owed to outfits like Club BodyTech and Amazon.com. . . .

I make the decision to move closer to Key West. First, because of the drive. Second and third, also because of the drive: gas is eating up $4 to $5 a day, and although Jerry's is as high-volume as you can get, the tips average only 10 percent, and not just for a newbie like me. Between the base pay of $2.15 an hour and the obligation to share tips with the busboys and dishwashers, we're averaging only about $7.50 an hour. Then there is the $30 I had to spend on the regulation tan slacks worn by Jerry's servers—a setback it could take weeks to absorb. (I had combed the town's two downscale department stores hoping for something cheaper but decided in the end that these marked-down Dockers, originally $49, were more likely to survive a daily washing.) Of my fellow servers, everyone who lacks a working husband or boyfriend seems to have a second job: Nita does something at a computer eight hours a day; another welds. Without the forty-five-minute commute, I can picture myself working two jobs and having the time to shower between them. . . .

I can do this two-job thing, is my theory, if I can drink enough caffeine and avoid getting distracted by George's ever more obvious suffering. . . .

Then it comes, the perfect storm. Four of my tables fill up at once. Four tables is nothing for me now, but only so long as they are obligingly staggered. As I bev table 27, tables 25, 28, and 24 are watching enviously. As I bev 25, 24 glowers because their bevs haven't even been ordered. Twenty-eight is four yuppyish types, meaning everything on the side and agonizing instructions as to the chicken Caesars. Twenty-five is a middle-aged black couple, who complain, with some justice, that the iced tea isn't fresh and the tabletop is sticky. But table 24 is the meteorological event of the century: ten British tourists who seem to have made the decision to absorb the American experience entirely by mouth. Here everyone has at least two drinks—iced tea and milk shake, Michelob and water (with lemon slice, please)—and a huge promiscuous orgy of breakfast specials, mozz sticks, chicken strips, quesadillas, burgers with cheese and without, sides of hash browns with cheddar, with onions, with gravy, seasoned fries, plain fries, banana splits. Poor Jesus! Poor me! Because when I arrive with their first tray of food—after three prior trips just to refill bevs—Princess Di refuses to eat her chicken strips with her pancake-and-sausage special, since, as she now reveals, the strips were meant to be an appetizer. Maybe the others would have accepted their meals, but Di, who is deep into her third Michelob, insists that everything else go back while they work on their "starters." Meanwhile, the yuppies are waving me down for more decaf and the black couple looks ready to summon the NAACP.

Much of what happened next is lost in the fog of war. . . .

I leave. I don't walk out, I just leave. I don't finish my side work or pick up my credit-card tips, if any, at the cash register or, of course, ask Joy's permission to go. And the surprising thing is that you can walk out without permission, that the door opens, that the thick tropical night air parts to let me pass, that my car is still parked where I left it. There is no vindication in this exit, no fuck-you surge of relief, just an overwhelming, dank sense of failure pressing down on me and the entire parking lot. I had gone into this venture in the spirit of science, to test a mathematical proposition, but somewhere along the line, in the tunnel vision imposed by long shifts and relentless concentration, it became a test of myself, and clearly I have failed. Not only had I flamed out as a housekeeper/server, I had even forgotten to give George my tips, and, for reasons perhaps best known to hardworking, generous people like Gail and Ellen, this hurts. I don't cry, but I am in a position to realize, for the first time in many years, that the tear ducts are still there, and still capable of doing their job. . . .

In one month, I had earned approximately $1,040 and spent $517 on food, gas, toiletries, laundry, phone, and utilities. If I had remained in my $500 efficiency, I would have been able to pay the rent and have $22 left over (which is $78 less than the cash I had in my pocket at the start of the month). During this time I bought no clothing except for the required slacks and no prescription drugs or medical care (I did finally buy some vitamin B to compensate for the lack of vegetables in my diet). Perhaps I could have saved a little on food if I had gotten to a supermarket more often, instead of convenience stores, but it should be noted that I lost almost four pounds in four weeks, on a diet weighted heavily toward burgers and fries.

How former welfare recipients and single mothers will (and do) survive in the low-wage workforce, I cannot imagine. Maybe they will figure out how to condense their lives—including child-raising, laundry, romance, and meals—into the couple of hours between full-time jobs. Maybe they will take up residence in their vehicles, if they have one. All I know is that I couldn't hold two jobs and I couldn't make enough money to live on with one. And I had advantages unthinkable to many of the long-term poor—health, stamina, a working car, and no children to care for and support. Certainly nothing in my experience contradicts the conclusion of Kathryn Edin and Laura Lein, in their recent book *Making Ends Meet: How Single Mothers Survive Welfare and Low-Wage Work*, that low-wage work actually involves more hardship and deprivation than life at the mercy of the welfare state. In the coming months and years, economic conditions for the working poor are bound to worsen, even without the almost inevitable recession. As mentioned earlier, the influx of former welfare recipients into the low-skilled workforce will have a depressing effect on both wages and the number of jobs available. A general economic downturn will only enhance these effects, and the working poor will of course be facing it without the slight, but nonetheless often saving, protection of welfare as a backup.

The thinking behind welfare reform was that even the humblest jobs are morally uplifting and psychologically buoying. In reality they are likely to be fraught with insult and stress. But I did discover one redeeming feature of the most abject low-wage work—the camaraderie of people who are, in almost all cases, far too smart and funny and caring for the work they do and the wages they're paid. The hope, of course, is that someday these people will come to know what they're worth, and take appropriate action.

KEY CONCEPTS

gender segregation   occupational segregation   working class

DISCUSSION QUESTIONS

1. What were the major lessons that Ehrenreich learned during her "experiment" as a low-wage worker and how do they inform your understanding of social class?

2. Having read Ehrenreich's article, how do you understand the relationship between work, stress, and health? What workplace policies could be implemented to improve worker health (mental and physical)?

57

The Power Elite

C. WRIGHT MILLS

C. Wright Mills's classic book, The Power Elite, *first published in 1956, remains an important analysis of the system of power in the United States. He argues that national power is located in three particular institutions: the economy, politics, and the military. An important point in his article is that the power of elites is derived from their institutional location, not their individual attributes.*

The powers of ordinary men are circumscribed by the everyday worlds in which they live, yet even in these rounds of job, family, and neighborhood they often seem driven by forces they can neither understand nor govern. "Great changes" are beyond their control, but affect their conduct and outlook none the less. The very framework of modern society confines them to projects not their own, but from every side, such changes now press upon the men and women of the mass society, who accordingly feel that they are without purpose in an epoch in which they are without power.

But not all men are in this sense ordinary. As the means of information and of power are centralized, some men come to occupy positions in American society from which they can look down upon, so to speak, and by their decisions mightily affect, the everyday worlds of ordinary men and women. They are not made by their jobs; they set up and break down jobs for thousands of others; they are not confined by simple family responsibilities; they can escape. They may live in many hotels and houses, but they are bound by no one community. They need not merely "meet the demands of the day and hour"; in some part, they create these demands, and cause others to meet them. Whether or not they profess their power, their technical and political experience of it far transcends that of the underlying population. What Jacob Burckhardt said of "great men," most Americans might well say of their elite: "They are all that we are not."

The power elite is composed of men whose positions enable them to transcend the ordinary environments of ordinary men and women; they are in positions to make decisions having major consequences. Whether they do or do not make such decisions is less important than the fact that they do occupy such pivotal positions: their failure to act, their failure to make decisions, is itself an act that is often of greater consequence than the decisions they do make. For they are in command of the major hierarchies and organizations of modern society. They rule the big corporations. They run the machinery of the state and claim its prerogatives. They direct the military establishment. They occupy the strategic command posts of the social structure, in which are now centered the effective means of the power and the wealth and the celebrity which they enjoy.

The power elite are not solitary rulers. Advisers and consultants, spokesmen and opinion-makers are often the captains of their higher thought and decision. Immediately below the elite are the professional politicians of the middle levels of power, in the Congress and in the pressure groups, as well as among the new and old upper classes of town and city and region. Mingling with them, in curious ways which we shall explore, are those professional celebrities who live by being continually displayed but are never, so long as they remain celebrities, displayed enough. If such celebrities are not at the head of any dominating hierarchy, they do often have the power to distract the attention of the public or afford sensations to the masses, or, more directly, to gain the ear of those who do occupy positions of direct power. More or less unattached, as critics of morality and technicians of power, as spokesmen of God and creators of mass sensibility, such celebrities and consultants are part of the immediate scene in which the drama of the elite is enacted. But that drama itself is centered in the command posts of the major institutional hierarchies.

The truth about the nature and the power of the elite is not some secret which men of affairs know but will not tell. Such men hold quite various theories about their own roles in the sequence of event and decision. Often they are uncertain about their roles, and even more often they allow their fears and their hopes to affect their assessment of their own power. No matter how great their actual power, they tend to be less acutely aware of it than of the resistances of others to its use. Moreover, most American men of affairs have learned well the rhetoric of public relations, in some cases even to the point of using it when they are alone, and thus coming to believe it. The personal awareness of the actors is only one of the several sources one must examine in order to understand the higher circles. Yet many who believe that there is no elite, or at any rate none of any consequence, rest their argument upon what men of affairs believe about themselves, or at least assert in public.

There is, however, another view: those who feel, even if vaguely, that a compact and powerful elite of great importance does now prevail in America often base that feeling upon the historical trend of our time. They have felt, for example, the domination of the military event, and from this they infer that generals and admirals, as well as other men of decision influenced by them, must be enormously powerful. They hear that the Congress has again abdicated to a handful of men decisions clearly related to the issue of war or peace. They know that the bomb was dropped over Japan in the name of the United States of America, although they were at no time consulted about the matter. They feel that they live in a time of big decisions; they know that they are not making any.

Accordingly, as they consider the present as history, they infer that at its center, making decisions or failing to make them, there must be an elite of power.

On the one hand, those who share this feeling about big historical events assume that there is an elite and that its power is great. On the other hand, those who listen carefully to the reports of men apparently involved in the great decisions often do not believe that there is an elite whose powers are of decisive consequence.

Both views must be taken into account, but neither is adequate. The way to understand the power of the American elite lies neither solely in recognizing the historic scale of events nor in accepting the personal awareness reported by men of apparent decision. Behind such men and behind the events of history, linking the two, are the major institutions of modern society. These hierarchies of state and corporation and army constitute the means of power; as such they are now of a consequence not before equaled in human history—and at their summits, there are now those command posts of modern society which offer us the sociological key to an understanding of the role of the higher circles in America.

Within American society, major national power now resides in the economic, the political, and the military domains. Other institutions seem off to the side of modern history, and, on occasion, duly subordinated to these. No family is as directly powerful in national affairs as any major corporation; no church is as directly powerful in the external biographies of young men in America today as the military establishment; no college is as powerful in the shaping of momentous events as the National Security Council. Religious, educational, and family institutions are not autonomous centers of national power; on the contrary, these decentralized areas are increasingly shaped by the big three, in which developments of decisive and immediate consequence now occur.

Families and churches and schools adapt to modern life; governments and armies and corporations shape it; and, as they do so, they turn these lesser institutions into means for their ends. Religious institutions provide chaplains to the armed forces where they are used as a means of increasing the effectiveness of its morale to kill. Schools select and train men for their jobs in corporations and their specialized tasks in the armed forces. The extended family has, of course, long been broken up by the industrial revolution, and now the son and the father are removed from the family, by compulsion if need be, whenever the army of the state sends out the call. And the symbols of all these lesser institutions are used to legitimate the power and the decisions of the big three.

The life-fate of the modern individual depends not only upon the family into which he was born or which he enters by marriage, but increasingly upon the corporation in which he spends the most alert hours of his best years; not only upon the school where he is educated as a child and adolescent, but also upon the state which touches him throughout his life; not only upon the church in which on occasion he hears the word of God, but also upon the army in which he is disciplined.

If the centralized state could not rely upon the inculcation of nationalist loyalties in public and private schools, its leaders would promptly seek to modify the decentralized educational system. If the bankruptcy rate among the top five hundred corporations were as high as the general divorce rate among the thirty-seven million married couples, there would be economic catastrophe on an international scale. If members of armies gave to them no more of their lives than do believers to the churches to which they belong, there would be a military crisis.

Within each of the big three, the typical institutional unit has become enlarged, has become administrative, and, in the power of its decisions, has become centralized. Behind these developments there is a fabulous technology, for as institutions, they have incorporated this technology and guide it, even as it shapes and paces their developments.

The economy—once a great scatter of small productive units in autonomous balance—has become dominated by two or three hundred giant corporations, administratively and politically interrelated, which together hold the keys to economic decisions.

The political order, once a decentralized set of several dozen states with a weak spinal cord, has become a centralized, executive establishment which has taken up into itself many powers previously scattered, and now enters into each and every cranny of the social structure.

The military order, once a slim establishment in a context of distrust fed by state militia, has become the largest and most expensive feature of government, and, although well versed in smiling public relations, now has all the grim and clumsy efficiency of a sprawling bureaucratic domain.

In each of these institutional areas, the means of power at the disposal of decision makers have increased enormously; their central executive powers have been enhanced; within each of them modern administrative routines have been elaborated and tightened up.

As each of these domains becomes enlarged and centralized, the consequences of its activities become greater, and its traffic with the others increases. The decisions of a handful of corporations bear upon military and political as well as upon economic developments around the world. The decisions of the military establishment rest upon and grievously affect political life as well as the very level of economic activity. The decisions made within the political domain determine economic activities and military programs. There is no longer, on the one hand, an economy, and, on the other hand, a political order containing a military establishment unimportant to politics and to moneymaking. There is a political economy linked, in a thousand ways, with military institutions and decisions. On each side of the world-split running through central Europe and around the Asiatic rimlands, there is an ever-increasing interlocking of economic, military, and political structures. If there is government intervention in the corporate economy, so is there corporate intervention in the governmental process. In the structural sense, this triangle of power is the source of the *interlocking directorate* that is most important for the historical structure of the present. . . .

At the pinnacle of each of the three enlarged and centralized domains, there have arisen those higher circles which make up the economic, the political, and the military elites. At the top of the economy, among the corporate rich, there are the chief executives; at the top of the political order, the members of the political directorate; at the top of the military establishment, the elite of soldier-statesmen clustered in and around the Joint Chiefs of Staff and the upper echelon. As each of these domains has coincided with the others, as decisions tend to become total in their consequence, the leading men in each of the three domains of power—the warlords, the corporation chieftains, the political directorate—tend to come together, to form the power elite of America. . . .

By the powerful we mean, of course, those who are able to realize their will, even if others resist it. No one, accordingly, can be truly powerful unless he has access to the command of major institutions, for it is over these institutional means of power that the truly powerful are, in the first instance, powerful. Higher politicians and key officials of government command such institutional power; so do admirals and generals, and so do the major owners and executives of the larger corporations. Not all power, it is true, is anchored in and exercised by means of such institutions, but only within and through them can power be more or less continuous and important.

Wealth also is acquired and held in and through institutions. The pyramid of wealth cannot be understood merely in terms of the very rich; for the great inheriting families, as we shall see, are now supplemented by the corporate institutions of modern society: every one of the very rich families has been and is closely connected—always legally and frequently managerially as well—with one of the multi-million dollar corporations.

The modern corporation is the prime source of wealth, but, in latter-day capitalism, the political apparatus also opens and closes many avenues to wealth. The amount as well as the source of income, the power over consumer's goods as well as over productive capital, are determined by position within the political economy. If our interest in the very rich goes beyond their lavish or their miserly consumption, we must examine their relations to modern forms of corporate property as well as to the state; for such relations now determine the chances of men to secure big property and to receive high income. . . .

If we took the one hundred most powerful men in America, the one hundred wealthiest, and the one hundred most celebrated away from the institutional positions they now occupy, away from their resources of men and women and money, away from the media of mass communication that are now focused upon them— then they would be powerless and poor and uncelebrated. For power is not of a man. Wealth does not center in the person of the wealthy. Celebrity is not inherent in any personality. To be celebrated, to be wealthy, to have power requires access to major institutions, for the institutional positions men occupy determine in large part their chances to have and to hold these valued experiences. . . .

KEY CONCEPTS

| interlocking directorate | power | power elite model |

DISCUSSION QUESTIONS

1. What evidence do you see of the presence of the power elite in today's economic, political, and military institutions? Suppose that Mills were writing his book today; what might he change about his essay?

2. Mills argues that the power elite use institutions such as religion, education, and the family as the means to their ends. Find an example of this from the daily news and explain how Mills would see this institution as being shaped by the power elite.

58

Has the Power Elite Become Diverse?

RICHARD L. ZWEIGENHAFT AND G. WILLIAM DOMHOFF

Richard Zweigenhaft and G. William Domhoff ask whether the power elite has changed by incorporating more diverse groups into the power structure of the United States. Although there is now more diversity in these circles of power, the "newcomers" tend to share the values of those already in power.

Injustices based on race, gender, ethnicity, and sexual orientation have been the most emotionally charged and contested issues in American society since the end of the 1960s, far exceeding concerns about social class and rivaled only by conflicts over abortion. These issues are now subsumed under the umbrella term *diversity*, which has been discussed extensively from the perspectives of both the aggrieved and those at the middle levels of the social ladder who resist any changes.

... [W]e look at diversity from a new angle: we examine its impact on the small group at the top of American society that we call the *power elite*, those who own and manage large banks and corporations, finance the political campaigns of conservative Democrats and virtually all Republicans at the state and national levels, and serve in government as appointed officials and military leaders. We ask whether the decades of civil disobedience, protest, and litigation by civil rights groups, feminists, and gay and lesbian rights activists have resulted in a more

SOURCE: Richard L. Zweigenhaft and G. William Domhoff. 2006. *Diversity in the Power Elite: How It Happened, Why It Matters.* Lanham, MD: Rowman & Littlefield.

diverse power elite. If they have, what effects has this new diversity had on the functioning of the power elite and on its relation to the rest of society?

We also compare our findings on the power elite with those from our parallel study of Congress to see whether there are differences in social background, education, and party affiliation for women and other underrepresented groups in these two realms of power. We explore the possibility that elected officials come from a wider range of occupational and income backgrounds than members of the power elite. In addition, we ask whether either of the major political parties has been more active than the other in advancing the careers of women and minorities.

According to many popular commentators, the composition of the higher circles in the United States had indeed changed by the late 1980s and early 1990s. . . .

Since the 1870s, the refrain about the new diversity of the governing circles has been closely intertwined with a staple of American culture created by Horatio Alger Jr., whose name has become synonymous with upward mobility in America. Far from being a Horatio Alger himself, the man who gave his name to upward mobility was born into a patrician family in 1832; his father was a Harvard graduate, a Unitarian minister, and a Massachusetts state senator. Horatio Jr. graduated from Harvard at the age of nineteen, after which he pursued a series of unsuccessful efforts to establish himself in various careers. Finally, in 1864, Alger was hired as a Unitarian minister in Brewster, Massachusetts. Fifteen months later, he was dismissed from this position for homosexual acts with boys in the congregation.

Alger returned to New York, where he soon began to spend a great deal of time at the Newsboys' Lodging House, founded in 1853 for footloose youngsters between the ages of twelve and sixteen and home to many youths who had been mustered out of the Union Army after serving as drummer boys. At the Newsboys' Lodging House, Alger found his literary niche and his subsequent claim to fame: writing books in which poor boys make good. His books sold by the hundreds of thousands in the last third of the nineteenth century, and by 1910, they were enjoying annual paperback sales of more than one million.[1]

The deck is not stacked against the poor, according to Alger. When they simply show a bit of gumption, work hard, and thereby catch a break or two, they can become part of the American elite. The persistence of this theme, reinforced by the annual Horatio Alger Award given to such well-known personalities as Ronald Reagan, Bob Hope, and Billy Graham (who might not have been so eager to accept the award had they known that Alger did not fit their fantasy of a straight, white, patriarchal, American male), suggests that we may be dealing once again with a cultural myth.

In its early versions, of course, the story concerned the great opportunities available for poor white boys willing to work their way to the top. More recently, the story has featured black Horatio Algers who started in the ghetto, Latino Horatio Algers who started in the barrio, Asian American Horatio Algers whose parents were immigrants, and female Horatio Algers who seem to have no class backgrounds, all of whom now sit on the boards of the country's largest corporations.[2]

Few people read Horatio Alger today, but they still believe in upward mobility in an era when real wages have been stagnant for the bottom 80 percent since the 1970s, the percentage of low-income people finishing college is decreasing, and the rate of upward mobility is declining.[3] . . .

But is any of the talk about Horatio Alger and upward mobility true? Can anecdotes, dime novels, and self-serving autobiographical accounts about diversity, meritocracy, and upward social mobility survive a more systematic analysis? Have very many women or members of other previously excluded groups made it to the top? Has class lost its importance in shaping life chances?

. . . [We] address these and related questions within the framework provided by the iconoclastic sociologist C. Wright Mills in his classic *The Power Elite*, published half a century ago in 1956 when the media were in the midst of what Mills called the "Great American Celebration," and still accurate today in terms of many of the issues he addressed. In spite of the Great Depression of the 1930s, Americans had pulled together to win World War II, and the country was both prosperous at home and influential abroad. Most of all, according to enthusiasts, the United States had become a relatively classless and pluralistic society, where power belonged to the people through their political parties and public opinion. Some groups certainly had more power than others, but no group or class had too much. The New Deal and World War II had forever transformed the corporate-based power structure of earlier decades.

Mills challenged this celebration of pluralism by studying the social backgrounds and career paths of the people who occupied the highest positions in what he saw as the three major institutional hierarchies in postwar America, the corporations, the executive branch of the federal government, and the military. He found that almost all members of this leadership group, which he called the power elite, were white, Christian males who came from "at most, the upper third of the income and occupational pyramids,"despite the many Horatio Algeresque claims to the contrary.[4] A majority came from an even narrower stratum, the 11 percent of U.S. families headed by businesspeople or highly educated professionals like physicians and lawyers. Mills concluded that power in the United States in the 1950s was just about as concentrated as it had been since the rise of the large corporations, although he stressed that the New Deal and World War II had given political appointees and military chieftains more authority than they had exercised previously.

It is our purpose, therefore, to take a detailed look at the social, educational, and occupational backgrounds of the leaders of these three institutional hierarchies to see whether they have become more diverse in terms of gender, race, ethnicity, and sexual orientation, and also in terms of socioeconomic origins. Unlike Mills, however, we think the power elite is more than a set of institutional leaders; it is also the leadership group for the small upper class of owners and managers of large, income-producing properties, the 1 percent of American households that owned 44.1 percent of all privately held stock, 58.0 percent of financial securities, and 57.3 percent of business equity in 2001, the last year for which systematic figures are available.(By way of comparison, the bottom 90 percent, those who work for hourly wages or monthly salaries, have a mere 15.5

percent of the stock, 11.3 percent of financial securities, and 10.4 percent of business equity.) Not surprisingly, we think the primary concern of the power elite is to support the kind of policies, regulations, and political leaders that maintain this structure of privilege for the very rich.[5]

We first study the directors and chief executive officers (CEOs) of the largest banks and corporations, as determined by annual rankings compiled by *Fortune* magazine. The use of *Fortune* rankings is now standard practice in studies of corporate size and power. Over the years, *Fortune* has changed the number of corporations on its annual list and the way it groups them. For example, the separate listings by business sector in the past, like "life insurance companies," "diversified financial companies," and "service companies," have been combined into one overall list, primarily because many large businesses are now involved in more than one of the traditional sectors. Generally speaking, we use the *Fortune* list or lists available for the time period under consideration.

Second, again following Mills, we focus on the appointees to the president's cabinet when we turn to the "political directorate," Mills's general term for top-level officials in the executive branch of the federal government. We also have included the director of the CIA...because of the increased importance of that agency since Mills wrote. Third, and rounding out our portrait of the power elite, we examine the same two top positions in the military, generals and admirals, that formed the basis for Mills's look at the military elite.

As we have noted, we also study Congress. In the case of senators, we do the same kind of background studies that we do for members of the power elite. For members of the House of Representatives, we concern ourselves only with party affiliation for most groups. We include findings on senators and representatives from underrepresented groups for two reasons. First, this allows us to see whether there is more diversity in the electoral system than in the power elite. Second, we do not think, as Mills did, that Congress should be relegated to the "middle level" of power. To the contrary, we believe that Congress is an integral part of the power structure in America. Similarly, unlike Mills, because we think that the Supreme Court is a key institution within the power elite, we have added information on Supreme Court appointments. . . .

In addition to studying the extent to which women and other previously excluded groups have risen in the system, we focus on whether they have followed different avenues to the top than their predecessors did, as well as on any special roles they may play. Are they in the innermost circles of the power elite, or are they more likely to serve as buffers and go-betweens? Do they go just so far and no farther? What obstacles does each group face?

We also examine whether or not their presence affects the power elite itself. Do they influence the power elite in a more liberal direction, or do they end up endorsing traditional conservative positions, such as opposition to trade unions, taxes, and government regulation of business?... We argue that the diversity forced on the power elite has had the ironic effect of strengthening it by providing it with a handful of people who can reach out to the previously excluded groups and by showing that the American system can deliver on its most important promise, an equal opportunity for every individual. It is as if the diversity efforts in the final thirty-five years of the twentieth century were scripted for the arrival of a George W. Bush, the most conservative and uncompromising president since Herbert Hoover, a president with the most diverse cabinet in the history of the country; in the first five years, its members have included six women, four African Americans, three Latinos, and two Asian Americans who also held a total of ten corporate directorships and two corporate law partnerships before joining the cabinet.

The issues we address are not simple. They involve both the nature of the American power structure and the way in which people's need for self-esteem and a coherent belief system mesh with the hierarchical social structure they face. Moreover, the answers to some of the questions we ask vary greatly depending on which previously disadvantaged group we are talking about. Nonetheless, in the course of our research, a few general patterns have emerged. . . .

1. The power elite now shows considerable diversity, at least as compared with its state in the 1950s, but its core group continues to consist of wealthy, white, Christian males, most of whom are still from the upper third of the social ladder. Like the white, Christian males, those who are newly arrived to the power elite have been filtered through the same small set of elite schools in law, business, public policy, and international relations.

2. In spite of the increased diversity of the power elite, high social origins continue to be the most important factor in making it to the top. There are relatively few rags-to-riches stories in the groups we studied, and those we did find tended to have received scholarships to elite schools or to have been elected to office, usually within the Democratic Party. In general, it still takes the few who make it at least three generations to rise from the bottom to the top in the United States.

3. The new diversity within the power elite is transcended by common values and a subjective sense of hard-earned and richly deserved class privilege. The newcomers to the power elite have found ways to signal that they are willing to join the game as it has always been played, assuring the old guard that they will call for no more than relatively minor adjustments, if that. There are few liberals and fewer crusaders in the power elite, despite its newly acquired diversity. Class backgrounds, current roles, and future aspirations are more powerful in shaping behavior in the power elite than gender, ethnicity, race, or sexual orientation.

4. Not all the groups we studied have been equally successful in contributing to the new diversity in the power elite. Women, African Americans, Latinos, Asian Americans, and openly homosexual men and women are all underrepresented, but to varying degrees and with different rates of increasing representation. There is a real possibility that Americans are replacing the old black-versus-white distinction with a black-versus-non-black dividing line, with "white" coming to be just another word for the in-group. . . .

5. There is greater diversity in Congress than in the power elite, especially in terms of class origins, and the majority of these more diverse elected officials

are Democrats, whose presence has forced the Republicans to play catch-up by including some candidates from the previously excluded groups.

6. Although the corporate, political, and military elites have accepted diversity only in response to pressure from social-movement activists and feminists, the power elite has benefited from the presence of new members. Some serve either a buffer or a liaison function with such groups and institutions as consumers, angry neighborhoods, government agencies, and wealthy foreign entrepreneurs. More generally, their simple presence at the top serves a legitimating function. Tokenism does work in terms of reassuring the general population that the system is fair. . . .

In what may be the greatest and most important irony of them all, the diversity forced upon the power elite may have helped to strengthen it. Diversity has given the power elite buffers, ambassadors, tokens, and legitimacy. This is an unintended consequence that few insurgents or social scientists foresaw. As recent social psychology experiments show and experience confirms, it often takes only a small number of upwardly mobile members of previously excluded groups, perhaps as few as 2 percent, to undermine an excluded group's definition of who is "us" and who is "them," which contributes to a decline in collective protest and disruption and increases striving for individual mobility. That is, those who make it are not only "role models" for individuals, but they are safety valves against collective action by aggrieved groups.

Tokens at the top create ambiguity and internal doubt for members of the subordinated group. Maybe "the system" is not as unfair to their group as they thought it was. Maybe there is something about them personally that keeps them from advancing. Once people begin to ponder such possibilities, the likelihood of any sustained group action declines greatly. Because a few people have made it, the general human tendency to think of the world as just and fair reasserts itself: since the world is fair, and some members of my group are advancing, then it may be my fault that I have been left behind. . . .

DO MEMBERS OF PREVIOUSLY EXCLUDED GROUPS ACT DIFFERENTLY?

Perhaps it is not surprising that when we look at the business practices of the members of previously excluded groups who have risen to the top of the corporate world, we find that their perspectives and values do not differ markedly from those of their white male counterparts. When Linda Wachner, one of the first women to become CEO of a *Fortune*-level company, the Warnaco Group, concluded that one of Warnaco's many holdings, the Hathaway Shirt Company, was unprofitable, she decided to stop making Hathaway shirts and to sell or close down the factory. It did not matter to Wachner that Hathaway, which started making shirts in 1837, was one of the oldest companies in Maine, that almost all of the five hundred employees at the factory were working-class women, or even that the workers had given up a pay raise to hire consultants to teach them to work more effectively and, as a result, had doubled their productivity. The bottom-line issue was that the company was considered unprofitable, and the average wage of the Hathaway workers, $7.50 an hour, was thought to be too high. (In 1995, Wachner was paid $10 million in salary and stock, and Warnaco had a net income of $46.5 million.) "We did need to do the right thing for the company and the stockholders," explained Wachner.[6]

Nor did ethnic background matter to Thomas Fuentes, a senior vice president at a consulting firm in Orange County, California, a director of Fleetwood Enterprises, and chairman of the Orange County Republican Party. Fuentes targeted fellow Latinos who happened to be Democrats when he sent uniformed security guards to twenty polling places in 1988 "carrying signs in Spanish and English warning people not to vote if they were not U.S. citizens." The security firm ended up paying $60,000 in damages when it lost a lawsuit stemming from this intimidation.[7] We also can recall that the Fanjuls, the Cuban American sugar barons, have no problem ignoring labor laws in dealing with their migrant labor force, and that Sue Ling Gin, one of the Asian Americans on our list of corporate directors, explained to an interviewer that, at one point in her career, she had hired an all-female staff, not out of feminist principles but "because women would work for lower wages." Linda Wachner, Thomas Fuentes, the Fanjuls, and Sue Ling Gin acted as employers, not as members of disadvantaged groups. That is, members of the power elite of both genders and of all ethnicities practice class politics.

CONCLUSION

The black and white liberals and progressives who challenged Christian, white, male homogeneity in the power structure starting in the 1950s and 1960s sought to do more than create civil rights and new job opportunities for men and women who had previously been mistreated and excluded, important though these goals were. They also hoped that new perspectives in the boardrooms and the halls of government would spread greater openness throughout the society. The idea was both to diversify the power elite and to shift some of its power to underrepresented groups and social classes. The social movements of the 1960s were strikingly successful in increasing the individual rights and freedoms available to all Americans, especially African Americans. As we have shown, they also created pressures that led to openings at the top for individuals from groups that had previously been ignored.

But as some individuals made it, and as the concerns of social movements, political leaders, and the courts gradually came to focus more and more on individual rights and individuals advancement, the focus on "distributive justice," general racial exclusion, and social class was lost. The age-old American commitment to individualism, reinforced by tokenism and reassurances from members of the power elite, won out over the commitment to greater equality of income and wealth that had been one strand of New Deal liberalism and a major emphasis of left-wing activism in the 1960s.

We therefore conclude that the increased diversity in the power elite has not generated any changes in an underlying class system in which the top 1 percent of households (the upper class) own 33.4 percent of all marketable wealth, and the next 19 percent (the managerial, professional, and small business stratum) have 51 percent, which means that just 20 percent of the people own a remarkable 84 percent of the privately owned wealth in the United States, leaving a mere 16 percent of the wealth for the bottom 80 percent (wage and salary workers).[8] In fact, the wealth and income distributions became even more skewed starting in the 1970s as the majority of whites, especially in the South and Great Plains, switched their allegiance to the Republican Party and thereby paved the way for a conservative resurgence that is as antiunion, antitax, and antigovernment as it is determined to impose ultraconservative social values on all Americans.

The values of liberal individualism embedded in the Declaration of Independence, the Bill of Rights, and American civic culture were renewed by vigorous and courageous activists in the years between 1955 and 1975, but the class structure remains a major obstacle to individual fulfillment for the overwhelming majority of Americans. The conservative backlash that claims to speak for individual rights has strengthened this class structure, one that thwarts advancement for most individuals from families in the bottom 80 percent of the wealth distribution. This solidification of class divisions in the name of individualism is more than an irony. It is a dilemma.

Furthermore, this dilemma combines with the dilemma of race to obscure further the impact of class and to limit individual mobility, simply because the majority of middle-American whites cannot bring themselves to make common cause with African Americans in the name of greater individual opportunity and economic equality through a progressive income tax and the kind of government programs that lifted past generations out of poverty. These intertwined dilemmas of class and race lead to a nation that celebrates individualism, equal opportunity, and diversity but is, in reality, a bastion of class privilege, African American exclusion, and conservatism.

NOTES

1. See Richard M. Huber, *The American Idea of Success* (New York: McGraw-Hill, 1971), 44–46; Gary Scharnhorst, *Horatio Alger, Jr.* (Boston: Twayne, 1980), 24, 29, 141. For a discussion of the general pattern by which the media eulogize tycoons as "self-made," see Richard L. Zweigenhaft, "Making Rags out of Riches: Horatio Alger and the Tycoon's Obituary," *Extra!*, January/February 2004, 27–28.

2. As C. Wright Mills wrote half a century ago, "Horatio Alger dies hard." C. Wright Mills, *The Power Elite* (Oxford University New York Press, 1956), 91.

3. See Harold R. Kerbo, *Social Stratification and Inequality* (New York: McGraw-Hill,

2006), ch. 12. See, also, the series on class in the *New York Times*, "A Portrait of Class in America," spring 2005.

4. Mills, *The Power Elite*, 279. For Mills's specific findings, see 104–105, 128–29, 180–81, 393–94, and 400–401.

5. Edward N. Wolff, "Changes in Household Wealth in the 1980s and 1990s in the U.S." (working paper 407, Levy Economics Institute, Bard College, 2004), at www.levy.org. See table 2, p. 30, and table 6, p. 34.

6. Sara Rimer, "Fall of a Shirtmaking Legend Shakes Its Maine Hometown," *New York Times*, May 15, 1996. See, also, Floyd Norris, "Market Place," *New York Times*, June 7, 1996; Stephanie Strom, "Double Trouble at Linda Wachner's Twin Companies," *New York Times*, August 4, 1996. Strom's article reveals that Hathaway Shirts "got a reprieve" when an investor group stepped in to save it.

7. Claudia Luther and Steven Churm, "GOP Official Says He OK'd Observers at Polls," *Los Angeles Times*, November 12, 1988; Jeffrey Perlman, "Firm Will Pay $60,000 in Suit over Guards at Polls," *Los Angeles Times*, May 31, 1989.

8. Edward N. Wolff, "Changes in Household Wealth in the 1980s and 1990s in the U.S." (working paper 407, Levy Economics Institute, Bard College, 2004), at www.levy.org.

KEY CONCEPTS

interlocking directorate   power

DISCUSSION QUESTIONS

1. To what extent do Zweigenhaft and Domhoff see diverse groups as becoming a part of the power elite? What factors do they identify as important in gaining entrance to the power elite and how does this challenge the myth of upward mobility typically symbolized by the Horatio Alger story?

2. At the heart of Zweigenhaft and Domhoff's analysis is the question, "Does having more women and people of color in positions of power change institutions?" Using empirical evidence, how would you answer this question?

59

Forever Seen as New

Latino Participation in American Elections

LOUIS DESIPIO AND RODOLFO O. DE LA GARZA

The authors here identify a number of changes that are likely to result in an increase in the political influence of Latinos in years ahead. Like other groups, however, the extent of voting participation among Latinos is also shaped by factors such as age, social class, and region of residence. The authors show how changes in the Latino population are linked to patterns of political participation and influence.

. . . The rapid growth of the Latino population in the past thirty years has raised popular expectations for its political impact. Now numbering more than 35 million and soon to overtake African Americans as the nation's largest minority, Latinos are often the subject of naive predictions for their imminent domination of politics and society (de la Garza 1996a). The electorate, of course, is much smaller than the total population: just one in six Latinos votes.

Much of the gap between rates of electoral participation for Latino and Anglo citizens can be explained by simple demographics. Among all populations, the young, the less well educated, and the low-income are less likely to vote. All of these groups are disproportionately represented among Latinos. Remedies for the impact on turnout of youth, limited education, and low-income are less clear. Community-wide mobilization, such as that which the African American community experienced as a result of the civil rights movement, can overcome these impediments to participation. Without such mobilization, however, Latinos are likely to continue to experience lower-than-average rates of participation, despite the steady growth in the number of Latino voters from election to election.

One of these traits—youth—has a natural remedy, aging. At present, Latinos are approximately nine years younger than the average non-Hispanic white.

SOURCE: DeSipio, Louis, and Rodolfo O. de la Garza. "Forever Seen as New: Latino Participation in American Elections." In *Latinos: Remaking America,* edited by Marcelo M. Suarez-Drozco and Mariela Paez, 398–409. Berkeley: University of California Press, 2002. Reprinted with permission.

Immigration and high birth rates will keep Latino voting rates low, but increasing numbers of Latinos will enter their forties and fifties, the ages that see peak voting in all populations. Today, approximately 6.3 million Latinos are forty-five or older. This number will increase to 11.0 million by 2010 and to 20.7 million by 2030 (U.S. Bureau of the Census 1996). Income and education levels are also rising, particularly among the U.S.-born (de la Garza 1996b). However, these slow gains could disappear if the U.S. economy were to deteriorate. For the time being, the steady aging of the Latino population and the growth of an educated middle class spur an increase of 10 to 15 percent every four years in the number of Latinos who turn out to vote.

Electoral growth, however, does not guarantee increased Latino influence. In 1996, for example, the Latino electorate had grown by 16 percent over 1992, and the impact of this growth in the Latino share of the electorate was magnified by a decline in the Anglo electorate. Yet Latino voters were influential in more states in 1992 when the election was closer and more states where many Latinos lived were crucial to the outcome. Thus, increasing the size of the electorate is an important goal, but it is not directly related to influence.

Over time, an increase in the size of the Latino electorate could bring people with different positions or interests into the electorate. This happened, for instance, as Cuban Americans began to vote, reducing the Democratic share of the Latino electorate. Looking to the future, electoral growth spurred by increasing incomes among Latinos would probably increase the Republican share of the electorate. On the other hand, as Latinos age, health care and Social Security could assume a central position in the issue agenda of Latino voters in a way that they do not today. These have long been Democratic issues, and their increased salience could strengthen Latino Democratic ties....

NATURALIZED CITIZENS

Whereas demographic limits on electoral participation affect all groups, a second characteristic disproportionately affects Latinos. Noncitizens total 39 percent of Latino adults and are the largest potential new electorate among Latinos (DeSipio 1996). Any effort to mobilize them, however, must begin by encouraging naturalization, because noncitizens are barred from voting in virtually all elections.

. . . Approximately 2.4 million Latinos became naturalized citizens between 1995 and 2001. (They were joined by approximately 2.6 million non–Latino new citizens.) These 2.4 million new Latino citizens are potentially influential for several reasons. First, if they were all to join the electorate (an unlikely scenario), they would add almost 50 percent to the existing Latino vote overnight. Second, the newly naturalized are concentrated in a few states, so their impact would be magnified. Finally, many became naturalized at least in part in response to government efforts to limit the rights of immigrants, so there was a political dimension to the decisions made by many such citizens....

NON-NATURALIZED IMMIGRANTS

What of the remaining (non-naturalized) Latino immigrants? Do they offer the foundation for an expanding new electorate? The answer is yes, but their impact will be felt slowly. Approximately 1.1 million Latinos (including 7.7 million adults) were not U.S. citizens in 1999. Of these, more than 4 million were undocumented, so there were approximately four million permanent residents. Of these 1.6 million immigrated in the past five years and thus are presently ineligible for naturalization. The remaining 2.4 million naturalization-eligible Latino immigrants are a pool for further growth in the Latino electorate.

In all likelihood, however, this remaining pool of Latino immigrants will be slow to become naturalized. A unique set of pressures and incentives encouraged most of the eligible who were immigrants interested in citizenship to seek naturalization in the late 1990s. Those who did not do so will require added encouragement to pursue citizenship now. . . .

CONCENTRATION

Latinos are more geographically concentrated than Anglos and are becoming even more concentrated. Some argue that this concentration boosts Latino empowerment. We find that its minuses may well outweigh its plusses. Let's consider both.

Instead of having a national election for the presidency, the electoral college reflects fifty state elections, in which most states award all of their delegates to the candidate who receives the most votes. Any concentrated electorate that can secure a state for a candidate exerts a form of influence that would be lost if each vote were tallied nationally. Latinos benefit from concentrating most of their numbers in just nine states and from having cohesive voting patterns in each. Latino advocates are quick to observe that Latinos are concentrated in states that elect three-quarters of the electors needed to win the presidency.

Concentration also entails a cost in national elections. No campaign runs equal efforts in all fifty states. Rather, campaigns calculate how they can best spend their money. Little is spent in states that are probable losses *or* in states where victory is very likely. Money and time are focused on the competitive states so that a winning margin of 270 electoral votes can be earned. . . .

In the short run, then, concentration is an advantage only in states or other electoral districts that are competitive. In the longer run, areas of concentration can become the sites of sustained multiyear mobilization efforts targeted not just at a specific campaign or election cycle. Efforts such as these, which make concentration an advantage, require leadership and resources that have been absent from most Latino outreach in recent years.

ELITE RESOURCES AND INSTITUTIONS

Although it has been little studied, one genuinely new phenomenon in Latino electoral participation is the rise of a new cadre of Latino elites and new institutions to shape candidate outreach to Latinos. The new elites are made up primarily of young, highly educated Latinos who have begun to populate campaigns. In the process, they have drawn attention to Latino issues and have educated non-Latino candidates about Latino communities. Institutional development has been slower but will probably expand in the next decade. These institutions include Latino political action committees (PACs) and Latino organizations within the national and state political parties. . . .

The increase in the number of Latino votes and the opportunities for these votes to prove influential in electoral outcomes pave the way for expansion in the role of elite resources and institutions. This growth will be spurred by the talent pool of skilled Latino politicos who seek both influence in campaigns and a voice on Latino issues. Thus, although the phenomenon of elite institutions and resources is not new, its potential for influence is, and it merits continuing scholarly appraisal. . . .

CONCLUSION

In the continual search for what is new in the Latino electorate, there is a tendency to neglect the incremental but steady changes that shape Latino politics. Latino votes have become increasingly sought by candidates and parties. These votes do occasionally determine outcomes, although when they do, they are usually determined by factors exogenous to the Latino community. Both parties have designed Latino-specific outreach strategies and dedicated resources to winning Latino votes, and Latinos have become more centrally positioned in campaigns and elite networks. As a result, candidates are somewhat less likely to speak from ignorance when they address Latino issues. Although we have not discussed it here, the number of Latino elected and appointed officials has grown. When the Voting Rights Act was extended to Latinos in 1975, they were not a nationally influential electorate; today they are. What we find, however, is not a single, dramatic change but incremental change and growth in the electorate. We expect this pattern to continue, but we do not anticipate a sudden burst of new Latino electoral empowerment.

This increase in the importance of the Latino electorate is not the result of mobilization among all Latinos or all Latino citizens. Instead, Latinos have followed the pattern of Anglos: voting is more common among the educationally and economically advantaged. For a truly new Latino electorate to emerge, this pattern would have to change, and mobilization would have to extend to all segments of the Latino electorate. Survey data indicate that even if this were to occur, such a broad-based mobilization would not appreciably change the issue focus of Latino electorates, just the likelihood of their influencing electoral outcomes.

Will Latinos play a central role in the outcomes of upcoming national elections? The answer to this question is found primarily outside of the Latino community. The incremental growth in Latino electorates makes it, all other factors remaining constant, slightly more likely each presidential election cycle. But the final answer will be determined each election year. Latinos will be important if the race comes down to the states where their numbers are concentrated, if one or both of the candidates see Latino votes as central to their ability to win those states, and if one or both of the candidates dedicate resources to winning Latinos' votes (or to preventing the other candidate from winning their votes). In this scenario, their votes could make the difference. The candidates and parties now possess the expertise to structure a campaign to win their votes. In this scenario, their strong partisanship and state-level cohesion make them an inviting target relative to other electorates.

REFERENCES

de la Garza, Rodolfo O. 1996a. "El Cuento de los Números and Other Latino Political Myths." In *Su Voto es Su Voz: Latino Politics in California*, ed. Aníbal Yáñez-Chávez, 11–32. San Diego: Center for U.S. Mexican Studies, University of California, San Diego.

de la Garza, Rodolfo O. 1996b. "The Effects of Primordial Claims, Immigration, and the Voting Rights Act on Mexican American Sociopolitical Incorporation." In *The Politics of Minority Coalitions: Race, Ethnicity, and Shared Uncertainties*, ed. Wilbur C. Rich, 163–76. Westport, CT: Praeger Publishers.

DeSipio, Louis. 1996. *Counting on the Latino Vote: Latinos as a New Electorate.* Charlottesville, VA: University Press of Virginia.

U.S. Bureau of the Census. 1996. *Population Projections of the United States by Age, Sex, Race, and Hispanic Origin: 1995 to 2050.* Current Population Reports, Series P-25, No. 1130.

KEY CONCEPTS

minority group   political economy

DISCUSSION QUESTIONS

1. What factors that are particular to the Latino population are likely to influence voting patterns in the future? How will the Latino vote be likely to shape election issues in the future?

2. How might the political participation of Latinos compare to other groups, such as African Americans, Whites, and Asian Americans? How does this show the influence of social factors on democratic participation?

60

Capturing the Youth Vote

DEBORAH CARR

Voter turnout among young people is notoriously low. Here, Deborah Carr examines patterns in youth voting since 1972, also asking how voter turnout drives aimed at young people might have affected voting behavior.

Senator John Kerry probably is not a fan of pro wrestling. And despite Jenna and Barbara Bush's valiant efforts to convince Americans that President George W. Bush grooves to Outkast, most hip-hop fans cannot envision W. passing the Courvoisier to rap star P. Diddy. To their surprise, Kerry and Bush found themselves defending their political views to young fans of pro wrestling and the hip-hop community during the heated 2004 presidential election campaigns. OutKast fans and SmackDown devotees were expected to play a bigger role than soccer moms or NASCAR dads in determining the election.

Why? Roughly 30 million "missing" youth votes were up for grabs. Just 42 percent of 18- to 24-year-old Americans voted in the 2000 presidential election, compared to 70 percent of people 25 and older (see Figure 1). The gender gap in youth voting has also increased slightly over time, with black and white women more likely than their male peers to vote. Among young Hispanics, however, men and women are equally likely to vote: just 30 percent voted in the 2000 presidential election.

This political apathy is a far cry from the late 1960s when 18- to 20-year-olds demanded to know why they were being called to fight in Vietnam yet could not vote until they were 21. In 1971, 18- to 20-year-olds won the right to vote with the ratification of the 26th Amendment. It is not clear whether novelty, the political zeitgeist, or a streak of youthful political activism was at work, but 55 percent of 18- to 21-year-olds voted in the 1972 presidential election, and that proportion has been inching downward ever since. Experts attribute declining voter turnout among youths to the distinctive characteristics of the Generation X and Y cohorts, including apathy, disillusionment, a lack of civics education, and high exposure to television ads that downplay candidates' ideological differences yet play up their personal transgressions.

SOURCE: Deborah Carr. 2005. "Capturing the Youth Vote." *Contexts 4*, no.1 (Winter): 47.

FIGURE 1 Voter Turnout among U.S. Citizens, 1972–2000

SOURCE: *Current Population Survey, November Supplements, 1972–2000*

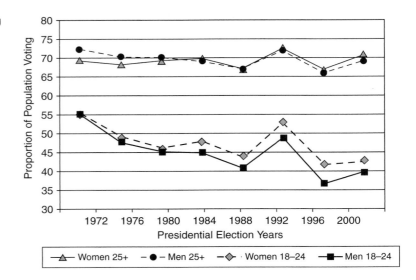

Recognizing the potential impact of young voters, innovative programs to increase the youth vote cropped up everywhere from the inner city to rural mobile-home parks in 2004. Rap star P. Diddy started the "Citizen Change" program, urging young adults—particularly urban African Americans and Latinos—to "Vote or Die!" World Wrestling Entertainment initiated the "Smack-Down Your Vote" program in 2002, a nonpartisan effort to encourage young Americans to become informed citizens. The centerpiece effort was "18–30 VIP" (vote.wwe.com), a Web-based forum where Kerry and Bush shared their views on jobs, the economy, terrorism, college loans, and other topics of interest to young voters.

We cannot determine with any certainty whether these programs are effective. While youthful voters did turn out in large numbers in the 2004 presidential election, they hardly drove the election outcome, as some political observers had predicted. According to exit-poll data from the Associated Press, 18- to 24-year-old voters accounted for just 10 percent of all voters in 2004, the exact same proportion as in 2000. The more uplifting news is that total number of voters increased from 105 million to an estimated 116 to 120 million, and the youth vote is believed to have increased proportionally. Experts attribute this less to young adults' heeding the call of their celebrity icons and more to the fact that issues critically important to young people—the war in Iraq, the economy, and funding for education—were at stake.

KEY CONCEPTS

democracy   voter turnout

DISCUSSION QUESTIONS

1. What social and historical factors do you think most influence young people and their participation (or lack thereof) in national elections?

2. How do popular figures, such as celebrities, influence the political participation of young people?

61

The Social Meanings of Illness

ROSE WEITZ

In this article, Rose Weitz outlines the sociological study of health and illness, summarizes the medical model of illness, and compares the two models. She explains that sociologists largely see illness as socially constructed and that race, class, and gender inequalities play a part in how we define health and sickness.

All Marco Oriti has ever wanted, ever imagined, is to be taller. At his fifth birthday party at a McDonald's in Los Angeles, he became sullen and withdrawn because he had not suddenly grown as big as his friends who were already five: in his simple child's calculus, age equaled height, and Marco had awakened that morning still small. In the six years since then, he has grown, but slowly, achingly, unlike other children. "Everybody at school calls me shrimp and stuff like that," he says.

"They think they're so rad. I feel like a loser. I feel like I'm nothing." At age 11, Marco stands 4 feet 1 inch— 4 inches below average—and weighs 49 pounds. And he dreams, as all aggrieved kids do, of a sudden, miraculous turnaround: "One day I want to, like, surprise them. Just come in and be taller than them."

Marco, a serious student and standout soccer player, more than imagines redress. Every night but Sunday, after a dinner he seldom has any appetite for, his mother injects him with a hormone known to stimulate bone growth. The drug, a synthetic form of naturally occurring human growth hormone (HGH) produced by the pituitary,

SOURCE: Rose Weitz. 2007. *The Sociology of Health, Illness, and Health Care,* 4th ed. Belmont, CA: Thomson Wadsworth, pp. 125–151.

has been credited with adding up to 18 inches to the predicted adult height of children who produce insufficient quantities of the hormone on their own—pituitary dwarfs. But there is no clinical proof that it works for children like Marco, with no such deficiency. Marco's rate of growth has improved since he began taking the drug, but his doctor has no way of knowing if his adult height will be affected. Without HGH, Marco's predicted height was 5 feet 4 inches, about the same as the Nobel Prize-winning economist Milton Friedman and . . . Masters golf champion, Ian Woosnam, and an inch taller than the basketball guard Muggsy Bogues of the Charlotte Hornets. Marco has been taking the shots for six years, at a cost to his family and their insurance company of more than $15,000 a year [$21,000 in 2005 dollars]. . . .

A Cleveland Browns cap splays Marco Oriti's ears and shadows his sparrowish face. Like many boys his age, Marco imagines himself someday in the NFL. He also says he'd like to be a jockey—making a painful incongruity that mirrors the wild uncertainty over his eventual size. But he is unequivocal about his shots, which his mother rotates nightly between his thighs and upper arms. "I hate them," he says.

He hates being short far more. Concord, the small Northern California city where the Oriti family now lives, is a high-achievement community where competition begins early. So Luisa Oriti and her husband, Anthony, a bank vice president, rationalize the harshness of his treatment. "You want to give your child that edge no matter what," she says, "I think you'd do just about anything." (Werth, 1991)

Does Marco have an illness? According to his doctors, who have recommended that he take an extremely expensive, essentially experimental, and potentially dangerous drug, it would seem that he does. To most people, however, Marco simply seems short. . . .

MODELS OF ILLNESS

The Medical and Sociological Models of Illness

What do we mean when we say something is an illness? As Marco's story suggests, the answer is far from obvious. Most Americans are fairly confident that someone who has a cold or cancer is ill. But what about the many postmenopausal women whose bones have become brittle with age, and the many older men who have bald spots, enlarged prostates, and urinary problems? Or the many young boys who have trouble learning, drink excessively, or enjoy fighting? Depending on who you ask, these conditions may be defined as normal human variations, as illnesses, or as evidence of bad character. As these questions suggest, defining what is and is not an illness is far from a simple task. In this section we explore the medical model of illness: what doctors typically mean when they say something is an illness. This medical model is not accepted in its entirety by all doctors—those in public health, pediatrics, and family practice are especially likely to question it—and is not rejected by all sociologists, but it is the dominant conception of illness in the medical world. The sociological model of illness summarizes critical sociologists' retort to the medical model of illness. This sociological model reflects sociologists' view of how the world currently operates, not how it ideally should operate. . . .

The medical model of illness begins with the assumption that illness is an *objective* label given to anything that deviates from normal biological functioning (Mishler, 1981). Most doctors, if asked, would explain that polio is caused by a virus that disrupts the normal functioning of the neurological system, that menopause is a "hormone deficiency disease" that, among other things, impairs the body's normal ability to regenerate bone, and that men develop urinary problems when their prostates grow excessively large and unnaturally compress the urinary tract. Doctors might further explain that, because of scientific progress, all educated doctors can now recognize these problems as illnesses, even though they were not considered as such in earlier eras.

In contrast, the sociological model of illness begins with the statement that illness (as the term is actually used) is a *subjective* label, which reflects personal and social ideas about what is normal as much as scientific reasoning (Weitz, 1991). Sociologists point out that ideas about normality differ widely across both individuals and social groups. A height of 4 feet 6 inches would be normal for a Pygmy man but not for an American man. Drinking three glasses of wine a day is normal for Italian women but could lead to a diagnosis of alcoholism in American medical circles. In defining normality, therefore, we need to look not only at individual bodies but also at the broader social context. Moreover, even within a given group, "normality" is a range and not an absolute. The median height of American men, for example, is 5 feet 9 inches, but most people would consider someone several inches taller or shorter than that as still normal. Similarly, individual Italians routinely and without social difficulties drink more or less alcohol than the average Italian. Yet medical authorities routinely make decisions about what is normal and what is illness based not on absolute, objective markers of health and illness but on arbitrary, statistical cutoff points—deciding, for example, that anyone in the fifth percentile for height or the fiftieth percentile for cholesterol level is ill. Culture, too, plays a role: Whereas the American Society of Plastic and Reconstructive Surgeons recommends breast enlargement for small breasts, which it considers a disease ("micromastia") and believes "results in feelings of inadequacy, lack of self-confidence, distortion of body image and a total lack of well-being due to a lack of self-perceived femininity" (1989: 4–5), in Brazil large breasts are denigrated as a sign of African heritage and breast *reduction* is the most popular cosmetic surgery (Gilman, 1999).

Because the medical model assumes illness is an objective, scientifically determined category, it also assumes there is no *moral* element in labeling a condition or behavior as an illness. Sociologists, on the other hand, argue that illness is inherently a moral category, for deciding what is illness always means deciding what is good or bad. When, for example, doctors label menopause a "hormonal deficiency disease," they label it an undesirable deviation from normal. In contrast, many women consider menopause both normal and desirable and enjoy the freedom from fear of pregnancy that menopause brings (E. Martin, 1987). In the same manner, when we define cancer, polio, or diabetes as illnesses, we judge the bodily changes these conditions produce to be both abnormal and undesirable, rather than simply normal variations in functioning, abilities, and life expectancies. (Conversely, when we define a condition as healthy, we judge it normal and desirable.)

Similarly, when we label an individual as ill, we also suggest that there is something undesirable about that *person*. By definition, an ill person is one whose actions, ability, or appearance do not meet social norms, or expectations within a given culture regarding proper behavior or appearance. Such a person will typically be considered less whole and less socially worthy than those deemed healthy. Illness, then, like virginity or laziness, is a moral status: a social condition that we believe indicates the goodness or badness, worthiness or unworthiness, of a person.

From a sociological stand, illness is not only a moral status but (like crime or sin) a form of deviance (Parsons, 1951). To sociologists, labeling something deviant does not necessarily mean that it is immoral. Rather, deviance refers to behaviors or conditions that socially powerful persons within a given culture *perceive*, whether accurately or inaccurately, as immoral or as violating social norms. We can tell whether behavior violates norms (and, therefore, whether it is deviant) by seeing if it results in *negative social sanctions*. This term refers to any punishment, from ridicule to execution. (Conversely, positive social sanctions refers to rewards, ranging from token gifts to knighthood.) These social sanctions are enforced by social control agents including parents, police, teachers, peers, and doctors. Later in this [article] we will look at some of the negative social sanctions imposed against those who are ill.

For the same reasons that the medical model does not recognize the *moral* aspects of illness labeling, it does not recognize the *political* aspects of that process. Although some doctors at some times are deeply immersed in these political processes—arguing, for example, that insurance companies should cover treatment for newly labeled conditions such as fibromyalgia or multiple chemical sensitivity—they rarely consider the ways that politics underlie the illness-labeling process in general. In contrast, sociologists point out that any time a condition or behavior is labeled as an illness, some groups will benefit more than others, and some groups will have more power than others to enforce the definitions that benefit them. As a result, there are often open political struggles over illness definitions (a topic we will return to later in this [article]). For example, vermiculite miners and their families who were constantly exposed to asbestos dust and who now have strikingly high rates of cancer have fought with insurance companies and doctors, in clinics, hospitals, and the courts, to have "asbestosis" labeled an illness; meanwhile, the mining companies and the doctors they employed have argued that there is no such disease and that the high rates of health problems in mining communities are merely coincidences (A. Schneder and McCumber, 2004).

In sum, from the sociological perspective, illness is a *social construction*, something that exists in the world not as an objective condition but *because we have defined it as existing*. This does not mean that the virus causing measles does not exist, or that it does not cause a fever and rash. It does mean, though, that when we talk about measles as an illness, we have organized our ideas about that virus, fever, and rash in only one of the many possible ways. In another place or time, people might conceptualize those same conditions as manifestations of witchcraft, as a healthy response to the presence of microbes, or as some other illness altogether. To sociologists, then, *illness*, like *crime* or *sin*, refers to biological, psychological, or social conditions subjectively defined as undesirable by those within a given culture who have the power to enforce such definitions.

In contrast, and as we have seen, the medical model of illness assumes that illness is an objective category. Based on this assumption, the medical model of health care assumes that each illness has specific features, universally recognizable in all populations by all trained doctors, that differentiate it both from other illnesses and from health (Dubos, 1961; Mishler, 1981). The medical model thus assumes that diagnosis is an objective, scientific process.

Sociologists, on the other hand, argue that diagnosis is a subjective process. The subjective nature of diagnosis expresses itself in three ways. First, patients with the same symptoms may receive different diagnoses depending on various social factors. Women who seek medical care for chronic pain, for example, are more likely to receive psychiatric diagnoses than are men who report the same symptoms. Similarly, African Americans (whether male or female) are more likely than whites are to have their chest pain diagnosed as indigestion rather than as heart disease (Hoffman and Tarzian, 2001; Nelson, Smedley, and Stith, 2002). Second, patients with the same underlying illness may experience different symptoms, resulting in different diagnoses. For example, the polio virus typically causes paralysis in adults but only flu-like symptoms in very young children, who often go undiagnosed. Third, different cultures identify a different range of symptoms and categorize those symptoms into different illnesses. For example, U.S. doctors assign the label of attention deficit disorder (ADD) to children who in Europe would be considered lazy troublemakers. And French doctors often attribute headaches to liver problems, whereas U.S. doctors seek psychiatric or neurological explanations (Payer, 1996). In practice, the American medical model of illness assumes that illnesses manifest themselves in other cultures in the same way as in American culture and, by extension, that American doctors can readily transfer their knowledge of illness to the treatment and prevention of illness elsewhere.

Finally, the medical model of illness assumes that each illness has not only unique symptoms but also a unique *etiology*, or cause (Mishler, 1981). Modern medicine assumes, for example, that *tuberculosis*, polio, HIV disease, and so on, are each caused by a unique microorganism. Similarly, doctors continue to search for limited and unique causes of heart disease and cancer, such as high-cholesterol diets and exposure to asbestos. Yet even though illness-causing microorganisms exist everywhere and environmental health dangers are common, relatively few people become ill as a result. By the same token, although cholesterol levels and heart disease are strongly correlated among middle-aged men many men eat high-cholesterol diets without developing heart disease, and others eat low-cholesterol diets but die of heart disease anyway. The doctrine of unique ecology discourages medical researchers from asking why individuals respond in such different ways to the same health risks and encourages researchers to search for magic bullets—a term first used by Paul Ehrlich, discoverer of the first effective treatment for syphilis, in referring to drugs that almost miraculously prevent or cure illness by attacking one specific etiological factor. . . .

MEDICINE AS SOCIAL CONTROL

Creating Illness: Medicalization

The process through which a condition or behavior becomes defined as a medical problem requiring a medical solution is known as medicalization (Conrad and Schneider, 1992; Conrad, 2005). For example, as social conditions have changed, activities formerly considered sin or crime, such as masturbation, homosexual activity, or heavy drinking, have become defined as illnesses. The same has happened to various natural conditions and processes such as uncircumsised penises, limited sexual desire, aging, pregnancy, and menopause (e.g., F. Armstrong, 2000; Barker, 1998; Figert, 1996; Rosenfeld and Faircloth, 2005). The term *medicalization* also refers to the process through which the definition of an illness is *broadened*. For example, when the World Health Organization (WHO) in 1999 lowered the blood sugar level required for diagnosis with diabetes, the number of persons eligible for this diagnosis increased in some populations by as much as 30 percent (Shaw, de Courten, Boyko, and Zimmet, 1999).

For medicalization to occur, one or more organized social groups must have both a vested interest in it and sufficient power to convince others (including doctors, the public, and insurance companies) to accept their new definition of the situation. Not surprisingly, doctors often play a major role in medicalization, for medicalization can increase their power, the scope of their practices, and their incomes. For example, during the first half of the twentieth century, improvements in the standard of living coupled with the adoption of numerous public health measures substantially reduced the number of seriously ill children. As a result, the market for pediatricians declined, and their focus shifted from treating serious illnesses to treating minor childhood illnesses and offering well-baby care. Pediatrics thus became less well-paid, interesting, and prestigious. To increase their market while obtaining more satisfying and prestigious work, some pediatricians have expanded their practices to include children whose behavior concerns their parents or teachers and who are now defined as having medical conditions such as attention deficit disorder or antisocial personality disorder (Halpern, 1990). Doctors have played similar roles in medicalizing premenstrual syndrome (Figert, 1996), drinking during pregnancy (E. Armstrong, 1998), impotence (Loe, 2004; Tiefer, 1994), and numerous other conditions. . . .

The Consequences of Medicalization   In some circumstances, medicalization can be a boon, leading to social awareness of a problem, sympathy toward its sufferers, and development of beneficial therapies. Persons with epilepsy, for example, lead far happier and more productive lives now that drugs usually can control their seizures and few people view epilepsy as a sign of demonic possession. But defining a condition as an illness does not necessarily improve the social status of those who have that condition. Those who use alcohol excessively, for example, continue to experience social rejection even when alcoholism is labeled a disease. Moreover, medicalization also can lead to new problems, known by sociologists as unintended negative consequences (Conrad and Schneider, 1992; Zola, 1972). . . .

The Rise of Demedicalization   The dangers of medicalization have fostered a counter movement of demedicalization (R. Fox, 1977). A quick lock at medical textbooks from the late 1800s reveals many "diseases" that no longer exist. For example, nineteenth-century medical textbooks often included several pages on the health risks of masturbation. One popular textbook from the late nineteenth century asserted that masturbation caused "extreme emaciation, sallow or blotched skin, sunken eyes, . . . general weakness, dullness, weak back, stupidity, laziness, . . . wandering and illy defined pains," as well as infertility, impotence, consumption, epilepsy, heart disease, blindness, paralysis, and insanity (Kellogg, 1880: 365). Today, however, medical textbooks describe masturbation as a normal part of human sexuality.

Like medicalization, demedicalization often begins with lobbying by consumer groups. For example, medical ideology now defines childbirth as an inherently dangerous process, requiring intensive technological, medical assistance. Since the 1940s, however, growing numbers of American women have attempted to redefine childbirth as a generally safe, simple, and natural process and have promoted alternatives ranging from natural childbirth classes, to hospital birthing centers, to home births assisted only by midwives (Sullivan and Weitz, 1988). Similarly . . . gay and lesbian activists have at least partially succeeded in redefining homosexuality from a pathological condition to a normal human variation. More broadly, in recent years, books, magazines, television shows, and popular organizations devoted to teaching people to care for their own health rather than relying on medical care have proliferated. For example, in the early 1970s, the Boston Women's Health Book Collective published a 35-cent mimeographed booklet on women's health. From this, they have built a virtual publishing empire that has sold to consumers worldwide millions of books (including the best-selling *Our Bodies, Ourselves*) on the topics of childhood, adolescence, aging, and women's health. . . .

Social Control and the Sick Role

. . . Medicine also can work as an institution of social control by pressuring individuals to *abandon* sickness, a process first recognized by Talcott Parsons (1951).

Parsons was one of the first and most influential sociologists to recognize that illness is deviance. From his perspective, when people are ill, they cannot perform the social tasks normally expected of them. Workers stay home, homemakers tell their children to make their own meals, students ask to be excused from exams. Because of this, either consciously or unconsciously, people can use illness to evade their social responsibilities. To Parsons, therefore, illness threatened social stability.

Parsons also recognized, however, that allowing some illness can *increase* social stability. Imagine a world in which no one could ever "call in sick." Over time, production levels would fall as individuals, denied needed recuperation time, succumbed to physical ailments. Morale, too, would fall while resentment would rise among those forced to perform their social duties day after day without relief. Illness, then, acts as a kind of pressure valve for society—something we recognize when we speak of taking time off work for "mental health days."

From Parsons's perspective, then, the important question was how did society control illness so that it would increase rather than decrease social stability? The author's emphasis on social stability reflected his belief in the broad social perspective known as functionalism. Underlying functionalism is an image of society as a smoothly working, integrated whole, much like the biological concept of the human body as a homeostatic environment. In this model, social order is maintained because individuals learn to accept society's norms and because society's needs and individuals' needs match closely, making rebellion unnecessary. Within this model, deviance—including illness—is usually considered dysfunctional because it threatens to undermine social stability.

Defining the Sick Role   Parsons's interest in how society manages to allow illness while minimizing its impact led him to develop the concept of the sick role. The sick role refers to social expectations regarding how society should view sick people and how sick people should behave. According to Parsons, the sick role as it currently exists in Western society has four parts. First, the sick person is considered to have a legitimate reason for not fulfilling his or her normal social role. For this reason, we allow people to take time off from work when sick rather than firing them for malingering. Second, sickness is considered beyond individual control, something for which the individual is not held responsible. This is why, according to Parsons, we bring chicken soup to people who have colds rather than jailing them for stupidly exposing themselves to germs. Third, the sick person must recognize that sickness is undesirable and work to get well. So, for example, we sympathize with people who obviously hate being ill and strive to get well and question the motives of those who seem to revel in the attention their illness brings. Finally, the sick person should seek and follow medical advice. Typically, we expect sick people to follow their doctors' recommendations regarding drugs and surgery, and we question the wisdom of those who do not.

Parsons's analysis of the sick role moved the study of illness forward by highlighting the social dimensions of illness, including identifying illness as deviance and doctors as agents of social control. It remains important partly because it was the first truly sociological theory of illness. Parsons's research also has proved important because it stimulated later research on interactions between ill people and others. In turn, however, that research has illuminated the analytical weaknesses of the sick role model.

Critiquing the Sick Role Model   Many recent sociological writings on illness—including this [article]—have adopted a conflict perspective rather than a functionalist perspective. Whereas functionalists envision society as a harmonious whole held together largely by socialization, mutual consent, and mutual interests, those who hold a conflict perspective argue that society is held together largely by power and coercion, as dominant groups impose their will on others. Consequently, whereas functionalists view deviance as a dysfunctional element to be controlled, conflict theorists view deviance as a necessary force for social change and as the conscious or unconscious expression of individuals who refuse to conform to an oppressive society. Conflict theorists therefore have stressed the need to study social control agents as well as, if not more than, the need to study deviants. . . .

In sum, the sick role model is based on a series of assumptions about both the nature of society and the nature of illness. In addition, the sick role model confuses the experience of *patienthood* with the experience of *illness* (Conrad, 1987). The sick role model focuses on the interaction between the ill person and the mainstream health care system. Yet interactions with the medical world form only a small part of the experience of living with illness or disability, as the next chapter will show. For these among other reasons, research on the sick role has declined precipitously; whereas *Sociological Abstracts* listed 71 articles on the sick role between 1970 and 1979, it listed only 7 articles between 1990 and 1999, even though overall far more academic articles were published during the 1990s than during the 1970s.

CONCLUSION

The language of illness and disease permeates our everyday lives. We routinely talk about living in a "sick" society or about the "disease" of violence infecting our world offhandedly labeling anyone who behaves in a way we don't understand or don't condone as "sick."

This metaphoric use of language reveals the true nature of illness: behaviors, conditions, or situations that powerful groups find disturbing and believe stem from internal biological or psychological roots. In other times or places, the same behaviors, conditions, or situations might have been ignored, condemned as sin, or labeled crime. In other words, illness is both a social construction and a moral status.

In many instances, using the language of medicine and placing control in the hands of doctors offers a more humanistic option than the alternatives. Yet, as this [article] has demonstrated, medical social control also carries a price. The same surgical skills and technology for cesarean sections that have saved the lives of so many women and children now endanger the lives of those who have cesarean sections unnecessarily. At the same time, forcing cesarean sections on women potentially threatens women's legal and social status. Similarly, the development of tools for genetic testing has saved many individuals from the anguish of rearing children doomed to die young and painfully, but has cost others their jobs or health insurance.

In the same way, then, that automobiles have increased our personal mobility in exchange for higher rates of accidental death and disability, adopting the language of illness and increasing medical social control bring both benefits and costs. These benefits and costs will need to be weighed carefully as medicine's technological abilities grow.

REFERENCES

Armstrong, Elizabeth M. 1998. "Diagnosing moral disorder: The discovery and evolution of fetal alcohol syndrome." *Social Science and Medicine* 47: 2025–2042.

———. 2000. "Lessons in control: Prenatal education in the hospital." *Social Problems* 47: 583–605.

Barker, K. K. 1998. "A ship upon a stormy sea: The medicalization of pregnancy." *Social Science and Medicine* 47: 1067–1076.

Conrad, Peter. 1987. "The experience of illness: Recent and new directions." *Research in the Sociology of Health Care* 6: 1–32.

———. 2005. "The shifting engines of medicalization." *Journal of Health and Social Behavior* 46: 3–14.

Conrad, Peter, and Joseph W. Schneider. 1992. *Deviance and Medicalization: From Badness to Sickness*. Philadelphia: Temple University Press.

Dubos, Rene. 1961. *Mirage of Health*. New York: Anchor.

Figert, Anne E. 1996. *Women and the Ownership of PMS: The Structuring of a Psychiatric Disorder*. New York: Aldine DeGruyter.

Fox, Renee. 1977. "The medicalization and demedicalization of American society." *Daedalus* 106: 9–22.

Gilman, Sander L. 1999. *Making the Body Beautiful: A Cultural History of Aesthetic Surgery*. Princeton, NJ: Princeton University Press.

Halpern, S. A. 1990. "Medicalization as a professional process: Postwar trends in pediatrics." *Journal of Health and Social Behavior* 31: 28–42.

Hoffman, Diane E., and Anita J. Tarzian. 2001. "The girls who cried pain: A bias against women in the treatment of pain." *Journal of Law, Medicine, and Ethics* 29: 13–27.

Kellogg, J. H. 1880. *Plain Facts for Young and Old*. Burlington, IA: Segner and Condit.

Loe, Meika. 2004. *The Rise of Viagra: How the Little Blue Pill Changed Sex in America*. New York: New York University Press.

Martin, Emily. 1987. *The Woman in the Body*. Boston: Beacon.

Mishler, Elliot G. 1981. "Viewpoint: Critical perspectives on the biomedical model." Pp. 1–23 in *Social Contexts of Health, Illness, and Patient Care*, edited by Elliot G. Mishler. Cambridge, UK: Cambridge University Press.

Nelson, Alan R., Brian D. Smedley, and Adreinne Y. Stith. 2002. *Unequal Treatment: Confronting Racial and Ethnic Disparities in Health Care*. Washington, DC: Institute of Medicine, National Academy Press.

Parsons, Talcott. 1951. *The Social System*. New York: Free Press.

Payer, Lynn. 1996. *Medicine and Culture*. Rev. ed. New York: Holt.

Rosenfeld, Dana, and Christopher A. Faircloth (eds.). 2005. *Medicalized Masculinities*. Philadelphia: Temple University Press.

Schneider, Andrew, and David McCumber. 2004. *An Air That Kills: How the Asbestos Poisoning of Libby, Montana Uncovered a National Scandal*. New York: Putnam's Sons.

Shaw, Jonathan E., Maximilian de Courten, Edward J. Boyko, and Paul Z. Zimmet. 1999. "Impact of new diagnostic criteria for diabetes on different populations." *Diabetes Care* 22: 762–766.

Sullivan, Deborah A., and Rose Weitz. 1988. *Labor Pains: Modern Midwives and Home Birth*. New Haven, CT: Yale University Press.

Tiefer, Leonore. 1994. "The medicalization of impotence: Normalizing phallocentrism." *Gender & Society* 8: 363–377.

Weitz, Rose. 1991. *Life with AIDS*. New Brunswick, NJ: Rutgers University Press.

Werth, Barry. 1991. "How short is too short? Marketing human growth hormone." *New York Times Magazine* June 16: 14+.

Zola, Irving K. 1972. "Medicine as an institution of social control." *Sociological Review* 20: 487–504.

KEY CONCEPTS

conflict theory	functionalism	medicalization of illness

DISCUSSION QUESTIONS

1. Compare and contrast the sociological and the medical models of illness. What are the pros and cons of each model?
2. What does the author mean by illness as a moral status?

62

What Do We Think of the U.S. Health Care System?

GRACE BUDRYS

*Grace Budrys discusses how the American health care system has been and should
be evaluated. There are many complaints about health care in America, and there
is evidence that we are lagging behind many other developed countries with regard
to the availability of affordable health care for all citizens. With evidence from
opinion polls and surveys, she outlines where our system falls short and suggests a
list of changes that may help to shape a better health care system for America.*

During the campaign preceding the 1992 election, President George Bush
was fond of saying that Americans enjoy the best health care system in the
world.[1] Many other Americans still say that. Sociologists would say this is an
example of "ethnocentrism." It means that we think "ours (whatever "ours"
refers to at the moment) is best" simply because it is "ours." When ethnocentrism
kicks in, people resent being asked to explain why they prefer what they prefer;
nor are they pleased to be asked to provide some evidence to show that it,
whatever it is, is, in fact, better in some way. No matter how much effort one is
willing to put into being "health-care-delivery-system-centric," saying that our
health care system is the best just does not stand up in the face of the evidence no
matter how much we would like it to.

Consider the most basic indicators. First, Americans don't live as long as
people in most industrialized countries; in fact, according to the World Health
Organization, twenty-two countries have a better male life expectancy and
women live longer in eighteen countries. (These figures have been fairly stable
over the past two decades or so.) Second, our infant mortality rate is higher than
that of twenty-four countries. This is important because infant mortality (death of
babies under one year of age) is considered the most sensitive measure of the
health of a society because infants are the most vulnerable. Third, we pay more
for health care than anyone else in the world! (We will explore the matter of
expenditures in greater detail later in this [article].)

SOURCE: Grace Budrys. 2001. *Our Unsystematic Health Care System.* Lanham,
MD: Rowman & Littlefield, pp. 23–35.

So is our technology what makes our system the best? We have the best medical technology in the world, and we have lots of it. After all, it is undeniable that people from all over the world come here to have operations. So, why isn't all that technology doing much for our life expectancy? Because people smoke, drink, eat the wrong things, don't exercise? Actually, people in other countries smoke a lot more, drink more, eat rich foods, don't exercise, and still live a lot longer. Look at the French!

Maybe it's because there is something wrong with the way our health system is organized. Some people, including former President George Bush, were, when pressed, willing to acknowledge that everyone does not necessarily benefit from our technology. Not because it isn't there, they point out, but because people aren't making the effort to take advantage of what is readily available. Almost everyone involved in studying the health care delivery system would say that's stretching it. Defenders of the system say that our technology is the best and available to everyone; public hospitals have it available and will provide it for free to those who really can't pay for it; if people don't use it, there is nothing you can do to make them use it.

There are a couple of assertions in these statements, which are not explicit, that deserve further exploration. For one, the statements suggest that the level of technology in other countries is not as advanced. That is not true in the case of the major industrialized European countries and Japan. They have the same kind of technology, only not as much of it, as we have. That brings us to the second built-in assumption, which is, more technology is always better. After all you can get a CT scan on every corner and that's great—right? In actuality, health policymakers in the United States and other countries discovered a couple of decades ago that the doctors and technicians who perform various "high-tech" procedures and tests must do a certain number of them on a regular basis to maintain a high level of proficiency. Having many machines simply means that you have people operating them, or worse yet interpreting the results, who do not have sufficient experience to do a good job—to recognize aberrations, to interpret the significance of subtle differences, and so on. This is difficult to hear and it strikes at the heart of the claim that lots of technology is good.

Where does that leave us in evaluating our health care system? It seems that when we talk to one another, we have many complaints about the system. For years, a majority of Americans have been saying that the health care system is in a state of crisis. How is it that we can say that we have the best health care system in the world one minute and say that the system is in crisis the next minute? In part, different people are saying these things. In my opinion, there is another dimension of ethnocentrism being played out here: I can criticize *my* family members, *my* ethnic group, *my* community, but don't *you* dare say something negative about them! For some reason, the health care system seems to inspire an "us" and "them" reaction, a "you must either be with us or you are against us" stance. Whatever the explanation for such inconsistency, it will be helpful to look at the evidence that indicates why so many of us feel that our system is far from perfect and needs to be changed.

EXACTLY WHAT HAVE WE BEEN SAYING
ABOUT THE SYSTEM?

... So how do we really know how satisfied or dissatisfied Americans are with current health care delivery system arrangements? Is it possible that all the talk about the need for reform is just another example of media hype, or an issue that is important within the Washington D.C. "beltway" but not other parts of the country, or maybe it is really of greater interest to academics than anyone else? Yet, it is hard to miss the increasing number of news reports, magazine stories, television specials that focus on some aspect of the current health care system. As we all know such reports would not be so common unless the media watchers had established that such material would attract a large audience.

On this point it is interesting to note that health care executives complain that public opinion is being shaped by the news media, which have been unduly critical of how health care organizations treat patients.[2] A recent analysis of the content of news publications and network news broadcasts between 1990 and 1997 indicates that the stories have become more critical over time.[3] Reporters tend to dramatize single incidents and end up identifying insurance companies and the health care organizations they sponsor as the "villains" in their stories.

There is, of course, no way to know how much of the dissatisfaction with the health care system expressed by Americans can be attributed to the negative reports found in the media. What we can do is continue to ask the public what it thinks about the workings of the health care system on a regular basis to trace shifts in public opinion even if we cannot be sure why those shifts are occurring. This suggests that we should be looking at the results of opinion polls, especially polls conducted by well-known survey research organizations and summarized by respected scholars.

Several features make opinion polls a reliable source of information, one being the fact that the results can be compared across time. That tells us whether satisfaction is increasing or decreasing. Occasionally polls compare our satisfaction level with the satisfaction level registered by people in other countries. The expense involved in conducting this type of poll prevents international surveys from being conducted on a regular basis, but those that have been done are particularly informative.

The results of a comparative international study, reported in 1990, are particularly striking. Scholars associated with Harvard University in cooperation with the Harris Associates polling organization, designed the survey.[4] A random sample of people in ten advanced, industrialized countries were asked to evaluate their health care arrangements. . . .

The comparison leaves little doubt that, at the time this survey was carried out, Americans, in comparison to people in all those other highly industrialized countries, were at the low end of the spectrum on satisfaction with our health care system. Given this information we should be able to identify what it is that Americans are so dissatisfied about and find a solution.

Indeed, the country, led by President Clinton, did take up the issue of health care reform shortly after these results were reported.... The health care reform effort captured the attention of all those who had any connection to the health sector as well those outside of this sector who would certainly be affected. A committee, headed by Hillary Rodham Clinton, produced a detailed reform plan, which virtually all interested parties played a part in shaping. Yet that plan failed to gain the support of many legislators let alone of the general public.

If so many of us were so clearly unhappy with prevailing arrangements, and could, if we made the effort, register those views and offer solutions, why did all that effort to alter the health care delivery system result in a reform plan that failed? Another way of putting the question—why is it so difficult to change arrangements that we are clearly dissatisfied with? Once structures (i.e., policies, programs, and organizations) come into existence, they resist change. Social institutions do change, but they change slowly. The Clinton health reform plan, which was pretty much designed to reorganize the whole health care system, turned out to be far too radical an approach for addressing dissatisfaction regardless of the fact that the dissatisfaction rate was as high as it was.

There was also one other major problem—the lapse between the *will* and the *way*. It is clear that the will to change current arrangements was there. Unfortunately, we got bogged down in trying to identify the way. In other words, there was not enough agreement on what should be changed or how it should be changed. What we, as a society, have difficulty agreeing on involves the most basic understandings about what this social institution should look like, such as who is entitled to receive health care services. Should everyone or only those who have earned the right because they work and contribute to society? What about children? Is it right to deprive children of access to health care services because their parents are not hard-working, respectable, and therefore "deserving" members of society? At another level, is it logical to deprive people of basic health care services knowing that they will be sicker and more expensive to treat when they come to the door of the hospital that cares for poor people? Who should make all these decisions? This brings us to the basic question that always bothers Americans—who is going to pay for it? That question tends to bring forth the following familiar refrain: Why should I pay for other people's health care? They probably are misusing it anyway.

PUBLIC ATTITUDES REGARDING HEALTH CARE ARRANGEMENTS

It is not that Americans aren't prepared to say what they think is wrong with the health care system; it's that not everyone sees the problem or the solution exactly the same way. (In this case, you can see how the "definition of the problem" is a reflection of the "social construction of reality.") While public opinion surveys on our views of the health care system date back many years, we cannot review all of them here. We will look at the ones that received the most attention during previous policy debates about health care reform as well as those receiving attention in the wake of failed health care reform efforts.

A particularly revealing assessment, published in 1984, reviewed fifteen national public opinion surveys conducted between 1981 and 1984.[5] The researchers captured the views expressed by Americans in five summarizing statements. Why consider what Americans thought about their health care arrangements that long ago? There are a couple of good reasons. It will tell us if the problems people complained about then were actually addressed. It may also help us understand how those complaints shaped the changes that did take place. Finally, we will see whether earlier solutions worked out the way we expected. The results of the fifteen surveys showed the following findings.

1. Americans see rising costs as the nation's number-one problem in health care, but do not see it as the most important issue on the list of problems facing the nation.

2. Americans are deeply disturbed by the sharply rising prices of their health care, particularly the increasing cost of a stay in the hospital or a visit to a physician. However, most are not troubled by the growing share of the nation's economy that is devoted to health care. In fact, many believe that our society currently spends too little rather than too much for these services.

3. Although the majority of Americans believe that the country's present health care arrangements are not satisfactory, they are not prepared to change their personal health care arrangements.

4. Because Americans do not see themselves as having created the problem of health care costs, they tend to support only the cost-containment proposals that do not affect their own health care arrangements.

5. The views of practicing physicians are more influential than the opinions of groups proposing health reforms, including government officials, business-people, and labor leaders.

After looking at these particular surveys more closely, the researchers identified a couple of themes that are especially noteworthy. For one, while Americans said that the health care system is seriously flawed, they also said that their own doctor was terrific. That has some interesting implications. If people do not have a doctor with whom they have a personal relationship, are they less likely to see the health care system favorably? The answer seems to be yes. This may also explain why the level of public dissatisfaction has been increasing in recent years, as more of us are finding that our health insurance plan requires us to rely on an organization of rotating doctors rather than a particular doctor of our own choosing.

A second observation is that, since Americans do not believe they are responsible for the problems affecting the health care system, they are not willing to make any personal sacrifices. In their view, if anyone should make sacrifices, it is other people, especially those who are getting free health care that they have not worked for and are using inappropriately anyway. Does this observation sound dated or pretty current? If more Americans felt that they might not be able to get health care when they need it, would that change how we all feel about how the health care system should be changed? Apparently so, and we will get back to this question shortly.

Given that the health care delivery system changed a great deal since those surveys were conducted, you might expect that the changes were made in response to the points of public dissatisfaction and that the public would respond accordingly, registering greater satisfaction. When some of the same researchers looked at public opinion regarding health care arrangements again ten years later, they found twenty-eight national surveys conducted between 1990 and 1993 to examine.[6] The survey results indicated that Americans were even more concerned about health care issues than they had been previously. Over the same period, it seems that the public's and the policy experts' respective assessments of the problem were becoming more divergent. The experts were becoming increasingly more troubled about spending too much on health care, while the public kept insisting that we, as a nation, were spending too little, even though the public indicated that health care was becoming too costly for individuals. In 1993, we were spending about 13.9 percent of the gross domestic product (GDP) on all goods and services related to health.... Considering the difference in how the problem was defined by the public in contrast to policy analysts, it is not surprising that the two ended up favoring very different solutions. While the experts were devising policies that would require some sacrifice on the part of the public, the public was leaning toward price controls on the rates that doctors, hospitals, and insurance companies could charge. One would think this suggests that the public wanted a much greater government role. It is interesting to find, therefore, that the public did not go so far as to accept a "ceiling" (a firm budgetary limit) on expenditures set by the government, fearing that such a ceiling could lead to rationing....

SATISFACTION AND DISSATISFACTION FACTORS

Researchers who have been following changes in public opinion recently conducted another international survey. We can again compare our level of satisfaction to that being registered by people in other countries at the end of a decade that saw so much attention devoted to health care system arrangements in the United States. In this more recent survey researchers surveyed people in five highly industrialized, English-speaking countries. (There are no comparative data on the other countries included in the 1990 survey.) The results indicate that we are no longer at the extreme end of the dissatisfaction scale....

Can this be interpreted as a significant decrease in dissatisfaction among Americans (you notice that I am not prepared to call it a "rise in satisfaction")? That is, of course, a difficult question to answer with any certainty. The researchers who spent a great deal of time reflecting on it say that the level of anxiety about being able to obtain health care when one needs it declined in the United States. However, we don't know what accounts for this.

Could it be that our economy has been so strong during the decade of the 1990s that Americans have a greater sense of confidence in their individual ability to pay for their health care? Or, is it that people are becoming more comfortable with the changes that have taken place in our health care delivery system and are truly more pleased with their own health care arrangements? Or, is it just that people have just become more resigned to the way things are and are satisfied with having any kind of health care coverage because they know that many other people are losing theirs? Or, is it that people are distracted from worrying about health care issues because they are worrying more about other things, for example, worrying about their own job security right in the middle of this period of national prosperity? There is no strong evidence to support any particular interpretation, and so the debate continues.

WHERE DO WE GO FROM HERE?

Clearly Americans are far from being completely satisfied with current health care arrangements. Is it possible to determine what alterations would increase our satisfaction? If we were to put together a list of possible changes we want to see implemented, a good place to begin would be identifying the objectives we hold forth for our health care system. These are the objectives that virtually all policymakers, no matter where they stand regarding solutions, see as the three guiding principles that have shaped the structure of our health care arrangements for at least the past five decades. They are not formally documented and written down anywhere, for good reason—we don't have "A System." Nevertheless, here they are: (1) attaining the highest *quality* care, (2) ensuring *access* to care, and (3) *cost containment*. Without a doubt these are great goals but they are impossible to define concretely and even more difficult to measure. What does quality mean? Who should define it—doctors, patients, or some group of experts authorized (by whom?) to monitor quality? How much access? Access for everyone? Assurance that we can have all the health care services we want? Or, that we can pay for? How do we measure cost containment? Do we want costs to drop or just not go up so fast? Are we prepared to cut anything?

While we can all argue about the answers to these questions, it is also abundantly clear that some people (the "haves"?) have had a greater impact on the design of the system than others. What we can say for certain is that this social institution, like all the others, is the product of human effort. As is always the case, the process of constructing it and rehabilitating it has not been particularly neat and orderly, so it is not always easy to identify the contribution made by any particular set of participants. . . .

NOTES

1. Victor Fuchs, "No Pain, No Gain," *Journal of the American Medical Association* 269 (Feb. 3, 1993) 631–633.
2. Karen Ignani, "Covering a Breaking Revolution: The Media and Managed Care," *Health Affairs* 17 (Jan./Feb. 1998): 26–34.
3. Mollyann Brodie, Lee Ann Brady, and Drew Altman, "Media Coverage of Managed Care: Is There a Negative Bias?" *Health Affairs* 17 (Jan./Feb. 1998): 9–25.
4. Robert Blendon, Robert Leitman, Ian Morrison, and Karen Donelan, "Satisfaction with Health Systems in Ten Nations," *Health Affairs* (Summer 1990): 185–192.
5. Robert Blendon and Drew Altman, "Public Attitudes about Health-Care Costs," *New England Journal of Medicine* 311 (Aug. 30, 1984): 613–616.
6. Robert Blendon, Tracey Stelzer Hyams, and John Benson, "Bridging the Gap Between Expert and Public Views on Health Care Reform," *Journal of the American Medical Association* 269 (May 19, 1993): 2573–2578.

KEY CONCEPTS

health maintenance organization (HMO)	life expectancy	managed care

DISCUSSION QUESTIONS

1. What are common complaints about the American health care system? What complaints have you or someone in your family had about health care?

2. How are *access* and *cost* of care different? How are they related?

63

Uninsured, Exposed and at Risk—But Not Powerless

LOURDES A. RIVERA

In this article, the author summarizes the recent statistics about the number of uninsured Americans. Women are particularly affected by this problem, and the current trends in access to health care show no signs of improvement. Low-income women, women of color, single mothers, and immigrant women all face inadequate health care and little opportunity to obtain the insurance needed to afford adequate health care. The article ends with information about what local, state, and federal agencies can do to help alleviate this national crisis.

Holes riddle the patchwork quilt that is the American health care system, leaving nearly 17 million women exposed to the dangers of being uninsured.[1] These dangers include a lower quality of care, preventable serious health problems, economic hardship and even death caused by their inability to access desperately needed health services.

Because they usually can't afford to pay for health care, uninsured women are less likely to access preventive care, such as mammograms, Pap tests and cervical screenings. They are more likely to postpone care and not fill prescriptions.[2] When they do seek out medical attention—often only when the condition has become very serious—it is likely to be of lower quality than the care delivered to insured women, more expensive and with worse health outcomes.[3,4] Some 18,000 uninsured American adults die each year because they can't get appropriate health care.[5,6] Being uninsured also creates financial insecurity for individuals and families who resort to borrowing money from family or friends or getting loans or mortgages to pay for their health care expenses. Some families also have had to declare bankruptcy, jeopardizing their future credit rating.[7,8]

Medicaid fills the insurance gap for nearly 16 million poor, non-elderly women. Millions more, however, do not meet the program's restrictive eligibility

SOURCE: Lourdes A. Rivera. 2004. "Uninsured, Exposed and at Risk—But Not Powerless." A prior version of this article was published in *Women's Health Activist*, the National Women's Health Network newsletter, September/October.

criteria, which also include categorical requirements such as being pregnant, having children or having a disability.[9] Also, while Medicaid is the largest public payer of family planning services, coverage for abortion is limited.[10] Under the Hyde Amendment, federal funds may be used to pay for abortions only in the case of rape, incest or to save the life of the mother.[11] Just 17 states use their own funds to provide medically necessary abortions; four do so voluntarily, and 13 states do so under court orders enforcing the state constitution.[12]

Among the 17 million women who neither qualify for Medicaid not have private health insurance, women of color, low-income women, single mothers and immigrant women are especially vulnerable.

INCOME STATUS

Low-income individuals and families are most likely to be uninsured. Almost a third of women ages 18 and older have family incomes that qualify as low income (below 200% of the federal poverty level).[13,14] Close to 40 percent of women who are poor (with family incomes below 100% of the federal poverty level) and almost 30 percent of near-poor women (100–199% of poverty) are uninsured.[15] Poor women are six times more likely to be uninsured and 3.5 times less likely to have employer-based health insurance than higher income women (with family income at or over 300% of poverty).[16] Low-income women also tend to be younger, less educated, less able to access resources and more likely to have childrearing responsibilities than more affluent women.[17] All of these factors present obstacles to accessing health care.

Low-income women in rural areas are disproportionately uninsured. One in five uninsured people lives in a rural area, and more than 70 percent of uninsured people in states like Montana and Maine are from rural areas.[18] Residents in remote rural areas have much higher uninsurance rates because they are far more likely to have low-wage jobs than urban residents, and are more likely to work in small businesses. Rural residents also have much less access to health care resources in their communities and few means of public transportation to reach service sites outside of their areas.[19]

Even more low-income women and families would be uninsured if it were not for Medicaid and SCHIP (State Children's Health Insurance Program). Medicaid expansions and the implementation of SCHIP between 1997 and 2002 significantly offset the drop in employer-sponsored health insurance during that same period, especially benefitting children. For children from higher-income families and for adults, the public expansions only somewhat offset the drop in private health insurance.[20,21] For adult women, for example, Medicaid covers only one-third of the lowest-income women, leaving most with few resources with no health coverage.[22]

EMPLOYMENT STATUS

Most Americans are insured through their employment, and women, as is well documented, tend to have jobs that pay less and offer fewer benefits. Simply being employed outside the home is no guarantee of having health insurance. Only 39 percent of working-age women have health coverage through their jobs, compared to 51 percent of men.[23] Women are less likely to be eligible for employer health plans because they are more likely to work part-time and to have lower incomes.[24] Even workers who are eligible for employer-sponsored health coverage can't always afford their portion of the premium. Overall, of the 45.5 million people who were uninsured in 2004, over 8 in 10 were in working families. The average employee's share of family premium for an employer-sponsored group plan increased by about $1,000 between 2000 and 2005, bringing the employee's family premium cost to an average of $2,713.[25]

Another factor behind the disparity between men and women is that women are more likely to be insured through their spouses' jobs. Twenty-five percent of women have job-based insurance through dependent coverage, compared to 13 percent of men.[26] This makes women more vulnerable to losing their coverage when they divorce or become widowed or if their spouse's employer drops dependent coverage. . . .

RACE AND ETHNICITY

Latinos and African Americans are much less likely to be insured than whites, and Latinas and African-American women have the highest rates of uninsurance of all. Thirty-eight percent of Latinas are uninsured, compared to 17 percent of African-American women and 13 percent of white women.[27] Seventy percent of white women have job-based health insurance, compared to 59 percent of African-American women and 39 percent of Latinas.[28]

Similar patterns extend to men and children. Approximately 35 percent of employed Latino adults are uninsured, compared to 18 percent of employed African Americans and 11 percent of working white adults.[29] White children are more likely to have access to private health insurance, while African-American and Latino children are more likely to be covered by the public programs Medicaid and SCHIP or to be uninsured.[30,31]

SINGLE PARENTS

Women are more likely to be single parents than men, limiting their ability to access job-based insurance. A California study found that 81 percent of the state's 1.4 million single parents in 1998 and 1999 were women.[32] That study showed single mothers in California to have higher uninsured rates (28 percent) than women in every other family situation except single women with no children. Only 19 percent of married mothers were uninsured.

IMMIGRATION STATUS

Women who are non-citizens, the majority of them in the United States legally, are disproportionately uninsured. They are as likely as citizens to have a full-time worker in their families, but their jobs tend to be low paying and not to offer health insurance. Thus in 2002, between 42 and 51 percent of non-citizens were uninsured, compared to 15 percent of U.S. citizens.[33]

Non-citizens also are less likely to qualify for Medicaid or SCHIP, especially after the passage of the Personal Responsibility and Work Opportunity Act of 1996 (the "welfare reform" law), which placed a five-year ban on public benefits for new immigrants.[34] In 2002, only 14 percent of low-income non-citizens residing in the U.S. for less than six years and 16 percent residing in the U.S. for six or more years received Medicaid or SCHIP coverage, compared to almost a third of low-income citizens.[35]

AFFORDABLE ACCESS FOR ALL

Even women who have health insurance often must pay out of pocket for many health needs. Insurance plans routinely exclude contraceptive drugs and devices, although they generally cover prescription drugs. Some states have responded to this inequity. According to the Alan Guttmacher Institute, twenty-three states, for instance, require health insurers that cover prescription drugs to also cover contraceptives and related services. But there are exemptions within these and other mandates. Fourteen states, for instance, exempt religious employers from including contraceptive coverage in their plans.[36] Some states, like California and New York, allow exemptions for churches and other narrowly-defined religious organizations, but requiring other employers, like religiously-sponsored hospitals that hire employees with varying beliefs and serving the general public, to provide contraceptive coverage.

Without a national response, millions of women, men and children will continue to lack health insurance and needlessly suffer poor health outcomes. Medicaid and SCHIP are the closest programs we have to national health insurance for low-income people. While the need and demand for these and other programs are increasing, they are under attack. Some states responding to their own budget crises already have imposed aggressive cost-cutting measures by freezing enrollment, rolling back eligibility and increasing out-of-pocket costs for those in the program. . . .

Moreover, in 2005 Congress passed the Deficit Reduction Act in which Medicaid is cut by $6.9 billion over the next five years and $28.3 billion over the next 10 years.[37] These cuts, which begin in fiscal year 2006, will be achieved partly by cutting benefits, including potentially family planning, imposing higher out-of-pocket costs, and making it more difficult for certain low-income citizens to apply for the program by requiring them to show a passport or birth certificate. The Bush Administration also proposed an additional $14 billion in cuts over the next five years, and $35 billion over the next 10 years in its fiscal year 2007 proposed budget.[38] The budget proposal continues to severely underfund the Title X Family Planning Block Grant, which provides access to comprehensive family planning services and reproductive health counseling, requesting increased funding for ineffective "abstinence-only" programs.

These policies fail to address comprehensive community health needs. A system that guarantees affordable access to quality health care, including comprehensive reproductive health care, is long overdue.

WHAT CAN WOMEN DO?

Current policy debates provide an opportunity to focus national attention on the lack of health access for so many women, their families and communities. Concerned women can take the important steps of calling and writing their elected state and Congressional representatives to voice their concerns about the dismantling of the health care safety-net, the need to restore funding for public programs, and the increasing number of people who are increasingly "sick and in debt" because of their medical bills.[39] Regardless of who is elected to Congress or the state houses, women can demand that elected officials preserve public programs that provide critical lifelines for so many people, and develop a system that extends comprehensive health services to all.

As part of their advocacy, women should learn more about the international human rights framework of the "right to health" which is inclusive of comprehensive reproductive health and demands a proactive role of government to protect, respect, and fulfill these rights.[40] The right to health also envisions services that are available, accessible, culturally and ethically appropriate, and of good quality.[41] By embracing this framework, we can achieve an equitable health care system in the U.S. that meets the needs of all communities.

NOTES

1. Henry J. Kaiser Family Foundation Fact Sheet. *Women's Health Insurance Coverage.* Nov. 2004.
2. Ibid.
3. Institute of Medicine. *Care Without Coverage: Too Little, Too Late.* May 2002.
4. Henry J. Kaiser Family Foundation. *The Cost of Not Covering the Uninsured.* June 2003.
5. Institute of Medicine. *Care Without Coverage: Too Little, Too Late.* May 2002.

6. Institute of Medicine. *Insuring America's Health: Principles and Recommendations*. Jan. 2004.

7. Duchon, Lisa, et al., "Commonwealth Fund." *Security Matters: How Instability in Health Insurance Puts U.S. Workers at Risk*. Dec. 2001.

8. Institute of Medicine. *Health Insurance is a Family Matter*. Sep. 2002.

9. Henry J. Kaiser Family Foundation Fact Sheet. *Women's Health Insurance Coverage*. Nov. 2004.

10. For an overview of Medicaid reproductive coverage, see National Health Law Program, *Medicaid Coverage of Reproductive Health Services*. June 2001. www.health-law.org/pubs/20010625reprofactsheet.html.

11. Ibid.

12. Alan Guttmacher Institute, *State Policies in Brief: State Funding of Abortion Under Medicaid*. March 1, 2006.

13. Henry J. Kaiser Foundation. *Women and Health Care: A National Profile*. July 2005.

14. The 2006 federal poverty level for a family of four is $20,000 per year. U.S. Dept. Health & Human Services. *2006 Federal Poverty Guidelines*, available at http://aspe.hhs.gov/poverty/06poverty.shtml.

15. Henry J. Kaiser Foundation. *Women and Health Care: A National Profile*. July 2005.

16. Ibid.

17. Ibid.

18. Henry J. Kaiser Family Foundation. *The Uninsured in Rural America*. April 2003.

19. Ibid.

20. Finegold K, Wherry I. "Race, Ethnicity and Health." *The Urban Institute; Snapshots of America's Families III*, No. 20 March 2004.

21. Holohan J. & Wang M. "Changes in health insurance coverage during the economic downturn 2000–2002." *Health Affairs*, January 28, 2004 (Web exclusive)

22. Henry J. Kaiser Foundation. *Women and Health Care: A National Profile*. July 2005.

23. Henry J. Kaiser Family Foundation Fact Sheet. *Women's Health Insurance Coverage*. Nov. 2004.

24. Ibid.

25. Henry J. Kaiser Family Foundation. *The Uninsured: A Primer*. Jan. 2006.

26. Henry J. Kaiser Family Foundation Fact Sheet. *Women's Health Insurance Coverage*. Nov. 2004.

27. Ibid.

28. Ibid.

29. Robert Wood Johnson Foundation, "Study Shows One in Three Working Hispanics Have No Health Coverage, Suffer Health Gaps as a Result" (press release). May 10, 2004.

30. Finegold K, Wherry I. "Race, Ethnicity and Health." *The Urban Institute; Snapshots of America's Families III*, No. 20 March 2004.

31. Holohan J. & Wang M. "Changes in health insurance coverage during the economic downturn 2000–2002". *Health Affairs* January 28, 2004 (Web exclusive)

32. Wyn R. & Ojeda V. *Single Mothers in California, Understanding Their Health Insurance Coverage.* UCLA Center for Health Policy Research, Policy Brief. May 2002.

33. Henry J. Kaiser Family Foundation & National Council of La Raza. *Immigrants and Health Coverage: A Primer.* June 2004.

34. For an overview of immigrant health access rules, see The Access Project & National Health Law Program. *Immigrant Access to Health Benefits: A Resource Manual.* Aug. 2002. For example, undocumented immigrants, if they meet the income and category requirements, can receive coverage of Medicaid for emergency services. Also, some states are using their funds to cover some services for otherwise unqualified immigrants (e.g. for unqualified immigrant pregnant women).

35. Henry J. Kaiser Family Foundation & National Council of La Raza. *Immigrants and Health Coverage: A Primer.* June 2004.

36. Alan Guttmacher Institute. *State Policies in Brief: Insurance Coverage of Contraceptives.* March 2006.

37. Congressional Budget Office. *Cost Estimate: S. 1932 Deficit Reduction Act of 2005.* Jan. 27, 2006.

38. Henry J. Kaiser Family Foundation. *The President's FY 2007 Budget Proposal: Overview and Briefing Charts.* Undated.

39. See, e.g., Health Consumer Alliance. *Sick and In Debt: Improper Practices that Cause Medical Debt for Low-Income Californians.* Sept. 2004.

40. Cook, Rebecca J., et al., *Reproductive Health and Human Rights.* Clarendon Press, 2003

41. Ibid.

KEY CONCEPTS

immigration   Medicaid   Medicare

DISCUSSION QUESTIONS

1. What factors contribute to the likelihood of having health insurance in America? Who, then, is left uninsured?

2. Why is the high number of uninsured Americans a problem? What impact does this have on society?

64

Beauty Myths and Realities and Their Impact on Women's Health

JANE SPRAGUE ZONES

The "beauty myth" defines a narrow range of social ideals for women's appearance. As Zones shows here, trying to attain such an ideal has consequences for women's physical and mental health.

Many women concur that personal beauty, or "looking good," is fostered from a very early age. It is probably true that the ways in which people assess physical beauty are not naturally determined but socially and culturally learned and therefore "in the eye of the beholder." However, we tend to discount the depth of our *common* perception of beauty, mistakenly assuming that individuals largely set their own standards. At any period in history, within a given geographic and cultural territory, there are relatively uniform and widely understood models of how women "should" look. . . .

The preoccupation with appearance serves to control and contain women's ambitions and motivations to gain power in larger political contexts. To the degree that many females feel they must dedicate time, attention, and resources to maintaining and improving their looks, they neglect activities to improve social conditions for themselves or others. Conversely, as women become increasingly visible as powerful individuals in shaping events, their looks become targeted for irrelevant scrutiny and criticism in ways with which men in similar positions are not forced to contend. . . .

BEAUTY MYTHS AND THE EROSION OF SELF-WORTH

Perhaps the biggest toll the "beauty myth" takes is in terms of women's identity and self-esteem. Like members of other oppressed groups of which we may also be part, women internalize cultural stereotypes and expectations, perpetuating

SOURCE: Zones, Jane Sprague. 1997. "Beauty Myths and Realities and Their Impact on Women's Health." In *Women's Health, Complexities and Differences*, edited by Sheryl Burt Ruzek, Virginia L. Oleson, and Adele E. Clark, 249–75. Columbus, OH: The Ohio State University. Reprinted with permission.

them by enforced acceptance and agreement. For women, this is intensified by the interaction of irrational social responses to physical appearance not only with gender but with other statuses as well—race, class, age, disability, and the like. Continuous questioning of the adequacy of one's looks drains attention from more worthwhile and confidence-building pursuits. . . .

QUANTIFYING BEAUTY: CONVENTIONALITY AND COMPUTER ENHANCEMENT

The predominant, nearly universal standard for beauty in American society is to be slender, young, upper-class, and white without noticeable physical imperfections or disabilities. To the extent that a woman's racial or ethnic heritage, class background, age, or other social and physical characteristics do not conform to this ideal, assaults on opportunities and esteem increase. Physical appearance is at the core of racism and most other social oppressions, because it is generally what is used to classify individuals. . . .

BEAUTY AND THE CHALLENGE OF SOCIAL DIVERSITY

Although significant beauty ideals appear to transcend cultural subgroup boundaries, appearance standards do vary by reference group. Clothing preferred by adolescents, for example, which experiences quick fashion turnover, is considered inappropriate for older people. Body piercing, a current style for young white people in urban areas of the United States, is repellent to most older adults and some ethnic minorities in the same age group. Religious and political ideologies are often identified through appearance. Islamic fundamentalist women wear clothing that covers body and face, an expression of religious sequestering; Amish women wear conservative clothing and distinctive caps; orthodox Jewish women wear wigs or cover their hair; African American women for many years wore natural hairdos to show racial pride; and Native American women may wear tribal jewelry and distinctive clothes that indicate their respect for heritage. In recent years, the disability rights movement has encouraged personal visibility to accompany the tearing down of barriers to access, resulting in a greater variety of appliances (including elegant streamlined wheelchairs) and functional clothing.

Although there are varying and conflicting standards of good looks and appropriate appearance that are held simultaneously by social subgroups, the dominant ideals prevail and are legitimated most thoroughly in popular culture. . . .

One major way that dominant social forces have dealt with those who diverge is to remove these expressions from view—through ghettoization, anti-immigration policies, special education programs, retirement policies, and so on.

The ultimate social insult is to render the oppressed invisible. Social barriers to visibility are expressed as well in pressures to avoid drawing attention to oneself. Those features that render us "different" are frequently the objects of harassment or unwanted attention. We learn to appear invisible. In the following sections, the gender effects of appearance in combination with other social statuses are described through personal accounts and social research. . . .

THE COMMERCIAL IMPERATIVE IN THE QUEST FOR BEAUTY

. . . As new standards of beauty expectations are created, physical appearance becomes increasingly significant, and as the expression of alternative looks are legitimized, new products are developed and existing enterprises capitalize on the trends. Liposuction, developed relatively recently, has become the most popular of the cosmetic surgery techniques. Synthetic fats have been developed, and there is now a cream claimed to reduce thigh measurements.

Weight Loss

Regardless of the actual size of their bodies, more than half of American females between ages ten and thirty are dieting, and one out of every six college women is struggling with anorexia and bulimia (Iazzetto 1992). The quest to lose weight is not limited to white, middle-class women. Iazzetto cites studies that find this pervasive concern in black women, Native American girls (75 percent trying to lose weight), and high school students (63 percent dieting). However, there may be differences among adolescent women in different groups as to how rigid their concepts of beauty are and how flexible they are regarding body image and dieting (Parker et al. 1995). Studies of primary school girls show more than half of all young girls and close to 80 percent of ten- and eleven-year-olds on diets because they consider themselves "too fat" (Greenwood 1990; Seid 1989). Analyses of the origins, symbolic meanings, and impact of our culture's obsession with thinness (Chernin 1981; Freedman 1989; Iazzetto 1988; Seid 1989) occupy much of the body-image literature.

Concern about weight and routine dieting are so pervasive in the United States that the weight-loss industry grosses more than $33 billion each year. Over 80 percent of those in diet programs are women. These programs keep growing even in the face of 90 to 95 percent failure rates in providing and maintaining significant weight loss. Congressional hearings in the early 1990s presented evidence of fraud and high failure rates in the weight-loss industry, as well as indications of severe health consequences for rapid weight loss (Iazzetto 1992). The Food and Drug Administration (FDA) has reviewed documents submitted by major weight-loss programs and found evidence of safety and efficacy to be insufficient and unscientific. An expert panel urged consumers to consider program effectiveness in choosing a weight-loss method but acknowledged lack of scientific data for making informed decisions (Brody 1992).

Fitness

Whereas in the nineteenth century some physicians recommended sedentary lifestyle to preserve feminine beauty, in the past two decades of the twentieth century, interest in physical fitness has grown enormously. Nowhere is this change more apparent than in the gross receipts of some of the major fitness industries. In 1987, health clubs grossed $5 billion, exercise equipment $738 million (up from $5 million ten years earlier), diet foods $74 billion, and vitamin products $2.7 billion (Brand 1988). Glassner (1989) identifies several reasons for this surge of interest in fitness, including the aging of the "baby boom" cohort with its attendant desire to allay the effects of aging through exercise and diet, and the institution of "wellness" programs by corporations to reduce insurance, absentee, and inefficiency costs. A patina of health, well-toned but skinny robustness, has been folded into the dominant beauty ideal.

Clothing and Fashion

For most of us, first attempts to accomplish normative attractiveness included choosing clothing that enhanced our self-image. The oppressive effects of corsets, clothing that interfered with movement, tight shoes with high heels, and the like have been well documented (Banner 1983). Clothing represents the greatest monetary investment that women make in their appearance. Sales for exercise clothing alone in 1987 (including leotards, bodysuits, warm-up suits, sweats, and shoes) totaled $2.5 billion (Schefer 1988). To bolster sales, fashion leaders introduce new and different looks at regular intervals, impelling women to invest in what is currently in vogue....

Cosmetics

The average person in North America uses more than twenty-five pounds of cosmetics, soaps, and toiletries each year (Decker 1983). The cosmetics industry produces over twenty thousand products containing thousands of chemicals, and it grosses over $20 billion annually (Becker 1991; Wolf 1991). Stock in cosmetics manufacturers has been rising 15 percent a year, in large part because of depressed petroleum prices. The oil derivative ethanol is the base for most products (Wolf 1991:82, 307). Profit margins for products are over 50 percent (McKnight 1989). Widespread false claims for cosmetics were virtually unchallenged for fifty years after the FDA became responsible for cosmetic industry oversight in 1938, and even now, the industry remains largely unregulated (Kaplan 1994). Various manufacturers assert that their goods can "retard aging," "repair the skin," or "restructure the cell." "Graphic evidence" of "visible improvement" when applying a "barrier" against "eroding effects" provides a pastiche of some familiar advertising catchphrases (Wolf 1991:109–10).

The FDA has no authority to require cosmetics firms to register their existence, to release their formulas, to report adverse reactions, or to show evidence of safety and effectiveness before marketing their products (Gilhooley 1978; Kaplan 1994). Authorizing and funding the FDA to regulate the cosmetics industry would allow some means of protecting consumers from the use of dangerous products.

Cosmetic Surgery

In interviews with cosmetic surgeons and users of their services, Dull and West (1991) found that the line between reconstructive plastic surgery (repair of deformities caused congenitally or by injury or disease) and aesthetic surgery has begun to blur. Doctors and their patients are viewing unimpaired features as defective and the desire to "correct" them as intrinsic to women's nature, rather than as a cultural imperative.

Because of an oversupply of plastic surgeons, the profession has made efforts to expand existing markets through advertising and by appeals to women of color. Articles encouraging "enhancement of ethnic beauty" have begun to appear, but they focus on westernizing Asian eyelids and chiseling African American noses. As Bordo (1993:25) points out, this technology serves to promote commonality rather than diversity.

Plastic surgery has been moving strongly in the direction of making appearance a bona fide medical problem. This has been played out dramatically in recent times in the controversy regarding silicone breast implants, which provides plastic surgeons with a substantial amount of income. Used for thirty years in hundreds of thousands of women (80 percent for cosmetic augmentation), the effects of breast implants have only recently begun to be studied to determine their health consequences over long periods (Zones 1992). . . .

HEALTH RISKS IN QUEST OF BEAUTY

Mental Health

For most women, not adhering to narrow, standardized appearance expectations causes insecurity and distraction, but for many, concerns about appearance can have serious emotional impact. Up until adolescence, boys and girls experience about the same rates of depression, but at around age twelve, girls' rates of depression begin to increase more rapidly. A study of over eight hundred high school students found that a prime factor in this disparity is girls' preoccupation with appearance. . . .

Physical Health

Perceived or actual variation from society's ideal takes a physical toll, too. High school and college-age females who were judged to be in the bottom half of their group in terms of attractiveness had significantly higher blood pressure than the young women in the top half. The relationship between appearance and blood pressure was not found for males in the same age group (Hansell, Sparacino, and Ronchi 1982). . . .

There are direct risks related to using commodities to alter appearance. According to the Consumer Products Safety Commission, more than 200,000 people visit emergency rooms each year as a result of cosmetics related health problems (Becker 1991). Clothing has its perils as well. In recent years, meralgia paresthetica, marked by sciatica, pain in the hip and thigh region, with tingling and itchy skin, has made an appearance among young women in the form of "skin-tight jean syndrome" (Gateless and Gilroy 1984). In earlier times, the same problems have arisen with the use of girdles, belts, and shoulder bags. . . .

Approximately 33 to 50 percent of all adult women have used hair coloring agents. Evidence over the past twenty-five years has shown that chemicals used in manufacturing hair dyes cause cancers in animals. Scientists at the National Cancer Institute (NCI) recently reported a significantly greater risk of cancers of the lymph system and of a form of cancer affecting bone marrow, multiple myeloma, in women who use hair coloring (Zahm et al. 1992). . . .

Because no cosmetic products require follow-up research for safety and effectiveness, virtually anything can be placed on the market without regard to potential health effects. Even devices implanted in the body, which were not regulated before 1978, can remain on the market for years without appropriate testing. During the decade of controversy over regulating silicone breast implants, the American Society of Plastic and Reconstructive Surgeons vehemently denied any need for controlled studies of the implant in terms of long-term safety. The society spent hundreds of thousands of dollars of its members' money in a public relations effort to avoid the imposition of requirements for such research to the detriment of investing in the expensive scientific follow-up needed (Zones 1992). . . .

Health consequences of beauty products extend beyond their impact on individuals. According to the San Francisco Bay Area Air Quality Management District, aerosols release 25 tons of pollution everyday. Almost half of that is from hairsprays. Although aerosols no longer use chlorofluorocarbons (CFCs), which are the greatest cause of depletion of the upper atmosphere ozone layer, aerosol hydrocarbons in hair sprays are a primary contributor to smog and ground pollution.

THE BEAUTY OF DIVERSITY

Both personal transformation and policy intervention will be necessary to allow women to present themselves freely. Governmental institutions, including courts and regulatory agencies, need to accord personal and product liability related to appearance products and services the attention they require to ensure public health and safety. The legal system must develop well-defined case law to assist the court in determining inequitable treatment based on appearance discrimination. . . .

The personal solution to individual self-doubt or even self-loathing of our physical being is to continuously make the decision to contradict the innumerable messages we are given that we are anything less than lovely as human beings. . . .

REFERENCES

Banner, Lois W. 1983. *American Beauty*. Chicago: University of Chicago Press.

Becker, Hilton. 1991. "Cosmetics: saving face at what price?" *Annals of Plastic Surgery* 26:171–73.

Bordo, Susan. 1993. *Unbearable Weight: Feminism, Western Culture and the Body*. Berkeley: University of California Press.

Brand, David. 1988. "A nation of health worrywarts?" *Time*, 25 July, 66.

Brody, Jane E. 1992. "Panel criticizes weight-loss programs." *New York Times*, 2 April, A10.

Chernin, Kim. 1981. *The Obsession: Reflections on the Tyranny of Slenderness*. New York: Harper Colophon Books.

Decker, Ruth. 1983. "The not-so-pretty risks of cosmetics." *Medical Self-Care* (Summer):25–31.

Dull, Diana, and Candace West. 1991. "Accounting for cosmetic surgery: the accomplishment of gender." *Social Problems* 38:54–70.

Freedman, Rita. 1986. *Beauty Bound*. Lexington, MA: Lexington Books.

———. 1989. Bodylove. New York: Harper and Row.

Gateless, Doreen, and John Gilroy. 1984. "Tight-jeans meralgia: hot or cold?" *Journal of the American Medical Association* 252:42–43.

Gilhooley, Margaret. 1978. "Federal regulation of cosmetics: an overview." *Food Drug Cosmetic Law Journal* 33:231–38.

Glassner, Barry. 1989. "Fitness and the postmodern self." *Journal of Health and Social Behavior* 30:180–91.

Greenwood, M. R. C. 1990. "The feminine ideal: a new perspective." *UC Davis Magazine* (July):8–11.

Hansell, Stephen, J. Sparacino, and D. Ronchi. 1982. "Physical attractiveness and blood pressure: sex and age differences." *Personality and Social Psychology Bulletin* 8:113–21.

Iazzetto, Demetria. 1988. "Women and body image: reflections in the fun house mirror." Pp. 34–53 in Carol J. Leppa and Connie Miller (eds.), *Women's Health Perspectives: An Annual Review*, Volcano, CA: Volcano Press.

———. 1992. "What's happening with women and body image?" *National Women's Health Network* News:1, 6, 7.

Kaplan, Sheila. 1994. "The ugly face of the cosmetics lobby." *Ms.* (Jan.–Feb.):88–89.

McKnight, Gerald. 1989. *The Skin Game: The International Beauty Business Brutally Exposed*. London: Sidgwick and Jackson.

Parker, Sheila, Mimi Nichter, Mark Nichter, Nancy Vuckovic, Colette Sims and Cheryl Ritenbaugh. 1995. "Body image and weight concerns among African Americans and white adolescent females: differences that make a difference." *Human Organization* 54(2):103–13.

Schefer, Dorothy. 1988. "Beauty: The real cost of looking good." *Vogue* (Nov.): 157–68.

Seid, Roberta Pollack. 1989. *Never Too Thin: Why Women Are at War with Their Bodies.* New York: Prentice-Hall.

Wolf, Naomi. 1991. *The Beauty Myth: How Images of Beauty Are Used against Women.* New York: William Morrow.

Zahm, Sheila Hoar, Dennis D. Weisenburger, Paula A. Babbitt, et al. 1992. "Use of hair coloring products and the risk of lymphoma, multiple myeloma, and chronic lymphocytic leukemia." *American Journal of Public Health* 82:990–97.

Zones, Jane Sprague. 1992. "The political and social context of silicone breast implant use in the United States." *Journal of Long-Term Effects of Medical Implants* 1:225–41.

KEY CONCEPTS

anorexia nervosa	bulimia	eating disorder

DISCUSSION QUESTIONS

1. What health risks does Zones identify for women who are overly concerned with the beauty ideals established for them? What alternatives are there?

2. Much of the discussion of beauty and health has focused on women. How do ideals regarding appearance affect men's health?

Applying Sociological Knowledge:
An Exercise for Students

Choose one of the institutions discussed in Part XII (family, religion, education, work, government, or health care). Go to www.democrats.org and www.gop.com and look at each political party's most recent platform. Search each document for the institution you have chosen. For example, you can enter the word "family" or "religion" in the search box and find the information about how the Democratic party plans to address problems with this social institution or to improve circumstances for this social institution.

Compare and contrast the two opposing political viewpoints.

1.   What does each party believe is the most important issue facing the institution?
2.   What does each party plan to do to improve circumstances for each institution?
3.   Which platform makes the most sense to you? Which do you agree with most?
4.   Would this information influence your voting behavior in the next election?

65

American Apartheid

DOUGLAS S. MASSEY AND NANCY A. DENTON

Douglas S. Massey and Nancy A. Denton argue that segregation, particularly residential segregation, is a fundamental dimension of race relations in the United States and is all too often ignored by policymakers and even scholars. It is a major cause of many of the ills of race relations in this country. They argue that it is the "missing link" in past attempts to understand the urban poor.

It is quite simple. As soon as there is a group area then all your uncertainties are removed and that is, after all, the primary purpose of this Bill [requiring racial segregation in housing].

Minister of the Interior,
Union of South Africa legislative debate on the Group Areas
Act of 1950

During the 1970s and 1980s a word disappeared from the American vocabulary. It was not in the speeches of politicians decrying the multiple ills besetting American cities. It was not spoken by government officials responsible for administering the nation's social programs. It was not mentioned by journalists reporting on the rising tide of homelessness, drugs, and violence in urban America. It was not discussed by foundation executives and think-tank experts proposing new programs for unemployed parents and unwed mothers. It was not articulated by civil rights leaders speaking out against the persistence of racial inequality; and it was nowhere to be found in the thousands of pages written by social scientists on the urban underclass. The word was segregation.

Most Americans vaguely realize that urban America is still a residentially segregated society, but few appreciate the depth of black segregation or the degree to which it is maintained by ongoing institutional arrangements and contemporary individual actions. They view segregation as an unfortunate holdover from a racist past, one that is fading progressively over time. If racial residential segregation persists, they reason, it is only because civil rights laws

SOURCE: Reprinted by permission of the publisher from *American Apartheid: Segregation and The Making of the Underclass* by Douglas S. Massey and Nancy A. Denton, pp. 1–7. Cambridge, Mass.: Harvard University Press. Copyright © 1993 by the President and Fellows of Harvard College.

passed during the 1960s have not had enough time to work or because many blacks still prefer to live in black neighborhoods. The residential segregation of blacks is viewed charitably as a "natural" outcome of impersonal social and economic forces, the same forces that produced Italian and Polish neighborhoods in the past and that yield Mexican and Korean areas today.

But black segregation is not comparable to the limited and transient segregation experienced by other racial and ethnic groups, now or in the past. No group in the history of the United States has ever experienced the sustained high level of residential segregation that has been imposed on blacks in large American cities for the past fifty years. This extreme racial isolation did not just happen; it was manufactured by whites through a series of self-conscious actions and purposeful institutional arrangements that continue today. Not only is the depth of black segregation unprecedented and utterly unique compared with that of other groups, but it shows little sign of change with the passage of time or improvements in socioeconomic status.

If policymakers, scholars, and the public have been reluctant to acknowledge segregation's persistence, they have likewise been blind to its consequences for American blacks. Residential segregation is not a neutral fact; it systematically undermines the social and economic well-being of blacks in the United States. Because of racial segregation, a significant share of black America is condemned to experience a social environment where poverty and joblessness are the norm, where a majority of children are born out of wedlock, where most families are on welfare, where educational failure prevails, and where social and physical deterioration abound. Through prolonged exposure to such an environment, black chances for social and economic success are drastically reduced.

Deleterious neighborhood conditions are built into the structure of the black community. They occur because segregation concentrates poverty to build a set of mutually reinforcing and self-feeding spirals of decline into black neighborhoods. When economic dislocations deprive a segregated group of employment and increase its rate of poverty, socioeconomic deprivation inevitably becomes more concentrated in neighborhoods where that group lives. The damaging social consequences that follow from increased poverty are spatially concentrated as well, creating uniquely disadvantaged environments that become progressively isolated—geographically, socially, and economically—from the rest of society.

The effect of segregation on black well-being is structural, not individual. Residential segregation lies beyond the ability of any individual to change; it constrains black life chances irrespective of personal traits, individual motivations, or private achievements. For the past twenty years this fundamental fact has been swept under the rug by policymakers, scholars, and theorists of the urban underclass. Segregation is the missing link in prior attempts to understand the plight of the urban poor. As long as blacks continue to be segregated in American cities, the United States cannot be called a race-blind society.

THE FORGOTTEN FACTOR

The present myopia regarding segregation is all the more startling because it once figured prominently in theories of racial inequality. Indeed, the ghetto was once seen as central to black subjugation in the United States. In 1944 Gunnar Myrdal wrote in *An American Dilemma* that residential segregation "is basic in a mechanical sense. It exerts its influence in an indirect and impersonal way: because Negro people do not live near white people, they cannot . . . associate with each other in the many activities founded on common neighborhood. Residential segregation . . . becomes reflected in uni-racial schools, hospitals, and other institutions" and creates "an artificial city . . . that permits any prejudice on the part of public officials to be freely vented on Negroes without hurting whites."

Kenneth B. Clark, who worked with Gunnar Myrdal as a student and later applied his research skills in the landmark *Brown v. Topeka* school integration case, placed residential segregation at the heart of the U.S. system of racial oppression. In *Dark Ghetto*, written in 1965, he argued that "the dark ghetto's invisible walls have been erected by the white society, by those who have power, both to confine those who have no power and to perpetuate their powerlessness. The dark ghettos are social, political, educational, and—above all—economic colonies. Their inhabitants are subject peoples, victims of the greed, cruelty, insensitivity, guilt, and fear of their masters."

Public recognition of segregation's role in perpetuating racial inequality was galvanized in the late 1960s by the riots that erupted in the nation's ghettos. In their aftermath, President Lyndon B. Johnson appointed a commission chaired by Governor Otto Kerner of Illinois to identify the causes of the violence and to propose policies to prevent its recurrence. The Kerner Commission released its report in March 1968 with the shocking admonition that the United States was "moving toward two societies, one black, one white—separate and unequal." Prominent among the causes that the commission identified for this growing racial inequality was residential segregation.

In stark, blunt language, the Kerner Commission informed white Americans that "discrimination and segregation have long permeated much of American life; they now threaten the future of every American." "Segregation and poverty have created in the racial ghetto a destructive environment totally unknown to most white Americans. What white Americans have never fully understood—but what the Negro can never forget—is that white society is deeply implicated in the ghetto. White institutions created it, white institutions maintain it, and white society condones it."

The report argued that to continue present policies was "to make permanent the division of our country into two societies; one, largely Negro and poor, located in the central cities; the other, predominantly white and affluent, located in the suburbs." Commission members rejected a strategy of ghetto enrichment coupled with abandonment of efforts to integrate, an approach they saw "as another way of choosing a permanently divided country." Rather, they insisted that the only reasonable choice for America was "a policy which combines ghetto enrichment with programs designed to encourage integration of substantial numbers of Negroes into the society outside the ghetto."

America chose differently. Following the passage of the Fair Housing Act in 1968, the problem of housing discrimination was declared solved, and residential segregation dropped off the national agenda. Civil rights leaders stopped pressing for the enforcement of open housing, political leaders increasingly debated employment and educational policies rather than housing integration, and academicians focused their theoretical scrutiny on everything from culture to family structure, to institutional racism, to federal welfare systems. Few people spoke of racial segregation as a problem or acknowledged its persisting consequences. By the end of the 1970s residential segregation became the forgotten factor in American race relations.

While public discourse on race and poverty became more acrimonious and more focused on divisive issues such as school busing, racial quotas, welfare, and affirmative action, conditions in the nation's ghettos steadily deteriorated. By the end of the 1970s, the image of poor minority families mired in an endless cycle of unemployment, unwed childbearing, illiteracy, and dependency had coalesced into a compelling and powerful concept: the urban underclass. In the view of many middle-class whites, inner cities had come to house a large population of poorly educated single mothers and jobless men—mostly black and Puerto Rican—who were unlikely to exit poverty and become self-sufficient. In the ensuing national debate on the causes for this persistent poverty, four theoretical explanations gradually emerged: culture, racism, economics, and welfare.

Cultural explanations for the underclass can be traced to the work of Oscar Lewis, who identified a "culture of poverty" that he felt promoted patterns of behavior inconsistent with socioeconomic advancement. According to Lewis, this culture originated in endemic unemployment and chronic social immobility, and provided an ideology that allowed poor people to cope with feelings of hopelessness and despair that arose because their chances for socioeconomic success were remote. In individuals, this culture was typified by a lack of impulse control, a strong present-time orientation, and little ability to defer gratification. Among families, it yielded an absence of childhood, an early initiation into sex, a prevalence of free marital unions, and a high incidence of abandonment of mothers and children.

Although Lewis explicitly connected the emergence of these cultural patterns to structural conditions in society, he argued that once the culture of poverty was established, it became an independent cause of persistent poverty. This idea was further elaborated in 1965 by the Harvard sociologist and then Assistant Secretary of Labor Daniel Patrick Moynihan, who in a confidential report to the President focused on the relationship between male unemployment, family instability, and the inter-generational transmission of poverty, a process he labeled a "tangle of pathology." He warned that because of the structural absence of employment in the ghetto, the black family was disintegrating in a way that threatened the fabric of community life.

When these ideas were transmitted through the press, both popular and scholarly, the connection between culture and economic structure was somehow lost, and the argument was popularly perceived to be that "people were poor because they had a defective culture." This position was later explicitly adopted by the conservative theorist Edward Banfield, who argued that lower-class culture—with its limited time horizon, impulsive need for gratification, and psychological self-doubt—was primarily responsible for persistent urban poverty. He believed that these cultural traits were largely imported, arising primarily because cities attracted lower-class migrants.

The culture-of-poverty argument was strongly criticized by liberal theorists as a self-serving ideology that "blamed the victim." In the ensuing wave of reaction, black families were viewed not as weak but, on the contrary, as resilient and well adapted survivors in an oppressive and racially prejudiced society. Black disadvantages were attributed not to a defective culture but to the persistence of institutional racism in the United States. According to theorists of the underclass such as Douglas Glasgow and Alphonso Pinkney, the black urban underclass came about because deeply imbedded racist practices within American institutions—particularly schools and the economy—effectively kept blacks poor and dependent.

As the debate on culture versus racism ground to a halt during the late 1970s, conservative theorists increasingly captured public attention by focusing on a third possible cause of poverty: government welfare policy. According to Charles Murray, the creation of the underclass was rooted in the liberal welfare state. Federal antipoverty programs altered the incentives governing the behavior of poor men and women, reducing the desirability of marriage, increasing the benefits of unwed childbearing, lowering the attractiveness of menial labor, and ultimately resulted in greater poverty.

A slightly different attack on the welfare state was launched by Lawrence Mead, who argued that it was not the generosity but the permissiveness of the U.S. welfare system that was at fault. Jobless men and unwed mothers should be required to display "good citizenship" before being supported by the state. By not requiring anything of the poor, Mead argued, the welfare state undermined their independence and competence, thereby perpetuating their poverty.

This conservative reasoning was subsequently attacked by liberal social scientists, led principally by the sociologist William Julius Wilson, who had long been arguing for the increasing importance of class over race in understanding the social and economic problems facing blacks. In his 1987 book *The Truly Disadvantaged*, Wilson argued that persistent urban poverty stemmed primarily from the structural transformation of the inner-city economy. The decline of manufacturing, the suburbanization of employment, and the rise of a low-wage service sector dramatically reduced the number of city jobs that paid wages sufficient to support a family, which led to high rates of joblessness among minorities and a shrinking pool of "marriageable" men (those financially able to support a family). Marriage thus became less attractive to poor women, unwed childbearing increased, and female-headed families proliferated. Blacks suffered disproportionately from these trends because, owing to past discrimination, they were concentrated in locations and occupations particularly affected by economic restructuring.

Wilson argued that these economic changes were accompanied by an increase in the spatial concentration of poverty within black neighborhoods. This new geography of poverty, he felt, was enabled by the civil rights revolution of the 1960s, which provided middle-class blacks with new opportunities outside the ghetto. The out-migration of middle-class families from ghetto areas left behind a destitute community lacking the institutions, resources, and values necessary for success in postindustrial society. The urban underclass thus arose from a complex interplay of civil rights policy, economic restructuring, and a historical legacy of discrimination.

Theoretical concepts such as the culture of poverty, institutional racism, welfare disincentives, and structural economic change have all been widely debated. None of these explanations, however, considers residential segregation to be an important contributing cause of urban poverty and the underclass. In their principal works, Murray and Mead do not mention segregation at all and Wilson refers to racial segregation only as a historical legacy from the past, not as an outcome that is institutionally supported and actively created today. Although Lewis mentions segregation sporadically in his writings, it is not assigned a central role in the set of structural factors responsible for the culture of poverty, and Banfield ignores it entirely. Glasgow, Pinkney, and other theorists of institutional racism mention the ghetto frequently, but generally call not for residential desegregation but for race-specific policies to combat the effects of discrimination in the schools and labor markets. In general, then, contemporary theorists of urban poverty do not see high levels of black-white segregation as particularly relevant to understanding the underclass or alleviating urban poverty.

The purpose of this [argument] is to redirect the focus of public debate back to issues of race and racial segregation and to suggest that they should be fundamental to thinking about the status of black Americans and the origins of the urban underclass. Our quarrel is less with any of the prevailing theories of urban poverty than with their systematic failure to consider the important role that segregation has played in mediating, exacerbating, and ultimately amplifying the harmful social and economic processes they treat.

We join earlier scholars in rejecting the view that poor urban blacks have an autonomous "culture of poverty" that explains their failure to achieve socioeconomic success in American society. We argue instead that residential segregation has been instrumental in creating a structural niche within which a deleterious set of attitudes and behaviors—a culture of segregation—has arisen and flourished. Segregation created the structural conditions for the emergence of an oppositional culture that devalues work, schooling, and marriage and that stresses attitudes and behaviors that are antithetical and often hostile to success in the larger economy. Although poor black neighborhoods still contain many people who lead conventional, productive lives, their example has been overshadowed in recent years by a growing concentration of poor, welfare-dependent families that is an inevitable result of residential segregation.

We readily agree with Douglas, Pinkney, and others that racial discrimination is widespread and may even be institutionalized within large sectors of American society, including the labor market, the educational system, and the welfare bureaucracy. We argue, however, that this view of black subjugation is incomplete without understanding the special role that residential segregation plays in enabling all other forms of racial oppression. Residential segregation is the institutional apparatus that supports other racially discriminatory processes and binds them together into a coherent and uniquely effective system of racial subordination. Until the black ghetto is dismantled as a basic institution of

American urban life, progress ameliorating racial inequality in other arenas will be slow, fitful, and incomplete.

We also agree with William Wilson's basic argument that the structural transformation of the urban economy undermined economic supports for the black community during the 1970s and 1980s. We argue, however, that in the absence of segregation, these structural changes would not have produced the disastrous social and economic outcomes observed in inner cities during these decades. Although rates of black poverty were driven up by the economic dislocations Wilson identifies, it was segregation that confined the increased deprivation to a small number of densely settled, tightly packed, and geographically isolated areas.

Wilson also argues that concentrated poverty arose because the civil rights revolution allowed middle-class blacks to move out of the ghetto. Although we remain open to the possibility that class-selective migration did occur, we argue that concentrated poverty would have happened during the 1970s with or without black middle-class migration. Our principal objection to Wilson's focus on middle-class out-migration is not that it did not occur, but that it is misdirected: focusing on the flight of the black middle class deflects attention from the real issue, which is the limitation of black residential options through segregation.

Middle-class households—whether they are black, Mexican, Italian, Jewish, or Polish—always try to escape the poor. But only blacks must attempt their escape within a highly segregated, racially segmented housing market. Because of segregation, middle-class blacks are less able to escape than other groups, and as a result are exposed to more poverty. At the same time, because of segregation no one will move into a poor black neighborhood except other poor blacks. Thus both middle-class blacks and poor blacks lose compared with the poor and middle class of other groups: poor blacks live under unrivaled concentrations of poverty and affluent blacks live in neighborhoods that are far less advantageous than those experienced by the middle class of other groups.

Finally, we concede Murray's general point that federal welfare policies are linked to the rise of the urban underclass, but we disagree with his specific hypothesis that generous welfare payments, by themselves, discouraged employment, encouraged unwed childbearing, undermined the strength of the family, and thereby caused persistent poverty. We argue instead that welfare payments were only harmful to the socioeconomic well-being of groups that were residentially segregated. As poverty rates rose among blacks in response to the economic dislocations of the 1970s and 1980s, so did the use of welfare programs. Because of racial segregation, however, the higher levels of welfare receipt were confined to a small number of isolated, all-black neighborhoods. By promoting the spatial concentration of welfare use, therefore, segregation created a residential environment within which welfare dependency was the norm, leading to the intergenerational transmission and broader perpetuation of urban poverty....

Our fundamental argument is that racial segregation—and its characteristic institutional form, the black ghetto—are the key structural factors responsible for the perpetuation of black poverty in the United States. Residential segregation is the principal organizational feature of American society that is responsible for the creation of the urban underclass....

KEY CONCEPTS

apartheid   culture of poverty   hypersegregation
segregation

DISCUSSION QUESTIONS

1. Regardless of your race or ethnicity, did you grow up in a racially segregated environment? How central in your life was this fact? What consequences did it have? If you know anyone who did, what in your estimation were the effects on their life?

2. What is the "culture of poverty" view? Do you agree with it? What do Massey and Denton have to say about it?

66

The Politics of Illegal Dumping

An Environmental Justice Framework

DAVID N. PELLOW

David Pellow provides an analytical framework for understanding how groups, especially those affected by environmental racism, can mobilize to address issues of environmental injustice.

Communities of color and low-income neighborhoods are disproportionately burdened with a range of environmental hazards, including polluting industries, landfills, incinerators, and illegal dumps. Researchers have supported this conclusion with analyses of census data and case studies of contaminated communities where poor persons and people of color are the residential majority. Significant attention has also focused on the environmental justice (EJ)

SOURCE: David N. Pellow. 2004. *Qualitative Sociology* 27, no. 4 (Winter): 511–525.

movement that has emerged with the aim of reducing the level of environmental risk in communities across the U.S. (Bullard 2000).

In this article, I argue that while these studies offer persuasive indicators of environmental inequality, we have little understanding of the complexity involved in the decision-making that produces or otherwise influences these unequal outcomes. Specifically, few if any studies have proposed a conceptual framework for understanding how environmental inequality emerges in communities. I propose an "environmental justice framework" to address these questions. In the remainder of the article, I review the relevant literature on environmental inequality/racism; I then develop an environmental justice framework, and apply it to a case study of illegal dumping in communities of color in Chicago, Illinois.

STUDIES OF ENVIRONMENTAL INEQUALITY: ORIGINS AND PROCESSES

Place-Specific Explanations for Environmental Inequality

While environmental racism and inequality are national, even global, phenomena, they unfold and impact people differently across space, depending on any number of factors. In their recent study of the state of Massachusetts, Faber and Krieg (2001) argue that a combination of white and middle-class flight to the suburbs, a rise in the number of people of color in the central cities, and increases in illegal dumping and the siting of incinerators produced environmental inequalities in that state. In a case study of the US Steel Corporation in Gary, Indiana, Hurley (1995) found that Latinos and African Americans faced disproportionately high levels of exposure to environmental toxins on the job and in their neighborhoods as a result of local racial discrimination.

Roberts and Toffolon-Weiss (2001) argue that the primary cause of environmental inequalities in the state of Louisiana is an alliance among business, the state, and other "growth machine" interests to create a "good business climate" that favors private profits over public and environmental health. The authors conclude that each particular EJ struggle may reveal driving factors that are unique to that specific context. For African Americans, Vietnamese, and Native American communities in the state, the process by which environmental inequalities emerged was distinct. For example, in the case of some African Americans in Louisiana, they face environmental racism not only as a result of contemporary racist environmental policies, but also because many polluting corporations occupy land that was once the site of slave plantations, land where the descendants of slaves and sharecroppers now live. Thus context, place, and history matter a great deal.

In a study of Torrance, California, Sidawi (1997) demonstrated that, as a result of racially biased urban planning during the last century, Chicanos faced the highest levels of exposure to industrial pollution in that city, when compared to Anglos. Similarly, Boone and Modarres (1999) argue that in Commerce, California, although hazardous industry was sited in close proximity to Latino populations, zoning and urban planning practices had as much, if not more, to do with industrial location decisions as did demographics and racial politics. As they state, "demographics alone are not responsible for the concentration of manufacturing in Commerce" (ibid., p. 165). They emphasize the importance of "place-specific analysis" to determine the root causes of environmental inequalities.

Macro-Level Explanations for Environmental Inequality

While discriminatory zoning practices and laws certainly contribute to environmental inequality, Bullard (1990, 1996, 2000) demonstrates that all of the city of Houston's municipal landfills are located in African American neighborhoods, *despite the absence* of any zoning laws. In this case, he maintains that there was *de facto* zoning by Houston's powerful white city leaders, who apparently viewed African American neighborhoods as appropriate sites for waste disposal. In other words, Bullard concludes that racism is a general organizing principle of city and county politics in the U.S.

Compounding the institutional racism in politics, Pastor, Sadd, and Hipp (2001, p. 19) argue that a lack of "pre-existing racially based social capital" also places communities of color at a disproportionately higher environmental risk than white communities. Communities with low levels of voting behavior, home ownership, wealth, and disposable income are more vulnerable to high concentrations of polluting facilities than are other communities. Unfortunately, for communities of color, these characteristics are often highly correlated with race. . . .

Thus, on one hand, we can cull from this literature particular factors that have produced environmental inequalities in certain case studies: local growth machine politics, discriminatory and insensitive zoning (or lack thereof), residential and occupational segregation, and white flight. On the other hand, a number of scholars point to a range of interrelated macro-level factors believed to be the general cause of environmental inequality: racism, social inequality, a lack of social capital, and the limited capacity to engage in collective action in the face of powerful political economic institutions. I argue that, while both groups of studies are helpful for understanding the nature of environmental racism, a new framework is needed to address both the generalities and complexities of this phenomenon, a framework that acknowledges both the macro-level themes of institutional inequality and the more micro-level motivations of particular organizations and actors involved in EJ struggles. In order to capture these factors that define many EJ struggles I propose an "environmental justice framework."

An Environmental Justice Framework

The framework I propose is organized around four dimensions relevant to most EJ conflicts. First is the importance of viewing environmental inequality as a sociohistorical process rather than as a discrete event. A focus on the history of an EJ conflict opens a window into the processes by which hazards are created and distributed. Second is a focus on the complex roles of the many people and organizations (stakeholders) involved. Environmental inequality impacts many actors and institutions with often contradictory and crosscutting allegiances (the state, workers, environmentalists, residents, private capital, and neighborhood organizations). These stakeholders are engaged in struggles for access to valuable resources (clean and safe working and living environments, natural resources, power, and profit). Third is the effect of social inequality on stakeholders. Institutional racism, class and gender inequalities, political hierarchies, and other forms of stratification play decisive roles in EJ struggles. Specifically, those populations of workers and residents lower on the social hierarchy are generally low-income and/ or people of color and are therefore more likely to suffer environmental injustices. Fourth is agency—the power of populations confronting environmental inequalities to shape the outcomes of these conflicts. That is, marginal groups can sometimes create openings in the political process to mitigate environmental inequality. Through resistance they can shape environmental inequalities.

I refer to this as an environmental *justice* framework because, although it underscores the causes and nature of environmental *inequalities*, I view the absence of environmental justice (EJ) as the central problem. In this study, I focus on a case in which a community of color faces an ecological threat. By drawing on the environmental justice framework, I seek to explain the social dynamics that produced this outcome.

METHODS

I collected data from 1997 through 1999—as part of a larger project—employing archival, interview, and participant observation methods. I collected numerous newspaper articles and government documents on EJ conflicts in Chicago during the 1990s. Data were also culled from hundreds of memos, reports, internal documents, and studies from various grassroots and environmental advocacy organizations involved in these struggles. I conducted forty face-to-face semi-structured interviews with residents, environmentalists, and government officials active in Chicago's solid waste conflicts. Interviews ranged from thirty minutes to two hours and were audiotaped, transcribed, and analyzed with a thematic coding system. Thus, through observation, interviews, and documentary analysis, I have triangulated data sources to provide the most accurate and complete presentation of the case study as possible (Denzin 1970).

HISTORY AND PROCESS: CHICAGO, A HOTBED OF ENVIRONMENTAL JUSTICE CONFLICT

Operation Silver Shovel was an illegal dumping scandal that occurred in Chicago's West Side Latino and African American communities during the 1990s. Thousands of tons of debris from construction, demolition, and residential remodeling projects were dumped in these neighborhoods, creating small mountains of waste, over the objections of local residents. Unfortunately, this case was not without precedent. . . .

Chicago is a city frequently rocked with environmental justice battles, as evidenced by its continual "garbage wars," wherein residents struggle to keep solid waste out of their communities and, by extension, inside other communities. More than eighty percent of the city's formal garbage disposal occurs in landfills on the mostly working-class, African American, Latino, and European ethnic Southeast side. Chicago hosts the most landfills per square mile in the nation. But formal waste disposal practices are only one dimension of Chicago's garbage wars. Less visible is the *informal* practice of illegal, or "fly," dumping.

Chicago has a long history of conflicts over illegal waste disposal. At the end of the nineteenth century in the immigrant enclave of the twenty-ninth ward, Alderman Tom Carey made a fortune creating large pits in the neighborhood to make bricks for his own construction company and then receiving payment from the city to fill them in with garbage (Hull House 1895, ch. VII). In 1910, the Superintendent for the Department of Health was dismissed after an investigation concluded that several incidents involving graft and extortion at a waste dump had occurred on his watch (Citizens' Association of Chicago 1910). In 1914 the City Council of Chicago passed an ordinance outlawing unregulated dumping. But since that time, there have been frequent scandals over the illicit trade in garbage for favors and cash.

This phenomenon also reflects the problem of environmental inequalities. A study during the 1990s by the Chicago Department of Streets and Sanitation revealed that, of the ten city neighborhoods with the most illegally dumped garbage, all are at least sixty percent African American or Latino. And wards where people of color are the majority account for seventy-nine percent of all illegally dumped garbage in the city (Cohen 1992). This is the historical context in which Operation Silver Shovel must be viewed.

OPERATION SILVER SHOVEL

A Multi-Stakeholder View of Illegal Dumping:
Politicians, Firms, and Residents

John Christopher was a businessman. Since the late 1980s, he was in the business of "recycling" construction and demolition (C&D) waste and finding places to dump it at the lowest possible cost. The vast majority of the waste Christopher was dumping originated from highway construction projects and remodeling firms across the mostly white North Side of the city and suburbs. John Christopher began dumping his waste in working-class and low-income African American and Latino communities on Chicago's West Side, particularly Lawndale and Austin. In order to ensure that he could commit these crimes without detection by the police or City Hall, he paid local aldermen cash bribes. . . .

Every community where Christopher dumped his waste was primarily African American or Latino, as was each alderman whom he bribed. The KrisJon company claimed to be recycling the C&D waste for use in future construction operations. However, KrisJon was not actually recycling the waste. Instead, the company was crushing large rocks and concrete blocks and simply piling them up—creating dumps in the neighborhoods. Local activists soon discovered that KrisJon had no permits for this operation and was therefore in violation of several city ordinances. KrisJon was engaged in illegal dumping.

Communities confronting environmental inequalities in the U.S. typically respond in a variety of ways. Public protests, petitioning and targeting political leaders, and negotiating directly with polluting firms are just a few methods commonly found in EJ groups' tactical repertoires (Gottlieb 1993; Hurley 1995). Resident stakeholders on Chicago's West Side were no different.

Local neighborhood groups in Lawndale and Austin protested against the illegal dumping operations early on. In 1990, residents held public hearings to discuss a site that was being proposed for a dump that KrisJon was to operate. A short while later, when the site was operational and producing a large volume of dust, an organization called Concerned Parents of Sumner, Frazier, and Webster Elementary School Children (schools located within blocks of the dump) sent letters to John Christopher requesting a number of environmental improvements. Christopher never responded. His dumpsites, however, were bustling and receiving waste from ninety-six different locations around metropolitan Chicago. The noise and vibrations from trucks delivering the waste was so extensive that it cracked the streets and sidewalks, damaged the foundations of nearby homes, and kept residents awake at night. Residents believed that the dust from the operation was linked to severe respiratory problems in these African American and Latino communities. One man reported that he had experienced

> . . . coughing, wheezing, short of breath, headache, sinus and I have been hospitalized six times since the dump been on Kildare [Street]. I had to remove carpet from floors because of dust and dust flares up my asthma (Interview 1997).

Another resident, Doreen Jenkins, wrote of her son's medical problems:

> . . . including a trip to the emergency room in January for headaches, dizziness and difficulty breathing. The emergency services report attached to Ms. Griffin's letter describes that the "community is quite polluted by trucks that dump dirt into the air." Ms. Griffin pleaded to "please stop this illegal operation in our neighborhood" (Vinik and Harley 1997).

While perhaps not scientifically conclusive, these residents' claims resonate strongly with the testimonies of other persons living near polluting facilities in numerous communities across the nation (Brown and Mikkelsen 1990; Kroll-Smith and Floyd 1997).

Although the Illinois State's Attorney has the authority to bring legal action against a company when there is "substantial danger to the environment or to the public health," the community received only a noncommittal letter from an Assistant State's Attorney. By this time, one of the dumpsites was measured at eighty feet high. . . . Despite the fact that Christopher's facility violated . . . regulations, the Illinois

Environmental Protection Agency issued no citations. This official sanctioning of direct violations of city and state laws in a community of color stunned local activists.

The community meetings and protests continued. In February 1992, more than six hundred people signed a petition demanding the closure of KrisJon's two major sites at Kildare and Kostner Avenues.... The resident-activist stakeholders were fighting what appeared to be either a case of near total insensitivity by the state or an alliance between government and polluters to allow the dumps to remain in these communities of color.

But not everyone in the community was opposed to the dumping operations. Some residents joined John Christopher and participated in neighborhood "beautification projects" wherein he provided them with grass, flower, and vegetable seeds for landscaping and gardening. He also received the support of one of the strongest institutions in the African American community—the church....

While local leaders and politicians were involved in this struggle, the federal government was also well aware of the illegal dumping operations and used them as an opportunity to launch an investigation of political corruption. In 1992 the Federal Bureau of Investigation (FBI) secretly secured John Christopher's cooperation in what became known as Operation Silver Shovel. Christopher became a "mole," working undercover for the U.S. Attorney's Office and the FBI in an effort to uncover political corruption associated with the disposal of solid waste in Chicago. So, just as he had done in the 1980s, Christopher bribed African American and Latino aldermen to allow him to dump waste in their wards; the only difference was that in 1992 he was secretly videotaping the transactions. The public was not informed about this sting operation until 1996 when the media broke the case.

Residents who had already been protesting the dumping were incensed that the government had sanctioned this process.... Despite this persistence on the part of activists, it would be many more months before any positive changes would result from their protests. This is in large part because these communities possess few economic and political resources.

Inequalities: Race, Class, and Political Power

...The two largest illegal dumpsites on the West Side were on Kostner and Kildare Avenues. The level of class and racial inequality evident at these sites is remarkable. For example, the neighborhood surrounding the Kostner dump (based on a one-square-mile radius from the site) had a median household income of $20,469 compared to a citywide median income of $26,301 (U.S. Census Bureau 1990). The majority of residents in this community are people of color, with Latinos comprising 46.3 percent and African Americans 40.2 percent. Citywide, African Americans comprise 39.1 percent of the population and Latinos 19.6 percent. Thus the median household income in this neighborhood is well below both the national poverty level and the citywide median, and the percentage of people of color is higher than the citywide percentage. The figures for the Kildare site are even starker. For example, the neighborhood surrounding the Kildare dump (based on a one-square-mile radius from the site) had a median household income of $15,113. The majority of people in this community are also people of color, with African Americans comprising 89.6 percent and Latinos 6.6 percent. Both of these communities are what William Julius Wilson (1996) has called "new poverty areas," where the majority of people live in deep poverty and most adults are either unemployed or underemployed. Thus, the majority of the residents in these neighborhoods experienced a significant degree of poverty, economic instability, and relative deprivation.

With regard to the role of corrupt elected officials, this dimension of inequality requires analysis but should also include a consideration of activities by residents who themselves accepted bribes and facilitated illegal dumping practices. The relative lack of status and political influence over citywide politics among Chicago's African American and Latino aldermen is rivaled only by the political powerlessness among their constituents. Poor and working-class residents of color in Chicago are not particularly influential in local politics and therefore offer polluters easy targets. The attendant lack of economic stability that characterizes many of these neighborhoods only reinforces their political powerlessness and the diminished status of their elected officials. This subjugated position also renders these groups particularly vulnerable to efforts by polluting firms to "divide and conquer" residents over potential economic benefits that may accompany industrial activity.

Frequently, when communities confront environmental threats, they are beset by internal fractures surrounding family conflicts, fear of job losses, and loyalties to various neighborhood institutions and firms involved. These tensions generally intensify the pain and anxiety that normally develop during conflicts over environmental contamination (Brown and Mikkelsen 1990; Hurley 1995; Roberts and Toffolon-Weiss 2001). Chicago's West Side was characterized by a particularly intense array of divisive wedges that rendered these working-class, polluted communities of color quite vulnerable.

Foremost among these fractures was the abuse—by aldermen—of their political positions to allow waste dumping in return for cash. But in addition to this, John Christopher had to build broader community support for his operation to ensure its survival. One strategy was to bribe *residents* who had complained about his facilities. . . .

Austin and Lawndale are two communities in desperate need of sustainable economic development, so politicians and residents were easy prey to bribes, temporary jobs, or a range of cash-producing activities associated with illegal waste dumping. This dynamic illustrates the depths of economic despair in many communities of color, which have become so desperate for development that garbage—or one's willingness to accept it—is viewed as one of the only marketable resources available. In the next section I consider these communities' ability to produce real change and challenge environmental injustices.

Agency: Shaping and Reducing Environmental Inequalities

West Side neighborhood activists were ultimately successful at urging the city, the state EPA, and the federal EPA to take action in the Operation Silver Shovel case. The three levels of government have begun to develop a comprehensive "strategy to address the problem of illegal solid waste disposal in Chicago" (Vinik and Harley 1997, p. 32). Some of the concrete steps taken include: developing a community policing program specifically for illegal dumping; training police officers on illegal dumping surveillance techniques; strengthening the city ordinance regulating dumping; inspecting dumpsites and testing them for hazardous and chemical waste; and beginning waste clean-ups at the sites. New or modified city ordinances in Chicago include penalties such as jail time for illegal dumpers, the seizure and impoundment of all vehicles used for fly dumping, and the barring of contractors convicted of illegal dumping from future eligibility for city contracts. . . .

The community fought back repeatedly and, despite years without official support, was able to initiate governmental action, regulatory enforcement, and a clean-up of the waste. Angry constituents also later voted out of office one of the aldermen who accepted bribes. At the national level, as a result of the Operation Silver Shovel scandal, U.S. Representative Cardiss Collins of Illinois (an African American woman) introduced a House bill intended to outlaw environmental racism. Although the bill later died, this was evidence that grassroots activists were able to put the issue of environmental racism on the public agenda. In this way, even the most dispossessed and disenfranchised communities on Chicago's West Side shaped the discourse around—and even reduced the impacts of—environmental racism.

DISCUSSION AND CONCLUSION

Illegal Dumping and the Environmental Justice Framework

Operation Silver Shovel was a case of environmental inequality/racism and can be analyzed using the framework I introduced earlier. First, a historical analysis reveals that Operation Silver Shovel was not an isolated instance of environmental racism without precedent. Illegal dumping in Chicago's communities of color and immigrant neighborhoods is more than a century old and it has been one of the key battlefronts in that city's garbage wars. Hence, the fly dumping in the Lawndale and Austin communities was part of a longstanding and larger pattern of environmental inequality around the city. Second, many stakeholders with a host of complex motives were involved in this conflict. While most residents opposed the waste dumpers, many locals were facilitating the waste trade, including aldermen, neighbors in need of cash, and gang members. The other key stakeholders were the various levels of government, each of which was complicit in allowing these acts of environmental injustice to go unchallenged, despite their obvious illegality. Third, Operation Silver Shovel is largely rooted in institutional racism and class inequalities in that this type of locally unwanted land use is likely to appear only in low-income neighborhoods and communities of color. KrisJon and the FBI targeted these communities because of their vulnerability, their lack of political power, and the presence of politicians willing to accept bribes for favors. Finally, this was a defining moment for the power of the grassroots to exercise agency—to challenge powerful actors and institutions. Community groups persisted in their efforts to bring public attention to the problem of illegal dumping and to get the courts and government agencies to enforce the law and, despite virtually no assistance, were able to implement real changes. These formerly powerless networks of ordinary citizens placed the issue of environmental racism on the city's agenda and successfully reduced the level of environmental hazards in their neighborhoods.

Environmental inequality/racism occurs, therefore, when historical and contemporary social forces intersect to position various stakeholders in a state of power imbalance with regard to environmental resources. These conflicts are, in turn, shaped by all affected groups, and can be mitigated or exacerbated when the power imbalances are reduced or increased. Drawing on the environmental justice framework and the case of illegal dumping, three things are clear: 1) Environmental inequality/racism is not just about correlations between hazards and populations. It is about the power dynamics that produce these inequalities and the power of the grassroots to challenge and reverse them; 2) environmental inequality/racism is not just about communities of color versus white communities. While racism may play a persistent role in these conflicts, the range of motivations among stakeholders— including the desire for political power—and intraracial divisions across a range of community interests matter a great deal; and 3) until we understand why certain interests in communities of color are willing to support environmentally harmful practices we will never truly understand environmental racism and will therefore be ill-equipped to move toward environmental justice. It is my hope that future studies will take these findings into account.

REFERENCES

Boone, C., & Modarres, A. (1999). Creating a toxic neighborhood in Los Angeles County. *Urban Affairs Review, 35,* 163–187.

Brown, P., & Mikkelsen, E. (1990). *No safe place: Toxic waste, leukemia and community action.* Berkeley: University of California Press.

Bullard, R. (1990). *Dumping in Dixie* (1st ed.). Boulder, Co: Westview Press.

Bullard, R. (1996). The legacy of American apartheid and environmental racism. *St. John's Journal of Legal Commentary, 9,* 445–474.

Bullard, R. (2000). *Dumping in Dixie* (3rd ed.). Boulder, Co: Westview Press.

Cohen, L. (1992, April). Waste dumps toxic traps for minorities. *The Chicago Reporter,* p. 2.

Denzin, N. (1970). *The research act.* Chicago: Aldine.

Gottlieb, R. (1993). *Forcing the spring.* Washington, DC: Island Press.

Hurley, A. (1995). *Environmental inequalities.* Chapel Hill: UNC Press.

Krieg, E. (1998). The two faces of toxic waste. *Sociological Forum, 13,* 3–20.

Kroll-Smith, S., & Floyd, H. H. (1997). *Bodies in protest: Environmental illness and the struggle over medical knowledge.* New York: New York University Press.

Pastor, M., Sadd, J., & Hipp, J. (2001). Which came first? Toxic facilities, minority move-in, and environmental justice. *Journal of Urban Affairs, 23,* 1–21.

Roberts, T., & Toffolon-Weiss, M. (2001). *Chronicles from the environmental justice frontline.* New York: Cambridge University Press.

Sidawi, S. (1997). Planning environmental racism. *Historical Geography, 25*, 83–99.

U.S. Census Bureau. (1990). *Census of the population: 1980*. Washington, DC.

Vinik, N., & Harley, K. (1997). *Environmental injustice*. January. Chicago: Chicago Legal Clinic.

Wilson, W. J. (1996). *When work disappears*. New York: Random House.

KEY CONCEPTS

environmental justice   environmental racism   social movement

DISCUSSION QUESTIONS

1. What factors does Pellow identify that produce environmental inequities?
2. What are the four factors in Pellow's environmental justice framework? What evidence do you see of these factors in the community where you live?

67

Comparative Public Opinion and Knowledge on Global Climatic Change and the Kyoto Protocol
The U.S. versus the World?

STEVEN R. BRECHIN

This article compares American concern for the environment with citizens of other nations. The author finds that public opinion polls show America worries much less about global climate change and environmental policy than many other

SOURCE: Brechin, Steven R. 2003. "Comparative Public Opinion and Knowledge on Global Climatic Change and the Kyoto Protocol: The U.S. versus the World?" *International Journal of Sociology and Social Policy* 23: 106–134.

nations. American understanding of the issues and the policies that currently exist is very limited. The Kyoto Protocol is an international legal agreement to reduce greenhouse gases. When President Bush pulled America out of the agreement, the response internationally was much more negative than among Americans themselves. The article ends with some discussion about what will need to happen for Americans to care more about the environment.

The world's scientists today speak with a near unified voice on the existence of a human induced greenhouse effect and in least in general ways on its potential dramatic impacts. They argue that the resulting rising temperatures will likely have serious consequences for humans and ecosystems alike. The citizens of various nations of the world, on the other hand, appear to possess wide-ranging views and levels of understandings about global climate change as a real or potential threat. The purpose of this paper is to explore comparatively the views, attitudes and knowledge of ordinary citizens from a number of countries where public opinion data exists on global climate change itself. In part it attempts to build upon an earlier effort at comparisons using data from the early 1990s (see Dunlap 1998) to see if there have been any dramatic changes over the past ten years. In particular, the author compares the views, attitudes and knowledge of the U.S. citizens to those of other countries, especially around the Kyoto Protocol, the international agreement created to regulate the release of greenhouse gases among the world's nations. . . .

GLOBAL CONCERN FOR THE ENVIRONMENT

One of the major social science findings on the environment in the 1990s was the discovery of "global environmentalism" or the expression of environmental concern by citizens in countries worldwide (Dunlap et al. 1993; Brechin and Kempton 1994). This finding came about, interestingly enough, as a consequence of some of the preparations for the 1992 United Nations Conference on Environment and Development (UNCED) in Rio de Janeiro, Brazil. . . .

Given established beliefs at the time and their theoretical underpinnings, it was assumed from the early studies on environmentalism in the U.S. and Europe that public concern for the environment was a consequence, in its broadest sense, of economic wealth. It was viewed as an outcome of both rapid industrialization and the financial means to address those problems. Consequently, environmentalism was considered a product unique to Western industrialized countries.

Environmentalism as a Western phenomenon also received theoretical support from Ronald Inglehart's Postmaterialist Values Thesis (Inglehart 1990, 1997). Based upon this theory, concern for the environment was thought to be the result more specifically from an intergenerational change in cultural values resulting from unprecedented political stability and economic welfare following World War II. Built on a notion of Maslow's Hierarchy of Needs (Maslow

1954), a postmaterialist concern for the environment could be characterized as a luxury good, an object "purchased" with the extra resources from greater wealth or similarly demanded by yet unattained personal desire, after other more basic needs had been met. In short, the rise of environmentalism in the West was considered as a consequence of the rise of postmaterialist values that was also occurring throughout the same world region.

This particular Western view on the genesis of environmentalism was shattered with new empirical data, coming from a number of sources, with the most important being the first cross-national public surveys that included poorer industrializing countries in their studies....

The Gallup study of 24 nations, with varying level of economic wealth and political systems and stability, showed that concern for a number of environmental issues were high among citizens of most nations, rich or poor. Even statistical tests based on results of on national probability sampling (country-wide representation) that on a wide number of environmental issues demonstrated that there were few meaningful differences in levels of concern by the citizens of richer or poorer nations. In fact only on two environmental items, 1) rating environmental problems as serious for the nation, and 2) air pollution and smog as a very serious personal concern, were there significant differences between the country groups. And in both cases greater concern rested with the citizens of poorer nations rather than wealthier ones (see Brechin & Kempton 1994). Given the findings for broad international concern for environmental problems, it would be reasonable to expect this would include concern about global climatic change as well.

A GLOBAL CONSENSUS ON GLOBAL WARMING?

Although now over ten years old, the largest publicly released crossnational study on the public attitudes toward global warming remains Gallup's HOP survey.... [T]he percentage of respondents who personally find global warming a serious problem in the world vary enormously.... [T]hose respondents who indicate that they believe global warming is a "very serious problem" range from twenty-six percent in Nigeria to seventy-three percent in Germany. Of the twenty-four countries in the study, however, more than half, thirteen of them have more than fifty percent of their respective populations who feel global warming is a very serious problem. The U.S. populace stands at forty-seven percent in this survey, the same as South Korea's, both of which are in the bottom half of all countries. However, if we were to place half of the countries in a separate group of industrialized countries and the other half in an industrializing nations group, analyses would show no significant differences amongst the two groups on the question of global warming, as well as other international environmental problems (see Brechin and Kempton 1994; Brechin 1999).

If we were to combine responses to include those who said global warming was a "very serious" problem with those who said it was a "somewhat serious" problem, that is, all those who find global warming at least somewhat serious, only one country of the twenty-four, Nigeria, with thirty-nine percent, would not fall within this category of broad support. Sixty-five percent or more of the populace in each of the remaining twenty-three nations are at least somewhat concerned about global warming. From these results it is clear to see that the majorities in most of these countries seem concerned about global warming. In sum, even a decade ago there appeared to be a global public consensus on global warming, years before the science was as certain as it is today. It could be argued then, that given the more recent scientific evidence on global climatic change, there would be even greater cross-national public consensus on the issue today. Unfortunately we have only limited cross-national data on recent public opinion on global warming. . . .

The U.S. Public and Global Climatic Change

One of the best longitudinal data on global warming that we have on U.S. public opinion today comes from fairly sporadic surveys by the Gallup Organization from 1989 to 2003. . . . The question used in this survey was worded somewhat differently than in the HOP study just discussed above. Instead of level of "seriousness of concern" about a global environmental problem, this question focused on level of "personal worry". It reads, "I am going to read you a list of environmental problems. As I read each one, please tell me if you *personally* worry about this problem a great deal, a fair amount, only a little, or not at all. How much do you personally worry about the 'greenhouse effect' or global warming?" From this data we can see that over the years 1989 to 2003 anywhere from twenty-four percent to forty percent of the U.S. public personally worried "a great deal" about global warming. This is compared to only twelve to seventeen percent of U.S. respondents who did not worry about it at all. Even with the two different wordings on testing public concern regarding global warming, it is quite obvious that a majority of U.S. respondents at least continue to worry a fair amount (50–72%) about the phenomenon and have done so over a number of years.

How Does Global Warming Rank to Other Environmental Problems in the U.S. Today?

Global warming, however, typically ranks considerably lower by the U. S. public when compared among other environmental concerns. In the same 2003 Gallup survey. . . respondents ranked global warming ninth out of ten problems with twenty-eight percent respectively saying they worried "a great deal" . . . ; only acid rain at twenty-four percent respectively was lower. Pollution of drinking water was ranked the highest with fifty-four percent of the respondents indicating that they were worried a great deal. This was followed by pollution of rivers, lakes, and reservoirs (fifty-one percent); contamination of soil and water by toxic waste (fifty-one percent); maintenance of the nation's supply of fresh water for household needs (forty-nine percent); air pollution (forty-two percent); the loss of tropical rain forests (thirty-nine percent); damage to the earth's ozone layer (thirty-five percent); and extinction of plant and animal species (thirty-four percent) (Carroll 2002; Saad 2003). Americans, it seems, in comparison to other environmental problems are relatively less concerned about global warming. This relatively low ranking (or issue salience) has remained fairly consisted over the years.

Comparative Global Warming Rankings

As we shall see, however, citizens of a number of the other countries have similarly low rankings for global warming as in the U.S.; only a few countries ranked it high among other environmental problems....

In the HOP survey data from 24 countries, we see more or less the same low ranking of global warming as in the U.S. Only Brazil, Japan, and West Germany of the twenty-four countries ranked global warming at or near the top of the list of environmental concerns.... Ten of the twenty-four countries ranked global warming at or near the bottom of their lists.[1] The remaining eleven countries ranked global warming in the middle to lower end of their environmental concerns. These findings suggest that many other countries ranked global warming similarly as the U.S., although there are some international differences. What is missing is a full explanation for this trend....

CROSS-NATIONAL KNOWLEDGE
ON THE SOURCES OF GREENHOUSE GASES

How well does the general public cross-nationally understand the anthropogenic sources of gases that cause the greenhouse effect? Previous research has shown that most citizens in the few countries studied do not, even in wealthy industrialized countries (Dunlap 1998; Kempton 1991, 1993; Kempton et al. 1995; Lofstedt 1991, 1992, 1993; Rudig 1995). Have the understandings about global warming by the world's populations improved over the last decade? This is potentially important, especially in more democratic countries where with proper knowledge citizen voices could more likely demand and support more effective policies to combat global climatic change. Better-informed citizens everywhere may more likely shape their own behaviors to contribute more positively in protecting their environments instead of threatening them.

Two very interesting sets of recent findings from the research group Environics International focused on the public's understandings of the causes of global warming. Although through the work of several American anthropologists (e.g. Kempton et al. 1995) and others, it has been known for a number of years now that the U.S. public so far lacks a clear understanding of how human activities actually contribute to the greenhouse effect. The data from Environics International in 1999 and 2001 clearly show that the lack of knowledge is shared cross-nationally as well, even among those countries such as Germany that have a strong, pro-environmental image, and Japan, home to the Kyoto Protocol. . . .

Many people cross-culturally also share the misconception on the role the thinning of the ozone layer of the upper atmosphere plays in encouraging global warming. This misconception too has been noted for some time in the social science literature (e.g. Kempton 1993, Kempton et al. 1995; see too Dunlap 1998). From the same 1999 data from Environics International, Japan, with twelve percent, had the lowest percentage of respondents who identified the loss of the ozone layer as the main source of global warming; Indonesia again had the largest with forty-eight percent. Twenty-six percent of U.S. respondents selected the ozone layer loss as the principal source of global warming, the same as Mexico's on this particular issue in this 1999 study. . . .

When comparing the 1999 results with those of 2001, only eleven of the original twenty-seven countries were the same and there were four new countries added. . . . Although the results are quite similar between the two different years, it could be argued that there has been a general trend in better understanding the importance of burning fossil fuels in creating greenhouse gases. Perhaps the most striking finding from these cross-cultural comparisons is how poorly people from a wide-ranging number of countries understand the anthropogenic causes of global warming. . . .

CROSS-NATIONAL REACTIONS TO THE BUSH ADMINISTRATION'S POSITION ON THE KYOTO PROTOCOL

The Kyoto Protocol is a legally binding international agreement that grew out of the 1992 United Nations Framework Convention on Climate Change as part of the larger United Nations Conference on Environment and Development in Rio de Janeiro, Brazil. It is specifically the product of a 1997 Conference of Parties (meeting of nations) in Kyoto Japan that committed industrialized nations, upon ratification, to reduce emissions of the six greenhouse gases: carbon dioxide, methane, nitrous oxide, hydrofluorocarbons, perfluorocarbons, and sulfur dioxide over a fixed period of time. Although the specific emissions level per country could vary, under the protocol the overall emissions from industrialized countries would be reduced 5.2 percent below 1990 levels over a five-year period, 2008-2012 (Kyoda News International, Ltd. 2002). Developing countries, such as China, are not legally bound by the requirements. The U.S., the world leader in emissions would be required to cut emissions by seven percent (e.g. Kleiner 2001). . . .

In a June 11, 2001 speech at the White House, President Bush called the 1997 Kyoto protocol "fatally flawed in fundamental ways" (White House 2001) and pulled the U.S. out of the agreement. The Bush Administration's rationale for the withdrawal from the agreement was that in their view it was unfair to

American businesses and would unnecessarily hurt the U.S. economy. The Bush Administration was particularly upset that both China and India, two of the leading producers of greenhouse gases, were exempted from the group of countries required to reduce their emissions. The developing countries have been exempted mainly due to the fact that they are not responsible for the mass accumulation of gases that have occurred over the past 200 years or more of industrialization. However, it was China and India's exemption from the Protocol as noted above that caused, or gave the excuse for, the Bush Administration to withdraw from the agreement, labeling Kyoto as unfair. . . .

European Perspectives on Kyoto Protocol

Unlike in the U.S., the Bush Administration's rejection of the Kyoto protocol in 2001 produced a very strong negative reaction internationally, especially in Europe. Even religious leaders from Europe as well as the U.S. rebuked President Bush for his decision (Doogue 2001). Both the citizens of Europe and their leaders were outraged, especially by Bush's claim that it would hurt American businesses too much. A public opinion poll from five European countries, conducted by the Pew Research Center, the International Herald Tribune, and the Council on Foreign Relations and released on August 15, 2001, responded to a number of Bush's Administration's foreign policies, including the Kyoto agreement (Pew Research Center 2001). The poll presented results of national samples from Great Britain, Italy, Germany, France, as well as the U.S. It showed a very consistent and overwhelming negative reaction from the citizens of these countries regarding Bush's decision on Kyoto. The populace of each of five European countries disapproved the policy decision by eighty percent or more—Great Britain, 83; Italy, 80; Germany, 87; and France 85. Their approval ratings for the Bush Administration's position were only 10% to 12% for each of these European countries. . . .

Equally striking about the survey, however, are the views of U.S. residents. Although generally disappointed by the President's announcement, the reaction was much less severe than it was among our European allies. Only forty-four percent of the U.S. public disapproved of the action in 2001 (Pew Research Center 2001). This is roughly half the amount of their European counterparts. Similarly, twenty-nine percent of Americans were supportive of the policy move, essentially three times more than the Europeans. This poll was taken about a month before September 11, 2001 terrorists' raids on New York and Washington, D.C., and hence the numbers discussed here were neither tainted by that event nor by the later disagreement over Iraq. A bit of information worth focusing on concerns the number of respondents who either refused to answer or did not know what to say about the President's decision. While the European percentages were single digit, ranging from three percent in Germany to eight percent in Italy, twenty-seven percent of the Americans surveyed did not have any opinion about the decision on Kyoto or refused to answer the question. This, along with the evidence presented earlier, may suggest that the U.S. citizens are not very knowledgeable about global warming and when in doubt they tend to support their party's position regarding the problem and any related policies (Krosnik and Visser 1998).[2] The action on Kyoto by the Bush Administration, as well as a few other foreign policy issues, placed the U.S. in direct disagreement with the European Union countries (Pew Research Center 2001). It was also part of a series of early positions by the Bush Administration that started labeling the president and his administration as unilateralists, especially by Europeans.

CONCLUSIONS

This article revisits the questions of cross-national public concern for global warming raised over a decade ago. Although the scientific community today speaks about global warming with a more unified voice concerning its anthropogenic causes and its potential devastating impacts across the globe, the citizens of a number of nations seem to still harbor some uncertainties about the problem itself and certainly lack a clear understanding of its sources. While majorities of publics across many nations speak of personal concern about global warming as well as on the seriousness of the problem, in all except a few countries, global warming seems to rank near or at the bottom of their list of environmental concerns. Although it could be argued that there has been a slight improvement over the last decade in the public's understanding regarding the anthropogenic causes of global warming, the people of all the nations studied remain largely uniformed about the problem.... The citizens of the U.S., among the most educated in the world, were somewhere in the middle of the pack, tied with the citizens of Brazil at fifteen percent. Even the Cubans, at seventeen percent, were slightly more informed than the American public....

It is clear from this research, however, that the U.S. public appears to be nearly as out of step with other nations on support on the Kyoto protocol as the Bush Administration itself. Although forty-four percent of the American people disapproved of President Bush's decision to withdraw from the protocol in 2001, this is essentially half the level of dissatisfaction found in Europe for the American President's action. Almost thirty percent of Americans supported his decision, approximately five to six times more so than Europeans.

There are many questions that remain. One of the more critical ones concerns the question of who is leading whom? Is the Bush Administration following the American public's uncertainty and ambivalence about the problem of global warming or is its stance on environmental issues generally, with global warming particularly, influencing American public opinion? Although this does not need to be an either or answer, that is, both arguments could be at least partially correct, what little evidence that exists may suggest that American citizens continue to be influenced by political party lines. However, with the current popularity of President Bush, the general U.S. public support for his unilateralist approach so far, and the general acquiescence to date of the American public in a post September 11th era, the American public still may be willing to follow their President on this concern, now nearly two years after his decision on

Kyoto. It is uncertain how long the American public will continue to let the Bush Administration to have its way on environmental issues. At the same time, it is unlikely that the American public will for long abandon its deep support for environmental protection (see Kempton et al 1995; Dunlap 2002, 2003). The questions that remain, however, are when will the American people demand more action on protecting the environment? And under what circumstances will that support emerge? On the particular topic of global warming, however, the international community, especially the Europeans and Japanese, may need to continue to serve as America's conscience.

NOTES

1. Please note that a question on the level of concern about acid rain, the lowest among U.S. respondents in the survey above, was not asked in the HOP study. If this question is removed then global warming is ranked last among a list of environmental concerns.

2. See Krosnik and Visser 1998 for more information on the political effects of the Clinton Kyoto campagin. The Clinton Administration's effort in 1997 to rally the American public support for combating global warming and to gain approval for the Kyoto Protocol seemed to have only further divided the country. In particular, the scientific controversy in the U.S. media seemed to deepen attachment to partisan positions if not political ideology, with Democrats more concern about global warming and supportive of Kyoto than Republicans.

REFERENCES

Brechin, S. R. 1999. "Objective Problems and Subjective Values: Evaluating the Post-materialist Argument and Challenging a New Explanation." *Social Science Quarterly*. 8/4: 793–809.

Brechin, S. R. and W. Kempton. 1994. "Global environmentalism: A challenge to the postmaterialist thesis." *Social Science Quarterly*. 75(2) 245–69.

Carroll, J. 2002. "Public Slightly More Negative than Positive about the Quality of the environment: Americans most concerned about pollution of their drinking water." The Gallup Organization. http:www.gallup.com/poll/releases/pr020419.asp?version=p. April 19, 2002. Accessed 9/3/2002.

Doogue, E. 2001. "Religious leaders rebuke Bush administration over Kyoto Protocol." Presbyterian News Service April 4, 2001. http://www.pcusua.org/pcnews/oldnews/2001/01115.htm. Accessed 2/25/2003.

Dunlap, R. 1998. "Lay perceptions of global risk: Public views of global warming in cross-national context." *International Sociology*, 13:4: 473–498.

———. 2002. An enduring concern. *Public Perspectives*. September/October. www.PublicPerspectives.org.

————. 2003. "No Environmental Backlash Against Bush Administration: Though most Americans favor strong environmental policies." The Gallup Organization. http://www.gallup.com/poll/releases/pr030421.asp?Version=p. Accessed 4/22/03.

Inglehart, R. 1990. *Cultural shift in advanced industrial society*. Princeton, N.J.: Princeton University Press.

————. 1997. *Modernization and postmodernization: Cultural, economic, and political change in 43 societies*. Princeton, N.J.: Princeton University Press.

Kempton, W. 1991. "Lay perspectives on global climate change," *Global Environmental Change* 1: 183–208.

————. 1993. "Will public environmental concern lead to action on global warming?" *Annual Review of Energy and Environment* 18: 217–45.

Kempton, W, J. S. Boster, and J. A. Hartley. 1995. *Environmental Values in American Culture*. Cambridge, MA: MIT Press.

Kleiner, K. 2001. "Heat is on." New Scientist.com. March 1. www.newscientist.com/news/print.jsp?id=ns9999566, Accessed 2/25/03.

Kronsik, J. and Visser P. 1998. "The impact of the fall 1997 debate about global warming on American public opinion." Weathervane, Resources for the Future's digital forum on global climate policy (www.weathervane.rff.org). Accessed 12/10/2002.

Kyodo News International, Inc. 2002. "70% feel global warming is immediate threat." Global NewsBank — Category. April 1. http://infoweb1.Newsbank.com/iw-search/we/InfoWeb?p_action+pring&p_docid+0F49222DA2F7BE3%. Accessed 1/22/2003.

Lofstedt, R. E. 1991. "Climate change perceptions and energy-use decisions in northern Sweden." *Global Environmental Change* 1: 321–324.

————. 1992. "Lay perspectives concerning global climate change in Sweden." *Energy and Environment* 3: 161–175.

————. 1993. "Lay perspectives concerning global climate change in Vienna, Austria." *Energy and Environment*, 4: 140–154.

Maslow, A. K. 1954. *Motivation and Personality*. New York: Harper & Row.

Pew Research Center for the People & the Press. 2001. "Bush unpopular in Europe, seen as unilateralist." August 15. http://people-press.org/reports/print.php3?ReportID=5. Accessed 1/22/03.

Saad, L. 2003. "Giving Global Warming the Cold Shoulder." The Gallup Organization: Government & Public Affairs. http://www.gallup.com/poll/tb/goverpublic/20030422.asp?Version=p. Accessed 4/28/03.

White House. 2001. "President Bush Discuses Global Climate Change." Press Release, Office of the Press Secretary, June 11. www.whitehouse.gov.news/releases/2001/06/print/20010611-2html. Accessed 3/17/03.

KEY CONCEPTS

global warming   industrialization

DISCUSSION QUESTIONS

1. Had you heard about the Kyoto Protocol? What is it? Do you think the United States should be part of the agreement?
2. Where do you learn about environmental concerns? What sources of information would you turn to if you wanted to find out more?

Applying Sociological Knowledge: An Exercise for Students

For one full week, keep a log in which you list everything that you discard. (For purposes of this exercise, you can disregard human body waste.) Note in your log the material that each object is made from, how much it weighs, and where and how you disposed of it. At the end of the week, tally how much weight in the various categories of materials you (as one individual) have discarded. Can you think of alternative methods of disposal that would be more environmentally friendly? What individual changes would you have to make to do this? What local changes would accommodate your making such changes? What changes at the broadest level of society would be necessary to be more environmentally healthy?

68

Generations X, Y, and Z: Are They Changing America?

DUANE F. ALWIN

What creates social change? Is it the values of young people who bring new perspectives and new issues to society? Or is it the influence of major historical events? In this article, Duane Alwin examines generational sources of social change, arguing that historical events and shifts in the individual lives because of aging are both sources of social change.

The Greatest Generation saved the world from fascism. The Dr. Spock Generation gave us rebellion and free love. Generation X made cynicism and slacking off the hallmarks of the end of the 20th century. In the media, generation is a popular and all-purpose explanation for change in America. Each new generation replaces an older one's zeitgeist with its own.

Generational succession is increasingly a popular explanation among scholars, too. Recently, political scientist Robert Putnam argued in *Bowling Alone* that civic engagement has declined in America even though individual Americans have not necessarily become less civic minded. Instead, he argues, older engaged citizens are dying off and being replaced by younger, more alienated Americans who are less tied to institutions such as the church, lodge, political party and bowling league.

Next to characteristics like social class, race, and religion, generation is probably the most common explanatory tool used by social scientists to account for differences among people. The difficulties in proving such explanations, however, are not always apparent and are often overshadowed by the seductiveness of the idea. Generational arguments do not always hold the same allure once they are given closer scrutiny.

Changes in the worldviews of Americans result not only from the progression of generations but also from historical events and patterns of aging. For example, generational replacement seems to explain why fewer Americans now than 30 years ago say they trust other people, but historical events seem to explain why fewer say they trust government. Similarly, historical events in interaction

SOURCE: Alwin, Duane F. 2002. "Generations X, Y, and Z: Are They Changing America?" *Contexts* 1 (Winter): 42–50.

with aging (or life cycle change) may explain lifetime changes in church attendance and political partisanship better than generational shifts.

EXPLAINING SOCIAL CHANGE

Some rather massive changes over the past 50 years in Americans' attitudes need explaining. Consider this short list of examples:

- In 1977, 66 percent said that it is better if the man works and the woman stays home; in 2000, only 35 percent did.
- In 1972, 48 percent said that sex before marriage is wrong; in 2000, 36 percent did.
- In 1972, 39 percent said that there should be a law against interracial marriage; in 2000, 12 percent did.
- In 1958, 78 percent said that one could trust the government in Washington to do right; in 2000 only 44 percent did.

Do changes in beliefs and behaviors reflect the experiences of specific generations, do they occur when Americans of all ages change their orientations, or do they result from something else? Although the idea of generational succession is promising and useful, it also has problems that may limit it as an all-purpose explanation of social change.

SOME PRELIMINARIES

Before we begin to deconstruct the idea of generational replacement we need to clarify a few issues. The first is that when sociologists use the term *generation* it can refer to one of three quite different things:

1. All people born at the same time.
2. A unique position within a family's line of descent (as in the second generation of Bush presidents).
3. A group of people self-consciously defined, by themselves and by others, as part of an historically based social movement (as in the "hippie" generation).

There are many examples in the social science literature of all three uses, and this can create a great deal of confusion. Here I refer mainly to the first use, measuring generation by year of birth. Demographers prefer to use the term cohort. Either way the reference is to the historical period in which people grow to maturity. I use the terms cohort and generation more or less interchangeably.

When sociologists are discussing social change in less precise terms, they may refer to generations in a somewhat more nuanced, cultural sense. Generations in this usage do not necessarily map neatly to birth years. Rather, the distinction between generations is a matter of quality, not degree, and their exact time boundaries cannot always be easily identified. It is also clear that statistically there is no way to identify cohort or generation effects unequivocally. The interpretation of generational differences depends entirely on one's ability or willingness to make some rather hefty assumptions about other processes, such as how aging affects attitudes, but as we shall see, we can nonetheless develop reasonable conclusions.

COHORT REPLACEMENT

Cohort replacement is a fact of social life. Earlier-born cohorts die off and are replaced by those born more recently. The question is: do the unique formative experiences of cohorts become distinctively imprinted onto members' worldviews, making them distinct generations over the course of their lifetimes, or do people of all cohorts adapt to change, remaining pliable in their beliefs throughout their lives?

When historical events mainly affect the young, we have the makings of a generation. Such an effect—labeled a cohort effect—refers to the outcomes attributable to having been born in a particular historical period. When, for example, people describe the Depression generation as particularly thrifty, they imply that the experience of growing up under privation permanently changed the economic beliefs and style of life of people who grew to maturity during that time.

Unique events that happen during youth are no doubt powerful. Certainly, some eras and social movements, like the Civil Rights era and the women's movement, or some new ideologies (e.g. Roosevelt's New Deal) provide distinctive experiences for youth during particular times. As Norman Ryder put it, "the potential for change is concentrated in the cohorts of young adults who are old enough to participate directly in the movements impelled by change, but not old enough to have become committed to an occupation, a residence, a family of procreation or a way of life."

To some observers, today's younger generations—Generation X and its younger counterpart—display a distinctive lack of social commitment. The goals of individualism and the good life have replaced an earlier generation's involvement in social movements and organizations. Is this outlook simply part of being young, or is it characteristic of a particular generation?

Each generation resolves issues of identity in its own way. In the words of analyst Erik Erikson, "No longer is it merely for the old to teach the young the meaning of life . . . it is the young who, by their responses and actions, tell the old whether life as represented by the old and presented to the young has meaning; and it is the young who carry in them the power to confirm those who confirm them and, joining the issues, to renew and to regenerate or to reform and to rebel." . . .

Before we accept this way of understanding change, however, we should consider other possibilities. One is that people change as they get older, which we call an effect of aging. The older people get, for example, the more intensely they may hold to their views. America, as a whole, may be becoming more politically partisan because the population is getting older—an age effect. Another possibility is that people change in response to specific historical events, what sociologists call period effects. The Civil Rights movement, for example, may have changed many Americans' ideas about race, not just the views of the generation growing up in the 1960s. The events of September 11, 2001, likely had an effect on the entire nation, not just those in the most impressionable years of youth.

A third possibility is that the change is located in only one segment of society. Members of the Roman Catholic faith, for example, may be the most responsive to the current turmoil over the sexual exploitation of youth by some priests in ways that hardly touch the lives of Protestants. Let us weigh these possibilities more closely, looking at the issues raised by Putnam in *Bowling Alone*.

CHANGES IN SOCIAL CONNECTEDNESS AND TRUST

It is often relatively easy to construct a picture of generational differences by comparing data from different age groups in social surveys and polls, but determining what produced the data is considerably more complex.

Take, for example, one of the key empirical findings of Putnam's analysis: the responses people give to the question of whether they trust their fellow human beings. The General Social Survey (administered regularly to a nationwide, representative sample of American residents since 1972) asks the following question: "Do you think most people would try to take advantage of you if they got the chance, or would they try to be fair?" Figure 1 . . . presents the percentage

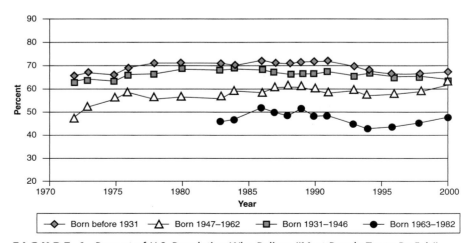

FIGURE 1   Percent of U.S. Population Who Believe "Most People Try to Be Fair"

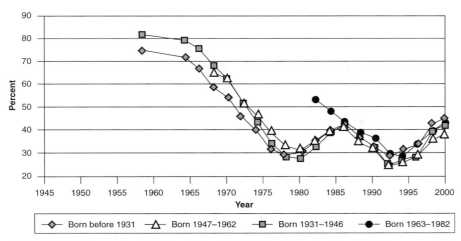

FIGURE 2   Percent of U.S. Population Who Believe "You Can Trust the Government in Washington to Do What Is Right"

of respondents in each set of cohorts who responded that people would try to be fair. The results show that birth cohorts were consistently different from one another, the recent ones being more cynical about human nature. There has been little change in this outcome over the years except insofar as new generations replaced older ones. These results reinforce the Putnam thesis, that the degree of social connectedness in the formative years of people's generation shapes their sense of trust.

Still, I would note some problems with these conclusions. First, generational experiences are not the only factors that differentiate these four groups, they also differ by age. Second, these data do not depict the young lives of the cohorts born before 1930 (who were 42 years of age or older in 1972), so we have little purchase on their beliefs before 1972. Third, there is remarkable growth in trust among the Baby Boom cohorts—those born from 1947 to 1962—over their midlife period, and in 2000 they had achieved a level of trust on a par with earlier cohorts. Finally, even the most recent cohorts (the lowest line in the figure) show some tendency to gain trust in recent years. The point is that while the data appear to show a pattern of generational differences—less trust among more recent cohorts—age might be just as plausible an explanation of the differences: trust goes up as people mature.

There may be more than one way to explain changes in Americans' trust of people, but generations do not explain changes in Americans' trust of government. In 1958 the National Election Studies (NES) began using the following question: "How much of the time do you think you can trust the government in Washington to do what is right—just about always, most of the time, or only some of the time?"

There are two important things to note about Figure 2. First, there are hardly any differences among birth cohorts who say most of the time or always in their responses to this question; the lines are virtually identical. Thus, generational replacement explains none of the very dramatic decline of trust in government. That decline may be better explained by historical events that affected all cohorts—the Vietnam War, the feminist movement, or the Watergate and Whitewater scandals—and there is little basis for arguing that more recent cohorts are more alienated from government than those born earlier. (Note that affirmations of trust in government rose dramatically right after 9/11.) . . .

GENERATIONS AND SOCIAL CHANGE

Society reflects, at any given time, the sum of its generations. Where one set of cohorts is especially large—like the Baby Boomers—its lifestyle dominates the society as it passes through the life course. Baby Boomers' taste in music and clothes, for example, disproportionately influence the whole culture. However, in cases where there are no major differences among generations (as in the example of trust in government), then generational succession cannot explain social change.

Where generations persistently differ, however, their succession will produce social change. Certainly, if the more recent generations have less affiliation and involvement with traditional religious groups, this will lead to social change, at least until they develop their own form of religiosity.

Because of the Baby Boomer generation's sheer size, its liberal positions on political and social issues will probably shape beliefs and behavior well into the new century, as Boomers replace the generations that came before. But even here, the Baby Boomers' distinctiveness may wane under the influence of historical events and processes of aging. Baby Boomers, for example, may be growing more conservative with age. This argues in favor of an alternative to the generational view: Generations do not necessarily differ in the same ways over time; individuals are not particularly consistent over their lives; and social change results as much from shifts in individual lives due either to aging or historical events. . . .

The existence of generation effects may depend very much on when one takes the snapshot of generational differences, and how generations differ may depend on which groups in society one examines. All fair warnings for the next essay you read on Generations X, Y or Z.

KEY CONCEPTS

age cohort   cohort effect   social change

DISCUSSION QUESTIONS

1. What is a cohort effect and how is it significant to the process of social change? What social changes are the result of cohort effects among your age generation?

2. As you imagine the future, what social changes do you think will result from the various sources of generational change that Alwin identifies?

69

Jihad vs. McWorld

BENJAMIN R. BARBER

The author compares and contrasts two countervailing sources of social change: Jihad and McWorld. Jihad is a traditionalist, anti-modern worldview (now especially familiar because of the post-9/11 social and political context). McWorld, on the other hand, is a world increasingly driven by market values and the commercialization of all forms of life. Though they seem like opposites, Barber shows how they produce each other and result from common forces of social change.

. . . [A]nyone who reads the daily papers carefully, taking in the front page accounts of civil carnage as well as the business page stories on the mechanics of the information superhighway and the economics of communication mergers, anyone who turns deliberately to take in the whole 360-degree horizon, knows that our world and our lives are caught between what William Butler Yeats called the two eternities of race and soul: that of race reflecting the tribal past, that of soul anticipating the cosmopolitan future. Our secular eternities are corrupted, however, race reduced to an insignia of resentment, and soul sized down to fit the demanding body by which it now measures its needs. Neither race nor soul offers us a future that is other than bleak, neither promises a polity that is remotely democratic.

The first scenario rooted in race holds out the grim prospect of a retribalization of large swaths of humankind by war and bloodshed: a threatened balkanization of nation-states in which culture is pitted against culture, people against people, tribe against tribe, a Jihad in the name of a hundred narrowly conceived faiths against every kind of interdependence, every kind of artificial social cooperation and mutuality: against technology, against pop culture, and against integrated markets; against modernity itself as well as the future in which modernity issues. The second paints that future in shimmering pastels, a busy portrait of onrushing economic, technological, and ecological forces that demand integration and uniformity and that mesmerize peoples everywhere with fast music, fast computers, and fast food—MTV, Macintosh, and McDonald's— pressing nations into one homogenous global theme park, one McWorld tied together by communications, information, entertainment, and commerce. Caught between Babel and Disneyland, the planet is falling precipitously apart and coming reluctantly together at the very same moment.

Some stunned observers notice only Babel, complaining about the thousand newly sundered "peoples" who prefer to address their neighbors with sniper rifles and mortars; others—zealots in Disneyland—seize on futurological platitudes and the promise of virtuality, exclaiming "It's a small world after all!" Both are right, but how can that be?

We are compelled to choose between what passes as "the twilight of sovereignty" and an entropic end of all history; or a return to the past's most fractious and demoralizing discord; to "the menace of global anarchy," to Milton's capital of hell, Pandemonium; to a world totally "out of control."

The apparent truth, which speaks to the paradox at the core of this argument, is that the tendencies of both Jihad *and* McWorld are at work, both visible sometimes in the same country at the very same instant. Iranian zealots keep one ear tuned to the mullahs urging holy war and the other cocked to Rupert Murdoch's Star television beaming in *Dynasty, Donahue* and *The Simpsons* from hovering satellites. Chinese entrepreneurs vie for the attention of party cadres in Beijing and simultaneously pursue KFC franchises in cities like Nanjing, Hangzhou, and Xian where twenty-eight outlets serve over 100,000 customers a day. The Russian Orthodox church, even as it struggles to renew the ancient faith, has entered a joint venture with California businessmen to bottle and sell natural waters under the rubric Saint Springs Water Company. Serbian assassins wear Adidas sneakers and listen to Madonna on Walkman headphones as they take aim through their gunscopes at scurrying Sarajevo civilians looking to fill family watercans. Orthodox Hasids and brooding neo-Nazis have both turned to rock music to get their traditional messages out to the new generation, while fundamentalists plot virtual conspiracies on the Internet.

. . . It is not Jihad and McWorld but the relationship between them that most interests me. For, squeezed between their opposing forces, the world has been sent spinning out of control. Can it be that what Jihad and McWorld have in common is anarchy: the absence of common will and that conscious and collective human control under the guidance of law we call democracy?

Progress moves in steps that sometimes lurch backwards; in history's twisting maze, Jihad not only revolts against but abets McWorld, while McWorld not only imperils but re-creates and reinforces Jihad. They produce their contraries and need one another. . . .

What then does it mean in concrete terms to view Jihad and McWorld dialectically when the tendencies of the two sets of forces initially appear so intractably antithetical? After all, Jihad and McWorld operate with equal strength in opposite directions, the one driven by parochial hatreds, the other by universalizing markets, the one re-creating ancient subnational and ethnic borders from within, the other making national borders porous from without. Yet Jihad and McWorld have this in common: they both make war on the sovereign nation-state and thus undermine the nation-state's democratic institutions. Each eschews civil society and belittles democratic citizenship, neither seeks alternative democratic institutions. Their common thread is indifference to civil liberty. Jihad forges communities of blood rooted in exclusion and hatred, communities that slight democracy in favor of tyrannical paternalism or consensual tribalism. McWorld forges global markets rooted in consumption and profit, leaving to an untrustworthy, if not altogether fictitious, invisible hand issues of public interest and common good that once might have been nurtured by democratic citizenries and their watchful governments. Such governments, intimidated by market ideology, are actually pulling back at the very moment they ought to be aggressively intervening. What was once understood as protecting the public interest is now excoriated as heavy-handed regulatory browbeating. Justice yields to markets, even though, as Felix Rohatyn has bluntly confessed, "there is a brutal Darwinian logic to these markets. They are nervous and greedy. They look for stability and transparency, but what they reward is not always our preferred form of democracy." If the traditional conservators of freedom were democratic constitutions and Bills of Rights, "the new temples to liberty," George Steiner suggests, "will be McDonald's and Kentucky Fried Chicken." . . .

Jihad is, I recognize, a strong term. In its mildest form, it betokens religious struggle on behalf of faith, a kind of Islamic zeal. In its strongest political manifestation, it means bloody holy war on behalf of partisan identity that is metaphysically defined and fanatically defended. Thus, while for many Muslims it may signify only ardor in the name of a religion that can properly be regarded as universalizing (if not quite ecumenical), I borrow its meaning from those militants who make the slaughter of the "other" a higher duty. I use the term in its militant construction to suggest dogmatic and violent particularism of a kind known to Christians no less than Muslims, to Germans and Hindis as well as to Arabs. The phenomena to which I apply the phrase have innocent enough beginnings: identity politics and multicultural diversity can represent strategies of a free society trying to give expression to its diversity. What ends as Jihad may begin as a simple search for a local identity, some set of common personal attributes to hold out against the numbing and neutering uniformities of industrial modernization and the colonizing culture of McWorld. . . .

. . . Jihad is then a rabid response to colonialism and imperialism and their economic children, capitalism and modernity; it is diversity run amok, multiculturalism turned cancerous so that the cells keep dividing long after their division has ceased to serve the healthy corpus.

Even traditionally homogenous integral nations have reason to feel anxious about the prospect of Jihad. The rising economic and communications interdependence of the world means that such nations, however unified internally, must nonetheless operate in an increasingly multicultural global environment. Ironically, a world that is coming together pop culturally and commercially is a world whose discrete subnational ethnic and religious and racial parts are also far more in evidence, in no small part as a reaction to McWorld....

THE SMALLING WORLD OF MCWORLD

. . . Every demarcated national economy and every kind of public good is today vulnerable to the inroads of transnational commerce. Markets abhor frontiers as nature abhors a vacuum. Within their expansive and permeable domains, interests are private, trade is free, currencies are convertible, access to banking is open, contracts are enforceable (the state's sole legitimate economic function), and the laws of production and consumption are sovereign, trumping the laws of legislatures and courts. In Europe, Asia, and the Americas such markets have already eroded national sovereignty and given birth to a new class of institutions—international banks, trade associations, transnational lobbies like OPEC, world news services like CNN and the BBC, and multinational corporations—institutions that lack distinctive national identities and neither reflect nor respect nationhood as an organizing or a regulative principle....

McWorld is a product of popular culture driven by expansionist commerce. Its template is American, its form style. Its goods are as much images as material, an aesthetic as well as a product line. It is about culture as commodity, apparel as ideology. Its symbols are Harley-Davidson motorcycles and Cadillac motorcars hoisted from the roadways, where they once represented a mode of transportation, to the marquees of global market cafés like Harley-Davidson's and the Hard Rock where they become icons of lifestyle. You don't drive them, you feel their vibes and rock to the images they conjure up from old movies and new celebrities, whose personal appearances are the key to the wildly popular international café chain Planet Hollywood. Music, video, theater, books, and theme parks—the new churches of a commercial civilization in which malls are the public squares and suburbs the neighborless neighborhoods—are all constructed as image exports creating a common world taste around common logos, advertising slogans, stars, songs, brand names, jingles, and trademarks. Hard power yields to soft, while ideology is transmuted into a kind of videology that works through sound bites and film clips. Videology is fuzzier and less dogmatic than traditional political ideology: it may as a consequence be far more successful in instilling the novel values required for global markets to succeed....

The dynamics of the Jihad–McWorld linkage are deeply dialectical. Japan has, for example, become more culturally insistent on its own traditions in recent years even as its people seek an ever greater purchase on McWorld. In 1992, the number-one restaurant in Japan measured by volume of customers was McDonald's, followed in the number-two spot by the Colonel's Kentucky Fried Chicken. . . .

In Russia, in India, in Bosnia, in Japan, and in France too, modern history then leans both ways: toward the meretricious inevitability of McWorld, but also into Jihad's stiff winds, heaving to and fro and giving heart both to the Panglossians and the Pandoras, sometimes for the very same reasons. The Panglossians bank on Euro-Disney and Microsoft, while the Pandoras await nihilism and a world in Pandemonium. Yet McWorld and Jihad do not really force a choice between such polarized scenarios. Together, they are likely to produce some stifling amalgam of the two suspended in chaos. Antithetical in every detail, Jihad and McWorld nonetheless conspire to undermine our hard-won (if only half-won) civil liberties and the possibility of a global democratic future. In the short run the forces of Jihad, noisier and more obviously nihilistic than those of McWorld, are likely to dominate the near future, etching small stories of local tragedy and regional genocide on the face of our times and creating a climate of instability marked by multimicrowars inimical to global integration. But in the long run, the forces of McWorld are the forces underlying the slow certain thrust of Western civilization and as such may be unstoppable. Jihad's microwars will hold the headlines well into the next century, making predictions of the end of history look terminally dumb. But McWorld's homogenization is likely to establish a macropeace that favors the triumph of commerce and its markets and to give to those who control information, communication, and entertainment ultimate (if inadvertent) control over human destiny. Unless we can offer an alternative to the struggle between Jihad and McWorld, the epoch on whose threshold we stand—postcommunist, postindustrial, postnational, yet sectarian, fearful, and bigoted—is likely also to be terminally postdemocratic.

KEY CONCEPTS

consumerism   modernization

DISCUSSION QUESTIONS

1. What does Barber mean when he says that what Jihad and McWorld may have in common is anarchy?

2. Why does Barber conclude that modern history leans both toward Jihad and McWorld and what does he see as the potential implications of this?

70

The Genius of the Civil Rights Movement

Can It Happen Again?

ALDON MORRIS

The civil rights movement was arguably the most influential movement in the United States during the twentieth century. Aldon Morris reviews the development of the civil rights movements and notes the products of this movement, including the mobilization of other national and international movements and the transformations in academic scholarship that the movement generated. By identifying the particular historical and social circumstances in which the civil rights movement developed, he also asks whether such a movement is possible again.

It is important for African Americans, as well as all Americans, to take a look backward and forward as we approach the turn of a new century, indeed a new millennium. When a panoramic view of the entire history of African Americans is taken into account, it becomes crystal clear that African American social protest has been crucial to Black liberation. In fact, African American protest has been critical to the freedom struggles of people of color around the globe and to progressive people throughout the world.

The purpose of this essay is: (1) to revisit the profound changes that the modern Black freedom struggle has achieved in terms of American race relations; (2) to assess how this movement has affected the rise of other liberation movements both nationally and internationally; (3) to focus on how this movement has transformed how scholars think about social movements; (4) to discuss the lessons that can be learned from this groundbreaking movement pertaining to future African American struggles for freedom in the next century.

It is hard to imagine how pervasive Black inequality would be today in America if it had not been constantly challenged by Black protests throughout each century since the beginning of slavery. The historical record is clear that slave resistance and slave rebellions and protest in the context of the Abolitionist movement were crucial to the overthrow of the powerful slave regime.

SOURCE: Morris, Aldon. "The Genius of the Civil Rights Movement: Can It Happen Again?" 2001. Northwestern University. Reprinted by permission of author.

The establishment of the Jim Crow regime was one of the great tragedies of the late nineteenth and early twentieth centuries. The overthrow of slavery represented one of those rare historical moments where a nation had the opportunity to embrace a democratic future or to do business as usual by reinstalling undemocratic practices. In terms of African Americans, the White North and South chose to embark along undemocratic lines.

For Black people, the emergence of the Jim Crow regime was one of the greatest betrayals that could be visited upon a people who had hungered for freedom so long; what made it even worse for them is that the betrayal emerged from the bosom of a nation declaring to all the world that it was the beacon of democracy.

The triumph of Jim Crow ensured that African Americans would live in a modern form of slavery that would endure well into the second half of the twentieth century. The nature and consequences of the Jim Crow system are well known. It was successful in politically disenfranchising the Black population and in creating economic relationships that ensured Black economic subordination. Work on wealth by sociologists Melvin Oliver and Thomas Shapiro (1995), as well as Dalton Conley (1999), are making clear that wealth inequality is the most drastic form of inequality between Blacks and Whites. It was the slave and Jim Crow regimes that prevented Blacks from acquiring wealth that could have been passed down to succeeding generations. Finally, the Jim Crow regime consisted of a comprehensive set of laws that stamped a badge of inferiority on Black people and denied them basic citizenship rights.

The Jim Crow regime was backed by the iron fist of southern state power, the United States Supreme Court, and white terrorist organizations. Jim Crow was also held in place by white racist attitudes. As Larry Bobo has pointed out, "The available survey data suggests that anti-Black attitudes associated with Jim Crow were once widely accepted . . . [such attitudes were] expressly premised on the notion that Blacks were the innately intellectual, cultural, and temperamental inferior group relative to Whites" (Bobo, 1997:35). Thus, as the twentieth century opened, African Americans were confronted with a powerful social order designed to keep them subordinate. As long as the Jim Crow order remained intact, the Black masses could breathe neither freely nor safely. Thus, nothing less than the overthrow of a social order was the daunting task that faced African Americans during the early decades of the twentieth century.

The voluminous research on the modern civil rights movement has reached a consensus: That movement was the central force that toppled the Jim Crow regime. To be sure, there were other factors that assisted in the overthrow including the advent of the television age, the competition for Northern Black votes between the two major parties, and the independence movement in Africa which sought to overthrow European domination. Yet it was the Civil Rights movement itself that targeted the Jim Crow regime and generated the great mass mobilizations that would bring it down.

What was the genius of the Civil Rights movement that made it so effective in fighting a powerful and vicious opposition? The genius of the Civil Rights movement was that its leaders and participants recognized that change could occur if they were able to generate massive crises within the Jim Crow order—crises of such magnitude that the authorities of oppression must yield to the demands of the movement to restore social order. Max Weber defined power as the ability to realize one's will despite resistance. Mass disruption generated power. That was the strategy of nonviolent direct action. By utilizing tactics of disruption, implemented by thousands of disciplined demonstrators who had been mobilized through their churches, schools, and voluntary associations, the Civil Rights movement was able to generate the necessary power to overcome the Jim Crow regime. The famous crises created in places like Birmingham and Selma, Alabama, coupled with the important less visible crises that mushroomed throughout the nation, caused social breakdown in Southern business and commerce, created unpredictability in all spheres of social life, and strained the resources and credibility of Southern state governments while forcing white terrorist groups to act on a visible stage where the whole world could watch. At the national level, the demonstrations and repressive measures used against them generated foreign policy nightmares because they were covered by foreign media in Europe, the Soviet Union, and Africa. Therefore what gave the mass-based sit-ins, boycotts, marches, and jailings their power was their ability to generate disorder.

As a result, within ten years—1955 to 1965—the Civil Rights movement had toppled the Jim Crow order. The 1964 Civil Rights Bill and the 1965 Voting Rights Act brought the regime of formal Jim Crow to a close.

The Civil Rights movement unleashed an important social product. It taught that a mass-based grass roots social movement that is sufficiently organized, sustained, and disruptive is capable of generating fundamental social change. In other words, it showed that human agency could flow from a relatively powerless and despised group that was thought to be backward, incapable of producing great leaders.

Other oppressed groups in America and around the world took notice. They reasoned that if American Blacks could generate such agency they should be able to do likewise. Thus the Civil Rights movement exposed the agency available to oppressed groups. By agency I refer to the empowering beliefs and action of individuals and groups that enable them to make a difference in their own lives and in the social structures in which they are embedded.

Because such agency was made visible by the Civil Rights movement, disadvantaged groups in America sought to discover and interject their agency into their own movements for social change. Indeed, movements as diverse as the Student movement, the Women's movement, the Farm Worker's movement, the Native American movement, the Gay and Lesbian movement, the Environmental movement, and the Disability Rights movement all drew important lessons and inspiration from the Civil Rights movement. From that movement other groups discovered how to organize, how to build social movement organizations, how to mobilize large numbers of people, how to devise appropriate tactics and strategies, how to infuse their movement activities with cultural creativity, how to confront and defeat authorities, and how to unleash the kind of agency that generates social change.

For similar reasons, the Black freedom struggle was able to effect freedom struggles internationally. For example, nonviolent direct action has inspired oppressed groups as diverse as Black South Africans, Arabs of the Middle East, and pro-democracy demonstrators in China to engage in collective actions. The sit-in tactic made famous by the Civil Rights movement, has been used in liberation movements throughout the third world, in Europe, and in many other foreign countries. The Civil Rights movement's national anthem "We Shall Overcome" has been interjected into hundreds of liberation movements both nationally and internationally. Because the Civil Rights movement has been so important to international struggles, activists from around the world have invited civil rights participants to travel abroad. Thus early in Poland's Solidarity movement Bayard Rustin was summoned to Poland by that movement. As he taught the lessons of the Civil Rights movement, he explained that "I am struck by the complete attentiveness of the predominantly young audience, which sits patiently, awaiting the translations of my words" (Rustin, undated).

Therefore, as we seek to understand the importance of the Black Freedom Struggle, we must conclude the following: the Black Freedom Struggle had provided a model and impetus for social movements that have exploded on the American and international landscapes. This impact has been especially pronounced in the second half of the twentieth century.

What is less obvious is the tremendous impact that the Black Freedom Struggle has had on the scholarly study of social movements. Indeed, the Black freedom struggle has helped trigger a shift in the study of social movements and collective action. The Black movement has provided scholars with profound empirical and theoretical puzzles because it has been so rich organizationally and tactically and because it has generated unprecedented levels of mobilization. Moreover, this movement has been characterized by a complex leadership base, diverse gender roles, and it has revealed the tremendous amount of human agency that usually lies dormant within oppressed groups. The empirical realities of the Civil Rights movement did not square with the theories used by scholars to explain social movements prior to the 1960s.

Previous theories did not focus on the organized nature of social movements, the social movement organizations that mobilize them, the tactical and strategic choices that make them effective, nor the rationally planned action of leaders and participants who guide them. In the final analysis, theories of social movements lacked a theory that incorporated human agency at the core of their conceptual apparatuses. Those theories conceptualized social movements as spontaneous, largely unstructured, and discontinuous with institutional and organizational behavior. Movement participants were viewed as reacting to various forms of strain and doing so in a non-rational manner. In these frameworks, human agency was conceptualized as reactive, created by uprooted individuals seeking to reestablish a modicum of personal and social stability. In short, social movement theories prior to the Civil Rights movement operated with a vague, weak vision of agency to explain phenomena that are driven by human action.

The predictions and analytical focus of social movement theories prior to the 1970s stood in sharp contrast to the kind of theories that would be needed to capture the basic dynamics that drove the Civil Rights movement. It became apparent to social movement scholars that if they were to understand the Civil Rights movement and the multiple movements it spun, the existing theoretical landscape would have to undergo a radical process of reconceptualization.

As a result, the field of social movements has been reconceptualized and this retheoritization will affect research well into the new millennium. To be credible in the current period any theory of social movements must grapple conceptually with the role of rational planning and strategic action, the role of movement leadership, and the nature of the mobilization process. How movements are gendered, how movement dynamics are bathed in cultural creativity, and how the interactions between movements and their opposition determine movement outcomes are important questions. At the center of this entire matrix of factors must be an analysis of the central role that human agency plays in social movements and in the generation of social change.

Thanks, in large part, to the Black freedom struggle, theories of social movements that grapple with real dynamics in concrete social movements are being elaborated. Intellectual work in the next century will determine how successful scholars will be in unraveling the new empirical and theoretical puzzles thrust forth by the Black freedom movement. Although it was not their goal, Black demonstrators of the Civil Rights movement changed an academic discipline.

A remaining question is: Will Black protest continue to be vigorous in the twenty-first century, capable of pushing forward the Black freedom agenda? It is not obvious that Black protest will be as sustainable and as paramount as it has been in previous centuries. To address this issue we need to examine the factors important to past protests and examine how they are situated in the current context.

Social movements are more effective when they can identify a clear-cut enemy. Who or what is the clear-cut enemy of African Americans of the twenty-first century? Is it racism, and if so, who embodies it? Is it capitalism, and if so, how is this enemy to be loosened from its abstract perch and concretized? In fact, we do not currently have a robust concept that grasps the modern form of domination that Blacks currently face. Because the modern enemy has become opaque, slippery, illusive, and covert, the launching of Black protest has become more difficult because of conceptual fuzziness.

Second, during the closing decades of the twentieth century the Black class structure has become more highly differentiated and it is no longer firmly anchored in the Black community. There is some danger, therefore, that the cross fertilization between different strata within the Black class structure so important to previous protest movements may have become eroded to the extent that it is no longer fully capable of launching and sustaining future Black protest movements.

Third, will the Black community of the twenty-first century possess the institutional strength required for sustaining Black protest? Black colleges have been weakened because of the racial integration of previously all white institutions of higher learning and because many Black colleges are being forced to integrate. The degree of institutional strength of the church has eroded because some of them have migrated to the suburbs in an attempt to attract affluent Blacks. In other instances, the Black Church has been unable to attract young people of the inner city who find more affinity with gangs and the underground economy. Moreover, a great potential power of the Black church is not being realized because its male clergy refuse to empower Black women as preachers and pastors. The key question is whether the Black church remains as close to the Black masses—especially to poor and working classes—as it once was. That closeness determines its strength to facilitate Black protest.

In short, research has shown conclusively that the Black church, Black colleges and other Black community organizations were critical vehicles through which social protest was organized, mobilized and sustained. A truncated class structure was also instrumental to Black protest. It is unclear whether during the twenty-first century these vehicles will continue to be effective tools of Black protest or whether new forces capable of generating protest will step into the vacuum.

In conclusion, I foresee no reason why Black protest should play a lesser role for Black people in the twenty-first century. Social inequality between the races will continue and may even worsen especially for poorer segments of the Black communities. Racism will continue to affect the lives of all people of color. If future changes are to materialize, protest will be required. In 1898 as Du Bois glanced toward the dawn of the twentieth century, he declared that in order for Blacks to achieve freedom they would have to protest continuously and energetically. This will become increasingly true for the twenty-first century. The question is whether organizationally, institutionally, and intellectually the Black community will have the wherewithal to engage in the kind of widespread and effective social protest that African Americans have utilized so magnificently. If previous centuries are our guide, then major surprises on the protest front should be expected early in the new millennium.

REFERENCES

Bobo, L. 1997. "The Color Line, the Dilemma, and the Dream: Race Relations in America at the Close of the Twentieth Century." In *Civil Rights and Social Wrongs: Black-White Relations since World War II*, edited by J. Higham, pp. 31–55. University Park, PA: Penn State University Press.

Conley, Dalton. 1999. *Being Black, Living in the Red: Race, Wealth, and Social Policy in America*. Berkeley: University of California Press.

Oliver, Melvin, and Thomas E. Shapiro. 1995. *Black Wealth/White Wealth: A New Perspective on Racial Inequality*. New York: Routledge.

Rustin, Bayard. no date. *Report on Poland*. New York: A. Philip Randolph Institute.

KEY CONCEPTS

civil rights   equal opportunity   social movement

DISCUSSION QUESTIONS

1. What does Morris mean by "the genius of the Civil Rights movement"? Can you imagine such a strategy being an effective means of combating the oppression of racial groups today? If so, how; if not, why not?

2. What does Morris identify as the products of the civil rights movement? What does this teach you about the connections between contemporary social movements and the civil rights movement?

Applying Sociological Knowledge:
An Exercise for Students

Previous generations in society have all been remembered for specific changes they brought about as they came of age. What types of things makes your generation distinctive compared to the previous one? How do you think your generation will be remembered in years to come? Do you think your defining achievements have already happened or are still to come?

Glossary

A

achieved status status attained by effort

acquaintance rape rape that occurs when someone is assaulted by someone they know; sometimes referred to as "date rape"

age cohort group of people born during the same time period

ageism institutionalized practice of age prejudice and discrimination

age prejudice negative attitude about an age group that is generalized to all people in that group

agency acting independently of constraints imposed by social systems

age stratification hierarchical ranking of age groups in society

alienation feelings of powerlessness and separation from one's group or society

anomie social structural condition existing when social regulations (norms) in a society break down

anorexia nervosa condition characterized by compulsive dieting, resulting in self-starvation

antimiscegenation laws laws outlawing the mixing of races through marriage

anti-Semitism hostility toward or discrimination against Jewish people

apartheid system in which different groups (typically racial groups) are completely segregated from each other in all aspects of public life

asceticism practice of promoting strict self-denial as a measure of spiritual discipline

ascribed status status determined by birth

assimilation process by which a minority becomes socially, economically, and culturally absorbed within the dominant society

attitude mental position with regard to a social phenomenon

authoritarian personality personality characterized by a tendency to rigidly categorize people, to submit to authority, to strictly conform, and to be intolerant of ambiguity

authority power that is perceived by others as legitimate

B

backstage behavior behavior normally not displayed in public

beliefs shared ideas held collectively by people within a given culture

bigotry state of mind of people who are intolerantly devoted to their own opinions and prejudices

block grant large amount of money allocated by the federal government to a state government for a specific purpose

bourgeoisie term used to loosely describe either the ruling or middle class in a capitalist society

brainwashing forcible indoctrination to induce someone to give up basic political, social, or religious beliefs and to accept contrasting, regimented ideas

breadwinner member of a family whose wages supply its livelihood

bulimia eating disorder that involves binge eating and then purging through vomiting and/or laxatives

bureaucracy type of formal organization characterized by an authority hierarchy, with a clear division of labor, explicit rules, and impersonality

C

capital accumulated goods devoted to the production of other goods

capitalism economic system based on the principles of market competition, private property, and the pursuit of profit

capitalist class those persons who own the means of production in a society

care work labor people (often women) do to nurture and take care of others

caste system system of stratification (characterized by low social mobility) in which one's place in the stratification system is determined and fixed by birth

charter school public school that has been granted exemption from some local or state regulations

chauvinism attitude of superiority toward members of the opposite sex; also behavior expressive of such an attitude

chi-square test statistic used to determine the statistical significance of a relationship between two variables

civil rights legal and political framework guaranteeing equal protection under the law

class see social class

clique narrow, exclusive circle or group of persons, especially one held together by common interests, views, or purposes

cohabitation living together as an intimate couple while not married

collective action (or collective behavior) behavior that occurs when the usual conventions are suspended and people collectively establish new norms of behavior in response to an emerging situation

collective consciousness body of beliefs that are common to a community or society and that give people a sense of belonging

collectivism political or economic theory advocating collective control over production and distribution of a system

coming-out process process of defining oneself as gay or lesbian

commodity chain network of production and labor processes by which a product becomes a finished commodity

communism economic system in which the state is the sole owner of the systems of production

conflict theory perspective that emphasizes the role of power and coercion in producing social order

consumerism preoccupation with and inclination toward the buying of consumer goods

consumption utilization of economic goods in the satisfaction of wants

contingent worker temporary worker

conversion experience associated with a decisive adoption of religion

corrections process by which convicted criminals are resocialized to be members of a society who behave acceptably

correlation statistical technique that analyzes patterns of association between pairs of sociological variables

counterculture subculture created as a reaction against the values of the dominant culture

crime behavior that violates criminal laws

criminal justice system institutions that a government develops to enforce standards of conduct and punish those who do not conform

cult religious group devoted to a specific cause or charismatic leader

cultural hegemony pervasive and excessive influence of one culture throughout society

cultural icon object of uncritical devotion within a society or culture

cultural imperialism forcing all groups in a society to accept the dominant culture as their own

cultural relativism idea that something can be understood and judged only in relationship to the cultural context in which it appears

culture complex system of meaning and behavior that defines the way of life for a given group or society

culture lag delay in cultural adjustments to changing social conditions

culture of poverty argument that poverty is a way of life and, like other cultures, is passed on from generation to generation

culture shock feelings of disorientation that can emerge when one encounters a new or rapidly changed cultural situation

custom long-established practice

cyberspace interaction sharing by two or more persons of a a virtual reality experience via communication and interaction with each other

D

data systematic information that sociologists use to investigate research questions

data set large collection of systematic information used in sociological research

de facto by practice or in fact, even if not in law

de jure by rule, policy, or law

decentralization redistribution of population and industry from urban centers to outlying areas

deindustrialization structural transformation of a society from a manufacturing-based economy to a service-based economy

democracy system of government based on the principle of representing all people through the right to vote

deterrence punishment intended to prevent or discourage the behavior being punished

deviance behavior that is recognized as violating expected rules and norms

diaspora people related by common historical descent who are now scattered far from their ancestral homeland

discourse interchange of ideas or opinions

discrimination overt negative and unequal treatment of members of some social group or stratum solely because of their membership in that group or stratum

distribution position, arrangement, or frequency of occurrence (as of the members of a group) over an area or throughout a unit of time

diversity variety of group experiences that result from the social structure of society

division of labor systematic interrelation of different tasks that develops in complex societies

divorce rate number of divorces per some number (usually 1,000) in a given year

doing gender analytical framework that interprets gender as an activity accomplished through everyday interaction

domestic violence exertion of physical or mental force by one household member against another

dominant culture culture of the most powerful group in society

double consciousness realizing that one is being viewed as the "other" in a social situation and simultaneously perceiving and understanding the dominant group

downsizing action by companies to eliminate job positions in order to cut the firm's operating costs

dramatic metaphor perspective used to suggest that social interaction has a likeness to dramas presented on a stage

dual labor market theoretical description of the occupational system as divided into two major segments: the primary and secondary labor markets

dual/shared custody arrangement of child-rearing responsibilities in which the child of divorced parents lives with each parent about the same percentage of time

dyad group consisting of two people

E

eating disorders illnesses that are both psychological and physical, involving dangerously unhealthy eating and dieting habits, such as bulimia nervosa or anorexia nervosa

economic restructuring contemporary transformations in the basic structure of work that are permanently altering the workplace, including the changing composition of the workplace, deindustrialization, the use of enhanced technology, and the development of a global economy

economy system on which the production, distribution, and consumption of goods and services is based

educational attainment total years of formal education

egalitarian societies or groups in which men and women share power

emotional labor (or management) work intended to produce a desired emotional effect on a client

Enlightenment period in eighteenth- and nineteenth-century Europe characterized by faith in the ability of human reason to solve society's problems

entitlement government program providing benefits to members of a specified group

environmental justice social movements that challenge the presence of greater pollution and toxic waste dumping in poor and minority communities

environmental racism disproportionate location of sources of toxic pollution in or very near communities of color

epistemology division of philosophy that investigates the nature and origin of knowledge

equal opportunity civil rights legislation that guarantees equality in employment, education, and other public institutions

ethnic enclave regional or neighborhood location that contains people of distinct cultural origins

ethnic group social category of people who share a common culture, such as a common language or dialect, a common religion, and common norms, practices, and customs

ethnocentrism belief that one's own group is superior to all others

ethnography descriptive account of social life and culture in a particular social system based on observations of what people actually do

ethnomethodology technique for studying human interaction by deliberately disrupting social norms and observing how individuals attempt to restore normalcy

exploitation unjust or improper use of another person for one's own profit or advantage

F

false consciousness idea that subordinated classes internalize the view of the ruling class

family primary group of people—usually related by ancestry, marriage, or adoption—who form a cooperative economic unit to care for any offspring (and each other) and who are committed to maintaining the group over time

feeling rules situational guidelines for emotional displays appropriate to a specific situation

feminism beliefs, actions and theories that attempt to bring justice, fairness, and equity to all women, regardless of their race, age, class, sexual orientation, or other characteristics

feminization of poverty trend in which a growing proportion of the poor are women and children

field research process of gathering data in a naturally occurring social setting

folkways general standards of behavior adhered to by a group

frequency number of individuals in a single class or category

functionalism theoretical perspective that interprets each part of society in terms of how it contributes to the stability of the whole society

G

game stage stage in childhood when children become capable of taking a multitude of roles at the same time

gemeinschaft German word meaning "community"; state characterized by a sense of fellow feeling among the members of a society, including strong personal ties, sturdy primary group memberships, and a sense of personal loyalty to one another; associated with rural life; compare gesellschaft

gender socially learned expectations and behaviors associated with members of each sex

gendered institution total pattern of gender relationships embedded in social institutions

gender identity one's definition of self as a woman or man

gender role learned expectations associated with being a man or a woman

gender segregation distribution of men and women in different positions in a social system

gender socialization process by which men and women learn the expectations associated with their sex

generalized other abstract composite of social roles and social expectations

genocide deliberate and systematic destruction of a racial, political or cultural group

gerontologist one who studies the branch of knowledge dealing with aging and the problems of the aged

gesellschaft German word meaning "society"; form of social organization characterized by a high division of labor, less prominence of personal ties, the lack of a sense of community among the members, and the absence of a feeling of belonging; associated with urban life; compare gemeinschaft

global care chain social network in which a series of personal links between mothers across the globe transfer caring (paid or unpaid) to children other than their own

global culture diffusion of a single culture throughout the world

global economy term acknowledging that all dimensions of the economy now cross national borders

globalization increased economic, political, and social interconnectedness and interdependence among societies in the world

global warming concept that the planet is experiencing a slow and steady change in temperature, resulting in various detrimental environmental consequences

grassroots activism political activity generated at the local level

group collection of individuals who interact and communicate, share goals and norms, and who have a subjective awareness as "we"

H

health maintenance organization (HMO) cooperative of doctors and other medical personnel who provide medical services in exchange for a set membership fee

hegemonic masculinity pervasivensss of culturally supported attributes associated with the dominant ideas of masculinity

hegemony ascendancy or dominance of one social group over another

heterogeneous containing dissimilar or diverse ingredients or constituents

heterosexism institutionalization of heterosexuality as the only socially legitimate sexual orientation

homogeneous containing the same or similar ingredients or constituents

homophobia fear and hatred of homosexuality

human nature inherent character of people

hypersegregation very high levels of segregation

hypothesis statement about what one expects to find when one does research

I

I preverbal part of the mind that experiences everything without the reflection and thought that language makes possible

ideology belief system that tries to explain and justify the status quo

immigration migration of people into one society from another society

impression management process by which people control how others perceive them

indicator something that represents an abstract concept

indirect effect variable that affects an outcome through another, intervening variable

individualism doctrine that states that the interests of the individual are and ought to be paramount

inductive reasoning logical process of building general principles from specific observations

industrialization sustained economic growth following the application of raw materials and other, more intellectual resources to mechanized production

in-group group with which one feels a sense of solidarity or community of interests

initiation rite ceremonies, ordeals, or instructions with which one is made a member of a sect or society or is invested with a particular function or status

institution see social institution

institutional privileges institutionalized benefits given to those of the dominant group that seem to "naturally" afford this group greater opportunity

instrumental ability emotionally neutral, task-oriented (goal-oriented) ability

interlocking directorate organizational linkages created when the same people sit on the boards of directors of a number of different corporations

internalization incorporation of values, patterns of culture, and so on within the self as conscious or subconscious guiding principles through learning or socialization

internalized oppression process by which members of a dominated social group accept the view of them held by the dominant group

issues problems that affect large numbers of people and have their origins in the institutional arrangements and history of a society

J

Jim Crow historical practice of completely segregating black and white Americans in all aspects of public life

K

kinship system pattern of relationships that define family members' relationships to one another

L

labeling theory theory that interprets the responses (or "labels") of others as most significant in determining the behavior of people

labor market available supply of jobs

language set of symbols and rules that, put together in a meaningful way, provides a complex communication system

legitimate conforming to recognized principles or accepted rules and standards

life chances opportunities people have in common by virtue of belonging to a particular class

life expectancy average number of years individuals in a particular group can expect to live

longitudinal study research design dealing with change within a specific group over a period of time

looking-glass self idea that people's conception of self arises through reflection about their relationship to others

M

managed care use of collective bargaining on the part of large collections of HMOs

marginalizing viewing social groups and their specific social processes as peripheral to the dominant group's processes

marriage rate number of marriages per some number (usually 1,000) in a given year

mass media channels of communication that are available to very wide segments of the population

McDonaldization process by which increasing numbers of services share the bureaucratic and rationalized processes associated with this food chain

me social being seen as "myself" in any social situation

means of production system by which goods are produced and distributed

means-testing method used to determine eligibility for government assistance or benefits based on income

measurement error inaccuracy due to flaws in a measurement instrument

median midpoint in a series of values that are arranged in numerical order

Medicaid government assistance program that provides health-care assistance for the poor

medicalization of deviance social process through which a norm-violating behavior is culturally defined as a disease and is treated as a medical condition

Medicare government assistance program that provides health-care assistance for the elderly

meritocracy system (as an educational system) presuming that the most talented are chosen and moved ahead on the basis of their talent and achievement

methodology practices and techniques used to gather, process, and interpret theories about social life

minority group any distinct group in society that shares common group characteristics and is forced to occupy low status in society because of prejudice and discrimination

modeling children's imitation of adults' behaviors

model minority minority group used as an example to suggest that social mobility is possible for minority groups (ignoring the fact that the model minority has only achieved partial success)

moral panic extreme response by a group of citizens to the behavior of other people whom that they see as deviant or undesirable

mores strict norms that control moral and ethical behavior

multiculturalism movement push to introduce more courses on different and diverse subcultures and groups, ethnic groups, and gender studies into the elementary, high-school, and college curricula

multivariate analysis any of several statistical or research methods for examining the effects of more than two variables at the same time

N

natural selection process in nature resulting in the survival and perpetuation of only those forms of animal and plant life having certain characteristics that best enable them to adapt to a specific environment

nonrepresentative sample sample not similar to the population from which it was drawn

norms specific cultural expectations for how to act in given situations

nuclear family social unit comprised of a man and a woman living together with their children

O

objectification process of treating a person or group of people as inanimate objects, devoid of their humanity

objectivity absence of bias in making or interpreting observations

occupational segregation pattern by which workers are separated into different occupations on the basis of social characteristics such as race and gender

oppression systematic, institutionalized mistreatment of one social group by another social group

organic metaphor similarity early sociologists saw between society and other organic systems

out-group group that is distinct from one's own and is usually an object of hostility or dislike

outsourcing hiring temporary workers outside of a firm to do certain jobs; such temporary workers seldom receive employee benefits

P

paradigm framework for understanding a phenomenon based on a particular and identifiable set of assumptions

participant observation method whereby the sociologist is both a participant in the group being studied and a scientific observer of the group

partner violence physical abuse of one person by an intimate significant other

pathology functional manifestations of a disease

patriarchy society or group in which men have power over women

play stage stage in childhood when children begin to take on the roles of significant people in their environment

pluralism state of society in which members of different ethnic, racial, religious, and social groups maintain their distinctive cultural traditions, respect the traditions of other groups, and share a common civilization

political economy interdependent workings and interests of governing and the material wealth production and distribution systems

popular culture beliefs, practices, and objects that are part of everyday traditions

positivism system of thought in which accurate observation and description is considered the highest form of knowledge

postmodernism theoretical perspective based on the idea that society is not an objective thing but is found in the words and images—or discourses—that people use to represent and describe behavior and ideas

poverty line figure established by the government to indicate the amount of money needed to support the basic needs of a household

power ability of a person or group to exercise influence and control over others

power elite model theoretical model of power positing a strong link between government and business

predictive validity extent to which a test accurately predicts later college grades or some other outcome, such as likelihood of graduating

prejudice negative evaluation of a social group and individuals within that group based on misconceptions about that group

presentation of self Erving Goffman's phrase referring to the way people display themselves during social interaction

prestige subjective value with which different groups or people are judged

primary group group characterized by intimate, face-to-face interaction and relatively long-lasting relationships

primary source firsthand accounts of events used as data

privatization change (as a business or industry) from public to private control or ownership

proletarian lowest social or economic class of a community; the laboring class, especially the class of industrial workers who lack their own means of production and hence sell their labor to live

proletarianization (1) process by which parts of the middle class become effectively absorbed into the working class; (2) process by which an occupational category is downgraded in occupational status and becomes more closely akin to working class jobs

proportion ratio of a part to the total

Protestant ethic belief that hard work and self-denial lead to salvation and success

Puritan member of a sixteenth- and seventeenth-century Protestant group in England and New England that opposed ceremonial worship

Q

qualitative method sociological research methodology based on interpretive and usually nonquantitative observation

R

race social category or social construction based on certain characteristics, some biological, that have been assigned social importance in the society

racial formation process by which groups come to be defined as a "race" through social institutions such as the law and the schools

racialization process whereby some social category, such as a social class or nationality, is assigned what are perceived to be racial characteristics

racial project organized effort to interpret and represent racial dynamics with the purpose of reorganizing and redistributing resources along particular racial lines

racism perception and treatment of a racial or ethnic group or member of that group as intellectually, socially, and culturally inferior to one's own group

random sample sample that gives everyone in the population an equal chance of being selected

rape myths commonly held beliefs about rape that are not true

rate parts per some number (e.g., per 10,000; per 100,000)

rationalization of society term used by Max Weber to describe the increasing organization of society around legal, empirical, and scientific forms of thought

real wages wage rate adjusted for the annual rate of inflation

recidivism relapse into a former pattern of criminal behavior

redlining discriminating against racial groups in housing, home-loan funds, or insurance

reliability likelihood that a particular measure would produce the same results if the measure were repeated

religion institutionalized system of symbols, beliefs, values, and practices by which a group of people interprets and responds to what they believe is sacred and that provides answers to questions of ultimate meaning

religiosity intensity and consistency of practice of a person's (or group's) faith

religious socialization process by which one learns a particular religious faith

ritual symbolic activities that express a group's spiritual convictions

role expected behavior associated with a given status in society

role conflict two or more roles associated with contradictory expectations

role negotiation working out for oneself the expectations of a specific role based on one's own beliefs and the social beliefs associated with the role

role strain conflicting expectations within the same role

S

sample any subset of units from a population that a researcher studies

sandwich generation people who are simultaneously caring for their elderly parents and their young children, including financial, emotional, and physical care

science body of thought about the natural world that rests on the idea that reliable knowledge must be based on systematic, observable facts that will lead anyone who considers them to the same conclusions

scientific method research process, including the steps of observation, hypothesis testing, analysis of data, and generalization

scientific objectivity see objectivity

script learned performance of a social role

secondary source secondhand account of an event used as data

second shift women's second round of work after wage labor, when they come home to manage their households

secular ordinary beliefs of daily life that are specifically not religious

self relatively stable set of perceptions of who we are in relation to ourselves, others, and the social system

self-fulfilling prophecy process by which the application of a label changes behavior and thus tends to justify the label

service sector part of the labor market composed of the nonmanual, nonagricultural jobs

sex biological identity as male or female

sex segregation distribution of men and women in any social group or society

sex trafficking international pattern of selling sex, often involving moving sex workers from one country to another

sex work paid labor involving sexual services

sexism system of practices and beliefs through which women are controlled and exploited because of the significance given to differences between the sexes

sexual assault term used in the criminal justice system for rape, a felony charge of sexual violence

sexuality sexual desire and behavior

sexual orientation manner in which individuals experience sexual arousal and pleasure

sexual politics the link feminists argue exists between sexuality and power and between sexuality and race, class, and gender oppression

sexual revolution widespread changes in men's and women's roles and greater public acceptance of sexuality as a normal part of social development

snowball sample subset of a population obtained by sampling one individual who recommends another to be included in the sample, and so forth

social class social structural position that groups hold relative to the economic, social, political, and cultural resources of society

social construction of reality process by which what individuals perceive as real is given objective meaning through a process of social interaction

social control process by which groups and individuals within those groups are brought into conformity with dominant social expectations

social facts social patterns that are external to individuals

social institution established, organized system of social behavior with a recognized purpose

social interaction behavior between two or more people that is given meaning

social mobility movement over time by a person from one social class to another

social movement group that acts with some continuity and organization to promote or resist change in society

social network set of links between individuals or other social units, such as groups or organizations

social order the stable social organization of the social world

social organization order established in social groups

social sanctions mechanisms of social control that enforce norms

social speedup having more to do in the same amount of time than was once the case

social stratification relatively fixed hierarchical arrangement in society by which groups have different access to resources, power, and perceived social worth; a system of structured social inequality

social structure patterns of social relationships and social institutions that comprise society

society system of social interactions that includes both culture and social organization

socioeconomic status (SES) measure of class standing, typically indicated by income, occupational prestige, and educational attainment

sociological imagination ability to see the societal patterns that influence individual and group life

sociology study of human behavior in society

standard deviation statistic showing the spread or dispersion of scores in a sample

standardized ability test tests given to large populations and scored with respect to population averages

state the organized system of power and authority in society

statistically significant relationship between variables that is larger than would be expected by chance alone

status established position in a social structure that carries with it a degree of prestige

status hierarchy power and prestige order of the group

steering practice of some real estate agents by which ethnic and racial minorities were influenced to buy houses that are not located in predominantly white communities in order to preserve the racial makeup of white communities

stereotype oversimplified set of beliefs about the members of a social group or social stratum that is used to categorize individuals of that group

stigma attribute that is socially devalued and discredited

subculture culture of groups whose values and norms of behavior are somewhat different from those of the dominant culture

subjectivity views of the subject rather those of an outside observer

survey to query (someone or a group or a sample) in order to collect data for the analysis of some aspect of a group or area

symbolic interaction theory perspective claiming that people act toward things because of the meaning things have for them

T

taboo prohibition imposed by social custom

time bind conflicts experienced by trying to integrate work and family responsibilities

tracking grouping, or stratifying, students in school on the basis of ability test scores

troubles privately felt problems that come from events or feelings in one individual's life

U

underemployment employment at a level below what would be expected, given a person's level of training or education

unemployment rate percentage of those not working but officially defined as looking for work

urban underclass grouping of people, largely minority and poor, who live at the absolute bottom of the socioeconomic ladder in urban areas

V

validity degree to which an indicator accurately measures or reflects a concept

values abstract standards in a society or group that define ideal principles

victimization experiences of people who have been victims of crimes or systematic oppression

voter turnout percentage of a given population that participates in elections

W

wealth monetary value of what someone owns, calculated by adding all financial assets (stocks, bonds, property, insurance, investments, etc.) and subtracting debts; also called net worth

welfare system public benefit system of a society designed to provide for the needs of those who cannot fully provide for themselves or their families

Westernization process by which non-Western peoples convert to or adopt Western culture

white privilege system in which white people benefit collectively from the social and economic history of racism

work productive human activity that produces something of value, either goods or services

working class those persons who do not own the means of production and therefore must sell their labor in order to earn a living

working families families characterized by having at least two members employed, typically husband and wife

working poor employed people whose wages are too low to bring their standard of living above the poverty level

world city coordination centers for the global economy where multinational corporations and large financial concerns house their headquarters, along with many other business service firms

Index

Student Reply Questionnaire

Dear Student,

I hope you enjoy reading *Understanding Society: An Introductory Reader*, Third Edition. With every book that I publish, my goal is to enhance your learning experience. If you have any suggestions that you feel would improve this book, I would be delighted to hear from you. All comments will be shared with the authors. My email address is Chris.Caldeira@thomson.com, or you can mail this form (no postage required). Thank you.

School and address: _____

Department: _____

Instructor's name: _____

1. What I like most about this book is: _____

_____

_____

2. What I like least about this book is: _____

_____

_____

3. I would like to say to the authors. . .

_____

_____

4. In the space below, or in an email to Chris.Caldeira@thomson.com, please write specific suggestions for improving this book and anything else you'd care to share about your experience using this book.

_____

_____

_____

DO NOT STAPLE. PLEASE SEAL WITH TAPE.

FOLD HERE

THOMSON

WADSWORTH

BUSINESS REPLY MAIL
FIRST-CLASS MAIL   PERMIT NO. 34   BELMONT CA

POSTAGE WILL BE PAID BY ADDRESSEE

Attn: Chris Caldeira, Sociology

Thomson Wadsworth
10 Davis Drive
Belmont, CA   94002-9801

FOLD HERE

OPTIONAL:

Your name: _____   Date: _____

May we quote you, either in promotion for *Understanding Society: An Introductory Reader*, Third Edition, or in future publishing ventures?

Yes: _____   No: _____

Sincerely yours,

Margaret L. Andersen
Kim A. Logio
Howard F. Taylor